STRUCTURE OF CRYSTALLINE POLYMERS

Structure of Crystalline Polymers

HIROYUKI TADOKORO
Department of Polymer Science
Osaka University
Toyonaka, Osaka, Japan

A Wiley-Interscience Publication

JOHN WILEY & SONS
New York · Chichester · Brisbane · Toronto

Library of Congress Cataloging in Publication Data:

Tadokoro, Hiroyuki, 1920–
Structure of crystalline polymers.

"A Wiley-Interscience publication."
Includes index.
1. Polymers and polymerization. 2. Crystals.
I. Title.
QD381.T33 547'.84 78–15024
ISBN 0–471–02356–6

Printed in the United States of America

10 9 8 7 6 5 4 3 2 1

To
Emeritus Professor Isamu Nitta
and
Akiko, my wife

'Tis a lesson you should heed,
Try, try, try again.
If at first you don't succeed,
Try, try, try again.
Then your courage should appear,
For if you will persevere
You will conquer, never fear.
Try, try, try again.

Hickson

Reprinted, by permission, from Edward Whymper, *Scrambles Amongst the Alps, In the Years 1860–69*, John Murray (Publishers) Ltd., London, 1871.

Foreword

The diversity of properties exhibited by polymeric substances, natural and synthetic, reflects the virtually endless variations in chemical constitution that can be incorporated in the long-chain macromolecules comprising members of this broad class of substances. Obviously, an understanding of the properties that distinguish one polymer from another requires an adequate knowledge of structure at the molecular level. Chemical formulas conventionally expressed in two dimensions do not suffice. The structure must be comprehended in three dimensions and in the detail necessary to portray the spatial locations of the constituent atoms in relation to one another. Required for attainment of this ambitious, but necessary, objective are bond lengths, bond angles and torsional rotations, or conformations, about the bonds of the structure.

The seeming complexities of the molecular architecture and conformations of polymers have engendered adoption of simplified models in the past. In adopting an artificial model, one should be cognizant of the hazards of applying it beyond the limits set by the fidelity of the model to the realities of the structure it is intended to represent. Abstraction is commonplace as an expedient in the scientific process, but secure foundations require collation and critical examination of relevent information in detail before making the step from reality to abstraction. Results of structural investigations pertaining to polymers stand at the forefront of essential information in this category.

X-ray crystallography and vibrational spectroscopy have proved to be the richest sources of structural data on macromolecular substances. These methods, supplemented by nuclear magnetic resonance, electron diffraction and other techniques, have yielded a wealth of information on bond lengths, bond angles and force constants for the principal chemical entities occurring in polymer chains. Additionally, they have provided important insights on conformational energies and on the preference for one conformation compared to another. Experimental methods have been supplemented in recent years by semi-empirical force field calculations, and also by molecular orbital calculations on simple molecular analogs.

The foregoing methods for the elucidation of structure and conformation in macromolecules are set forth in this volume at a level and in a manner that should meet the needs of students of the subject and of research workers who, though aware of the techniques and some of the results obtained through their application, often have not aquired an appreciation of them in the depth that should be desired. The author has taken the pains to lay forth the foundations underlying the diffraction of X-rays by crystals, commencing with molecular and crystal symmetry and culminating in the Fourier transformation of the crystal lattice into its representation in reciprocal space. He has avoided belaboring the reader with a compendium of formulas, superficially explained within the confines of a single volume. Infrared absorption and Raman spectroscopy are presented with corresponding regard for the uninitiated reader.

In devoting principal attention to the crystalline state, the author fulfills dual objectives. Crystalline polymers are important subjects of inquiry in their own right. Hence, information pertinent thereto immediately aids elucidation of the properties of polymers in this state. Additionally, crystallography has proved to be an especially fruitful source of structural and conformational information applicable to polymers in the amorphous state and in solution. Lucid descriptions of experimental methods, including critical assessment of their capabilities, illumine the interplay of experiment and theory. Systematic presentation of experimental results, including results of recent investigations, provides an abundance of illustrative material which, moreover, enhances the potential value of the book as a reference source.

This book should be warmly received by students and research investigators who face the increasingly formidable task of transcending the inevitable limitations of the current literature as an avenue to acquisition of a knowledge of the subject in the depth that is essential for their needs. Additionally, it can be counted upon to stimulate research in areas of enduring significance.

PAUL J. FLORY

Department of Chemistry
Stanford University

Preface

Few books that cover the field of polymer structure are suitable both for beginners, that is, chemistry undergraduate and graduate students, and for research workers. Although the structure of crystalline polymers is a broad field of study, this book is limited to my areas of speciality: X-ray analysis of crystal and molecular structures, infrared and Raman spectroscopy, and energy calculation, with recent reviews and aspects.

In 1951 when I started to study the structures of crystalline polymers, there was no suitable textbook covering this field. I studied by "running" between basic textbooks on X-ray diffraction, infrared spectroscopy, group theory, and so on, as well as the books, reviews, and original papers of polymer chemistry. I intend this book to be the sort that I would have desired to read at that time. I think I have given the reader a basis for understanding the current literature on the structure of polymers, along with materials useful for advanced studies in the field.

Infrared spectrophotometers are now in wide use, and many of them are used in the analysis of polymeric materials. I hope this book will also be helpful to people working in this field.

Information on the crystal structure and molecular conformation of polymers may often be needed by polymer scientists working in fields other than those treated in the present book, such as synthesis, morphology, and properties. The tables and references in Chapter 7 will, I hope, supply this information.

To keep this volume a reasonable size, I confine myself to the subjects that are especially important for polymer studies. Other topics, which are discussed in the standard texts on X-ray crystallography and infrared and Raman spectroscopy are treated only briefly here.

In any field it is very important to grasp the fundamentals of a given problem and to acquire a deep understanding of the essential features of the phenomena. However, most chemists seem to be more or less unfamiliar with the mathematical techniques important for theoretical treatments in physics. For this reason, I provide derivations and explanations of the theories and formulas in much detail. Where the derivation of a formula has been omitted, I have given the relevant references. In the process of writing the manuscript, I found that my personal knowledge, even of this narrow field, is limited. To complete the coverage I studied the basics over again, taking this opportunity to refresh my understanding.

We are now in a state of transition between the traditional system of units of measure and the SI. Nonetheless I have written this book using mainly the cgs units. This is in conformity with the original literature. A conversion table between the two systems appears in Appendix H.

The quotation by Hickson on page vi is from my favorite book, Edward Whymper's *Scrambles Amongst the Alps*. This poem includes a precept which I have always tried to follow. It was not intended as arrogance toward my readers. Mr. Whymper tried six times to climb the Matterhorn along the southwest ridge, then on the seventh attempt he tried the east wall. He finally succeeded in the first ascent on July 14, 1865, his eighth trial, by changing the route to the northeast ridge. I understand clearly what was in Mr. Whymper's mind when he cited this poem at the beginning of the chapter "Renewed Attempts to Ascend the Matterhorn." Although this lesson could be applied to any field, I think it fits the field of study presented in this book exactly. I will be very pleased if my readers share this feeling with me.

HIROYUKI TADOKORO

January 1979

Acknowledgments

First of all I am most grateful to Professor Herman F. Mark for his suggesting that I write this book. This book would not have been completed without the heartful cooperation of my colleagues, Drs. Y. Chatani, M. Kobayashi, Y. Takahashi, and K. Tashiro. It is also a great pleasure that many graduate students from my laboratory read and criticized earlier versions of the manuscript and spotted many errors. I express my cordial gratitude to Dr. C. W. Bunn, Professor C. H. Wang, and Dr. E. M. Barrall II who read through the manuscript and provided many valuable comments for its improvement. Thanks are also made to Professor L. E. Alexander, Professor R. N. Jones, Professor P. H. Geil, Professor J. L. Koenig, Dr. R. L. Miller, and Dr. A. J. Hopfinger for reading parts of the manuscript and giving many important suggestions. Finally my sincere thanks are due to my three teachers, Emeritus Professor I. Nitta, Emeritus Professor S. Murahashi, and Professor S. Seki, who have guided and encouraged me throughout my research work.

It is a great honor for me to have Professor Paul J. Flory, one of the most distinguished pioneers of polymer science, write the foreword to this book. I express my heartfelt thanks to him.

Thanks are also due to the following publishers and authors for their courtesy in granting permission for the use of copyrighted material. The names of authors and references are cited as they occur. Mr. E. Whymper, Mr. W. E. Hickson, John Murray, Ltd., Prof. A. Keller, The Chemical Society, U. K., Prof. J. D. Watson, MacMillan Journals, Ltd., Prof. K. Tomita, Dr. C. W. Bunn, The Royal Society, U. K., Cornell University Press, International Union of Crystallography, Prof. E. B. Wilson, Jr., McGraw-Hill Book Co., Prof. L. Pauling, National Academy of Sciences, U. S., Prof. S. C. Nyburg, Academic Press, G. Bell & Sons, John Wiley & Sons, Inc., IPC Business Press Ltd., Prof. R. Hosemann, American Institute of Physics, Prof. P. Corradini, Societa Italiana di Fisica il Nuovo Cimento, Dr. H. W. Starkweather, Jr., Springer-Verlag, Akademische Verlagsgesellschaft, Prof. E. S. Clark, Institute of Physics, U. K., D. Van Nostrand Co., Inc., Dr. D. Z. Robinson, The American Chemical Society,

Prof. R. N. Jones, Prof. S. Krimm, Prof. M. Shen, Dr. V. B. Carter, Prof. J. K. Koenig, Dr. R. Zbinden, Dr. M. C. Tobin, Prof. M. Tasumi, Prof. T. Shimanouchi, Kyoritsu Publishing Co., Ltd., Prof. T. Miyazawa, Dr. W. Myers, Pergamon Press, Prof. T. Yamamoto, Prof. G. Allegra, Prof. M. Farina, Hutig & Wepf Verlag, and Dr. Y. Kinoshita.

I wrote a book entitled *Kobunshi no Kozo* (*Structure of Macromolecules*) in Japanese which was published in 1976 by Kagaku Dojin, Co. Ltd., Kyoto, Japan. The present monograph has been written on the basis of the Japanese text in a new and revised form. I am deeply indebted to Kagaku Dojin for their courtesy in permitting me to reproduce many figures as well as their encouragement in preparing the English manuscript of the present book.

H. T.

Contents

APPENDICES

STRUCTURE OF CRYSTALLINE POLYMERS

Chapter 1

Introduction

1.1 STRUCTURE OF SOLID POLYMERS

A polymer can be partially crystalline, such as polyethylene (PE) and poly(ethylene terephthalate) (PET), or noncrystalline, such as commercially available poly(methyl methacrylate) and polystyrene. Partially crystalline polymers (customarily called crystalline polymers) consist of crystalline and amorphous regions as shown in Fig. 1.1; in the former the molecular chains are packed regularly and in the latter they are distributed randomly. This concept is based on clear evidence that the X-ray pattern of an oriented sample (Fig. 1.2) is the superposition of diffraction spots originating from the crystalline regions onto halos due to the amorphous regions. There are also intermediate regions, and the transition from noncrystalline to crystalline regions is considered to be continuous. A crystalline region is called a crystallite and is several hundred angstroms long. The usual length of the molecular chain is, in general, far greater than the size of the crystallites. For example, the length of the extended planar zigzag of a PE molecule of molecular weight 50,000 is about 4500 Å. Hence one molecular chain is considered to pass through many crystalline and noncrystalline regions successively. The molecular chains are arranged in the crystalline regions, with the crystal structure characteristic of each chemical structure of the polymer. If such a specimen is stretched under suitable conditions, the molecular chains and crystallites are oriented as shown in Fig. 1.1*b*. Crystallization of polymer samples with low crystallinity is accelerated by heat treatments under suitable conditions. The weight fraction of the crystalline regions gives the degree of crystallinity, for which many methods of estimation have been developed. The term "fine texture" is used to indicate the state of aggregation of polymers, such as the size, shape, and orientation of crystallites, and their distribution, as well as the structure of the amorphous regions. The concept of a paracrystal has been proposed for treating disordered crystalline regions (Section 4.10.2).

Single crystals of dimensions suitable for electron microscopy have been

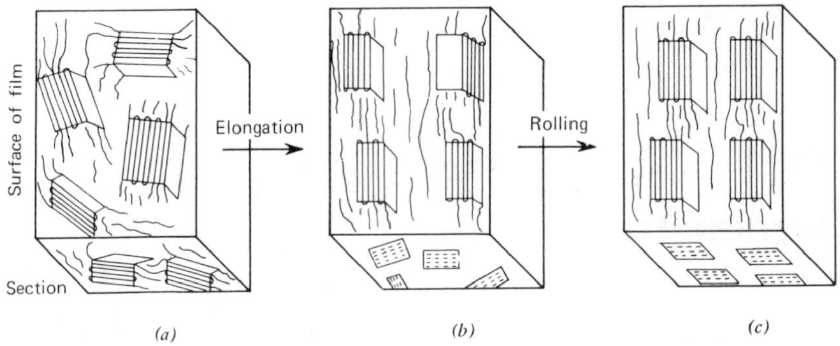

Fig. 1.1 Simplified models for the fine structure of crystalline polymers: (*a*) nonoriented crystalline and (*b*) uniaxially oriented crystalline. The axes of the molecular chains are nearly parallel to the elongation direction, but the orientation of the molecular plane in the cross section perpendicular to the elongation direction remains completely random. (*c*) Doubly oriented crystalline. A certain plane parallel to the molecular axis [e.g., in the case of poly (vinyl alcohol) the plane of the carbon zigzag chain] is also oriented parallel to the rolled plane.

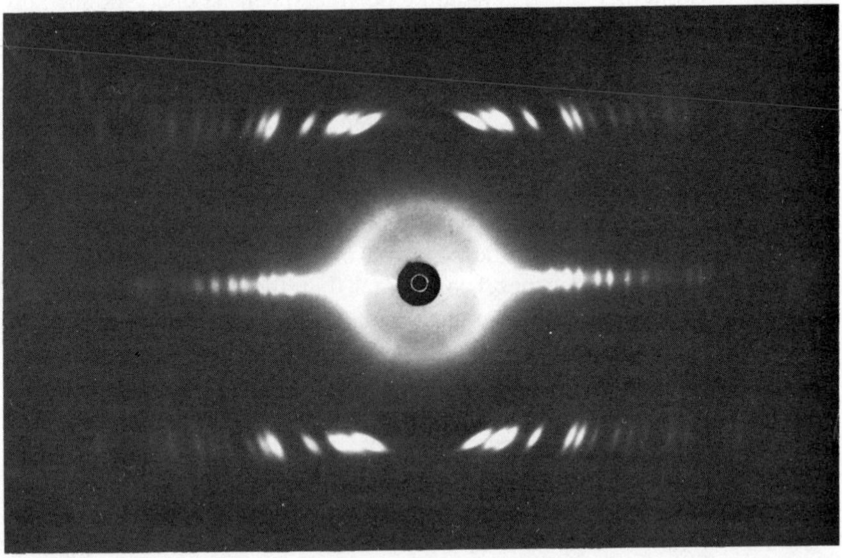

Fig. 1.2 X-Ray fiber diagram of a uniaxially oriented polyethylene sample obtained with a cylindrical camera.

obtained from dilute solutions of PE (Fig. 1.3) and many other polymers. Electron diffraction studies clearly show that in a single crystal the molecular chains are perpendicular to the crystal plane with a thickness of about 100 Å, and fold on the crystal surfaces.[1] This is known as the "folding model" or "lamellar model." Such folded lamellar structures have also been found in

Fig. 1.3 Electron micrograph of polyethylene (Marlex) crystallized from trichloroethylene solution and shadowed by Au-Pd on a carbon film. × 8800. (From Keller and O'Connor[1]; photograph courtesy of Professor A. Keller.)

film samples cast from solutions or those prepared from melts. Thus it has now been recognized that the interpretation of the microstructure of polymers in terms of the fringed micelle model, which has been used for a long time, is not really appropriate. Most likely the structures containing the fringed micelle and/or the lamellar structure can be prepared, depending on the conditions used. Also it may be presumed that the lamelar structure partially unfolds and changes to the fringed micellelike structure on stretching. The lamellar model and the fringed micelle model may be regarded as the two extreme cases. The highly crystalline polymers, such as high-density PE, may be close to the lamellar model, whereas stretched crystallized rubber, viscose rayon, and poly(*p*-phenylene terephthalamide) (Kevlar) (Section 7.10.2) may be represented by the fringed micelle model. Figure 1.1 shows an intermediate structure that describes most crystalline polymers.

Spherulites are an important product of crystallization from the melt. A spherulite consists of folded chain lamellae developing radially outwards. According to X-ray studies, a particular crystalline direction (e.g., the *b* axis in the case of PE) coincides with the radial direction. Under a microscope with crossed polaroids, maltese crosses are observed.

For polymer fine texture or morphology, see for example, Refs. 2–5.

1.2 SIGNIFICANCE OF STRUCTURAL STUDIES OF POLYMERS

The field of polymer science includes the study of syntheses, structure, properties, and processing. Several examples of important structural studies are given below.

1. Discovery of Stereoregular Polymers. Natta et al.[6] polymerized crystalline poly-α-olefins with heterogeneous catalysts. Using mainly X-ray analyses, they found that these polymers have a stereoregular structure, which they called "isotactic" (Section 2.1). They reported their discovery in 1955, and since then the field of "stereoregular polymers" has commenced its great development.

Seven years before this discovery, Schildknecht et al.[7] were the first to prepare a stereoregular polymer. In 1948 they prepared poly(vinyl isobutylether) $(\text{—}CH_2\text{—}\underset{\displaystyle OCH_2CH(CH_3)_2}{\underset{|}{CH}}\text{—})_n$, using $BF_3{\cdot}O(C_2H_5)_2$ as the catalyst, and obtained its crystalline X-ray pattern. Schildknecht et al. suggested that this crystallinity was due to a stereoisomeric structure, but they could not determine the actual structure of the polymer. In 1956 Natta et al.[8] found, using X-ray and electron diffraction methods, that this polymer has a structure similar to the threefold helix of isotactic polypropylene (PP) (see Fig. 4.86). Prior to this work Bunn[9] had proposed the threefold helix structure as one of the possible models for the carbon chain (see Fig. 2.5*c*). Therefore, if Schildknecht et al. had examined the X-ray patterns of poly(vinyl isobutylether) in more detail in relation to Bunn's study, an isotactic polymer (the name would be different) might have been discovered as early as 1948. The great development of the new field of stereoregular polymers in Natta's school is attributed to the wider range of their studies, covering syntheses, structure, and so forth.

2. Double Helix Structure of DNA and Genetic Information. Deoxyribonucleic acid (DNA) exists in chromosomes. It consists of a main chain having ester linkages of 2-deoxy-D-ribose and phosphoric acid, and four kinds of bases as side chains. Watson and Crick[10] found the right-hand double strand helix structure for DNA (Fig. 1.4) by X-ray analysis of the fiber diagram shown in Fig. 1.5 (for a discussion of the fiber diagram see Section 4.1). A pair of the DNA chains are linked by hydrogen bonds, which can only form between the special pairs of bases, that is, adenine and thymine or guanine and cytosine. The sequence of amino acids in the process of protein formation is controlled by the sequence of the four bases in DNA; in other words, the genetic information is merely the sequence of the bases in DNA. This very important discovery was based on X-ray analysis of the biopolymer that gives the fiber diagram.

3. Structure and Physical Properties. Crystal structures of various polymers have been determined mainly by X-ray analyses, and their physical properties have always been investigated in relation to their crystal structures. The most extensively studied example is PE, whose crystal structure was determined by Bunn[11] in 1939. PE molecules are branched, the extent depending on the condition of preparation, and the branching is closely

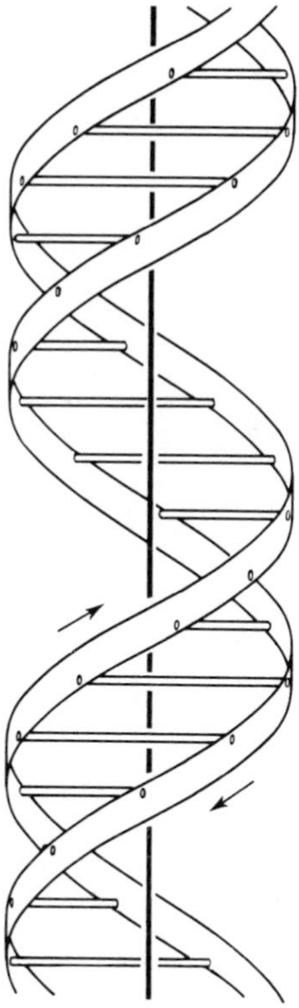

Fig. 1.4 Schematic model of double helix of DNA (From Watson and Crick[10]).

related to the properties of the polymer. Infrared spectroscopy is used to estimate the degree of branching.

The crystallite modulus also can be measured by the X-ray diffraction method[12,13] under the assumption of a series model, in which molecular chains are assumed to pass through the crystalline and amorphous regions along the fiber axis, so that the crystalline and amorphous regions are subjected to the same magnitude of stress. The crystallite modulus can also be calculated theoretically, if the geometrical structure is known and if suitable force constants can be assumed.[14–17] According to Sakurada et al.,[13] the force required to stretch a molecule by 1% in the direction of the molecular axis is 3×10^{-5} to 6×10^{-5} dyn for the planar zigzag-type polymers (PE, etc.), about 1.5×10^{-5} dyn for the glide-type polymers [poly(vinylidene fluoride) form II, etc.] (for the definition of glide plane see Section 3.5), and 0.2×10^{-5} to 1.5×10^{-5} dyn for the helical polymers (isotactic PP, etc.). Thus the force required to stretch a molecular chain is sensitive to the molecular conformation (Section 2.2).

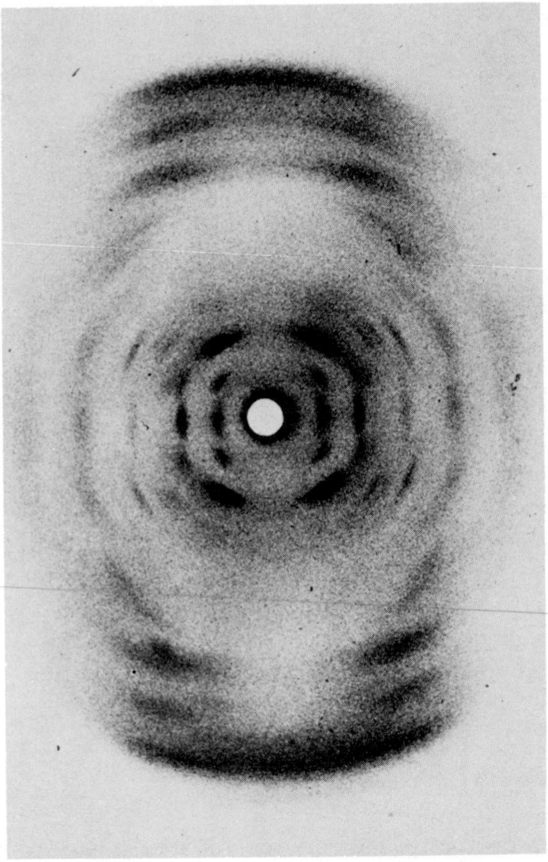

Fig. 1.5 X-Ray fiber diagram of DNA. (Photograph courtesy of Professor K. Tomita.)

1.3 METHODS OF STRUCTURAL STUDIES

This book deals essentially with the determination of molecular and crystal structures by X-ray diffraction, infrared and Raman spectroscopic methods, and energy calculations. The following methods are also important for structural studies of polymers: small-angle X-ray scattering, neutron scattering, nuclear magnetic resonance, optical and electron microscopy, electron diffraction, light scattering, density, differential thermal analysis, and differential scanning calorimetry.

It should be mentioned here that the analysis of crystal structure is made, even at present, essentially by a trial-and-error method, and it is not possible to proceed directly from the experimental data to the final structure. The

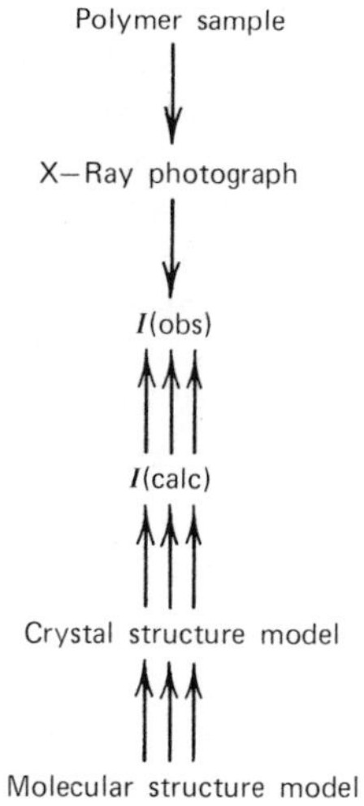

Fig. 1.6 Process of crystal structure analysis of high polymers.

process of X-ray analysis is schematically shown in Fig. 1.6. First the X-ray diffraction photographs are taken and the positions and intensities of diffraction maxima I (obs) are measured. Then molecular and crystal structure models are assumed, taking into account the empirically accumulated knowledge and information obtained from infrared and Raman spectroscopic methods, as well as the results of calculation of intra- and intermolecular interaction energy. To this is added intuition. The theoretical X-ray diffraction intensities I (calc) are then calculated for the assumed model and compared with the observed values I(obs). The calculation is repeated until a model is obtained that gives the best agreement between the observed and calculated intensities. In Fig. 1.6 a large number of upward-pointing arrows would have better represented the actual procedure, but for simplicity only three arrows are used.* It should be emphasized that when the true structure is of an unexpected new type, the analysis cannot be successful until the researcher is inspired with an idea of a suitable structure model.

REFERENCES

1. A. Keller and A. O'Connor, *Discuss. Faraday Soc.*, **25,** 114 (1958).
2. P. H. Geil, *Polymer Single Crystals*, Interscience, New York, 1963.

* For single crystals of low-molecular-weight compounds, there has been an increase in the number of cases in which trials of assumed molecular and crystal structure models are not necessarily needed, since the isomorphous replacement method and many kinds of statistical direct methods can be applied. For high polymers (except for proteins forming single crystals), however, the analyses can be made only by trial-and-error methods.

3. B. Wunderlich, *Macromolecular Physics*, Vol. 1, *Crystal Structure, Morphology, Defects*, Academic, New York, 1973.
4. B. Wunderlich, *Macromolecular Physics*, Vol. 2, *Crystal Nucleation, Growth, Annealing*, Academic, New York, 1976.
5. J. M. Schultz, *Polymer Materials Science*, Prentice-Hall, Engelewood Cliffs, New Jersey, 1974.
6. G. Natta, P. Pino, P. Corradini, F. Danusso, E. Mantica, G. Mazzanti, and G. Moraglio, *J. Am. Chem. Soc.*, **77,** 1708 (1955).
7. C. E. Schildknecht, S. T. Gross, H. R. Davidson, J. M. Lambert, and A. O. Zoss, *Ind. Eng. Chem.*, **40,** 2104 (1948).
8. G. Natta, I. Bassi, and P. Corradini, *Makromol. Chem.*, **18/19,** 455 (1956).
9. C. W. Bunn, *Proc. Roy. Soc. (Lond.)*, **A180,** 67 (1942).
10. J. D. Watson and F. H. C. Crick, *Nature*, **171,** 737 (1953).
11. C. W. Bunn, *Trans. Faraday Soc.*, **35,** 482 (1939).
12. W. J. Dulmage and L. E. Contois, *J. Polym. Sci.*, **28,** 275 (1958).
13. I. Sakurada and K. Kaji, *J. Polym. Sci. C*, **31,** 57 (1970).
14. L. R. G. Treloar, *Polymer*, **1,** 279 (1960).
15. T. Shimanouchi, M. Asahina, and Satoru Enomoto, *J. Polym. Sci.*, **59,** 93 (1962).
16. H. Sugeta and T. Miyazawa, *Polym. J.*, **1,** 266 (1970).
17. K. Tashiro, M. Kobayashi, and H. Tadokoro, *Macromolecules*, **10,** 731 (1977); **11,** 908, 914 (1978).

Chapter 2

Configuration and Conformation of Polymer Chains

The term "configuration" is used for describing an arrangement of atoms that cannot be altered by mere rotation of groups or atoms around single bonds. Configuration is determined during the polymerization process and cannot be altered except by breaking chemical bonds and forming new ones. Examples of different configurations are the D or L asymmetric carbon atom, the stereoregular sequences, such as isotactic (abbreviated as *it*) or syndiotactic (*st*), and the cis or trans isomers about the C=C double bond of diene polymers. Although the term configuration is also used by some authors in a wider sense, which includes the conformation and the shape of molecule as a whole, configuration is used here only according to the definition as given above.

On the other hand, "conformation" refers to the relative steric arrangement of atoms or groups that can be altered by rotation of the atoms or groups around a single bond. For example, conformation includes the trans and gauche arrangements of consecutive C—C single bonds and the helical arrangement of several polymers in the crystalline state. Figure 2.1 shows schematically the trans and two gauche conformations of the 1, 2-dihaloethane molecule. The C—C bond is perpendicular to the paper. The trans form is a staggered conformation in which the internal rotation angle X—C—C—X is 180°. This form is energetically the most stable. There are two types of gauche forms as shown in the figure. The gauche form is the next most stable conformation. In this book the trans and the two types of gauche forms are indicated by the notations T, G, and $\bar{G}$ (minus G), respectively, as shown in Fig. 2.1. Figure 2.2 shows the relationship between the potential energy and the internal rotation angle C—C—C—C of *n*-butane.

The cis conformation (denoted by C) corresponds to the internal rotation angle 0° and is the least stable form.

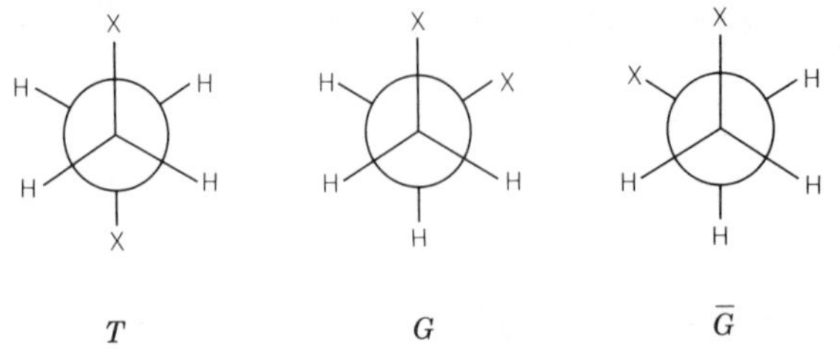

Fig. 2.1 The trans form and two types of gauche forms.

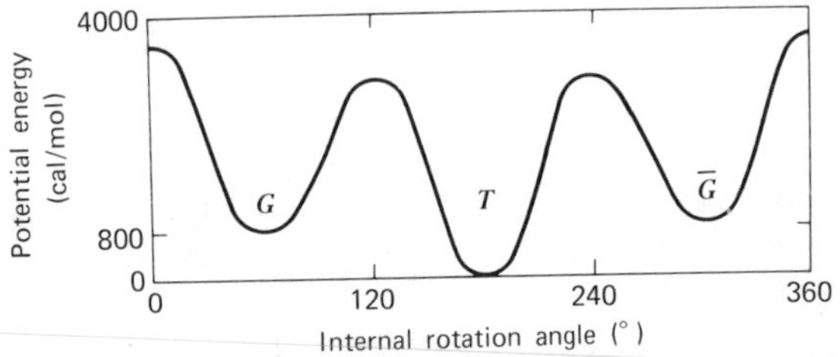

Fig. 2.2 Potential energy curve for internal rotation around the C—C bond of *n*-butane.

2.1 CONFIGURATION OF CHAIN POLYMERS

As early as 1932, Staudinger[1] recognized that the polymers of monosubstituted olefins $(\text{—CH}_2\text{—C*HR—})_n$ should contain asymmetric carbon atoms (denoted by an asterisk) along the chain. Thereafter the possibility of synthesizing such polymers in stereoregular forms was discussed by several authors (for instance, see Huggins[2]). The structure of such stereoregular polymers was really clarified for the first time by Natta et al. in 1955.[3] The definition of a stereoregular polymer is briefly discussed here according to the report on nomenclature proposed by Huggins et al.[4] See also the new nomenclature.[5]

An *it*-polymer is one in which the chemical structural unit possesses at least one main-chain carbon atom with two different substituent atoms or groups, and the configurations around the carbon atoms are all the same along the chain. If an *it*-polymer is shown with a Fischer projection, the substituents of each type in successive units are all placed on the same side of the line representing the main chain.

Two examples of the hypothetically extended zigzag chain of *it*-polymers are shown below with the corresponding Fischer projections:

it-polypropylene $[-CH_2CH(CH_3)-]_n$

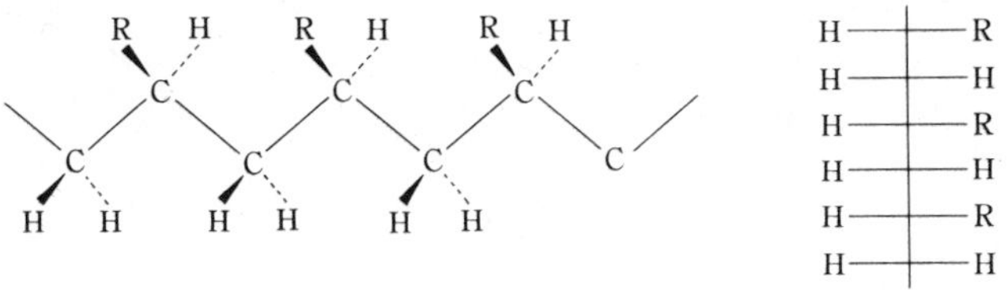

it-poly(propylene oxide) $[-CH_2CH(CH_3)O-]_n$

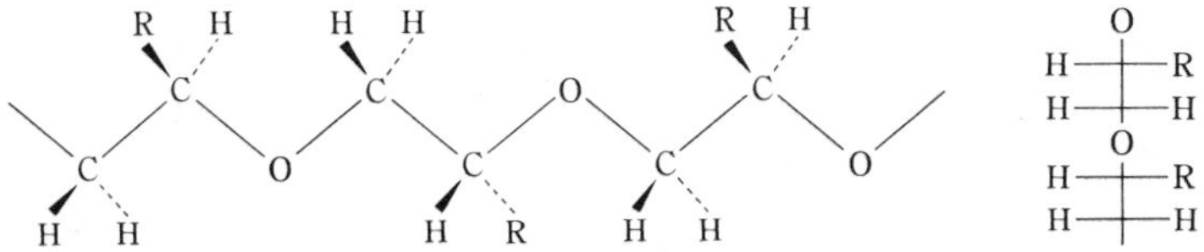

it-PP (there are two main-chain atoms per structural unit) has no true asymmetric carbon atom, while *it*-poly(propylene oxide) (three main-chain atoms) has true asymmetric carbon atoms and gives rise to two kinds of optical isomers. These are indicated by *R* (rectus) and *S* (sinister) according to the rules proposed by Cahn et al.[6] The above example belongs to *R* in this notation.

A *st*-polymer is a polymer in which the chemical structural unit possesses at least one main-chain carbon atom with two different substituents, and the configurations around the carbon atoms are opposite in successive structural units; for example,

st-polypropylene

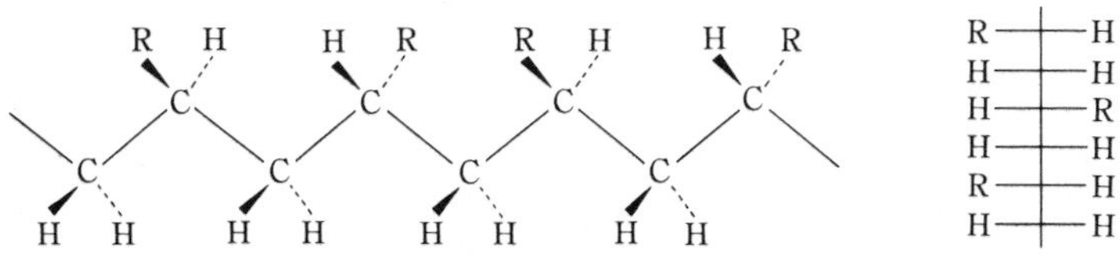

It should be clear from the description above that if a molecular chain is assumed to be extended as a planar zigzag, in the case of polymers composed of an even number of chain atoms per structural unit (for example, vinyl polymers), the substituents of *it*-polymers are all on the same side of the zigzag plane and those of *st*-polymers are located alternately on both sides. On the other hand, in the case of polymers composed of an odd number of chain atoms per structural unit, the situation is reversed; for example, in the case of *it*-poly(propylene oxide) the methyl groups alternate on both sides.

An atactic (*at*) polymer is one in which there is complete randomness with regard to the configurations at all the main-chain sites of steric isomerism. Polymers with a low degree of tactic order, having macroscopic properties practically indistinguishable from those of a strictly *at*-polymer, are also customarily designated as atactic.

In a diisotactic (*dit*) or disyndiotactic (*dst*) polymer, the structural unit possesses two carbon atoms, each having two different substituents, with configurations around each carbon atom in successive units making the molecule isotactic or syndiotactic, respectively. There are two types of *dit*-polymers; erythro-*dit*- and threo-*dit*-polymers in which the configurations at the two main-chain atoms in the structural unit are alike or opposite, respectively. The threo–erythro terminology originates from the names of D-threose and D-erythrose.

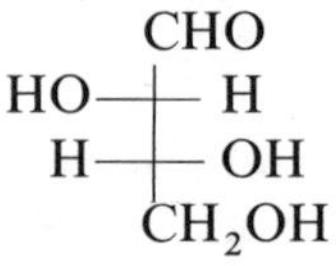

D-Threose

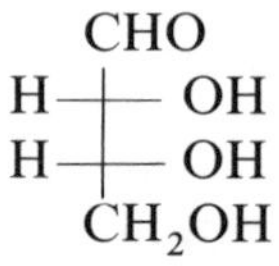

D-Erythrose

Threo-*dit*-polymer

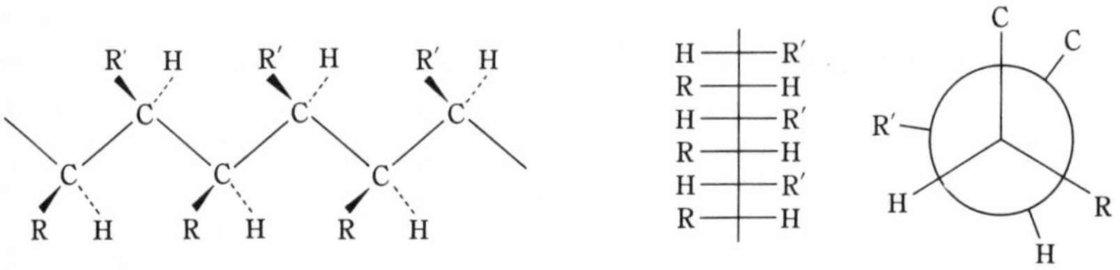

Erythro-*dit*-polymer

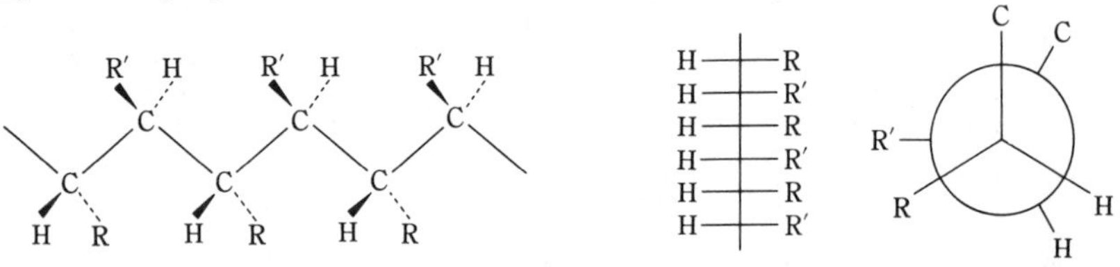

dst-Polymer

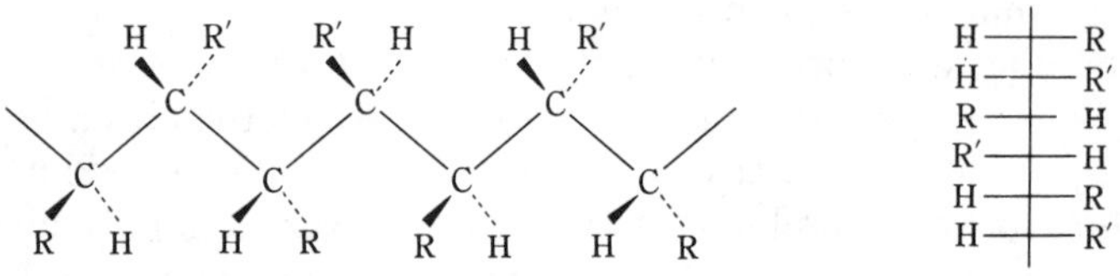

The Newman representations[7] are also given here. Only one type of *dst*-

polymer exists in the case with the structural unit composed of two main-chain atoms, but there are erythro- and threo-*dst*-polymers in the case of polymers composed of more than two main-chain atoms.

The structural unit of a polymer of 1,4-disubstituted butadiene contains two kinds of asymmetric carbon atoms and a C=C double bond. Such a polymer is called a "tritactic" polymer and is described in terms of cis or trans, erythro or threo, and *dit* or *dst*.

2.2 CONFORMATION OF CHAIN POLYMERS

Chain polymer molecules have their most stable conformations generally in the crystalline regions depending on the chemical structures, configurations, and so on of the molecules. There can be two or more types of conformations among the polymers, depending on the conditions of crystallization, which can result in various crystal modifications.

Figure 2.3 shows several possible conformations of single-bonded carbon chains. The nomenclature used here is a modified form of that first introduced by Bunn.[8] Although Bunn used the notations *A*, *B*, and *C* for the trans and two types of gauche forms, the notations *T*, *G*, and $\bar{G}$ are used here as described in the preceding section (Fig. 2.1). For example, $(TG)_3$ denotes a helical chain containing three alternations of *T* and *G* in the fiber identity period. This gives rise to a left-hand helix, while $(T\bar{G})_3$ gives a right-hand helix. The fiber identity period, or fiber period, is the special repeat length of the polymer chain in the crystalline region. This term originated from the fact that for high polymers the molecular axis (among the three axes of a crystal lattice) coincides in general with the fiber axis or elongated direction.

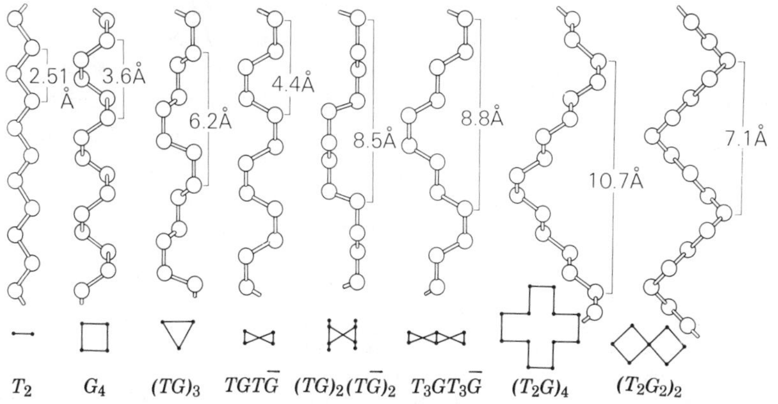

Fig. 2.3 Several possible conformations of a single-bonded carbon chain. Except for $(T_2G)_4$ and $(T_2G_2)_2$, these are reproduced from Bunn's paper.[8]

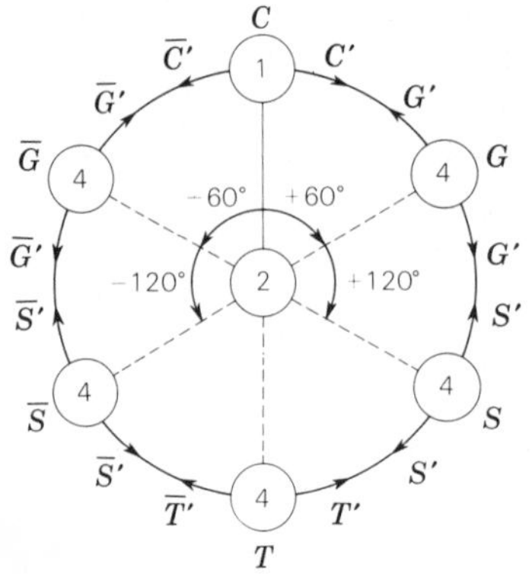

Fig. 2.4 Notations for the conformations and the internal rotation angles.

Figure 2.4 shows the notation of the conformation associated with the internal rotation angle. Here the sign of the internal rotation angle is defined as positive if, viewing the atoms along bond 2, 3 with 2 nearer the observer than 3, the angle from the projection of 2, 1 to the projection 3, 4 is traced in the clockwise direction.* The conformations deviating more or less from the exact G (60°), $\bar{G}$ (− 60°), S (skew, 120°), and $\bar{S}$ (− 120°) are denoted by a prime, if necessary. Those deviating from T and C are also indicated by T', $\bar{T}'$, and so on, depending on the signs of the internal rotation angles. The notation (u/t) gives a conformation in which the fiber identity period comprises u chemical units in t turns of the helix†: the same notation is used both for the right- and left-hand helices. For example, the conformation of *it*-PP is denoted by (3/1), since the fiber identity period comprises three chemical units in one turn of the helix.[9] The nonhelical polymers are indicated by $t = 0$.

The simplest conformation is the planar zigzag: the repetition of the trans forms $(TT\cdots)$. PE is a typical example and is expressed by T_2-(2/1).[10] A repetition of G or $\bar{G}$ gives rise to a right- or left-hand helix, G_4 or $\bar{G}_4$; the end view of this helix is a square as shown in Fig. 2.3. Orthorhombic polyoxymethylene (POM),[11] polyallene[12] [both G_4-(2/1)], and trigonal POM[13] [G'_{18}-(9/5)] are examples of this type of conformation. $TGT\bar{G}$ is the conformation of poly(vinylidene fluoride) form II.[14] An actual example of $(TG)_2(T\bar{G})_2$ has not been found. Examples of the $T_3GT_3\bar{G}$, $(T_2G)_4$, and $(T_2G_2)_2$ types are rubber hydrochloride,[15] poly(ethylene oxide) (PEO) [$(T_2'\ G')_7$-(7/2) helix],[16] and *st*-PP,[17] respectively.

* In X-ray crystallography and infrared and Raman spectroscopy the internal rotation angle of T is taken as 180°, while in the field of statistical thermodynamics it is 0°. Attention should be paid to this point, since no unified rule has yet been fixed.

† Although there is a different notation in which the polymer helix conformations are denoted by u_t, for instance, 3_1 for *it*-PP, (u/t) is used in this book as indicated here, since the notation u_t is used with a different meaning in crystallography (see Chapter 3).

2.3 FACTORS GOVERNING CONFORMATIONS (QUALITATIVE CONSIDERATION)

Conformations are governed by intramolecular interactions and intermolecular iteractions in the crystallite.

The intramolecular interactions include: (*a*) potential energy hindering the internal rotation around a single bond, (*b*) repulsive force and van der Waals attraction between nonbonded atoms or groups, (*c*) electrostatic interaction, and (*d*) hydrogen bonding. The intermolecular interactions also include *b*, *c*, and *d*, as well as the efficient packing in the crystal lattice as an integrated effect.

Although it may not be necessary to explain interaction *a* further, it should be noted again that the trans and two gauche forms are usually the first and the next most stable conformations, respectively. An example of interaction *b* may be seen in the case of polytetrafluoroethylene.[18] This molecule is a helix slightly distorted from a planar zigzag, in contrast to PE, which is a true planar zigzag as shown in Fig. 2.5*c*. In the case of PE, twice the van der

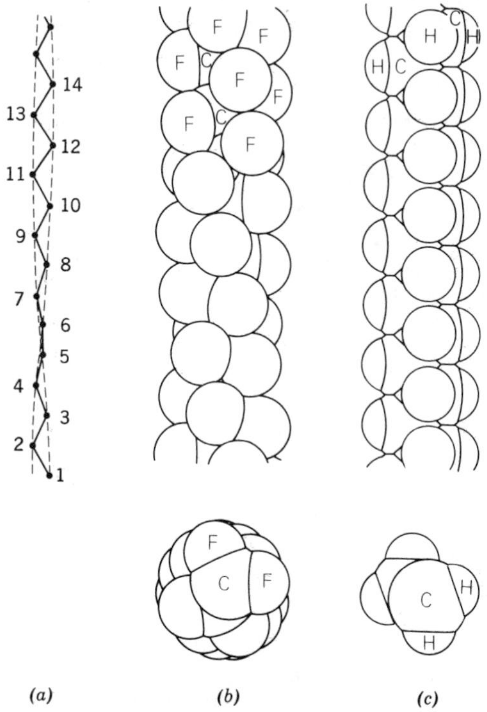

Fig. 2.5 (*a* and *b*) Molecular structure of polytetrafluoroethylene. (*c*) Molecular structure of PE (From Bunn and Howells[18]).

Waals radius of a hydrogen atom (2.2–2.4 Å) is shorter than the H···H distance in the planar zigzag (fiber period of PE 2.534 Å), so there is no appreciable repulsion. On the other hand, twice the van der Waals radius of the fluorine atom (2.7 Å) is considerably larger than the distance calculated for the planar zigzag form, and strong repulsion is expected. This is the main reason that the polytetrafluoroethylene molecule assumes a helical conformation instead of a planar zigzag. The polytetrafluoroethylene molecule is a (13/6) helix with a fiber period of 16.9 Å below 19°C and untwists a little above 19°C to form a (15/7) helix. *it*-PP, and so on assume helical conformations as a result of the repulsion between bulky side chains. The van der Waals radii of the atoms and groups important in polymer studies are given in Table 2.1. The distance of nonbonded atoms is determined by the van der Waals attraction and the repulsion due to the overlapping of the electron clouds. The nonbonded closest atomic distances have been shown to be approximately constant by structure analyses of various substances and may be estimated roughly by the sums of van der Waals radii. These are called van der Waals distances.

POM is used to illustrate interaction *c*. If only interactions *a* and *b* are taken into account the planar zigzag T_2 and the (2/1) helix G_4 are considered

Table 2.1 Van der Waals Radii (from Pauling[19])

Atom or group	Radius (Å)
H	1.2
N	1.5
O	1.40
F	1.35
P	1.9
S	1.85
Cl	1.80
Br	1.95
I	2.15
CH_2	2.0
CH_3	2.0
Half of the thickness of the benzene ring[20]	1.75–1.80

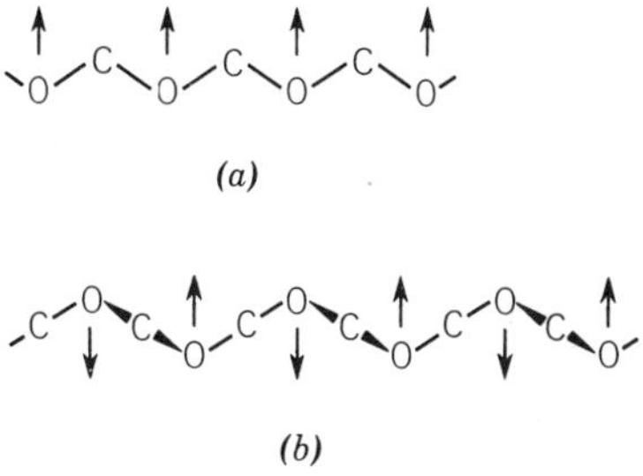

Fig. 2.6 Orientation of the dipole moments of the POM molecule: (*a*) planar zigzag conformation; (*b*) (2/1) helix conformation.

to be stable (Fig. 2.6). But there exist strong dipole moments at the COC groups of this molecule. If the molecule is in the planar zigzag conformation, the dipole moment vectors are parallel. On the other hand, in the case of a helix, the dipole moments are alternately antiparallel, and this orientation is far more favorable than the former one. Thus one of the important factors causing the helical structure may be the intramolecular dipole interaction.* The reason that the orthorhombic form with the (2/1) helix[11] is less stable experimentally than the trigonal form with the (9/5) helix[13] is not yet clear, but it may be due to the intermolecular interactions in the crystal.

Examples of interaction *d*, include the intramolecular hydrogen bonds of the α-helix of polypeptides[21] and the intermolecular hydrogen bonds of the β-form of polypeptides,[22] polyamides,[23] poly(vinyl alcohol) (PVA),[24] and so on.

The stability of molecular conformations and crystal modifications is discussed in detail in Chapter 6.

REFERENCES

1. H. Staudinger, *Die Hochmolekularen Organischen Verbindungen, Kautschuk und Cellulose*, Springer-Verlag, Berlin, 1932.
2. M. L. Huggins, *J. Am. Chem. Soc.*, **66,** 1991 (1944).
3. G. Natta, *J. Polym. Sci.*, **16,** 143 (1955).
4. M. L. Huggins, G. Natta, V. Desreux, and H. Mark, *J. Polym. Sci.*, **56,** 153 (1962); *Makromol. Chem.*, **81,** 1 (1965).
5. IUPAC Commission on Macromolecular Nomenclature, *Pure Appl. Chem.*, **40,** 479 (1974).
6. R. S. Cahn, C. K. Ingold, and V. Prelog, *Experientia*, **12,** 81 (1956).
7. M. S. Newman, *Steric Effects in Organic Chemistry*, Wiley, New York, 1956, p. 10.
8. C. W. Bunn, *Proc. Roy. Soc. (Lond.)*, **A180,** 67 (1942).
9. G. Natta and P. Corradini, *Nuovo Cimento, Suppl.*, **15,** 40 (1960).

* According to energy calculations, both the T_2 and G_4 conformations give potential minima only if interactions *a* and *b* are taken into account, but the G_4 is more stable. In addition to *a* and *b*, interaction *c* also stabilizes the G_4 form.

10. C. W. Bunn, *Trans. Faraday Soc.*, **35,** 482 (1939).
11. G. Carazzolo and M. Mammi, *J. Polym. Sci.*, **A1,** 965 (1963).
12. H. Tadokoro, Y. Takahashi, S. Otsuka, Kan Mori, and F. Imaizumi, *J. Polym. Sci. B*, **3,** 697 (1965).
13. H. Tadokoro, T. Yasumoto, S. Murahashi, and I. Nitta, *J. Polym. Sci.*, **44,** 266 (1960).
14. R. Hasegawa, Y. Takahashi, Y. Chatani, and H. Tadokoro, *Polym. J.*, **3,** 600 (1972).
15. C. W. Bunn and E. V. Garner, *J. Chem. Soc.*, **1942,** 654.
16. H. Tadokoro, Y. Chatani, T. Yoshihara, S. Tahara, and S. Murahashi, *Makromol. Chem.*, **73,** 109 (1964).
17. G. Natta, I. Pasquon, P. Corradini, M. Peraldo, M. Pegoraro, and A. Zambelli, *Atti Accad. Naz. Lincei, Rend., Cl. Sci. Fis. Mat. Nat.*, **28,** 539 (1960).
18. C. W. Bunn and E. R. Howells, *Nature*, **174,** 549 (1954).
19. L. Pauling, *The Nature of the Chemical Bond*, 3rd ed., Cornell Univ. Press, Ithaca, N. Y., 1960.
20. A. Bondi, *J. Phys. Chem.*, **68,** 441 (1964).
21. L. Pauling, R. B. Corey, and H. R. Branson, *Proc. Natl. Acad. Sci. U.S.A.*, **37,** 205 (1951).
22. R. E. Marsh, R. B. Corey, and L. Pauling, *Acta Crystallogr.*, **8,** 710 (1955).
23. C. W. Bunn and E. V. Garner, *Proc. Roy. Soc. (Lond.)*, **A189,** 39 (1947).
24. C. W. Bunn, *Nature*, **161,** 929 (1948).

Chapter 3

Symmetry of Molecules and Crystals

3.1 SYMMETRY OF MOLECULES

A chloroform molecule has the geometrical shape shown in Fig. 3.1. If the molecule is rotated by $2\pi/3$ about the C—H bond axis, it assumes a position indistinguishable from its original one, unless some kind of label is attached to the chlorine atoms. Such positions, which leave the molecule invariant, are said to be equivalent; by the operation the molecule is transformed into itself. A subsequent rotation of $2\pi/3$ also brings the molecule to another equivalent position. Such an axis of rotation is called the threefold axis or the threefold rotation axis, which is one example of symmetry elements. In general, a molecule with an n-fold axis can assume a position equivalent to the original one by a $2\pi/n$ rotation about the axis, and it has n equivalent positions. There are only two-, three-, four-, and sixfold axes in crystals,* but no such limitation exists for the number n in free molecules.

The operation of rotating a molecule from one equivalent position to another is a symmetry operation. A symmetry operation should be carefully distinguished from a symmetry element. Two types of symbols are used, to describe symmetry; those of Schoenflies and those of Hermann-Mauguin. At present, Schoenflies' symbols are mainly used in spectroscopy, and Hermann-Mauguin's in crystallography. As shown in Table 3.1, the rotation axes are indicated by C_n according to Schoenflies' symbols, and by Arabic figures 2, 3,··· according to Hermann-Mauguin's symbols. The various kinds of symmetry are now explained in terms of these two sets of symbols.[2,3]

A mirror plane or a plane of symmetry is another symmetry element, indicated by σ or m (Table 3.2). The corresponding symmetry operation is reflection through the plane. The chloroform molecule has three mirror planes, each coinciding with the plane H—C—Cl. If a straight line is drawn from one atom through a point within a molecule so that the extension meets

* For this reason see, e.g., Ref. 1.

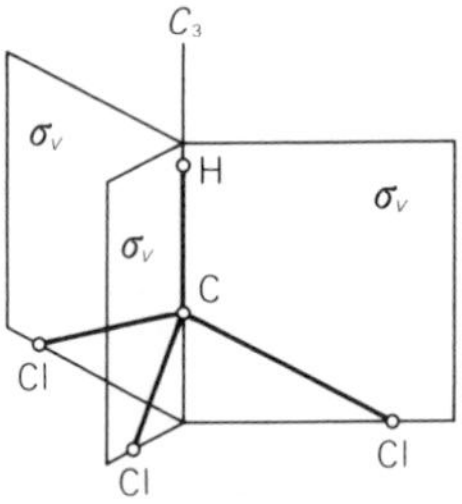

Fig. 3.1 Molecular symmetry of chloroform, C_{3v}.

an equivalent atom at the same distance from the point, this point is a center of symmetry and is denoted by i or $\bar{1}$. The corresponding symmetry operation is known as inversion. When the center of symmetry is taken as the origin of the coordinate system, one atom with the coordinates (x, y, z) is always accompanied by an equivalent atom at $(-x, -y, -z)$. The operation of inversion combined with that of rotation produces a new type of symmetry element known as the rotatory-inversion axis; these axes are represented by symbols $\bar{1}$, $\bar{2}$, $\bar{3}$, $\bar{4}$, and $\bar{6}$. Obviously $\bar{1}$ is the operation of inversion itself, and $\bar{2}$ is equivalent to the reflection across a plane perpendicular to the twofold axis. In spectroscopy another type of symmetry element is used, namely, the rotatory-reflection axis, usually designated by S_n. In a molecule having such an axis, a rotation by $2\pi/n$ about the axis followed by a reflection at a plane perpendicular to the axis transforms the molecule into itself. Of the symmetry elements appearing in the crystal, 1, 2, 3, 4, and 6 are called symmetry elements of the first kind, and $\bar{1}$, $\bar{2}$ $(=m)$ and $\bar{4}$ are those of the second kind.* If a symmetry operation of the first kind (rotation) is applied to a

Table 3.1 Axes of Symmetry[2].

Symmetry	No symmetry	Twofold axis	Threefold axis	Fourfold axis	Sixfold axis
Schoenflies' symbol	C_1	C_2	C_3	C_4	C_6
Hermann-Mauguin's symbol	1	2	3	4	6
Graphical symbol	None	⬮ (normal to paper) ⟶ (parallel to paper)	▲	◆	⬢

Reproduced by courtesy of the International Union of Crystallography.

* $\bar{3}$ and $\bar{6}$ are not independent symmetry elements, since $\bar{3}$ is the combination of 3 and $\bar{1}$, and $\bar{6}$ is a combination of 3 and $\bar{2}$.

Table 3.2 Rotatory-Inversion Axes[2]

Symmetry	Center of symmetry	Mirror plane	Threefold rotatory inversion axis (sixfold rotatory reflection axis)	Fourfold rotatory inversion axis (fourfold rotatory reflection axis)	Sixfold rotatory inversion axis (threefold rotatory reflection axis)
Schoenflies' symbol	i	σ	S_6	S_4	S_3
Hermann-Mauguin's symbol	$\bar{1}$	$m(=\bar{2})$	$\bar{3}$	$\bar{4}$	$\bar{6}$
Graphical symbol	○	—[a]	▲	◆	⬢

Reproduced by courtesy of the International Union of Crystallography.
[a]Normal to plane of projection.

geometrical pattern without the symmetry elements of the second kind (center of symmetry, mirror plane, etc.), the patterns before and after the operation are equivalent. On the other hand, a symmetry operation of the second kind (inversion, reflection, etc.) in this case gives the mirror image of the configuration before the operation, that is, the enantiomer. A geometrical pattern having no symmetry elements of the second kind is called chiral, while one having these elements is called achiral. In other words, if a molecule is equivalent to its mirror image, the molecule is achiral; otherwise it is chiral.

3.2 POINT GROUPS

The set of symmetry operations possessed by a molecule forms a group in the mathematical sense discussed in Chapter 5. All groups associated with molecules are called point groups because the molecules are all left with one point unaltered by any of the symmetry operations. All molecules can be classified into a small number of groups according to the number and nature of the symmetry elements. The reader is referred to Ref. 3 for examples of molecules belonging to each point group.

The simplest point group is C_1, to which molecules with no symmetry belong. It consists of only the identity E, which holds all the atoms in their original positions. A rotation by $2\pi k/n$ around an n-fold rotation axis C_n is represented by the symbol $C_n{}^k$. The collection of such operations $E\,(=C_n{}^0)$, $C_n{}^1, C_n{}^2, \cdots, C_n{}^{n-1}$ forms the point group C_n. Another group having n twofold axes at right angles to the principal symmetry axis C_n is called the group D_n (from the German *Diedergruppe*). Group D_2 has three equivalent twofold axes at right angles to each other and is also called V(from *Vierergruppe*).

Adding n vertical planes σ_v to a vertical axis C_n gives rise to a new group C_{nv}. The chloroform molecule shown in Fig. 3.1 belongs to the group C_{3v}. Adding a horizontal plane σ_h to a vertical axis C_n results in group C_{nh}. The special case with $n=1$ contains the identity E and the plane σ_h and is usually called C_s (the subscript s originates from *Spiegel*). D_n plus the mirror plane σ_h perpendicular to the C_n axis results in another group D_{nh}. D_{2h} is also called V_h.

For the groups S_n whose elements are $E, S_n{}^1, S_n{}^2, \cdots, S_n{}^{n-1}$, new groups are obtained only when n is even, since when n is odd the groups are identical to C_{nh}. When $n=2$, $S_2=C_i$, which is a group consisting of E and $S_2^1=i$.

If a set of vertical planes σ_d through the axis C_n bisecting the angles between the twofold axes is added to D_n, another series of groups D_{nd} arises (the subscript d originates from *diagonal*). For example, the ethane molecule with the staggered conformation belongs to D_{3d}.

There remain seven other possible point groups (T, T_d, T_h, O, O_h, I and I_h),

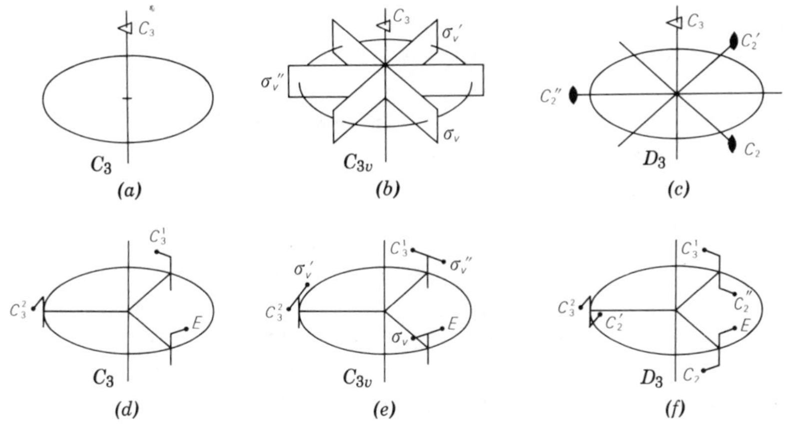

Fig. 3.2 Symmetry elements of point groups: (*a*) C_3 (*b*) C_{3v}; and (*c*) D_3. In *d–f*, the symbol labeling each point indicates the operation that carries the point *E* to that position (From E. B. Wilson, Jr, J. C. Decius, and P. C. Cross, *Molecular Vibrations*, Copyright McGraw-Hill Book Co., New York, 1945. Used with permission of McGraw-Hill Book Co.)

but these are not important for polymer study. Figure 3.2 shows the symmetry elements of three point groups and the points to which the original point (indicated by *E*) is shifted by the symmetry operations of the point groups.

The symmetries of crystals are divided into 32 crystal classes, which correspond to 32 crystallographic point groups. According to Hermann-Mauguin's symbols, $2/m$ means the combination of a twofold axis with a mirror plane normal to it, that is, C_{2h}. The symbol $2mm$ denotes the combination of a twofold axis and two mirror planes passing through the axis, C_{2v}. Symmetry elements concerning the main axis are written first. For the point groups of high symmetry a short symbol is used instead of the full symbol, since only some of the symmetry elements are sufficient to represent the point symmetry unambiguously. For instance, *mmm* denotes three mirror planes each perpendicular to the *x*, *y*, or *z* axis, and by this combination the *x*, *y*, and *z* axes should all be twofold axes. Consequently *mmm* is the short symbol of $2/m\ 2/m\ 2/m$ and is identical to D_{2h}. The symbol 32 is the same as D_3 (trigonal) and means that the main axis is a threefold axis and the subaxes are twofold, while 23, which is the same as *T* (cubic), means that the main axis is a twofold axis and the subaxes are threefold axes.

3.3 SYMMETRY OF POLYMER CHAINS (ONE-DIMENSIONAL SPACE GROUPS)

In crystalline regions the polymer molecule has a regular sequence of repeating units along the molecular axis. Although the size of the crystalline region is generally 100–150 Å, for simplicity an idealized system composed of a perfectly straight and infinitely long polymer chain, that is, a one-dimensional crystal, may be assumed. This idealized chain molecule has translational symmetry; it is invariant on translation along the chain axis in either direction by any multiple of the fiber identity period or unit cell length. Figure 3.3 shows the symmetry of a fully extended PE molecule. The distance between $C_I(j)$ and $C_I(j+1)$ is the fiber identity period. The above-mentioned translation is a symmetry operation and is denoted by T_n ($n = -\infty, \cdots, -1, 0, +1, \cdots, +\infty$). T_0 is equal to the identity operation E.

Many polymer molecules have some symmetry elements besides translation. In the PE molecule (Fig. 3.3), there are two kinds of twofold axes, $C_2(y)$ passing through the carbon atoms in the y direction, and $C_2(z)$ passing through the centers of the C—C bonds in the z direction. There are centers of symmetry i at the centers of the C—C bonds. The molecular plane containing all the carbon atoms $\sigma(xy)$ and the set of planes perpendicular to the chain axis at the carbon atoms $\sigma(yz)$ are mirror planes. The axis denoted by $C_2^s(x)$ is a twofold screw axis in which the symmetry operation is a 180° rotation about the molecular axis followed by a translation of half the length of the cell dimension along the molecular axis. The plane $\sigma_g(xz)$ is a glide plane, which is a combination of a mirror plane and a translation of half the length of the cell dimension. The whole set of symmetry operations corresponding to these symmetry elements and the translations including the identity E form a one-dimensional space group or a line group. (For group theory see Chapter 5.)

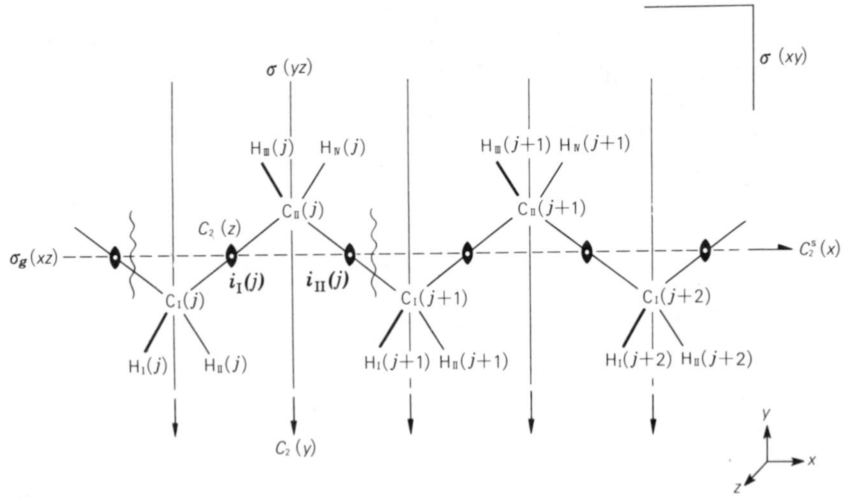

Fig. 3.3 Symmetry of an extended PE molecule.

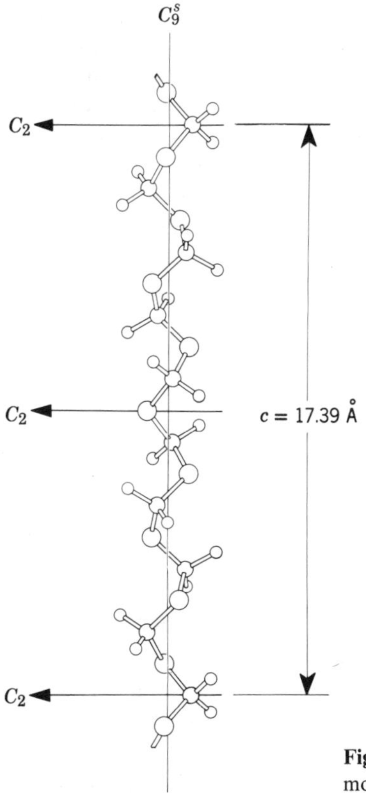

Fig. 3.4 A (9/5) helix symmetry of a trigonal POM molecule.[5]

The POM molecule in the trigonal form assumes a (9/5) helix in which nine monomeric units turn five times to form a fiber identity period of 17.39 Å, as shown in Fig. 3.4.[5] There is a ninefold screw axis $C_9{}^s$, which produces symmetry operations such as a rotation through $(5 \times 2\pi)/9$ followed by a translation along the axis by $\frac{1}{9}$ of the identity period. A rotation of $10\ \pi k/9$ followed by a translation of $k/9$ of the period is a symmetry operation, where $k = 1, 2, \cdots, 9$, with the special case of $k = 9$ corresponding to a translation of one identity period, T_1. Twofold axes passing through the oxygen or carbon atoms and crossing the chain axis at right angles are symmetry element of another kind.

3.4 SYMMETRY OF CRYSTALS (SPACE GROUPS)

A three-dimensional crystal is built up by contiguous repetition of unit cells in three dimensions. The unit cell is described by specifying the lengths

Table 3.3 Crystal Systems and Bravais Space Lattices

Crystal system	Axial length	Interaxial angle	Symbol of space lattice
Triclinic	$a \neq b \neq c$	$\alpha \neq \beta \neq \gamma$	P
Monoclinic	$a \neq b \neq c$	$\alpha = \gamma = 90° \neq \beta$	P, $A(C)$
Orthorhombic	$a \neq b \neq c$	$\alpha = \beta = \gamma = 90°$	P, $A(B, C)$, F, I
Tetragonal	$a = b \neq c$	$\alpha = \beta = \gamma = 90°$	P, I
Rhombohedral (trigonal)	$a = b = c$	$\alpha = \beta = \gamma < 120°$ ($\neq 90°$)	R
Hexagonal	$a = b \neq c$	$\alpha = \beta = 90°$, $\gamma = 120°$	P
Cubic	$a = b = c$	$\alpha = \beta = \gamma = 90°$	P, F, I

of cell edges a, b, and c and interaxial angles α, β, and γ (e.g., the angle between the b and c axes is denoted by α). These six parameters are called lattice constants.

By combining the 14 types of Bravais space lattices (Table 3.3) with the symmetry elements of the first and second kinds, as well as with screw axes and glide planes, the 230 space groups can be derived. In Table 3.3 P denotes the primitive lattice, in which the points are only at the corners of the lattice. The others are nonprimitive lattices and are built up by combining two or more primitive lattices translated parallel to each other. For convenience the nonprimitive lattices are sometimes used instead of the primitive lattices, although the former can be expressed by the latter in terms of different choices of axes. A, B, or C is the base-centered lattice (centered on the *bc*, *ac*, or *ab* face), F is the face-centered lattice, and I is the body-centered lattice (I originates from *innenzentriertes Gitter*). For some trigonal crystals the symbol R (rhombohedral lattice) is used, representing a primitive lattice with the shape of a rhombohedron.

In Hermann-Mauguin's symbols, the screw axes are denoted by p_q. This means a rotation of $2\pi/p$ of a right-hand turn combined with a translation of q/p the length of the identity period. In a crystal, the kinds of screw axes are subjected to the same limitation as the kinds of rotation axes, and so only the following 11 kinds of screw axes are possible: $2_1, 3_1, 3_2, 4_1, 4_2, 4_3, 6_1, 6_2, 6_3, 6_4$, and 6_5. For example, 3_1 and 3_2 are the right- and left-hand threefold screw axes, respectively, and 3_1 is the mirror image of 3_2. The glide planes are given in Table 3.4.

Of the space groups for 141 examples of polymer crystal structures, the number of those belonging to $P2_1/c\text{-}C_{2h}{}^5$ (including $P2_1/b$, $P2_1/a$, and $P2_1/n$) is 24, by far the largest amount. The next largest amounts belong to $P2_12_12_1\text{-}D_2{}^4$ and $P\bar{1}\text{-}C_i{}^1$, both 14; followed by $P1\text{-}C_1{}^1$, 9; $Pnam\text{-}D_{2h}{}^{16}$ (including $Pcmn$), 7; $R3c\text{-}C_{3v}{}^6$ (or $R\bar{3}c\text{-}D_{3d}{}^6$), 6; $P2_1cn\text{-}C_{2v}{}^9$, 6; and so on.

Table 3.4 Glide Planes[a]

		Graphical symbol		
Symbol	Glide plane	Normal to plane of projection[b]	Parallel to plane of projection[c]	Length of glide translation
a	Axial glide plane	Translation parallel to one of the axes on the plane of projection		$\frac{a}{2}$
b		- - - - - - - - - - - - - -		$\frac{b}{2}$
c		Translation perpendicular to the plane of projection · · · · · · · · · · · · · · · · · ·		$\frac{c}{2}$
n	Diagonal glide plane (net)	-·-·-·-·-·-·-·-·-·-·-·-		$(a+b)/2$, $(b+c)/2$, or $(c+a)/2$; or $(a+b+c)/2$ (tetragonal and cubic)

Reproduced by courtesy of the International Union of Crystallography.

[a]Diamond glide plane is omitted because it has no importance in polymers.

[b]In Ref. 2, p. 49, dashed and dotted lines are defined so as to be used only for the c axis projection. Here the definition is modified to be applicable to the projection along any axis.

[c]  denotes the existence of two glide planes, e.g., along the a and b axes in the C base-centered lattice.

The *International Tables for X-Ray Crystallography*[2] are explained here for the cases of $P2_1/c$-$C_{2h}{}^5$ and $Pnam$-$D_{2h}{}^{16}$. The space group is represented by combining the symbols of Hermann-Mauguin and Schoenflies. The Hermann-Mauguin symbol consists of the lattice symbol and symmetry elements; only the minimum elements necessary for representing the space group are usually given. The Schoenflies symbol for a space group is the crystal class symbol belonging to the space group with a superscript of an Arabic number.

3.4.1 Example 1: $P2_1/c$-$C_{2h}{}^5$ (Full Symbol $P\,1\,2_1/c\,1$)

Figure 3.5 is a reproduction of p. 99 of the *International Tables*, Vol. I. In row 1, the crystal system, crystal class, and full symbol of the space group

(1) Monoclinic $2/m$ $P1\,2_1/c\,1$ No. 14 $P2_1/c$ C_{2h}^5

(2)

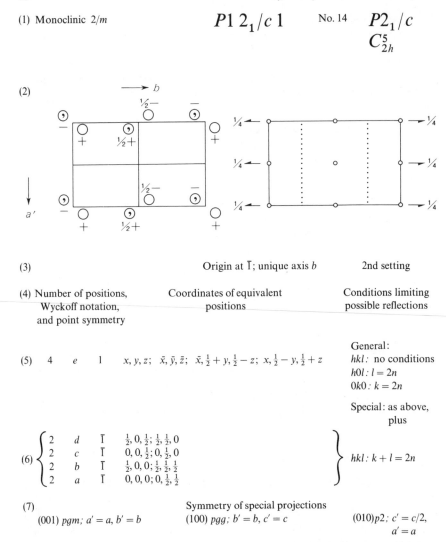

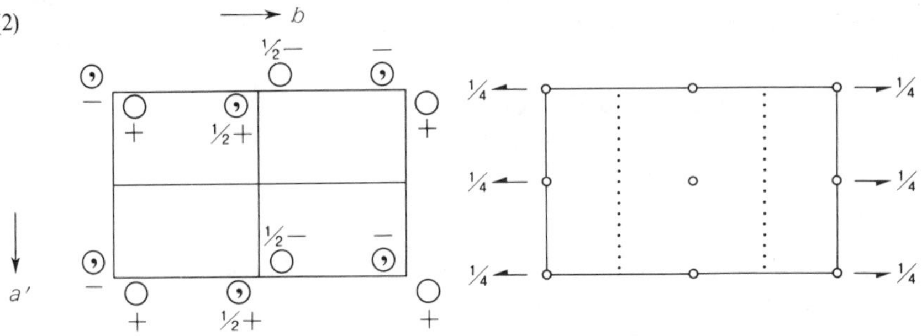

(3) Origin at $\bar{1}$; unique axis b 2nd setting

(4) Number of positions, Wyckoff notation, and point symmetry			Coordinates of equivalent positions	Conditions limiting possible reflections
				General:
(5) 4	e	1	$x, y, z;\ \bar{x}, \bar{y}, \bar{z};\ \bar{x}, \frac{1}{2}+y, \frac{1}{2}-z;\ x, \frac{1}{2}-y, \frac{1}{2}+z$	hkl: no conditions $h0l$: $l=2n$ $0k0$: $k=2n$
				Special: as above, plus
(6) 2	d	$\bar{1}$	$\frac{1}{2}, 0, \frac{1}{2};\ \frac{1}{2}, \frac{1}{2}, 0$	hkl: $k+l=2n$
2	c	$\bar{1}$	$0, 0, \frac{1}{2};\ 0, \frac{1}{2}, 0$	
2	b	$\bar{1}$	$\frac{1}{2}, 0, 0;\ \frac{1}{2}, \frac{1}{2}, \frac{1}{2}$	
2	a	$\bar{1}$	$0, 0, 0;\ 0, \frac{1}{2}, \frac{1}{2}$	

(7) Symmetry of special projections

(001) pgm; $a' = a$, $b' = b$ (100) pgg; $b' = b$, $c' = c$ (010) $p2$; $c' = c/2$, $a' = a$

Fig. 3.5 $P2_1/c$-C_{2h}^5 (second setting). (Reproduced by courtesy of the International Union of Crystallography.[2])

are given. No. 14 refers the space group number. The full symbol $1\ 2_1/c\ 1$ denotes the symmetry associated with the a, b, and c axes, respectively.

In the diagrams in row 2, the c axis is perpendicular to the page, the positive direction being upward from the page (a right-hand coordinate system). The upper left corner is taken as the origin, the projection of the a axis (along the *c* axis) points downward, and the *b* axis points in the right-hand direction.* The left-hand diagram shows the equivalent general positions produced by operation of the symmetry elements of the space group on one initial position chosen arbitrarily. The symbols +, $\frac{1}{2}-$, and so forth mean $+z$, $\frac{1}{2}-z$, and so on, where z is the fractional coordinate of the position along the *c* axis expressed in terms of the length *c*. The two positions denoted by ○ and ⊙ are those related by the symmetry operations of the second kind, that is, enantiomorphous to each other.

The right-hand diagram shows the positions of the symmetry elements, the 2_1 screw axis; *c* glide plane, and center of symmetry being denoted by a single-edged arrow (⇀), dotted line (·····), and small circle ∘, respectively. Although the 2_1 screw axes are at heights one quarter and three quarters along the *c* axis, the notation $\frac{3}{4}$ is omitted for simplicity. No indication is given for the center of symmetry, which is at heights 0 and $\frac{1}{2}$. The *b* axis projection of the lattice $P\ 1\ 2_1/c\ 1\text{-}C_{2h}{}^5$ (second setting) is shown in Fig. 3.6. The positive direction of the *b* axis points upward from the page. ◗ denotes the 2_1 screw axis perpendicular to the page, and $\frac{1}{4}$ denotes the glide plane parallel to the page with the glide translation along the axis indicated by the arrow. $\frac{1}{4}$ has the same meaning as explained above.

In row 3, "origin at $\bar{1}$" indicates that the position of the center of symmetry is taken as the origin. In the monoclinic space groups there are two kinds of setting for the axes; in the first setting, γ is not equal to 90° and the *c* axis is called the unique axis, while in the second setting the *b* axis is the unique axis. *International Tables*, Vol. I, shows the first and second settings on the

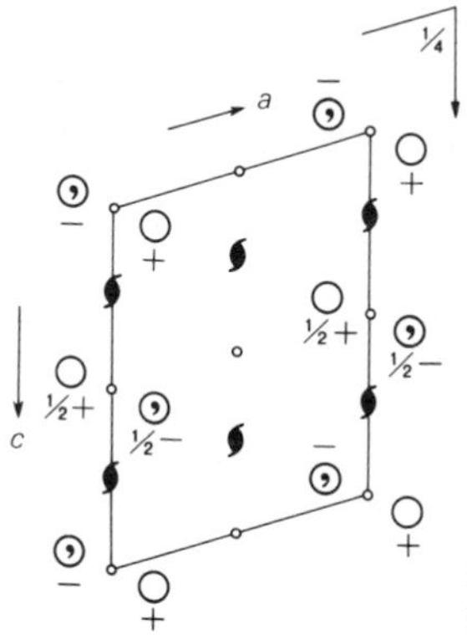

Fig. 3.6 The *b* axis projection of the lattice $P\ 1\ 2_1/c\ 1\text{-}C_{2h}{}^5$ (second setting).

* In the triclinic and in the second setting of the monoclinic system, the *a* axis (e.g., for Fig. 3.5) is not on the plane of projection, but is inclined to it, since the diagrams are clinographic projections. In the first setting of the monoclinic system, and in the systems with higher symmetries, the diagrams are orthographic projections. In all systems, except for the rhombohedral, the projection is made in the same way as shown here.

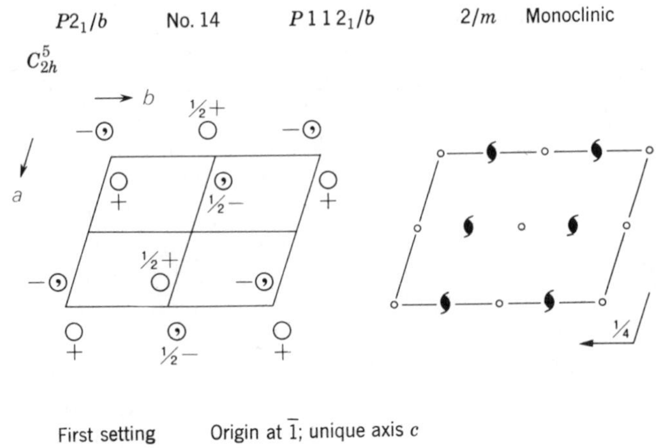

Fig. 3.7 $P2_1/b$-$C_{2h}{}^5$ (first setting). (Reproduced by courtesy of the International Union of Crystallography.[2])

left- and right-hand pages for the monoclinic space groups. Figure 3.7 shows parts of the description on p. 98 of the *International Tables*, Vol. I.

In row 5 of Figure 3.5 four is the number of the general equivalent positions. If the equivalent positions are located on a symmetry element, the number of positions decreases. In row 6 such equivalent positions are reduced to two and are called the special equivalent positions. *a*, *b*, *c*, *d*, and *e* in rows 5 and 6 denote the sets of equivalent points and are called the Wyckoff notation. 1, $\bar{1}$, and so forth indicate the symmetry of the equivalent points. A point in the crystal has one of the symmetries corresponding to the 32 crystallographic point groups and is usually called the point symmetry. The point symmetry of the general equivalent positions is 1 (that is, no symmetry), and those of the special equivalent positions shown in rows 6 are $\bar{1}$, that is, on the center of symmetry. The coordinates of the general equivalent positions are x, y, z; $\bar{x}, \bar{y}, \bar{z}$; $\bar{x}, \frac{1}{2}+y, \frac{1}{2}-z$; $x, \frac{1}{2}-y, \frac{1}{2}+z$. $\bar{x}, \bar{y}, \bar{z}$ denote $-x, -y, -z$, respectively. If, for instance, x, y, z is set equal to 0, 0, 0, the position $\bar{x}, \bar{y}, \bar{z}$ also becomes 0, 0, 0 and the other two positions become $0, \frac{1}{2}, \frac{1}{2}$, giving the special equivalent positions *a*. The systematic absences are given on the right-hand side of rows 4–6 (see section 4.2.7. F). There is no limiting condition for the appearance of the reflections *hkl*,* but there are conditions $l = 2n$ for $h0l$ and $k = 2n$ for $0k0$. On the right-hand side of rows 6, additional conditions are given for a crystal consisting of only the atoms at the special equivalent positions *a*, *b*, *c*, and *d*; the reflections *hkl* can be observed only for $k + l = 2n$.

Row 7 gives the notations for the two-dimensional space groups (plane groups) produced by projecting the lattice along the *a*, *b*, and *c* axes, denoted

* *hkl* is used for the indices of a reflection and the coordinates of a reciprocal-lattice point.

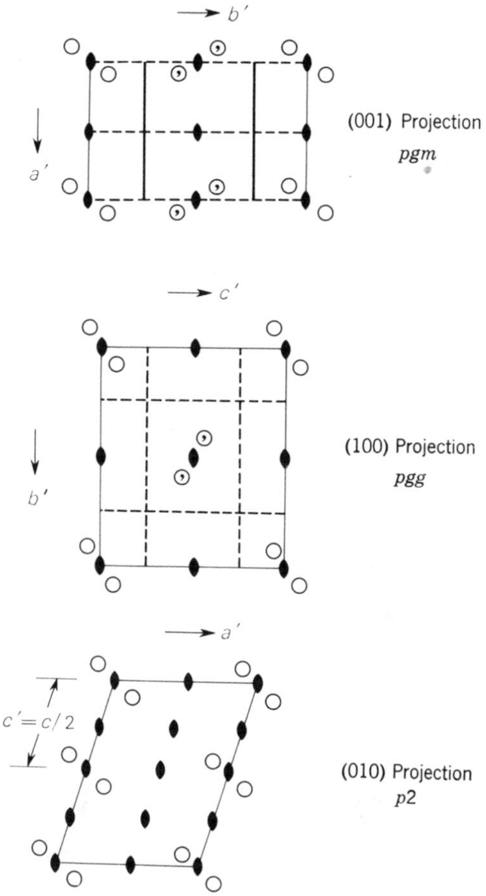

Fig. 3.8 Plane groups produced by projecting the lattice $P2_1/c\text{-}C_{2h}^{5}$ along the three axes.

by (100),† and so on. Such projections are shown in Fig. 3.8. There are 17 two-dimensional space groups, which are represented by using the Hermann-Mauguin symbols with lower-case letters. The glide planes are commonly denoted by *g*. The two-dimensional space groups are shown in the *International Tables*, Vol. I (pp. 57–72), and are useful for trial-and-error procedures for determining projected structures from equatorial reflections. In the plane groups the symmetry elements for the space groups change as shown on the next page. On the (010) projection in Fig. 3.8, the upper and lower halves are the same, the lattice having the axial length $c' = c/2$.

† (*hkl*) is used for indices of a crystal face, a single plane, or a set of parallel planes.

Space group		Plane group
n-Fold screw axis parallel to the axis of projection	⟶	*n*-Fold rotation axis (including twofold rotation axis, if *n* is even)
Center of symmetry	⟶	Twofold rotation axis
Glide plane along the axis of projection	⟶	Mirror plane
Twofold screw axis perpendicular to the axis of projection and diagonal glide plane parallel to the axis of projection	⟶	Glide plane

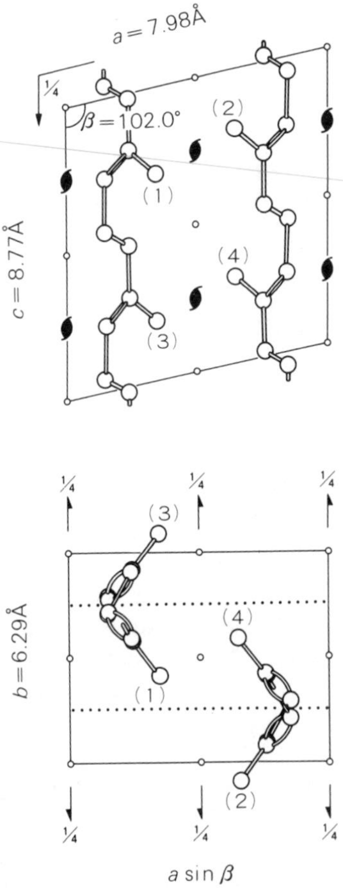

Fig. 3.9. Crystal structure of α-gutta percha ($P2_1/c$-C_{2h}^5).[6]

Figure 3.9 shows the crystal structure of α-gutta percha (*trans*-1,4-polyisoprene)[6] [—CH_2—$C(CH_3)$=CH—CH_2—]$_n$ as an example of $P2_1/c$. Methyl group 1 transforms to methyl group 2 by 2_1, to methyl group 3 by the *c* glide plane, and to methyl group 4 by the center of symmetry. These methyl groups are at the general equivalent positions. Rubber hydrochloride[7] [—CH_2 – $CH(CH_3)$—CHCl—CH_2—]$_n$ also belongs to $P2_1/c$, but is a special case with $\beta = 90°$.

If the *a* and *c* axes in $P\,1\,2_1/c\,1$ are exchanged, $P\,1\,2_1/a\,1$ is obtained; for example, see the crystal structure of poly(pentamethylene sulfide)[8] [—$(CH_2)_5$—S—]$_n$ shown in Fig. 3.10. In α-gutta percha the molecular chain has a glide symmetry and is on the *c* glide plane, while in poly(pentamethylene sulfide) the molecular chains are related by the *a* glide planes. PEO,[9] *trans*-1,4-polybutadiene,[10] and poly(ethylene adipate)[11] [—$O(CH_2)_2OCO(CH_2)_4$CO—]$_n$ are examples of $P2_1/a$.

For polymers the *c* axis is usually taken as parallel to the molecular

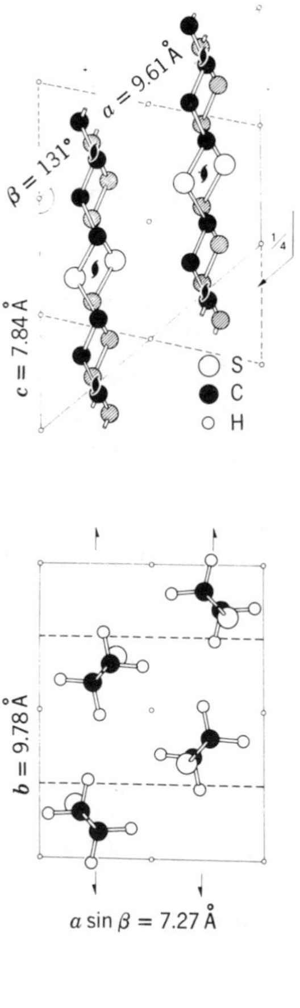

Fig. 3.10 Crystal structure of poly (pentamethylene sulfide) ($P\,1\,2_1/a\,1$-C_{2h}^5).[8] Dashed line in the upper figure indicates the alternative choice of space group $P2_1/n$.

axis. At the General Assembly of the International Union of Crystallography in 1951 it was agreed that the second setting of the monoclinic space groups should be accepted as standard, but that the first setting could also be used when there is a special reason.[2] In the γ-form of nylon 6[12] the b axis is taken as the unique axis according to this agreement, and so it belongs to $P2_1/a$, the molecular axis being parallel to the b axis instead of the c axis; the location of the symmetry elements is the same as in Fig. 3.10. On the other hand, the first setting ($P\ 1\ 1\ 2_1/b$, Fig. 3.7) is used for *it*-poly(3-methyl-1-butene)[13] (—CH—CH$_2$—)$_n$ (Fig. 7.2), polypivalolactone[14] [—CH$_2$C(CH$_3$)$_2$
 |
CH(CH$_3$)$_2$
COO—]$_n$, and so forth, in which the c axis is parallel to the molecular axis.

$P2_1/n$ is a special case of $C_{2h}{}^5$. If the unit cell is taken as shown by the broken line in the upper part of Fig. 3.10, the space group is given by $P2_1/n$. Using $P2_1/n$ makes the angle β closer to 90° and more convenient to deal with; in some cases, such as for poly(tetramethylene succinate)[15] [—O(CH$_2$)$_4$OCO(CH$_2$)$_2$CO—]$_n$, $\beta = 124°$ for $P2_1/n$ but $\beta = 149°$ for $P2_1/a$, which is too large.

3.4.2 Example 2: *Pnam*-$D_{2h}{}^{16}$ (Fig. 3.11)

This space group belongs to the orthorhombic system and is described as *Pnma* in the *International Tables*. By exchanging the b and c axes, *Pnam* is obtained. Orthorhombic PE belongs to this group.[16]

In the orthorhombic system there are six settings for the same unit cell under the right-hand coordinate system. The notation of the face center, the order of the axes or planes, and the direction of translation of the glide planes vary according to the setting of the axes. The notation **abc** at the head of the columns in the *International Tables*, Vol. I (e.g., p. 548), is the standard setting, and *Pnma* corresponds to it for No. 62, $D_{2h}{}^{16}$. The notation $\mathbf{a\bar{c}b}$ means that the original **a**, $\bar{\mathbf{c}}$, and **b** axes should be taken as the **a**, **b**, and **c** axes, respectively. Hence, if the coordinate axes are taken as shown in the right-hand figure below,

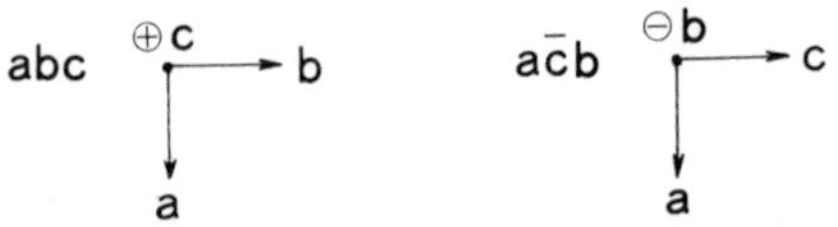

and $\mathbf{a\bar{c}b}$ is applied to the diagram *Pnma*-$D_{2h}{}^{16}$, then *Pnam* is obtained. The symbols ⊕ and ⊖ mean that the positive direction points upward or downward from the page, respectively.

In Fig. 3.11 the diagonal glide planes (—·—·—·) are normal to the a axis

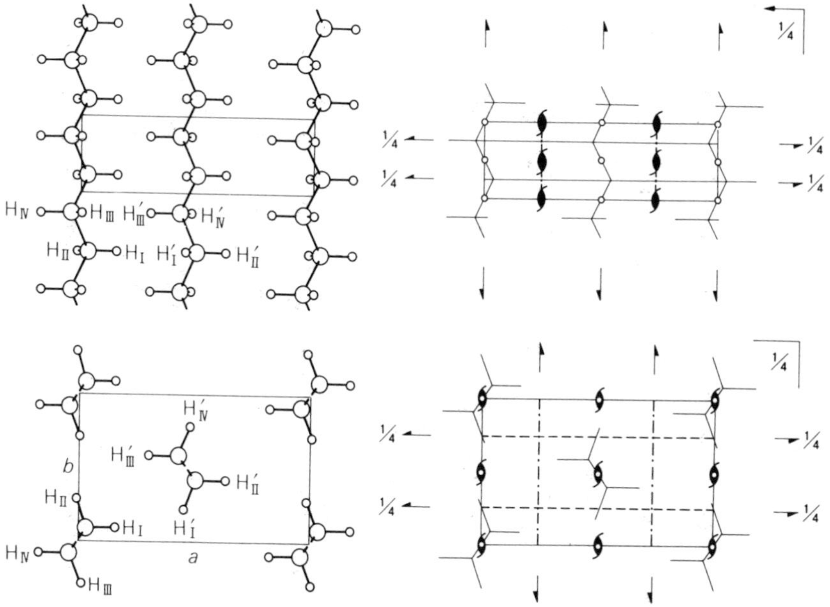

Fig. 3.11 Crystal structure of PE ($Pnam$-$D_{2h}{}^{16}$). (From Bunn.[16])

and translate by $(b+c)/2$. The a glide planes (---- *or* [corner symbol with arrow]) and the mirror planes (—— or [corner symbol]) are normal to the b and c axes, respectively. There are also the symmetry elements $\bar{1}$ and 2_1.

The molecular axis of PE coincides with the twofold screw axis. The symmetry elements of the molecular group $C_2{}^s(x)$, i, and $\sigma(yz)$ as shown in Fig. 3.3 also exist in the space group: 2_1, $\bar{1}$, and m. But the other symmetry elements of the single chain $C_2(y)$, $C_2(z)$, $\sigma(xy)$, and $\sigma_g(xz)$ are not in the crystal. Thus the symmetry of the molecule in the crystal lattice is often different from the symmetry of the isolated single molecule. The symmetry of the specific position of the molecular chain in this case is called the site symmetry. It is convenient to consider the site symmetry in addition to the point symmetry in the case of high polymers.

REFERENCES

1. C. W. Bunn, *Crystals: Their Role in Nature and in Science*, Academic, New York, 1964.
2. N. F. M. Henry and K. Lonsdale, Eds., *International Tables for X-Ray Crystallography*, Vol. I. *Symmetry Groups*, 2nd ed., Kynoch Press, Birmingham, 1965.
3. G. Herzberg, *Infrared and Raman Spectra of Polyatomic Molecules*, Van Nostrand, New York, 1945.

4. E. B. Wilson, Jr., J. C. Decius, and P. C. Cross, *Molecular Vibrations*, McGraw-Hill, New York, 1955.
5. H. Tadokoro, T. Yasumoto, S. Murahashi, and I. Nitta, *J. Polym. Sci.*, **44,** 266 (1960).
6. Y. Takahashi, T. Sato, H. Tadokoro, and Y. Tanaka, *J. Polym. Sci., Polym. Phys. Ed.*, **11,** 233 (1973).
7. C. W. Bunn and E. V. Garner, *J. Chem. Soc.*, **1942,** 654.
8. Y. Gotoh, H. Sakakihara, and H. Tadokoro, *Polym. J.*, **4,** 68 (1973).
9. Y. Takahashi and H. Tadokoro, *Macromolecules*, **6,** 672 (1973).
10. S. Iwayanagi, I. Sakurai, T. Sakurai, and T. Seto, *J. Macromol. Sci., Phys.*, **2,** 163 (1968).
11. A. Turner-Jones and C. W. Bunn, *Acta Crystallogr.*, **15,** 105 (1962).
12. H. Arimoto, *J. Polym. Sci.*, **A2,** 2283 (1964).
13. P. Corradini, P. Ganis, and V. Petraccone, *Eur. Polym. J.*, **6,** 281 (1970).
14. G. Perego, A. Melis, and M. Cesari, *Makromol. Chem.*, **157,** 269 (1972).
15. Y. Chatani, R. K. Hasegawa, and H. Tadokoro, Annual Meeting of the Society of Polymer Science, Japan, Tokyo, Preprint p. 28B07, 1971.
16. C. W. Bunn, *Trans. Faraday Soc.*, **35,** 482 (1939).

Chapter 4

X-Ray Diffraction Method

4.1 CHARACTERISTIC FEATURES OF X-RAY DIFFRACTION OF POLYMERS

In the X-ray analyses of polymers the diffraction data are usually less abundant than in the case of single crystals of low-molecular-weight substances. (1) The polymer samples are usually uniaxially oriented and give fiber diagrams that correspond to the rotation photographs of single crystals. Accordingly it is difficult to obtain three-dimensional reflection data uniquely. Three-dimensional data can be obtained only for doubly oriented samples (Section 4.3.2) and for solid-state polymerization products. (2) The diffraction spots are broad and diminish rapidly with increasing diffraction angle since the crystalline regions coexist with the amorphous ones and, furthermore, the former regions contain a considerable number of irregularities; the size of the crystallites is of the order of several hundred angstroms. (3) Incomplete orientation of the crystallites results in broadening of the spots along the arcs at a constant diffraction angle. (4) The number of independent observable reflections is small (at most 200, usually 40–100) compared to that of single crystals (usually more than 1000). Because of these factors, X-ray analyses of high polymers have, up to the present time, been made mainly by trial-and-error procedures. However several additional methods have recently been developed that give successful results.

Figure 4.1 shows the fiber diagrams of PEO (*a*) and of a PEO–urea complex (*b*) prepared by immersing an oriented film of PEO in a methanolic solution of urea, and a rotation photograph (*c*) around the *c* axis of a single crystal of the PEO–urea complex obtained by using PEO of low molecular weight (about 1000).[1] Photographs *a* and *b* show the characteristic features of the fiber diagram mentioned above. Photograph *c* corresponds to *b*, giving reflections with essentially the same positions and intensities.

In this chapter a general discussion of the basic principles of X-ray diffraction is given and then those specific to polymers are explained in detail. The *International Tables for X-Ray Crystallography*[2–5] are of basic importance, and there are also many good textbooks on X-ray crystallography.[6–18]

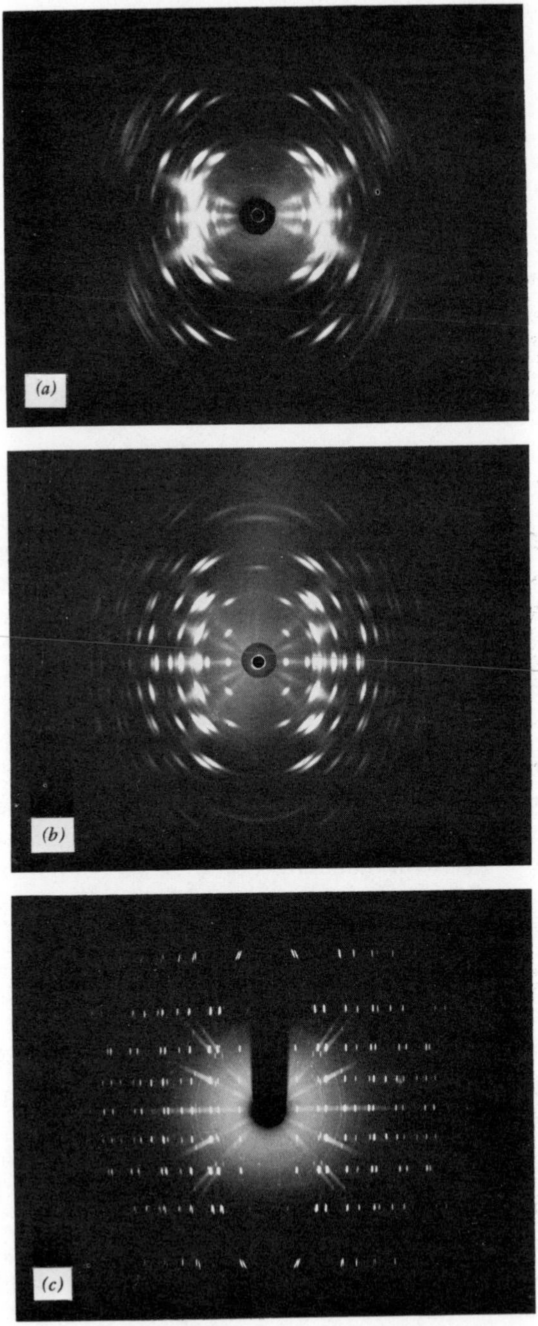

Fig. 4.1 Fiber diagrams of (*a*) PEO and of (*b*) PEO–urea complex, and (*c*) a rotation photograph of a single crystal of PEO–urea complex.[1]

4.2 BASIC PRINCIPLES OF X-RAY DIFFRACTION

When a monochromatic X-ray beam is incident on a crystal, the scattered X-rays from the atoms interfere with each other, giving strong diffractions in some special directions, since atomic distances are of the same order as the X-ray wavelength and the crystal consists of a lattice of regularly arrayed atoms. The direction of diffraction is related to the shape and size of the unit cell of the lattice, and the intensity depends on the atomic arrangement in the lattice.

4.2.1 Bragg's Condition and Polanyi's Formula

In a crystal the parallel planes of regularly arranged atoms are called atomic net planes. Diffracted X-rays can be treated as reflections by such atomic net planes, in which Bragg's condition is necessary (see Fig. 4.2). Then

$$2d \sin \theta = n\lambda \qquad n = 1,2,\cdots \tag{4.1}$$

where d is the spacing of atomic net planes, θ is the angle between the X-ray beam and the planes, and λ is the wavelength of the X-ray. Reflection occurs only when Bragg's condition is satisfied.

Considered next is a beam of X-rays with a wavelength λ perpendicularly incident on a one-dimensional linear lattice consisting of points equally spaced by a distance I. The X-rays scattered by the lattice points give rise to strong diffractions in the directions in which Polanyi's formula is fulfilled (see Fig. 4.3).

$$I \sin \phi_m = m\lambda \qquad m = 0, 1, 2, \cdots \tag{4.2}$$

For particular values of m and λ, ϕ_m is constant, that is, the diffracted rays are on a cone with the one-dimensional lattice as the axis. This is the reason for the occurrence of layer lines (equatorial, first, second, $\cdots$) on fiber photographs taken with a cylindrical camera (Section 4.2.6. B). Equation 4.2 is used for obtaining the fiber period I from a fiber diagram. The value ϕ_m is

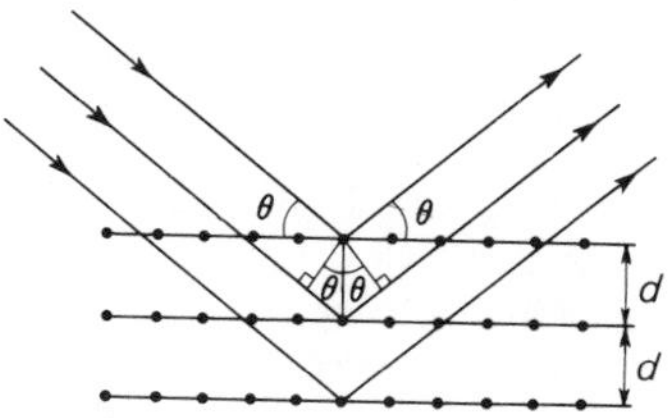

Fig. 4.2 Bragg's condition.

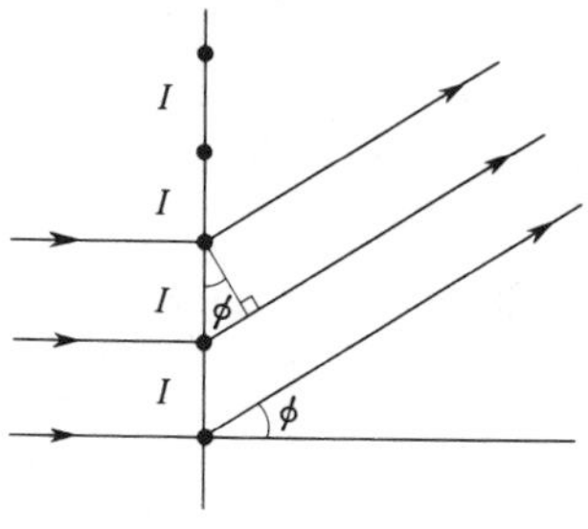

Fig 4.3 Derivation of Polanyi's formula.

calculated by

$$\tan\phi_m = \frac{S_m}{r} \tag{4.3}$$

where r is the radius of the cylindrical film and S_m is the distance between the zeroth and mth layer lines.

4.2.2 Scattering of X-Rays

Figure 4.4 shows the X-rays scattered by two electrons at the points O and R separated by the vector $\mathbf{r}$. The directions of the incident and scattered rays are indicated by unit vectors $\mathbf{s}_0$ and $\mathbf{s}$, respectively. The phase difference between the scattered waves at a point P far away compared to $\mathbf{r}$ is now examined. The incident path at R is longer by $\mathbf{s}_0 \cdot \mathbf{r}$ than at O, whereas the path of the ray scattered by R is shorter by $\mathbf{s} \cdot \mathbf{r}$. Here $\mathbf{s} \cdot \mathbf{r}$ is a scalar product whose length is $|\mathbf{s}|\,|\mathbf{r}|\cos\alpha$, α being the angle between the two vectors. As a whole, the path length of the X-rays passing through R is shorter by $(\mathbf{s} \cdot \mathbf{r} - \mathbf{s}_0 \cdot \mathbf{r})$ than the other. Accordingly the phase of the wave passing through R is in advance of that of the wave passing through O by $(2\pi/\lambda)(\mathbf{s} - \mathbf{s}_0) \cdot \mathbf{r}$. A scattering vector is defined by

$$\mathbf{S} = \frac{\mathbf{s} - \mathbf{s}_0}{\lambda} \tag{4.4†}$$

as shown in Fig. 4.5. The phase difference is $2\pi\mathbf{S} \cdot \mathbf{r}$. The scattered intensity

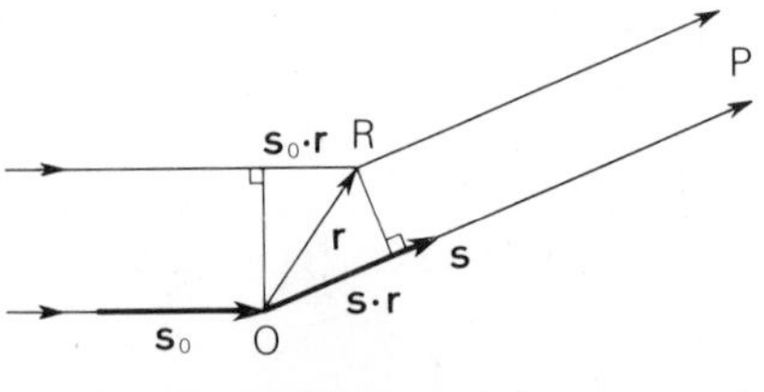

Fig. 4.4 Phase difference of the waves scattered by two electrons.

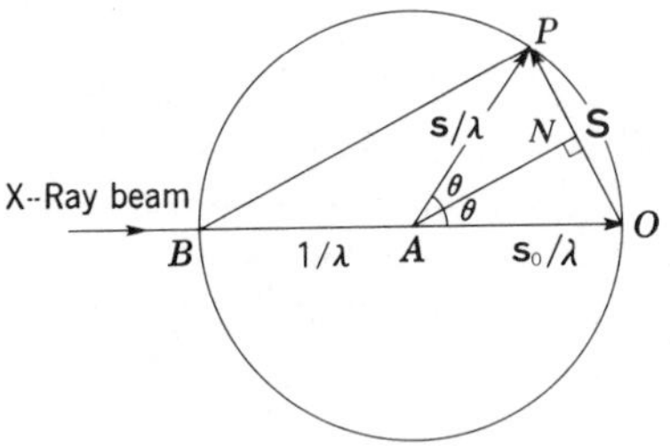

Fig. 4.5 Ewald's sphere of reflection.

† According to the present definition, $\mathbf{S}$ has the dimension $[L^{-1}]$, and the radius of the sphere of reflection is $1/\lambda$. In Ref. 8 the scattering vector is defined as $\mathbf{S} = \mathbf{s} - \mathbf{s}_0$, so that the phase difference is $(2\pi/\lambda)\mathbf{S} \cdot \mathbf{r}$, $|\mathbf{S}| = 2\sin\theta$, $\mathbf{S}$ is a dimensionless quantity, and the radius of the sphere of reflection is 1. In Sections 4.2.6.B, C, 4.5.1.A, and 4.5.1.B only, the treatment used in Ref. 8 is applied for the convenience of indexing, the dimensions of reciprocal space being multiplied by λ.

observed at P is given by

$$I(\mathbf{S}) = I_e F(\mathbf{S}) F^*(\mathbf{S}) = I_e |F(\mathbf{S})|^2 \tag{4.5}$$

I_e is the intensity of X-rays scattered by an electron and is given by the equation

$$I_e = I_0 \frac{e^4}{m^2 c^4 R^2} \frac{1 + \cos^2 2\theta}{2} \tag{4.6}$$

I_0 is the nonpolarized incident X-ray intensity, 2θ is the diffraction angle, R is the distance from the sample, m is the electronic mass, e is the electronic charge, and c is the velocity of light. $(1 + \cos^2 2\theta)/2$ is the polarization factor. The scattered X-rays are generally polarized, and the extent of polarization depends on the diffraction angle and affects the intensity. In structure analyses all parts except the polarization factor in Eq. 4.6 can be treated as constants and so are omitted hereafter.

$F(\mathbf{S})$ is called the structure amplitude or the structure factor and is given by the following equation summing up exponential functions expressing phase differences for a system of N electrons.

$$F(\mathbf{S}) = \sum_{j=1}^{N} \exp(2\pi i \mathbf{S} \cdot \mathbf{r}_j) \tag{4.7}$$

$F(\mathbf{S})$ is a complex function and $F^*(\mathbf{S})$ is its complex conjugate.

Since the electron is distributed as an electron cloud, and the electron density at a point indicated by vector $\mathbf{r}$ is denoted by $\rho(\mathbf{r})$, Eq. 4.7 can be rewritten as

$$F(\mathbf{S}) = \int \rho(\mathbf{r}) \exp(2\pi i \mathbf{S} \cdot \mathbf{r})\, dv \tag{4.8}$$

where dv is the volume element of scattering matter. Equation 4.8 is the most important equation in X-ray diffraction. It gives the atomic, molecular, or crystal structure factor, if the range of integration is taken over an atom, a molecule, or a crystal, respectively.

In Fig. 4.5 the sample is located at A, from which vectors $\mathbf{s}_0/\lambda$ and $\mathbf{s}/\lambda$ to the directions of the incident and scattered X-rays are drawn, the ends of the vectors being O and P, respectively. The scattering vector $\mathbf{S}$ corresponds to OP. Since the triangle APO is isosceles, a perpendicular AN drawn from A to PO bisects $\angle PAO$. Then $\angle BPO$ is 90°, B being the intersection of the incident beam and a sphere of radius $1/\lambda$ with the center located at A. The length of $\mathbf{S}$ is

$$|\mathbf{S}| = \frac{2 \sin \theta}{\lambda} \tag{4.9}$$

The sphere is known as Ewald's sphere of reflection.

Equation 4.8 is a Fourier transform as defined in mathematics (Appendix A). The space where vector **S** exists is called reciprocal space, while the space where $\rho(\mathbf{r})$ is defined and in which we live is called real, direct, or physical space. If $\rho(\mathbf{r})$ is known, $F(\mathbf{S})$ can be calculated uniquely. On the other hand, $\rho(\mathbf{r})$ can also be obtained if the amplitude and the phase angle of $F(\mathbf{S})$ are given (Section 4.8). Here it should be mentioned that the experimentally measurable quantity is only $I(\mathbf{S}) = |F(\mathbf{S})|^2$, which does not include the phase angle. Hence the most difficult problem is that the phase angle must be estimated by various methods.

4.2.3 Scattering by Atoms

From Eq. 4.8 the atomic scattering factor, or atomic structure factor $f(\mathbf{S})$, is given in terms of the scattering power of an electron:

$$f(\mathbf{S}) = \int_{\text{atom}} \rho(\mathbf{r}) \exp(2\pi i \mathbf{S}\cdot\mathbf{r})\, dv \tag{4.10}$$

The f values calculated by assuming spherical symmetry of the electron clouds are given in Ref. 4 as functions of $(\sin\theta)/\lambda$. Several examples are shown in Fig. 4.6. The value of f coincides with the number of electrons in the atom for $\theta = 0°$ because of the absence of phase differences, and it decreases with increasing θ.

4.2.4 Scattering by Molecules

The molecular scattering factor, given by Eq. 4.8, can be rewritten in terms

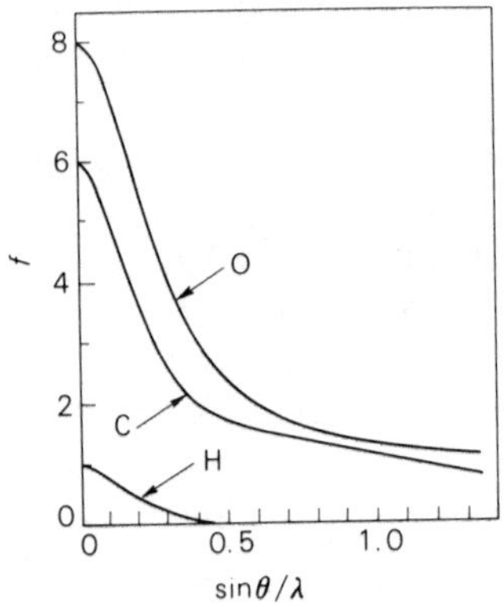

Fig. 4.6 Examples of the atomic scattering factor.

of the scattering by atoms.

$$F_M(\mathbf{S}) = \int_{\text{molecule}} \rho(\mathbf{r}) \exp(2\pi i \mathbf{S}\cdot\mathbf{r})\, dv \tag{4.11}$$

$$= \sum_n f_n \exp(2\pi i \mathbf{S}\cdot\mathbf{r}_n)$$

Here f_n and $\mathbf{r}_n$ are the atomic scattering factor and the coordinate vector of the nth atom, respectively.

A right-hand Cartesian coordinate system (x, y, z) is used here for designating atomic positions, and another right-hand Cartesian coordinate system (X, Y, Z) is used for $\mathbf{S}$ as shown in Fig. 4.7. Since $\mathbf{r}_n$ and $\mathbf{S}$ are represented by the components (x_n, y_n, z_n) and (X, Y, Z), Eq. 4.11 may be written as

$$F_M(\mathbf{S}) = \sum_n f_n \exp[2\pi i (X x_n + Y y_n + Z z_n)] \tag{4.12}$$

As an example of the simplest polymer molecule a one-dimensional lattice consisting of identical atoms equally spaced at intervals of c along the z axis is considered. The numbering of the atoms is given in Fig. 4.7*a*. The coordinate of the nth atom is

$$x_n = y_n = 0, \quad z_n = nc \tag{4.13}$$

$F_M(\mathbf{S})$ is given by

$$F_M(\mathbf{S}) = f \sum_{n=-N}^{N} \exp(2\pi i Z n c) \tag{4.14}$$

Since this equation is a series of geometrical progressions, using the formula

$$1 + a + a^2 + \cdots + a^{m-1} = \frac{1 - a^m}{1 - a} \tag{4.15}$$

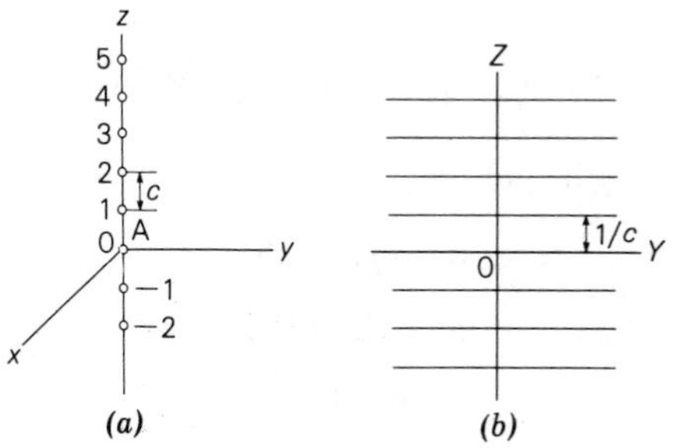

Fig. 4.7 (*a*) One-dimensional lattice and (*b*) its transform into reciprocal space (viewed along the *X* axis).

gives

$$F_M(\mathbf{S}) = f \exp(-2\pi i Z N c) \frac{1 - \exp[2\pi i Z(2N+1)c]}{1 - \exp(2\pi i Z c)}$$

$$= f \frac{\exp[-2\pi i Z(N+\frac{1}{2})c] - \exp[2\pi i Z(N+\frac{1}{2})c]}{\exp(-\pi i Z c) - \exp(\pi i Z c)} \qquad (4.16)$$

Furthermore, since

$$e^{i\omega} - e^{-i\omega} = 2i \sin \omega \qquad (4.17)$$

Eq. 4.16 becomes

$$F_M(\mathbf{S}) = f \frac{\sin[\pi Z(2N+1)c]}{\sin(\pi Z c)} \qquad (4.18)$$

Now for simplicity $2N + 1$ is replaced by N. The sine part of Eq. 4.18, or its square,

$$G(\mathbf{S})G^*(\mathbf{S}) = \frac{\sin^2(\pi N Z c)}{\sin^2(\pi Z c)} \qquad (4.19)$$

is called the Laue function. If N is large, G is nonzero on the planes of $Z = l/c$ (l is an integer) and nearly zero elsewhere.* Accordingly Eq. 4.14, shown in reciprocal space in Fig. 4.7*b*, has a value Nf only on the planes normal to the Z axis with equal separations $1/c$ and whose intersections with the sphere of reflection are coaxial circles. This result corresponds to Polanyi's formula (Eq. 4.2) and shows that when there is a periodicity only along the z direction in physical space, a periodicity exists only along the Z direction in reciprocal space.

4.2.5 Diffraction by Crystals

According to Eq. 4.8 the diffraction by a crystal is

$$F_{cr}(\mathbf{S}) = \int_{\text{crystal}} \rho(\mathbf{r}) \exp(2\pi i \mathbf{S}\cdot\mathbf{r})\, dv \qquad (4.20)$$

The origin of a unit cell is given by

$$\mathbf{R}_k = p\mathbf{a} + q\mathbf{b} + t\mathbf{c} \qquad (4.21)$$

and the position of a given atom in the lattice is denoted by

$$\mathbf{r} = \mathbf{R}_k + \mathbf{r}_j \qquad (4.22)$$

* When m is an integer, $\lim_{x\to m} (\sin \pi N x)/\sin \pi x = \lim_{x\to m} N(\cos \pi N x)/\cos \pi x = \pm N$. If N becomes infinite, Eq. 4.18 becomes a δ function (Eq. A. 11).

where p, q, and t are integers and $\mathbf{r}_j$ is a vector indicating the position of the jth atom in the kth unit cell. Using a method similar to that used in Eq. 4.11 gives for the crystal structure factor

$$F_{cr}(\mathbf{S}) = \sum_k \exp(2\pi i \mathbf{S}\cdot\mathbf{R}_k) \sum_j f_j \exp(2\pi i \mathbf{S}\cdot\mathbf{r}_j) \tag{4.23}$$

Next the reciprocal lattice defined by vectors **a***, **b***, and **c*** in reciprocal space is introduced.

$$\mathbf{a}^* = \frac{\mathbf{b}\times\mathbf{c}}{V}, \quad \mathbf{b}^* = \frac{\mathbf{c}\times\mathbf{a}}{V}, \quad \mathbf{c}^* = \frac{\mathbf{a}\times\mathbf{b}}{V} \qquad (4.24)^\dagger$$

$\mathbf{b}\times\mathbf{c}$ is a vector product. The vector product is a vector with the length $|\mathbf{b}||\mathbf{c}|\sin\alpha$ and is normal to the plane determined by **b** and **c**; the directions **b**, **c**, and $\mathbf{b}\times\mathbf{c}$ have the same relations as the coordinates x, y, and z in a right-hand coordinate system. V is the volume of the unit cell and is equal to $\mathbf{a}\cdot(\mathbf{b}\times\mathbf{c}) = \mathbf{b}\cdot(\mathbf{c}\times\mathbf{a}) = \mathbf{c}\cdot(\mathbf{a}\times\mathbf{b})$. The relationship between the direct and reciprocal lattices is shown in Fig. 4.8. The relationships between dimensions of the direct and reciprocal unit cells are given in Table 4.1. According to the properties of scalar products,

$$\begin{gathered}\mathbf{a}\cdot\mathbf{a}^* = \mathbf{b}\cdot\mathbf{b}^* = \mathbf{c}\cdot\mathbf{c}^* = 1 \\ \mathbf{a}\cdot\mathbf{b}^* = \mathbf{a}\cdot\mathbf{c}^* = \mathbf{b}\cdot\mathbf{a}^* = \mathbf{b}\cdot\mathbf{c}^* = \mathbf{c}\cdot\mathbf{a}^* = \mathbf{c}\cdot\mathbf{b}^* = 0\end{gathered} \tag{4.25}$$

Thus the a^*, b^*, and c^* axes are perpendicular to the bc, ca, and ab planes, respectively. In reciprocal space a set of parallel planes of the direct lattice is represented by a point.

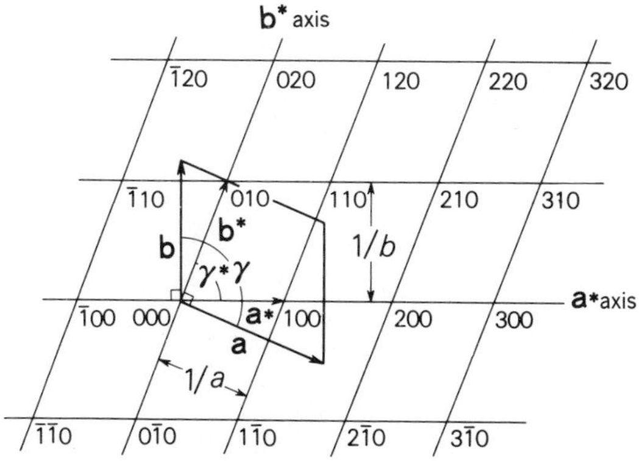

Fig. 4.8 Relationship between the direct (heavy lines) and reciprocal lattices (thin lines). An example of a monoclinic lattice of the first setting. Here $a^* = 1/(a \sin\gamma)$ and $b^* = 1/(b \sin\gamma)$.

† In several books, such as Ref. 8, **a*** is defined as $\mathbf{a}^* = \lambda\mathbf{b}\times\mathbf{c}/V$, etc., where **a*** is dimensionless.

Table 4.1 Relationships between Dimensions of Direct and Reciprocal Lattices[2]

$a^* = \dfrac{bc \sin \alpha}{V}$	$a = \dfrac{b^*c^* \sin \alpha^*}{V^*}$
$b^* = \dfrac{ca \sin \beta}{V}$	$b = \dfrac{c^*a^* \sin \beta^*}{V^*}$
$c^* = \dfrac{ab \sin \gamma}{V}$	$c = \dfrac{a^*b^* \sin \gamma^*}{V^*}$
$\cos \alpha^* = \dfrac{\cos \beta \cos \gamma - \cos \alpha}{\sin \beta \sin \gamma},$	$\cos \alpha = \dfrac{\cos \beta^* \cos \gamma^* - \cos \alpha^*}{\sin \beta^* \sin \gamma^*}$
$\cos \beta^* = \dfrac{\cos \gamma \cos \alpha - \cos \beta}{\sin \gamma \sin \alpha},$	$\cos \beta = \dfrac{\cos \gamma^* \cos \alpha^* - \cos \beta^*}{\sin \gamma^* \sin \alpha^*}$
$\cos \gamma^* = \dfrac{\cos \alpha \cos \beta - \cos \gamma}{\sin \alpha \sin \beta},$	$\cos \gamma = \dfrac{\cos \alpha^* \cos \beta^* - \cos \gamma^*}{\sin \alpha^* \sin \beta^*}$
$V = abc\,(1 - \cos^2 \alpha - \cos^2 \beta - \cos^2 \gamma + 2 \cos \alpha \cos \beta \cos \gamma)^{1/2}$	
$V^* = \dfrac{1}{V}$	

Reproduced by courtesy of the International Union of Crystallography.

If a given scattering vector **S** is denoted by

$$\mathbf{S} = \zeta \mathbf{a}^* + \eta \mathbf{b}^* + \zeta \mathbf{c}^* \tag{4.26}$$

taking Eq. 4.25 into account gives

$$\mathbf{S}\cdot\mathbf{a} = \xi, \qquad \mathbf{S}\cdot\mathbf{b} = \eta, \qquad \mathbf{S}\cdot\mathbf{c} = \zeta \tag{4.27}$$

For a parallelepiped crystal with the edges $L\mathbf{a}$, $M\mathbf{b}$, and $N\mathbf{c}$, the first summation of Eq. 4.23 is

$$\sum_k \exp(2\pi i \mathbf{S}\cdot\mathbf{R}_k) = \sum_{p=0}^{L-1} \sum_{q=0}^{M-1} \sum_{t=0}^{N-1} \exp[2\pi i \mathbf{S}\cdot(p\mathbf{a} + q\mathbf{b} + t\mathbf{c})]$$

$$= \sum_p \exp(2\pi i \xi p) \sum_q \exp(2\pi i \eta q) \sum_t \exp(2\pi i \zeta t) \tag{4.28}$$

These functions are the Laue functions already given in Eqs. 4.14 and 4.18. Since the values of L, M, and N are large in crystals, the diffraction intensity has the values given by Eq. 4.29 only at the points with $\xi = h$, $\eta = k$, and

$\zeta = l(h, k, l$: integers), namely, at the reciprocal lattice points, and zero elsewhere. Thus

$$F_{cr}(hkl)\, F_{cr}{}^{*}(hkl) = L^2M^2N^2F(hkl)F^{*}(hkl) \qquad (4.29)^{\dagger}$$

where $F(hkl)$ is the unit cell structure factor, or simply structure factor. Laue's equations are obtained from Eqs. 4.4 and 4.27 and the above condition:

$$\mathbf{S}\cdot\mathbf{a} = h, \qquad \mathbf{S}\cdot\mathbf{b} = k, \qquad \mathbf{S}\cdot\mathbf{c} = l \qquad (4.30)$$

The equivalence between Bragg's law and Laue's equations is discussed in many books, for example, Ref. 7.

For the reflection *hkl*, **S** is

$$\mathbf{S} = h\mathbf{a}^{*} + k\mathbf{b}^{*} + l\mathbf{c}^{*} \qquad (4.31)$$

and the coordinates of the *j*th atom in the unit cell are denoted by

$$\mathbf{r}_j = \mathbf{a}x_j + \mathbf{b}y_j + \mathbf{c}z_j \qquad (4.32)$$

Then $F(hkl)$ of Eq. 4.29 is given as follows using Eqs. 4.23 and 4.25.

$$F(hkl) = \sum_j f_j \exp(2\pi i \mathbf{S}\cdot\mathbf{R}_j)$$

$$= \sum_{j=1}^{n} f_j \exp[2\pi i(hx_j + ky_j + lz_j)] \qquad (4.33)$$

where x_j, y_j, and z_j are the fractional coordinates of the *j*th atom, the units being *a*, *b*, and *c*. If the atomic positions all assumed, the structure factors can be calculated by using this equation. A crystal structure analysis involves mainly the assumption and modification of the values of atomic coordinates so as to obtain good agreement between the observed and calculated structure factors. When the unit cell contains only one chain, the diffraction intensity at the reciprocal lattice points corresponds to the molecular transform (Sections 4.5.1.C and 4.6.4.A).

4.2.6 Mechanism of X-Ray Diffraction Measurements

Although the number of reciprocal lattice points *hkl* is infinite, only those intersecting the surface of the sphere of reflection with the radius $1/\lambda$ can be observed. In other words, only the inside of the sphere with radius $2/\lambda$ and origin *O* is observable. This is called a limiting sphere. For the spacing d_{hkl} shorter than $\lambda/2$, no reflection occurs, since $|\mathbf{S}| = 1/d_{hkl}$ would be larger

† In the case of Eq. 4.28, the Laue function is accompanied by the phase factor $\exp\{\pi i[(L-1)\xi + (M-1)\eta + (N-1)\zeta]\}$.

than the radius of the sphere of reflection. The number of reciprocal lattice points in the sphere of reflection depends on the wavelength; thus it becomes eight times as great if the wavelength is halved, because the volume of the limiting sphere increases eightfold, while the size of reciprocal lattice remains constant.

A. *Unoriented Polymer Samples* (*Debye-Scherrer Rings*). If crystallites or crystalline powders have random orientations, the reciprocal lattice points are distributed on the surfaces of concentric spheres. Their intersections with the sphere of reflection are circles, and a pattern of concentric circles is observed if a flat film is used. These circles are called Debye-Scherrer rings, from which the spacings d can be measured. Debye-Scherrer rings can also be recorded with cylindrical cameras (see next Section 4.2.6.B). If the distance between the sample and film is r, and the diameter of a ring is $2x$, then $\tan 2\theta = x/r$ for a flat film and $2\theta = x/r$ for a cylindrical film.

B. *Uniaxially Oriented Samples* (*Cylindrical Camera*). Figure 4.9 shows the measurement of a uniaxially oriented sample by a cylindrical camera. The sample is set at A with the orientation axis coinciding with the axis of the cylindrical film. The X-ray beam is incident from the left-hand side normal to the axis. For uniaxially oriented polymer samples, the axes of the molecular chains are nearly parallel to the elongation direction, but the orientation of a particular plane of crystallites (e.g., corresponding to the molecular plane of PVA) in the cross section perpendicular to the elongation direction remains completely random. Consequently the reciprocal lattice points are distributed uniformly on their loci, obtained by rotating the reci-

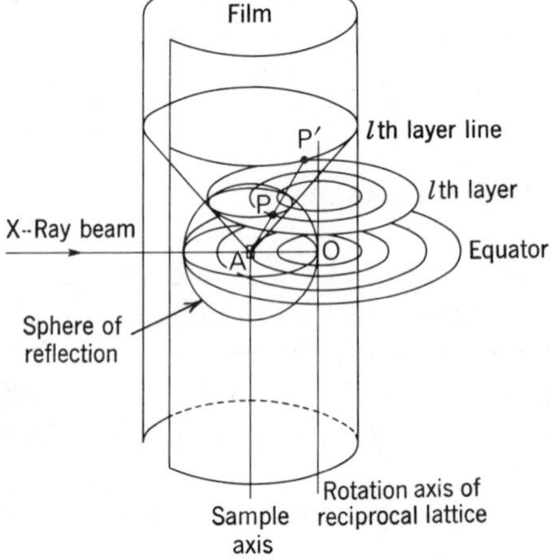

Fig. 4.9 Measurement of uniaxially oriented sample by a cylindrical camera.

procal lattice by 360° around the axis passing through the origin of the reciprocal lattice and parallel to the orientation direction of the sample. Such loci form concentric circles on every layer. In the figure only the equator and the lth layer are shown. If an intersecting point of such a circle and the sphere of reflection is denoted by P, diffraction occurs in the direction AP and is recorded at P' on the film. The X-rays diffracted by the lth layer lie on the surface of a cone, which intersects the film in a circle and gives rise to the lth layer lines on the film if it is opened out flat. Thus photographs of uniaxially oriented samples (e.g., Figs. 4.1*a* and 4.1*b*) are essentially equivalent to rotation photographs of low-molecular-weight single crystals (Fig. 4.1*c*). The spots on the vertical line passing through the point at which the primary beam would have struck the film had the beam stop not been in the way are called meridional reflections. In cases where the c and c^* axes are coincident (in other words, the c axis is normal to the ab plane), the reciprocal lattice points $00l$ cannot touch the surface of the sphere of reflection if the orientation in complete and the reciprocal lattice point is represented by a point (see Fig. 4.14). Actually the $00l$ reflections often appear on the meridional line of the fiber diagram because of the broadening of the reciprocal lattice points, due mainly to incomplete orientation, as a result of which the lattice points can intersect the sphere of reflection.

Hereafter in this section and also in Sections 4.2.6.A, 4.2.6.C, 4.5.1, and 4.5.2, the reciprocal lattice is considered to be multiplied by λ, so that the radius of the sphere of reflection is one. The cylindrical coordinates of the reciprocal lattice points (ξ, ω, ζ) are defined as shown in Fig. 4.10, the ζ axis being parallel to the cylindrical axis (the c axis). The reciprocal lattice points with the same ζ value give rise to reflections on the same layer line. The reciprocal lattice points with the same ξ are on the same cylindrical surface whose axis coincides with the rotation axis of the reciprocal lattice. In Fig. 4.9 if the intersection of the cylindrical surface and the sphere of reflection is projected from A to the cylindrical film, a curve with constant ξ is obtained. When the equator is taken as the x axis on the film opened out flat, and the direction of the cylinder (the meridian) is taken as the y axis, ξ and ζ are

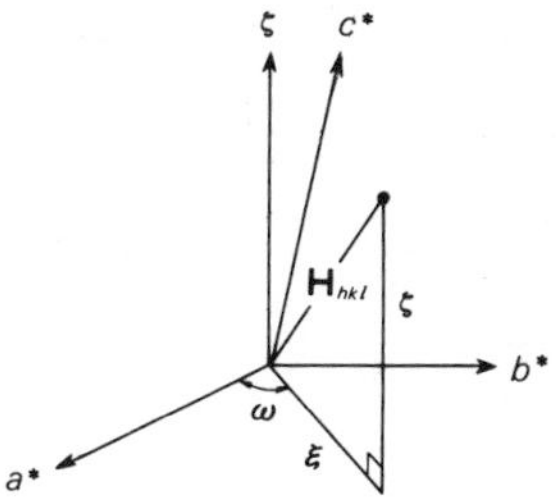

Fig. 4.10 Cylindrical coordinates of a reciprocal lattice showing the general case where the c and c^* axes do not coincide.

given by

$$\left.\begin{aligned}\xi &= \left[1+\left(1+\frac{y^2}{r^2}\right)^{-1} - 2\left(1+\frac{y^2}{r^2}\right)^{-1/2}\cos\left(\frac{x}{r}\right)\right]^{1/2}\\ \zeta &= \frac{y}{(r^2+y^2)^{1/2}}\end{aligned}\right\} \tag{4.34}$$

where r is the radius of the cylindrical film. Using Eq. 4.34, sets of curves with constant ξ and constant ζ can be drawn; the resulting figure is called Bernal's chart (Fig. 4.11).[19] Since on the equator $y = 0$, then

$$\xi = \left[2 - 2\cos\left(\frac{x}{r}\right)\right]^{1/2} \tag{4.35}$$

and $\xi = 2.0$ for $x = \pi r$. For cylindrical film of radius r mm the width of Fig. 4.11 is magnified to $2\pi r$ mm. When such a chart printed on a transparent paper or film is superposed on the fiber diagram, the ξ and ζ coordinates of each reflection can be read.

As an example of a uniaxially oriented sample, the reciprocal lattice of poly-1,3-dioxolane $[-OCH_2O(CH_2)_2-]_n$ form II is shown in Fig. 4.12.

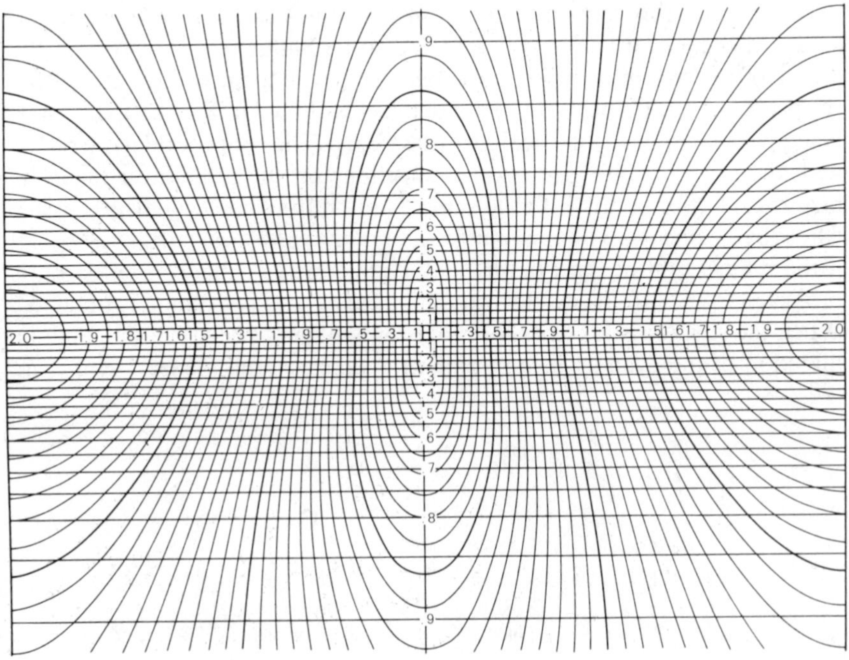

Fig. 4.11 Bernal's chart.[19]

The lattice is orthorhombic, $a = 9.07$ Å, $b = 7.79$ Å, c (fiber axis) $= 9.85$ Å, and the space group is $Pbca\text{-}D_{2h}^{15}$. The c and c^* axes coincide, and the ξ values of the reciprocal lattice points with the same hk are the same for different l. In Figs. 4.12a and 4.12b, the open circles show the reciprocal lattice points not observed because of the systematic absences (Section 4.2.7.F), and the closed circles represent the reciprocal lattice points that give reflections experimentally. The systematic absences in this case are

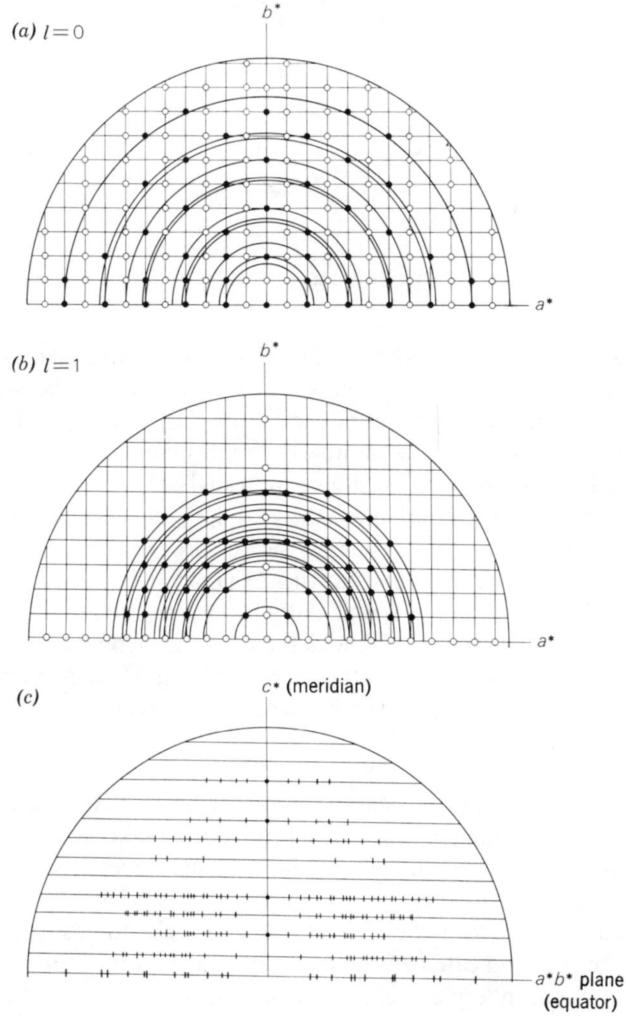

Fig. 4.12 The reciprocal lattice of poly-1, 3-dioxolane form II.[20] (a and b) the equatorial and first layer planes, respectively. (c) A section including the c^* axis.

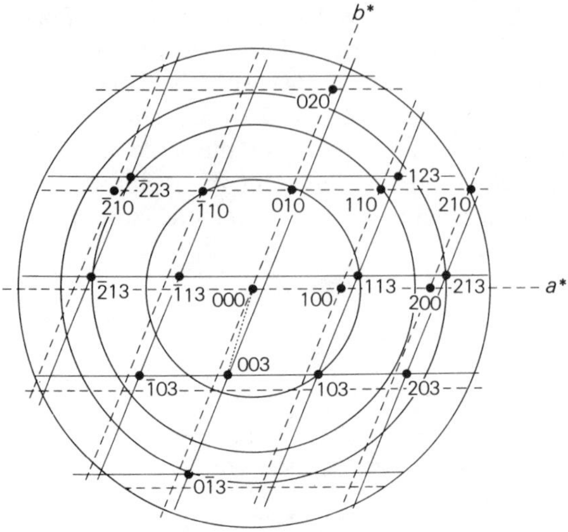

Fig. 4.13 Reciprocal lattice of the planar zigzag form of PEO[21]: (———) third layer, (-----) equator.

$h \neq 2n$ for $hk0$, $l \neq 2n$ for $h0l$, and $k \neq 2n$ for $0kl$. Figure 4.12*c* shows a section including the c^* axis, where the points on the layer line are the intersecting points of the section and the concentric circles shown in Fig. 4.9. The fiber diagram of a uniaxially oriented sample essentially corresponds to Fig. 4.12*c*, but it is deformed appreciably as shown in Bernal's chart.

As an example of the case where the c and c^* axes do not coincide (in the triclinic and monoclinic systems), the reciprocal lattice of the planar zigzag form of PEO (triclinic) is shown in Fig. 4.13. The concentric circles of 110, 210, 113, and 213 for a uniaxially oriented sample are shown, and their centers are all on the c axis, which passes through 000 and is normal to the page. The direction joining the points 000 and 003 in reciprocal space is the c^* axis. Since ξ is the distance from the c axis, it is clear that the ξ values of $hk0$ and $hk3$ are different.

C. *Uniaxially and Doubly Oriented Samples* (*Weissenberg Camera*).* For uniaxially oriented samples it is normal practice to take fiber diagrams by the above method. As shown in Fig. 4.14, the intersections of the sphere of reflection with the layer planes are circles, the radii of which are 1 for the equator and $(1 - \zeta_l^2)^{1/2}$ for the lth layer. Accordingly, the reciprocal lattice points in the part indicated by hatching cannot be observed on the fiber diagrams. A Weissenberg camera is useful to observe the reflections for all

* For doubly oriented samples a usual and simple method is to take photographs using a plate camera with the incident X-ray beam parallel to the rolled direction (Section 4.3.2).

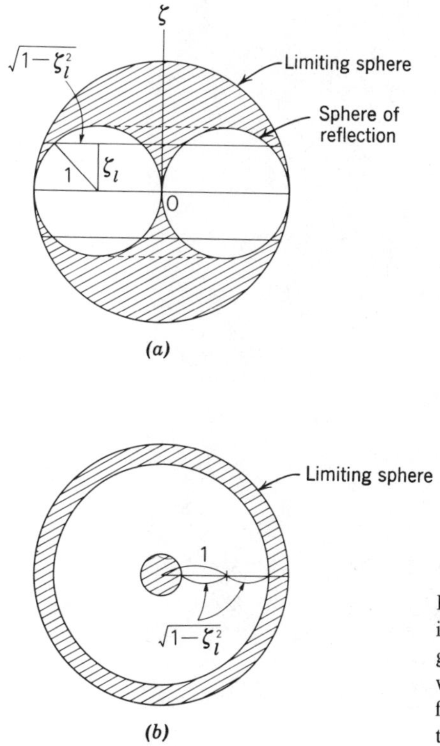

Fig. 4.14 (*a*) A section of reciprocal space including the ζ axis for the rotation photograph (see Figs. 4.9 and 4.10) and (*b*) the section with $\zeta = \zeta_l$. The cylindrical camera gives reflections only for the reciprocal lattice points in the part without hatching.

reciprocal lattice points in the limiting sphere. As the sample is rotated in the Weissenberg camera (Fig. 4.15), a cylindrical film is moved bodily along the axis of rotation. A complete to-and-fro cycle synchronized with the rotation of the sample (in practice oscillation within 200°) takes place. A slotted metal screen can be adjusted to permit the passage of any selected cone of reflection (see Fig. 4.9). For the Weissenberg camera a cassette of 57.3 mm diameter is usually available as a standard accessory. If the distance

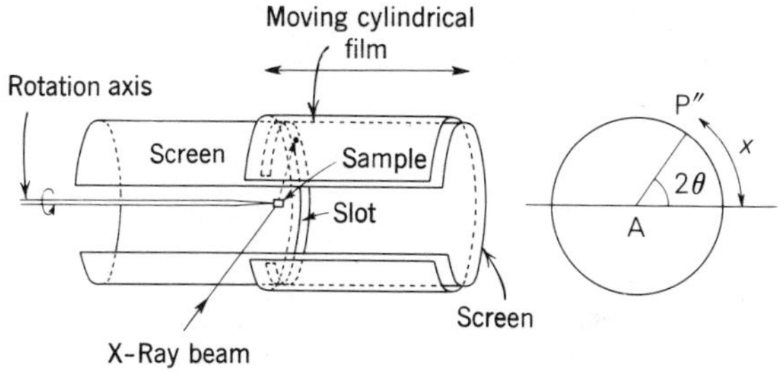

Fig. 4.15 Weissenberg camera.

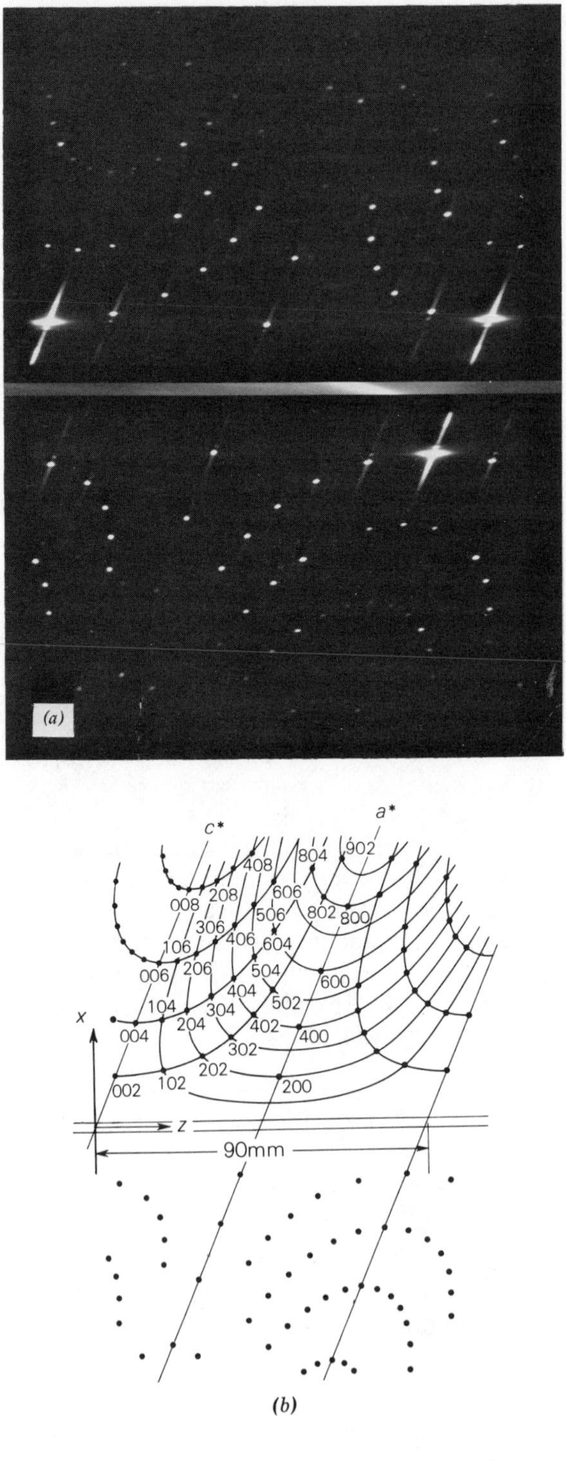
(a)
c*
a*
902
804
408
606
008
208
506
802
800
306
406
604
106
006
206
504
600
404
104
304
502
204
402
400
004
302
202
002
102
200
x
z
90mm

(b)

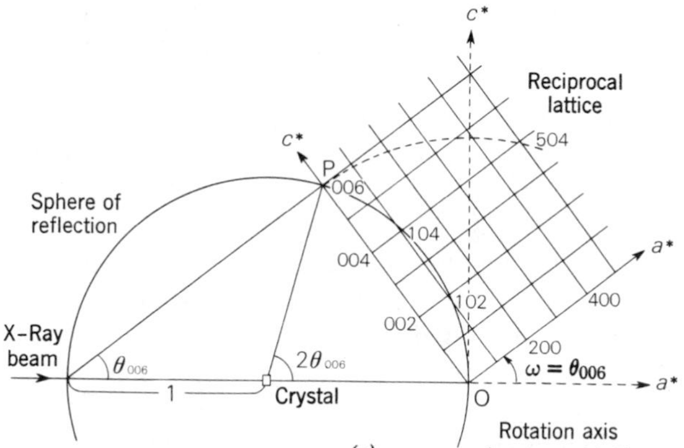

(c)

Fig. 4.16 (*a*) Weissenberg photograph of pentoxane, (*b*) indexing, and (*c*) the relationship between the reciprocal lattice and the sphere of reflection. The b and b^* axes are perpendicular to the page. The reflections $h0l$ with $l = 2n$, $h00$ with $h = 2n$, and $00l$ with $l = 2n$ are the only ones observed.[22]

between the spot of diffraction angle 2θ and the central line on the film (Fig. 4.15) is x mm, then x gives θ in degrees, since $x = 2r\theta$ rad $= 2r(\theta/57.3)^\circ$ and $2r = 57.3$ mm by using the relation 1 rad $= 180^\circ/\pi = 57.3^\circ$. The rotation angle of the sample $\theta = 180^\circ$ usually corresponds to the 90 mm movement of the film in commercially available cameras. Since the standard 57.3 mm cassette is rather small for polymer studies, the author uses specially made cassettes of 70 and 90 mm diameters. For the 70 mm cassette, x, when multiplied by a factor of $57.3/70 = 0.819$, gives θ in degrees.

Figure 4.16*a* shows the Weissenberg photograph of the $h0l$ reflections of pentoxane $(—CH_2O—)_5$ taken using CuKα radiation ($\lambda = 1.542$ Å) with a 70 mm cassette by rotating the sample around the b axis. Here the slot of the screen is fixed so as to transmit only the $h0l$ reflections. This sample is not a polymer, but a single crystal of a ring oligomer, and is orthorhombic, $Pbcn$-$D_{2h}{}^{14}$, with $a = 8.194$ Å, $b = 10.691$ Å, and $c = 7.682$ Å. Figure 4.16*b* shows the indexing of reflections. The reflections on the straight lines correspond to the reciprocal lattice points on the lines passing through the rotation axis (point O in Fig. 4.16*c*), $h00$ and $00l$. The other reflections, namely, those on the curves, are due to the points not on the lines passing through the rotation axis $h0l$. The abscissa, or the z axis, is parallel to the rotation axis and crosses the incident beam at right angles.

In Fig. 4.16*c* the solid line shows the orientation of the reciprocal lattice when the 006 reflection is recorded where $\sin\theta_{006} = 6\lambda c^*/2$. When the crystal rotates from the original orientation (the a^* and c^* axes are indicated by broken lines) by the rotation angle $\omega = \theta_{006} = \sin^{-1}(6\lambda c^*/2) = 37.0^\circ$, the reciprocal lattice point 006 just touches the surface of the sphere of reflection. Since a rotation of ω by 2° corresponds to 1 mm movement in the z axis direction on the film, the spot 006 would correspond to the point shifted by $z = 18.5$ mm from the original position on the film after ω is rotated by 37.0°. The $00l$ reflections appear as a straight line on the film, since the rotation angle ω from the original position coincides with the reflection angle θ for the points on the c^* axis and θ is proportional to both x and z. The slope of the line for a standard cassette is equal to $\tan^{-1}(x/z) = \tan^{-1} 2 = 63.4^\circ$, since x in millimeters equals θ in degrees and z in millimeters corresponds to $\theta/2$ in degrees. For a 70 mm cassette the slope is $\tan^{-1}(2 \times 70/57.3) = 67.7^\circ$. The situation is the same for the a^* axis except $\omega = \theta + 90^\circ$. When the crystal is rotated by $\theta_{006} = 37.0^\circ$, the reciprocal lattice points 102 and 104 also touch the surface of the sphere of reflection and the corresponding reflections are recorded on the points at $z = 18.5$ mm. The 504 reflection has a diffraction angle similar to that of 006 reflection, but it is recorded after further rotation when the reciprocal lattice point touches the sphere of reflection. Therefore 504 and 006 have similar x values, but they appear at positions with different z. All reflections other than $h0l$ are not recorded, as they are intercepted by the screen.

A square lattice with 0.1 unit spacings in the reciprocal space (a layer of the reciprocal lattice) is given by intersecting points of curves in the Weissenberg photograph as shown in Fig. 4.17. This figure is called the Weissenberg chart, which is a deformed representation of an orthogonal coordinate plane. When the abscissa of the chart is superposed on the central line (z axis) of the photograph and one of the principal axes of the chart is superposed on a series of spots on a straight line, the coordinate values of the

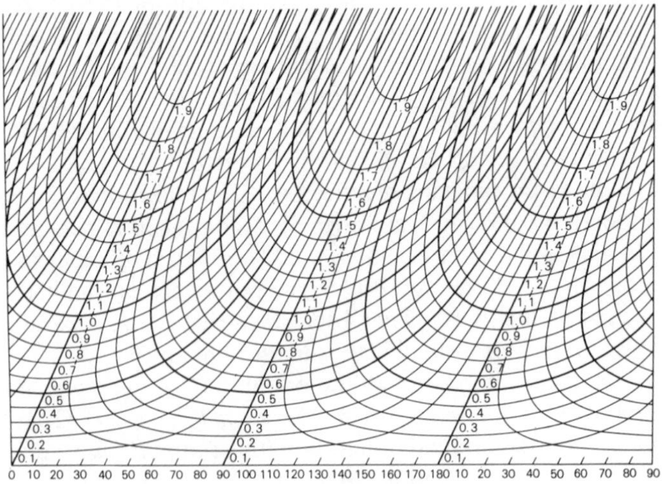

Fig. 4.17 Weissenberg chart.

reciprocal lattice points corresponding to the reflection spots can be read. Thus the reflection that appears only on a layer line when a cylindrical camera is used can be obtained in two-dimensions by the Weissenberg camera. The chart in Fig. 4.17 is for a standard 57.3 mm cassette. For cassettes of other sizes, the chart should be redrawn, because the ratios of the *x* and *z* axes are different. If the main axes recorded on the film are not perpendicular to each other, the distance between the main axes on the chart should be changed, depending on the interaxial angle.

For the equatorial reflections, the normal beam method is used, in which the X-ray beam is incident normal to the rotation axis. The layer reflections are measured by using the equi-inclination method, in which the camera is set so that the incident and reflected X-rays have the same angle to the rotation axis. Reflection data for doubly oriented samples and data for three dimensionally oriented samples obtained by the radiation polymerization of single crystals can be obtained by following the same procedure as that for single crystals and using a Weissenberg camera (Section 4.3.2).

Norman's method,[23] which is frequently used for uniaxially oriented samples and can provide more abundant reflection data that cannot be obtained by a cylindrical camera, is now explained. The sample is set at the goniometer head so that the orientation axis is normal to the rotation axis, and the slot is fixed at the plane including the incident beam. In Fig. 4.18 the X-ray beam is incident on the sample *A* from the *BA* direction, the rotation axis being normal to the page. When the orientation axis (*c* axis) is vertical in this figure, the *OC* direction (*O* is the origin of the reciprocal lattice) is parallel to the *c* axis. In the case of a uniaxially oriented sample as is described earlier (Fig. 4.9), the reciprocal lattice points are distributed uniformly on coaxial circles around the *c* axis. Any point *P* on these circles and also on the page is designated by ξ and ζ. When the sample rotates counterclockwise and *P* comes to *P*′, diffraction occurs along the direction *AP*′. The reciprocal

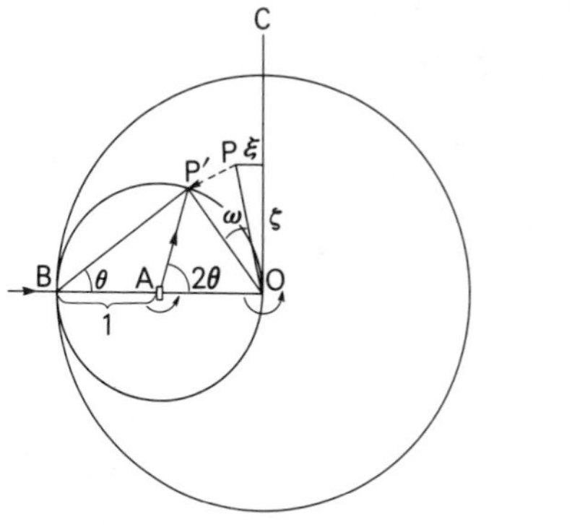

Fig. 4.18 Norman's method.

lattice has rotated at this moment by an angle ω ($\angle POP'$) and the film has moved simultaneously by $z = (\omega/2°)$ mm. Since the slot is fixed at the position including the sample axis, the reflection at P' is recorded on the film. Although there are many circles intersecting the sphere of reflection at the same time, the reflections other than those intersecting the circle $BP'O$ (on the page plane) are not recorded because of interception by the screen.

A Weissenberg photograph of poly-1,3-dioxolane form II taken by Norman's method is shown in Fig. 4.19. In this case, the structure is orthorhombic, and so the c and c^* axes coincide. The spots $00l$ on the photograph taken for a uniaxially oriented sample just correspond to the points on the c^* axis in Fig. 4.12c and appear on a straight line. ζ can be read by using the Weissenberg chart. The $hk0$ reflections, corresponding to the equatorial reflections of the fiber diagram, are also recorded on a straight line shifted by 45 mm (rotation angle 90°) with respect to the z direction. This line corresponds to the equator in Fig. 4.12c and ξ can be read off. The other reflections appear between these two lines, and their ξ and ζ can be read by using the values of the curves on the Weissenberg chart. This situation can be understood by the correspondence of the c^* axis and the equator of the Weissenberg photograph to the c^* and a^* axes of Fig. 4.16b, respectively. The statement given above for the case of coincidence of the c and c^* axes is also true in the case of the noncoincidence of the c and c^* axes, except for the $00l$ reflections appearing on two curves deviated from a line to both sides.

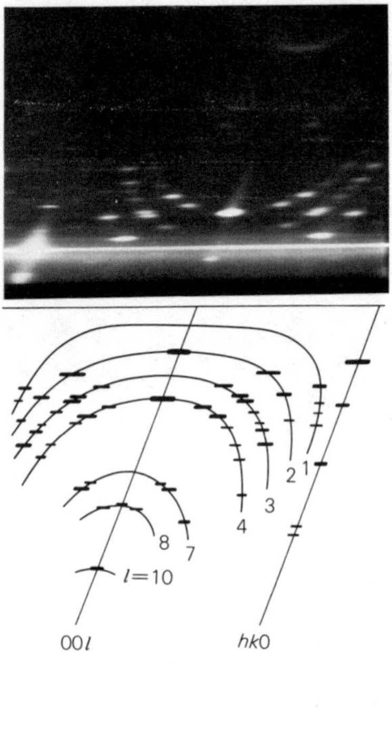

Fig. 4.19 Weissenberg photograph of poly-1, 3-dioxolane form II taken by Norman's method.[20]

4.2.7 Diffraction Intensity

To compare the crystal structure factor with the observed intensity $I(hkl)$, the following correction is necessary.

$$I = KmpLA|F|^2 \tag{4.36}$$

where K is the scale factor, m is the multiplicity, p is the polarization factor, L is the Lorentz factor, and A is the absorption factor.

A. *Structure Factor.* The structure factor, already derived in Eq. 4.33, gives the relationship between the diffraction intensity and the atomic arrangements in the unit cell. The main purpose of X-ray crystal structure analysis is to clarify the atomic coordinates included in the structure factor. Equation 4.33 can be rewritten as

$$\begin{aligned} F(hkl) &= A(hkl) + iB(hkl) \\ A(hkl) &= \sum_{j=1}^{N} f_j \cos[2\pi(hx_j + ky_j + lz_j)] \\ B(hkl) &= \sum_{j=1}^{N} f_j \sin[2\pi(hx_j + ky_j + lz_j)] \end{aligned} \tag{4.37}$$

and

$$\begin{aligned} |F(hkl)| &= [A^2(hkl) + B^2(hkl)]^{1/2} \\ A(hkl) &= |F(hkl)| \cos\alpha(hkl) \\ B(hkl) &= |F(hkl)| \sin\alpha(hkl) \end{aligned} \tag{4.38}$$

where $\alpha(hkl)$ is the phase angle of $F(hkl)$.

When the crystal structure has centers of symmetry, and if one of them is chosen at the origin, F becomes a real function

$$F(hkl) = 2\sum_{j=1}^{N/2} f_j \cos[2\pi(hx_j + ky_j + lz_j)] \tag{4.39}$$

since x, y, z are always accompanied by $-x$, $-y$, $-z$. Here the formula

$$\exp(\pm i\omega) = \cos\omega \pm i\sin\omega \tag{4.40}$$

has been used.

B. *Temperature Factor.* Each atom oscillates around its equilibrium position, and the amplitude increases with temperature. Because of the thermal vibrations, the atomic coordinates and hence the electron density

ρ should be statistically averaged. The atomic scattering factor diminishes by a function of $(\sin \theta)/\lambda$, as shown in Fig. 4.6, because of the spreading of the electron cloud around the nucleus. The effect of thermal vibrations gives rise to further spreading, on the average resulting in a more marked decrease in intensity. The atomic scattering factors given in Ref. 4 are for atoms at rest. If these atomic scattering factors at rest are denoted by f_0, then those to be used in practice are given by

$$f = f_0 \exp\left[-B\frac{\sin^2\theta}{\lambda^2}\right] \tag{4.41}$$

When anisotropic thermal vibrations are considered, the following types of temperature factors are used:

$$f = f_0 \exp\left[-(h^2B_{11} + k^2B_{22} + l^2B_{33} + 2hkB_{12} + 2klB_{23} + 2lhB_{31})\right] \tag{4.42}$$

B in Eq. 4.41 is called the isotropic temperature parameter, and B_{11}, and so on are the anisotropic temperature parameters. B_{11}, and so on are dimensionless, while B has a dimension of $[\mathrm{L}^2]$. Therefore for the reflections $h00$, h^2B_{11} corresponds to $(B \sin^2 \theta)/\lambda^2 = B/4d_{100}{}^2$. B has the following relation to the mean square amplitude $\langle u^2 \rangle$ of the atomic vibrations normal to the net plane:

$$B = 8\pi^2 \langle u^2 \rangle \tag{4.43}$$

C. *Polarization Factor and Lorentz Factor.* As is explained earlier (Eq. 4.6), the polarization factor is given by

$$p = \frac{1 + \cos^2 2\theta}{2} \tag{4.44}$$

When a monochromator is used, $p = (1 + \cos^2 2\theta_M \cos^2 2\theta)/2$ should be applied for equatorial reflections, where θ_M is the Bragg angle of the reflection from the monochromator. For the layer reflections, see Ref. 24.

The Lorentz factor L is the correction due to the relative fraction in which a given family of reflection planes exist within a narrow angular range over which reflection occurs (in the case of a rotation photograph this is the relative time that the reflecting planes spend within the angular range), and is given by the same formula for both fiber and rotation photographs:

$$L = \frac{1}{\sin 2\theta} \frac{\cos \theta}{(\cos^2 \phi - \sin^2 \theta)^{1/2}} \tag{4.45}$$

where ϕ is the angle between the reflection plane and the elongation axis (the rotation axis for a rotation photograph). The corrections for p and L can be made by using Cochran's chart.[25]

For Weissenberg photographs of uniaxially oriented samples (Section 4.2.6.C), a rough method of correction is often used for the equatorial

Lorentz and polarization factors (Eq. 4.45 with $\phi = 0$ is used) and the intensity values are further divided by ξ (defined by Eq. 4.34). However, since an accurate method for this case has not yet been established, the intensity data obtained by using this method should be taken into account only qualitatively, and they are not suitable for refinements by the least-squares method.

D. *Multiplicity.* The multiplicity of a reflection is the number of equivalent types of crystal plane contributing to one observed reflection; it is proportional to the number of crystallites that are suitably oriented for reflection by a particular type of plane. The orthorhombic system is considered as an example. The plane of type (100) is equivalent to $(\bar{1}00)$, and (210) is equivalent to $(\bar{2}10)$, $(2\bar{1}0)$, and $(\bar{2}\bar{1}0)$, the multiplicities being 2 and 4, respectively. The plane (211) is equivalent to $(\bar{2}11)$, $(2\bar{1}1)$, $(\bar{2}\bar{1}1)$, $(21\bar{1})$, $(\bar{2}1\bar{1})$, $(2\bar{1}\bar{1})$, and $(\bar{2}\bar{1}\bar{1})$, the multiplicity being 8. These numbers, applicable to powder samples and nonoriented polymer samples, are listed in Ref. 2 (p. 32). In the case of fiber diagrams or single crystal rotation photographs, the situation is different. The planes parallel to the fiber axis (or rotation

Table 4.2 Multiplicities for Fiber Diagrams

Reflection		Relative number of equivalent types of planes	Number of appearing reflections	Multiplicity m
Orthorhombic (c axis ∥ elongation axis)				
Equator	$h00$	2	2	1
	$0k0$	2	2	1
	$hk0$	4	2	2
Layer	$00l$	2	2	1
	$h0l$	4	4	1
	$0kl$	4	4	1
	hkl	8	4	2
Monoclinic (second setting, c axis ∥ elongation axis)				
Equator	$h00$	2	2	1
	$0k0$	2	2	1
	$hk0$	4	2	2
Layer	$00l$	2	4	0.5
	$h0l$	2	4	0.5
	$\bar{h}0l$	2	4	0.5
	$0kl$	4	4	1
	hkl	4	4	1
	$\bar{h}kl$	4	4	1

axis) give reflections in two places on the equator of the photograph, one on each side of the meridian, while the reflections from other planes are distributed among four positions, one in each quadrant. Therefore multiplicity m is obtained for fiber diagrams or rotation photographs by dividing the multiplicity for nonoriented samples by the number of positions given above. Table 4.2 gives the multiplicities for fiber diagrams for the orthorhombic and monoclinic systems. For the latter, the second setting with the c axis parallel to the elongation axis (the c and c^* axes do not coincide) is taken. In this case the reflections $00l$ are not located on the meridian but appear on the layer lines as four spots.

The quantity $[I(\text{obs})]^{1/2}$ may be compared with $[mF^2(\text{calc})]^{1/2}$ after the p and L corrections are made, m being taken from Table 4.2. If there are nonequivalent reflections that are accidentally overlapped, then $[\Sigma_i m_i F_i^2(\text{calc})]^{1/2}$ may be used.

Table 4.2 is also applicable to Weissenberg photographs taken by Norman's method. However m is one for the case where a moving film camera, such as a Weissenberg camera and a diffractometer, is used for a single crystal, because the intensity is measured only once for one reciprocal lattice point hkl. The procedure for Weissenberg photographs of the equator or layer lines of a doubly oriented polymer sample is the same as that for a single crystal, although the existence of two or more twinned components should be noticed (Section 4.3.2).

E. *Absorption Factor.* The absorption factor A depends on the shape, size, and density of the crystal, the elements composing its substance, and the wavelength of the X-ray (Ref. 4). Special correction for absorption is not necessary for most polymer samples containing no heavy atoms. It is preferable to use a cylindrical sample. Caution is needed for polymers containing heavy atoms such as chlorine (Section 4.4.2.B).

F. *Systematic Absences.* Systematic absences, or *Auslöschung* (German), occur for nonprimitive lattices and lattices containing screw axes or glide planes. Symmetry elements such as the center of symmetry, mirror plane, rotation axis, and rotatory inversion axis give no extinctions.

For example, in the base-centered lattice C, a point at (x, y, z) is always accompanied by another at $(\frac{1}{2}+x, \frac{1}{2}+y, z)$. Therefore the structure factor is

$$F(hkl) = \sum_{j=1}^{N/2} f_j \left\{ \exp[2\pi i(hx_j + ky_j + lz_j)] + \exp\left[2\pi i\left(hx_j + ky_j + lz_j + \frac{h}{2} + \frac{k}{2}\right)\right]\right\}$$

$$= \{1 + \exp[\pi i(h+k)]\} \sum f_j \exp[2\pi i(hx_j + ky_j + lz_j)] \quad (4.46)$$

The first factor in braces, being independent of the atomic coordinates, can have only two values, 2 for $h + k = 2n$, and 0 for $h + k \neq 2n$. Hence there are no reflections with $h + k \neq 2n$. It is supposed next that the lattice has a c glide plane passing through the origin and normal to the b axis. Since the points at (x, y, z) and $(x, \bar{y}, \frac{1}{2} + z)$ are equivalent, the following equation is derived in the same way as that above:

$$F(h0l) = [1 + \exp(\pi i l)] \sum f_j \exp[2\pi i(hx_j + lz_j)] \tag{4.47}$$

Hence the $h0l$ reflections are absent if l is odd.

For a 2_1 screw axis passing through the origin and parallel to the c axis, since the points at (x, y, z) and $(\bar{x}, \bar{y}, \frac{1}{2} + z)$ are equivalent,

$$F(00l) = [1 + \exp(\pi i l)] \sum f_j \exp(2\pi i l z_j) \tag{4.48}$$

and so, the $00l$ reflections are absent if l is odd.

The atomic arrangements in the primitive lattice of the rhombohedral system R (rhombohedral-R, indicated by the hexagon in Fig. 4.20) may also be represented by a hexagonal lattice. The latter corresponds to the non-primitive lattice (R_{hex}, hexagonal-R) produced from the hexagonal primitive lattice (hexagonal-P, the large rhomboid in Fig. 4.20) by adding two hexagonal-P lattices having origins at $(\frac{1}{3}, \frac{2}{3}, \frac{2}{3})$ and $(\frac{2}{3}, \frac{1}{3}, \frac{1}{3})$. The result leads to systematic absences for hkl reflections with $-h + k + l \neq 3n$. The simplest case is assumed, that is, $R3\text{-}C_3^4$ in which the equivalent points are $(0, 0, 0)$, $(\frac{1}{3}, \frac{2}{3}, \frac{2}{3})$, and $(\frac{2}{3}, \frac{1}{3}, \frac{1}{3})$ combined with (x, y, z), $(\bar{y}, x - y, z)$, and $(y - x, \bar{x}, z)$.

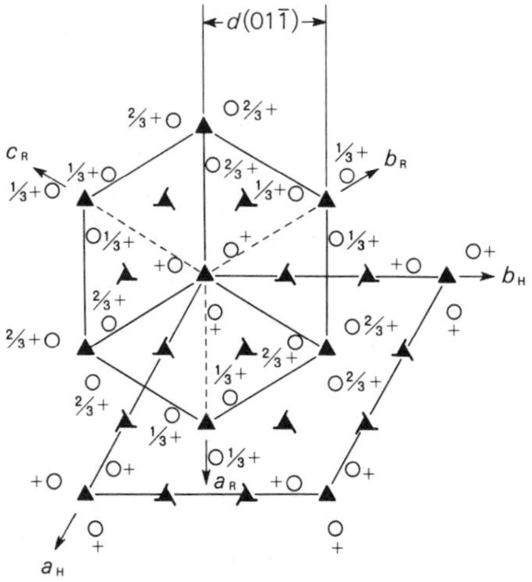

Fig. 4.20 Unit lattice of $R3\text{-}C_3{}^4$. In rhombohedral-R (denoted by hexagon), solid and dashed lines indicate the upper and lower edges, respectively. a_{H} and b_{H} are on the page, while a_{R}, b_{R}, and c_{R} point obliquely upward. The two ways to set rhombohedral axes to hexagonal axes are obverse and reverse, the former being the standard. (Reproduced by courtesy of the International Union of Crystallography.[2])

F is given by

$$F(hkl) = \sum f_j \left\{ 1 + \exp\left[2\pi i\left(\frac{h}{3} + \frac{2k}{3} + \frac{2l}{3}\right)\right] + \exp\left[2\pi i\left(\frac{2h}{3} + \frac{k}{3} + \frac{l}{3}\right)\right]\right\}$$
$$\times \{\exp[2\pi i(kx_j + hy_j + lz_j)] + \exp[2\pi i(kx_j + iy_j + lz_j)]$$
$$+ \exp[2\pi i(ix_j + hy_j + lz_j)]\} \qquad (4.49)^*$$

The first factor is calculated as follows.

$$\{\ \} = 1 + \exp\left[2\pi i\left(\frac{h}{3} - \frac{k}{3} - \frac{l}{3}\right)\right] + \exp\left[2\pi i\left(-\frac{h}{3} + \frac{k}{3} + \frac{l}{3}\right)\right]$$
$$= 1 + 2\cos\left[\frac{2\pi(-h + k + l)}{3}\right] \qquad (4.50)$$

This is equal to 3 for $-h + k + l = 3n$, and 0 for $-h + k + l = 3n + 1$ and $3n + 2$, resulting in the foregoing systematic absences.

Table 4.3 shows the systematic absences that often appear in high polymers.

Table 4.3 Examples of Systematic Absences

Symbol	Reflection	Condition of absence
P		Nonabsence
C	hkl	$h + k \neq 2n$
R_{hex}		$-h + k + l \neq 3n$
Screw Axis Parallel to c Axis		
$2_1, 4_2, 6_3$		$l \neq 2n$
$3_1, 3_2, 6_2, 6_4$	$00l$	$l \neq 3n$
$4_1, 4_3$		$l \neq 4n$
$6_1, 6_5$		$l \neq 6n$
Glide Plane Perpendicular to b Axis		
a		$h \neq 2n$
c	$h0l$	$l \neq 2n$
n		$h + l \neq 2n$

* In the hexagonal system the third axis is considered to make an angle of 120° with the a and b axes on the ab plane and the index $hkil$ is used, where $i = -(h + k)$. Then $-hy + k(x - y) = kx + iy$, and $h(y - x) - kx = ix + hy$. This i should be distinguished from the imaginary number i.

4.3 PREPARATION OF SAMPLES

For the purpose of X-ray analyses it is important to prepare samples as highly oriented and crystalline as possible. Doubly oriented samples should be prepared as well as uniaxially oriented samples.

4.3.1 Uniaxially Oriented Samples

Fiber samples are used as they are. Starting from a pellet or powder sample, films are cast from solution or prepared from melt, or model filaments are made by using an electrically heated syringe or a flow tester. Next such films or filaments are stretched or rolled under suitable conditions and are then heat treated. Examples of a device for elongation and a roller are shown in Figs. 4.21 and 4.22. Clamps can be made to grip a specimen without slipping by using a piece of sandpaper. A number of examples for the preparation of oriented crystalline samples are given below.

A. *Elongation of Melt-Quenched Samples.* By quenching molten samples in Dry Ice–methanol or ice water, amorphous or low-crystalline samples can be obtained, and such samples may be easily elongated.

Polyamides: quenching, and elongation followed by heat treatment at 10–20°C below the melting point.[26–28] PEO: quenching, stretching and no heat treatment.[29] Polyglycolide $(—CH_2COO—)_n$: quick quenching of melt in ice water followed by cold drawing and subsequent annealing at 180°C *in vacuo*; better oriented samples are easily obtainable from the copolymer of mole ratio glycolide 92% and lactide 8% compared with the homopolymer.[30] Poly(β-ethyl-β-propiolactone) [—CH-

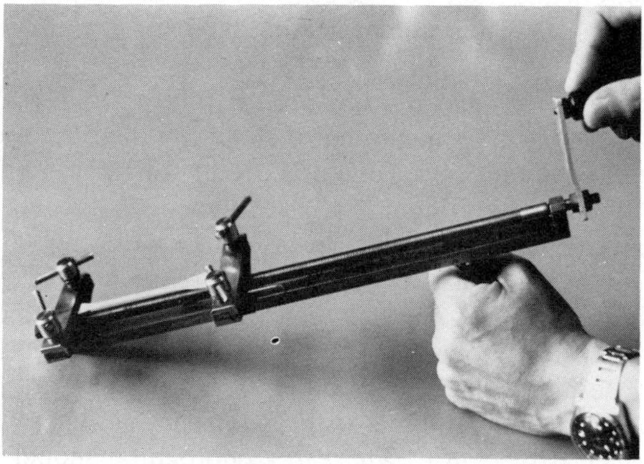

Fig. 4.21 Elongation device. The extension ratio can be read on a millimeter scale.

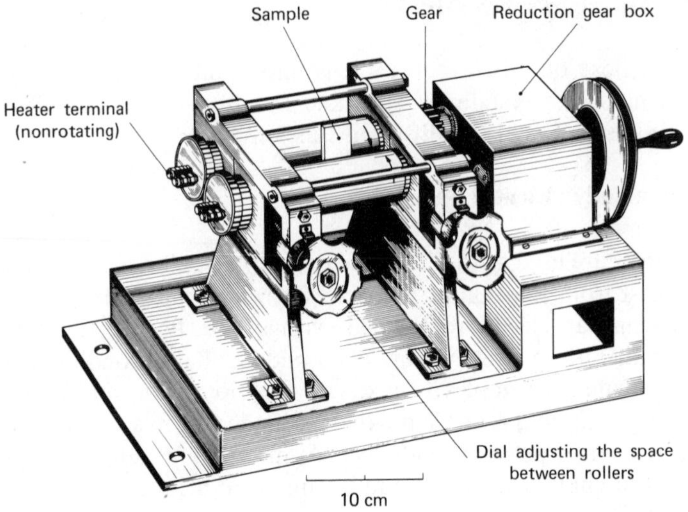

Fig. 4.22 Roller, which can be operated slowly manually or electrically. Heating devices are inside the cavities of rollers.

$(C_2H_5)CH_2COO—]_n$: quick drawing of melt during cooling to room temperature, followed by further elongation and heat treatment at 92°C under tension in *vacuo* for 1 day.[31]

B. *Elongation or Rolling at Room Temperature.* Polyisobutylene: Stretching a strip of the amorphous rubberlike sample to about 11 times its original length at room temperature, both ends being clamped.[32] *it*-Poly(methyl methacrylate): Casting from toluene solution and stretching to about 5–10 times its original length, followed by heating at 90°C for 6 hr *in vacuo*.[33]

C. *Elongation or Rolling at Elevated Temperatures.* *it*-Polystyrene: Stretching very slowly in boiling water to about five times its original length by using the elongation device (Fig. 4.21), followed by heating in Wood's alloy at about 180°C for 30 min.[34] *it*-Poly(*m*-methyl styrene): Drawing at 90°C to about 20 times its original length and heating at 100°C for 3 hr under tension.[35] *it*-Poly-1-butene: Preparation of oriented samples of pure form II has been reported to be very difficult because of instability, but it is possible according to the following procedure. The sample is cooled gradually in poly(ethylene glycol) from 140°C to 110–115°C at a rate of 1°C/min, slowly stretched to about 10 times its original length at the same temperature, and quenched to −78°C. The X-ray photograph should be taken at this temperature.[36]

D. *Elongation after Linking Chain Ends or Crosslinking.* The molecular weight can be increased by treating the polymer with diisocyanate, which links the molecules end-to-end by reaction with both hydroxyl and carboxyl end groups, forming carbamate and amide groups, respectively. This method has been successfully used for poly(ethylene adipate),[37] poly(ethylene suberate),[37] poly(ethylene succinate),[38] and poly-1,3-dioxolane.[20]

Another method involves crosslinking the polymer by irradiation. α-Gutta percha is obtained by slow cooling of the melt, but this does not yield an oriented sample. Although oriented samples can be prepared by stretching at 60–70°C, the α-form tears when it is highly stretched at such temperatures. To overcome this difficulty, the specimen can be crosslinked by γ-irradiation (0.7×10^6 rad/hr for 24 hr) and cooled slowly after being stretched in hot water at about 70°C.[39]

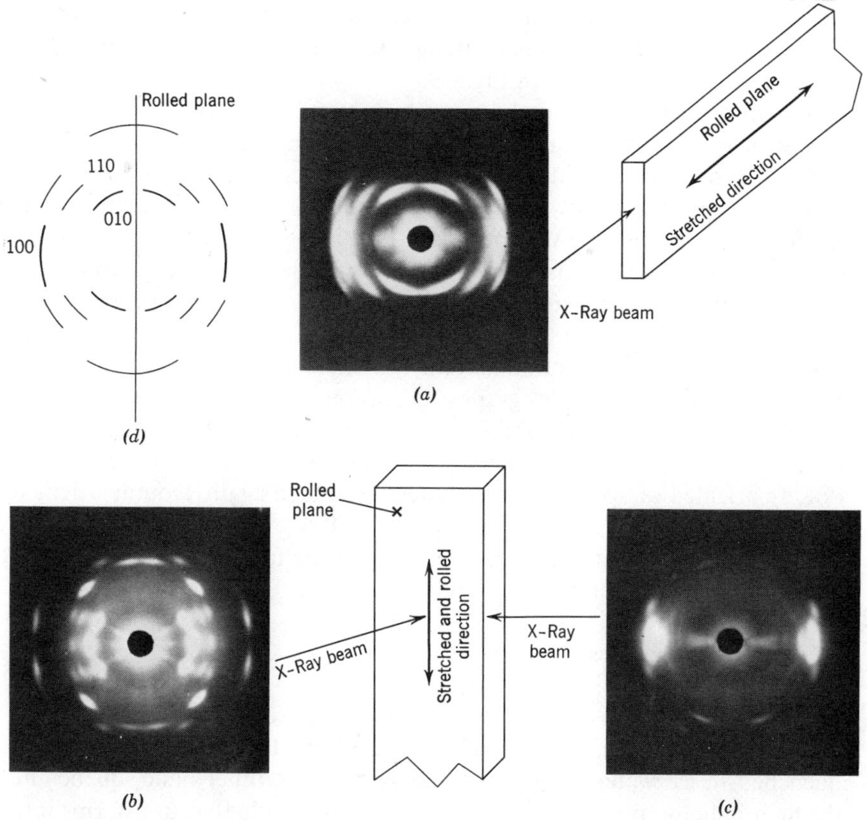

Fig. 4.23 (*a–c*) Tridirectional X-ray patterns of a highly doubly oriented PET specimen; (*d*) reflection indices for pattern *a*.[40] According to Statton and Godard,[41] the direction *a* is termed end, *b* is through, and *c* is edge.

4.3.2 Doubly Oriented Samples

When the uniaxially oriented sample is rolled under suitable conditions, one of the planes parallel to the molecular axis (in the case of PVA, the plane of the zigzag carbon chain) also becomes oriented parallel to the plane of rolling, as shown in Figure 1.1*c*. Such an orientation is called a double orientation.[26]* Figure 4.23 shows the X-ray patterns of a doubly oriented PET sample taken, using a plate camera, with beams parallel to three mutually perpendicular directions. Pattern *a* was obtained with the beam parallel to the rolling direction. If the sample was oriented uniaxially rather than doubly, the photograph taken by this method would not exhibit the arcs or spots as shown in *a*, but only Debye-Scherrer rings. Moreover, patterns *b* and *c* would be identical for a uniaxially oriented sample. In the case of doubly oriented samples, the reciprocal lattice points concentrate more or less in narrow ranges along the arcs, differing from uniaxially oriented samples, as shown in Fig. 4.9.

Figure 4.24*a* is an X-ray photograph of a highly doubly oriented PVA specimen, in which the beam is parallel to the rolling direction and the rolled plane is horizontal. The indices of the spots are given in Fig. 4.24*b*. Figure 4.24*c* is presented to clarify the relation between unit cell and the rolled plane. In the doubly oriented PVA specimens the (101) plane is oriented parallel to the rolled plane and is shared by two differently oriented lattices, as indicated by solid and broken lines in Figs. 4.24*b* and 4.24*c*. Therefore the 101 reflection appears at two points, while $10\bar{1}$ appears at four. Doubly oriented nylon 66 has four differently oriented lattices sharing the (010) plane.[26] Weissenberg photographs taken on the layer lines of doubly oriented samples are very useful for indexing, confirmation of systematic absences, and separation of intensity data. Therefore trials for preparing the doubly oriented samples are strongly recommended again. Doubly oriented samples of nylon 6,[43] poly(ethylene sulfide),[44] polyoxacyclobutane $[-(CH_2)_3-O-]_n$ form I,[45] polyglycolide,[30] α-gutta percha,[39] polyallene,[46] and so on have been prepared by rolling.

4.3.3 Special Samples

A. *Crystal Modifications.* Polyoxacyclobutane can be obtained in three crystal modifications, as shown in Fig. 4.25. If a molten sample is quenched in ice water, form I is obtained. On the other hand, quenching the melt in liquid nitrogen gives an amorphous sample that transforms into form II on stretching. The amorphous sample crystallizes to form III when

* The term "biaxial orientation" has a different meaning than double orientation. See Ref. 17, p. 254.

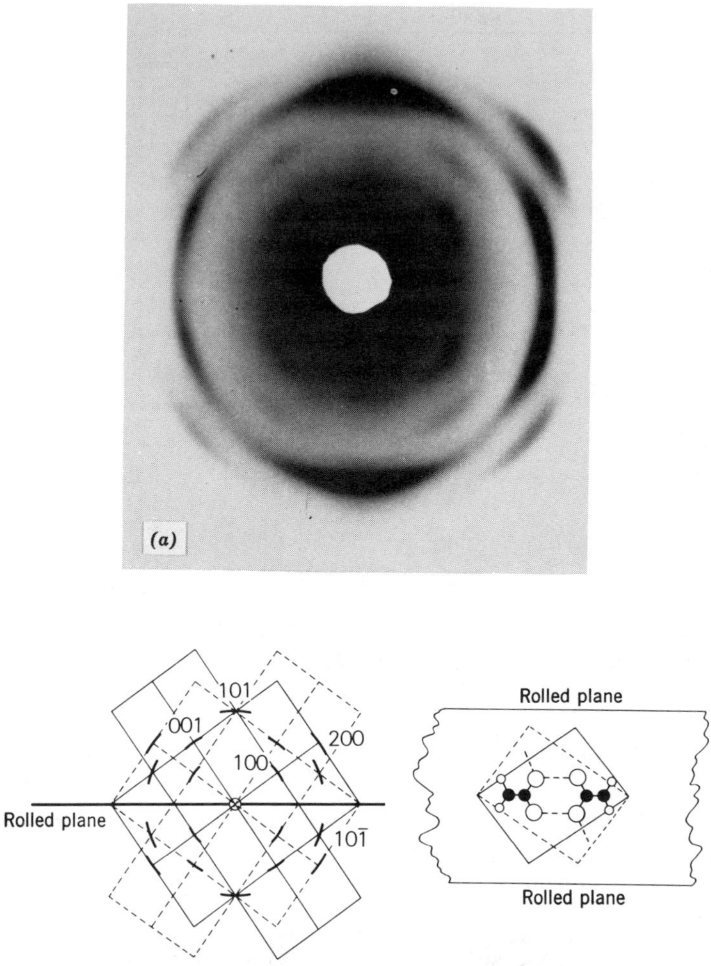

(b) (c)

Fig. 4.24 (*a*) The end directional X-ray photograph of a highly doubly oriented PVA. (*b*) Indices of the spots of photograph *a* and the position of the rolled plane. (*c*) Section of the specimen showing the relation of the unit cell to the rolled plane.[42]

it is kept at room temperature. Form III also transforms to form II on stretching. In the presence of water, forms II and III transform into form I. When form I is dried or kept *in vacuo*, form II or III is obtained.[45,47] Another crystal modification (form IV) has been found by further stretching of a specimen of form II or III. Form IV consists of planar zigzag molecular chains.[48]

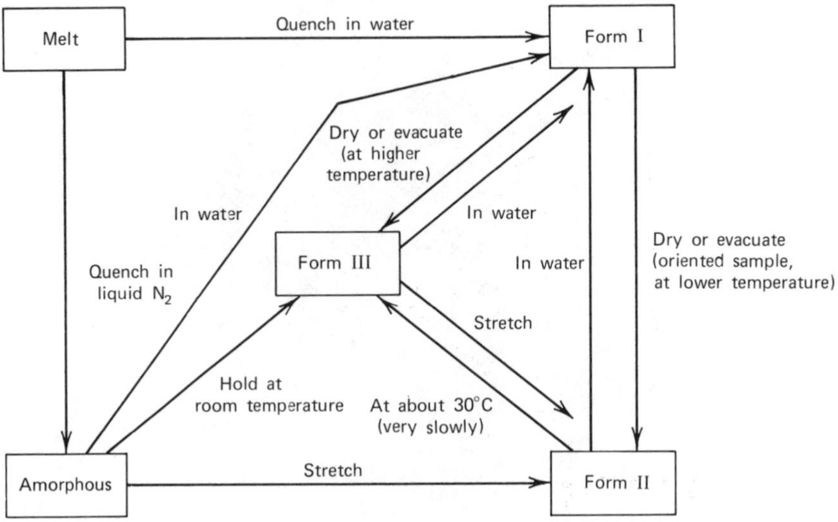

Fig. 4.25 Transitions among three modifications of polyoxacyclobutane.[47]

α-Nylon 6 forms an iodo complex on immersion in an aqueous KI solution of I_2. The complex transforms into the γ form if I_2 is removed in an aqueous $Na_2S_2O_3$ solution. If the original sample is doubly oriented, the orientation of the methylene portion does not change after transformation into the γ form.[43]

In PEO samples the molecules usually have a (7/2) helix conformation. The planar zigzag form is prepared by stretching a necked sample about twofold at room temperature. The X-ray photograph shown in Fig. 4.26 was taken by clamping the stretched sample in a metal holder under tension.[21] Arrows indicate reflections due to the planar zigzag form (cf. Fig. 4.1*a*).

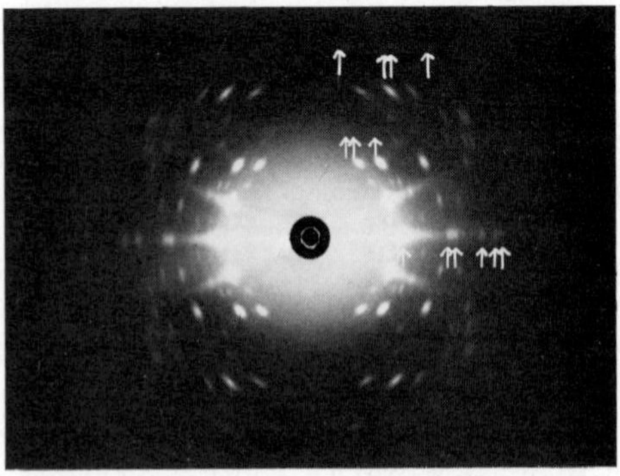

Fig. 4.26 Fiber diagram of PEO including the planar zigzag form.[21]

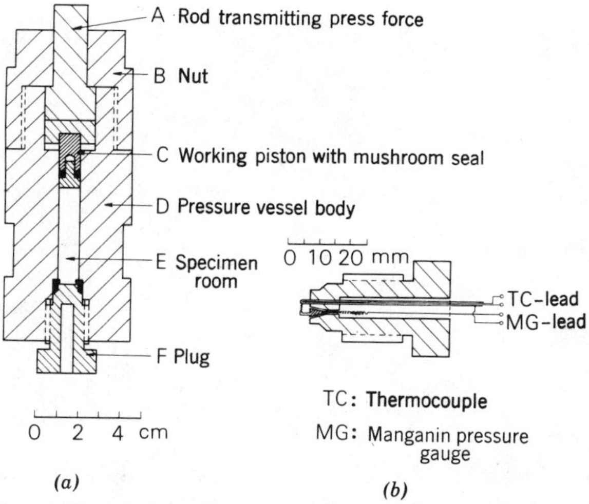

Fig. 4.27 High pressure treatment apparatus. (*a*) High pressure vessel and (*b*) plug assembly for measurement of temperature and pressure.[49]

The planar zigzag form is stable only under tension; when the sample is removed from the holder, the planar zigzag form disappears rapidly.

B. *Heat Treatments Under High Pressure.* Figure 4.27 shows an apparatus for high-pressure heat treatment.[49] *E* is the specimen chamber (diameter 10 mm) in which a sample, clamped in a metal holder (Fig. 4.33), is inserted with silicone oil. The vessel works as an intensifier by itself if a hydraulic handpress is used. After the nut *B* is fixed at a suitable pressure, the vessel is removed from the press and put into an oil heating bath. Formation of three crystal modifications of poly(vinylidene fluoride) has been confirmed with this apparatus.[50]

It has been found for *it*-poly(4-methyl-1-pentene) that the density of the crystalline regions (0.830 g/cm^3) is lower than that of the noncrystalline regions (0.838 g/cm^3) at 20°C.[51] This situation is analogous to that of ice and water. The ordinary crystallized sample transforms into a new crystal form of higher density (0.859 g/cm^3 at 20°C) by heat treatment under high pressure (4500 atm, 200–270°C).[49]

C. *Radiation Polymerization of Single Crystals.* A highly crystalline and three dimensionally oriented POM sample is obtained from a single crystal of tetraoxane by irradiation with ^{60}Co γ-rays. The polymerization temperature is 100°C and the irradiation dose is 1.4×10^5 rad/hr for 1 day; the conversion is 98%. The *c* axis rotation photograph of this sample is compared with the fiber diagram of Derlin-Acetal Resin (du Pont) in Fig. 4.28.[52]

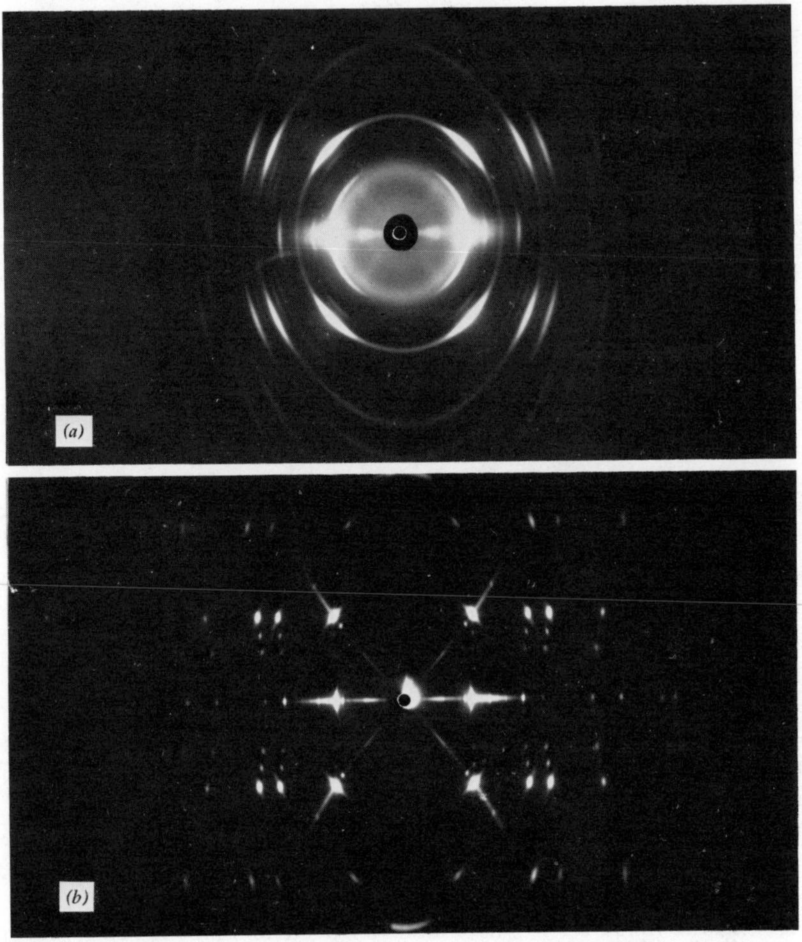

Fig. 4.28 (*a*) Fiber diagram of an oriented sample of Delrin-Acetal resin and (*b*) rotation photograph around the *c* axis of a POM sample polymerized from a tetraoxane single crystal.[52]

The same method can be used for poly-β-propiolactone, but the resultant polymer is paracrystalline and shows streaks on the layer lines (Section 4.10.4).[53]

D. *Radiation Polymerization of Adducts.* A mixture of 2 ml of nearly saturated methanol solution of thiourea and 0.1 ml of 2, 3-dichlorobutadiene at $-30°C$ after 1 week forms needlelike single crystals of the monomer thiourea complex. Irradiation by ^{60}Co γ-rays is carried out at $-78°C$; the total dose being 2.4×10^6 rad. After the thiourea is removed by washing with

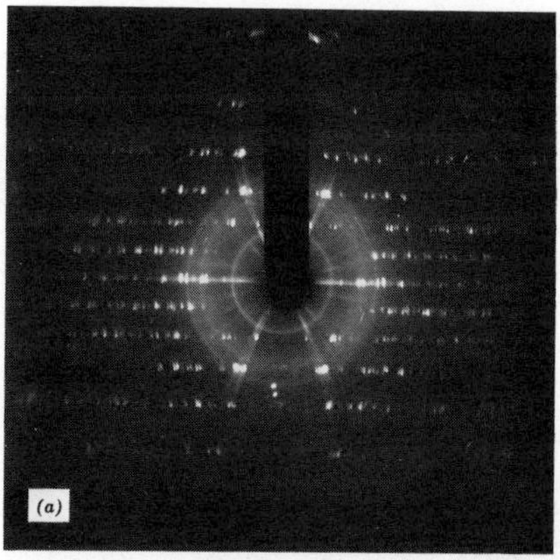

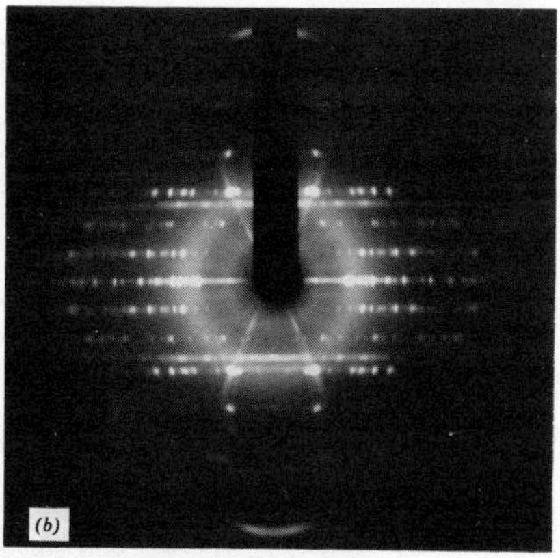

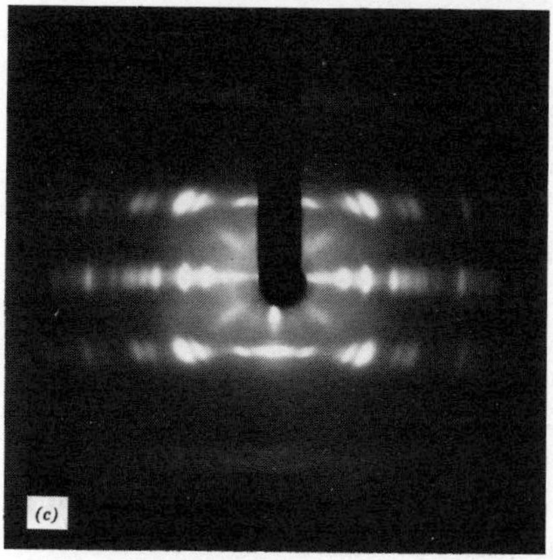

Fig. 4.29 X-Ray rotation photograph of (*a*) 2,3-dichlorobutadiene–thiourea complex and (*b*) of poly-2,3-dichlorobutadiene–thiourea complex, and (*c*) fiber diagram of poly-2,3-dichlorobutadiene.[54]

boiling methanol, poly-2,3-dichlorobutadiene is formed in a uniaxial orientation. Figure 4.29 shows the X-ray patterns corresponding to this process.[54] The same method is used for vinylidene chloride[55] and others.

4.3.4 Samples Difficult to Obtain by Usual Methods

A. *Polymer Single Crystal Mats.* Polyethers $[-(CH_2)_m-O-]_n$ with $m = 6$–12 are prepared by condensation polymerization of the corresponding diols $HO(CH_2)_mOH$ in the presence of concentrated sulfuric acid and $BF_3 \cdot Et_2O$ at 160–210°C. These samples cannot be oriented by drawing because of the low degree of polymerization. However single crystal mats, prepared by slow crystallization from hot methanol solution and subsequent sedimentation by centrifuge, give rise to X-ray patterns as shown in

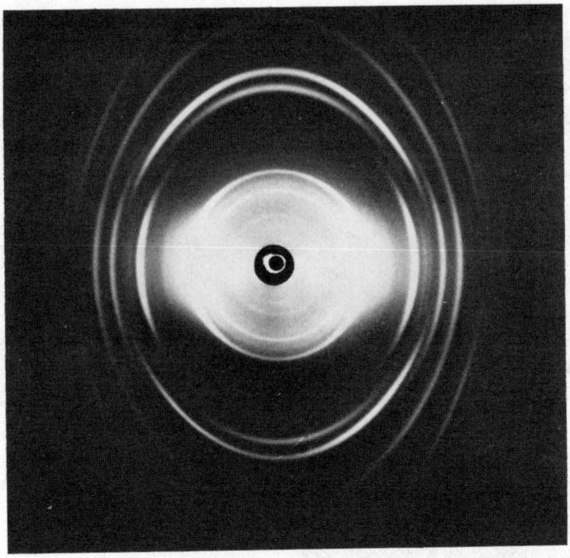

Fig. 4.30 X-Ray photograph of single crystal mat of poly (hexamethylene oxide).[56]

Fig. 4.30. This was taken by setting the mat plane parallel to the incident X-ray beam. Since the molecular chains in the mat are perpendicular to the mat plane, this photograph is quite similar to a fiber diagram of a uniaxially oriented sample.[56]

B. *Use of Nonoriented Data.* Structure analysis of orthorhombic POM was made by Carazzolo and Mammi[57] using powder sample data because of difficulties in obtaining an oriented sample. They used $FeK\alpha$ radiation in addition to $CuK\alpha$ radiation to obtain a better separation of reflections (Section 4.4.1). The lattice dimensions and space group were determined successfully by starting from a model of the orthohexagonal cell of trigonal POM. The C centered orthorhombic cell obtained by the alternative choice of axes of a hexagonal or trigonal cell is called an orthohexagonal cell (Fig. 4.31).

4.3.5 Polymer Complexes

PEO forms a crystalline complex with urea as mentioned earlier (Fig. 4.1). It also forms two types of complexes with mercuric chloride.[58,59] When a rolled film of PEO is soaked in a saturated ether solution of mercuric chloride at room temperature for 1 week, complex type I with a mole ratio of $CH_2CH_2O/HgCl_2 = 4:1$ is obtained. A sample of type II (mole ratio = 1:1) is prepared by soaking the powdered sample of type I in a saturated boiling

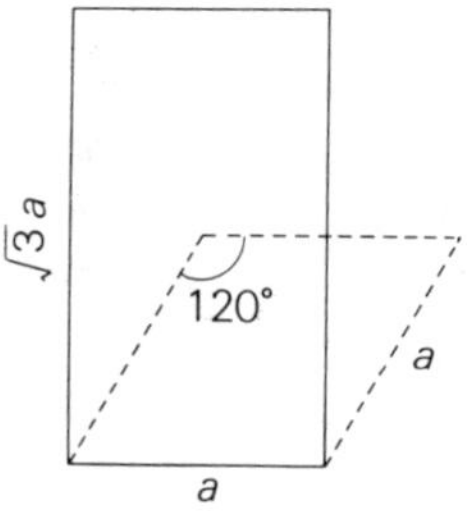

Fig. 4.31 Hexagonal cell (------) and orthohexagonal cell (———)

ether solution of mercuric chloride for about 2 days. An oriented film of type II is made by rolling.

4.4 DIFFRACTION MEASUREMENTS

4.4.1 Selection of Wavelengths

For polymer studies CuKα radiation (wavelength 1.542 Å)* is usually employed. The Kβ line (1.392 Å) is removed (or more accurately the intensity is reduced) by a nickel filter. Monochromatic beams are obtained by diffraction through lithium fluoride, graphite, and so on. The diffraction angle due to any given spacing depends on the wavelength of the X-rays as given by Bragg's condition (Eq. 4.1): the larger the wavelength, the larger the diffraction angle. Therefore X-rays with wavelengths different from those of CuKα are more effectively used in some specific cases. CoKα (1.790 Å) and FeKα (1.937 Å) radiations were utilized for the better resolution of diffraction data of nonoriented samples of a PEO–mercuric chloride complex[58] and orthorhombic POM,[57] respectively. MoKα (0.711 Å) is rarely used because of rapid diminution of the intensity with diffraction angle in polymers, although it is useful for collecting shorter spacing data for single crystals. Smaller absorption of MoKα [the mass absorption coefficient (cm^2/g) for graphite is 0.70 for MoKα and 5.50 for CuKα] is helpful in reducing the effect of sample shape. Photographs prepared with continuous X-rays (Laue photographs) are not employed for polymer studies at present.

4.4.2 Photographic Methods

A. *Cameras.* Flat cameras are used for taking the end directional photographs of doubly oriented samples (Section 4.3.2), as well as obtaining preliminary photographs of nonoriented or uniaxially oriented samples. The flat

*Customary weighted mean of Kα_1 and Kα_2, Kα_1 being given the weight 2 with respect to Kα_2.

camera is not suitable for measuring the fiber period, since the layer lines appear as hyperbolas instead of straight lines as in those photographs obtained with cylindrical cameras. For the cylindrical camera, the type of cassette in which the film is held tightly against the inner surface of the cylinder is convenient, because the construction is such that the film is held securely in position (yielding good precision) and only a small area near the beam is covered (small dead angle). Cassettes with radii of 35, 50 mm, and so on are usually available. The author and his co-workers have used a vacuum cylindrical camera with a radius of 100 mm. This camera gives good resolution of the reflections and also gives very clear photographs with air scattering reduced as shown in Fig. 4.32.

The author and his co-workers have used various sizes of collimators (beam defining systems): pinholes of diameters 1.0, 0.5, and 0.2 mm (distances between the first and second pinhole diaphragms are 25–95 mm) and rectangular slits of 0.2 mm × 2.5 mm (distance = 95 mm). The latter collimator is useful for obtaining a good resolution of the equatorial reflections by separating the sample and the film at a large distance (about 10–15 cm).

The sample is fixed on a brass holder (Fig. 4.33*a*), which can be set on a goniometer head with modeling clay. For a small sample a glass rod with

Fig. 4.32 Fiber diagram of polytetrafluoroethylene obtained with a cylindrical vacuum camera of radius 100 mm.

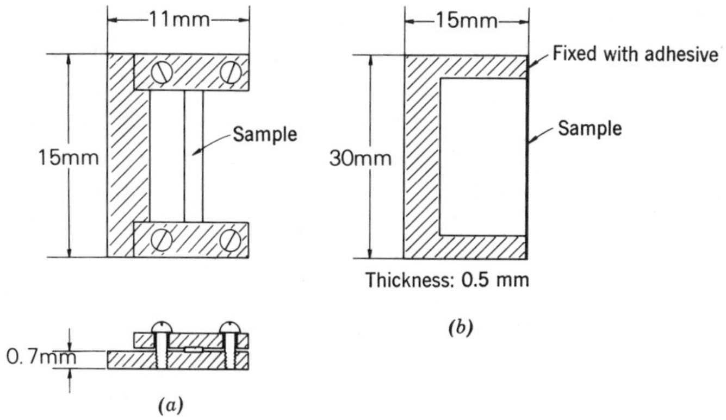

Fig. 4.33 Sample holders for X-ray measurements: (*a*) for general use and (*b*) for Norman's method.

a diameter of about 0.2 mm is used. The samples are fixed to the rod with nail enamel. A thin brass holder, Fig. 4.33*b*, is used for taking Weissenberg photographs by Norman's method, as it minimizes the shadowed area.

The usual operating conditions for CuKα tubes of 1.2 kW power are 30–35 kV and 20–25 mA; low voltage is used to reduce the intensity of continuous X-rays. A fiber diagram of a uniaxially oriented sample needs about 5 hr of exposure time. A Weissenberg photograph (Norman's method) needs about 100 hr if a collimator with a 0.3 mm diameter and a cylindrical cassette with a 35 mm radius are used.

Accurate determination of the spacings is made with a sample smeared with powder of a standard crystal; for example, the lattice constant of silicon at 25°C is 5.4301 Å, that of aluminum is 4.0494 Å at 25°C, that of sodium chloride is 5.6402 Å at 26°C, and so forth.[4]

B. *Size of the Sample.* A suitable sample size can be roughly determined from the reciprocal of the linear absorption coefficient of the material μ in centimeters. μ is obtained from the mass absorption coefficient (μ/ρ) by multiplying the density ρ. Using the mass absorption coefficients of the elements given in Ref. 4, (μ/ρ) of the sample can be calculated by the formula

$$\frac{\mu}{\rho} = \sum_i g_i \left(\frac{\mu}{\rho}\right)_i \tag{4.51}$$

where g_i is the weight fraction of the ith element of the material. For example, for CuKα μ is 3.8 cm^{-1} for PE, 11.6 cm^{-1} for POM, and 142 cm^{-1} for poly(vinylidene chloride). For ordinary polymers not containing heavy atoms such as chlorine, films about 0.5 mm thick are suitable, and no correction is needed for the absorption factor. It is preferable to use cylindrical samples for fiber diagrams and spherical or cubic samples for Norman's method, the diameter being smaller than that of the beam. In the case of polymers having a large μ, such as poly(vinylidene chloride), cylindrical samples of smaller diameter, about 0.2 mm, should be used. The PEO–mercuric chloride complex is an example of a very highly absorbing sample for which scintillation counter data (reflection method) on a powder sample were used.

C. *Measurement of the Diffraction Intensity.* In the author's laboratory photographs for intensity measurements are always taken using the multiple-film method, for which about five films are inserted together into the cassette. Measurements of the reflection intensity are made using a microphotometer and by visual comparison with a standard intensity scale. The 100 reflection of POM prepared by solid-state radiation polymerization is used for preparing a standard intensity scale for polymers. About 30 spots with diffrent relative intensities of 1 (very weak, barely visible) to 40 (very strong, a little saturated) are recorded on a film by using a Weissenberg camera, the intensity ratio being adjusted by counting the oscillational motions of the camera.

4.4.3 Diffractometric Methods

For polymers use of a diffractometer without the photographic method is not adequate, as the diffractions of high polymers are generally weak and broad. Thus it is recommended that both photographic data and intensity data as measured by the diffractometer equipped with a scintillation counter be used. The latter can be used to obtain accurate data for strong reflections. It is very important to increase the accuracy of the intensity data, especially for applying the least-squares method (Section 4.9).

In the case of polymer samples, if a CuKα tube at 30 kV and 20 mA is used, strong reflections give about 1000 cps (counts per second), medium ones give about 100 cps, and weak ones give less than 10 cps.

PSPC (the position-sensitive proportional counter) has been applied to small-angle X-ray scattering, since this is suitable for measurements of broad and weak reflections. Recently further application to the wide-angle diffraction of high polymers also has been found to be very useful based on experiments on various PE specimens.[60]

4.4.4 Measurements at High and Low Temperatures or under High Pressure

References 61 and 62 can be consulted for measurements at high and low temperatures. The author and his co-workers have used the apparatus in which air or nitrogen kept at a constant temperature is blown on the specimen set on the goniometer head.

Apparatus for measurements under high pressure can be roughly classified into two categories: piston type and anvil type. Using the former, measurements can be carried out under hydrostatic pressure. Ito and Marui have measured the spacings of PE (110, 200, etc.) at various pressures up to 3000 kg/cm^2.[63] Nakafuku and Takemura took fiber diagrams of polytetrafluoroethylene at 5500 kg/cm^2 and proposed an orthorhombic crystal structure.[64] The anvil-type apparatus is commercially available (High Pressure Diamond Optics, Inc.). Flack[65] has used this apparatus for polytetrafluoroethylene.

4.5 DETERMINATION OF LATTICE CONSTANTS AND SPACE GROUP

4.5.1 Indexing of Reflections

A. *The Orthorhombic System.* The fiber diagram of a uniaxially oriented sample of polyallene $[—CH_2—C(=CH_2)—]_n$ obtained with a cylindrical camera of 35 mm radius (Fig. 4.34)[46,66] is now considered. From the distances between the equator and the layer lines, $c = 3.87$ Å is obtained using Polanyi's formula (Eqs. 4.2 and 4.3). It is fortunate that the fiber period can be uniquely determined from fiber diagrams. The determination of molecular structure is often possible from the fiber period in the case of polymers having simple molecular conformations such as the planar-zigzag type.

The values ξ and ζ can be read from Bernal's chart, but ω (Fig. 4.10) is not known, because the fiber diagram corresponds to the full rotation photograph (not oscillation photographs over limited angular ranges). Thus fiber diagrams cannot be indexed uniquely. The ξ values read from Bernal's chart are not sufficiently accurate. Therefore ξ should be calculated by using Eq. 4.34 from the x and y values on the film measured with a comparator or a glass scale.

For indexing a circle of radius 20 cm (the limiting sphere) is drawn as shown in Fig. 4.35, 10 cm corresponding to $\xi = 1$. A straight line OQ is then drawn from the origin, and the ξ values for the equatorial reflections are plotted on it. Circles centered on the origin are drawn with the radii of all the ξ values. The values a^*, b^*, and γ^* are found by trial and error so that at least one reciprocal lattice point fits on each circle. Trials may be started by adjusting a few reflections of small ξ to the reciprocal lattice points of low diffraction order, such as 100, 010, 110, 200, 020. The unit cells of the orthorhombic (or tetragonal), monoclinic (or hexagonal), and triclinic systems should be examined in this order.

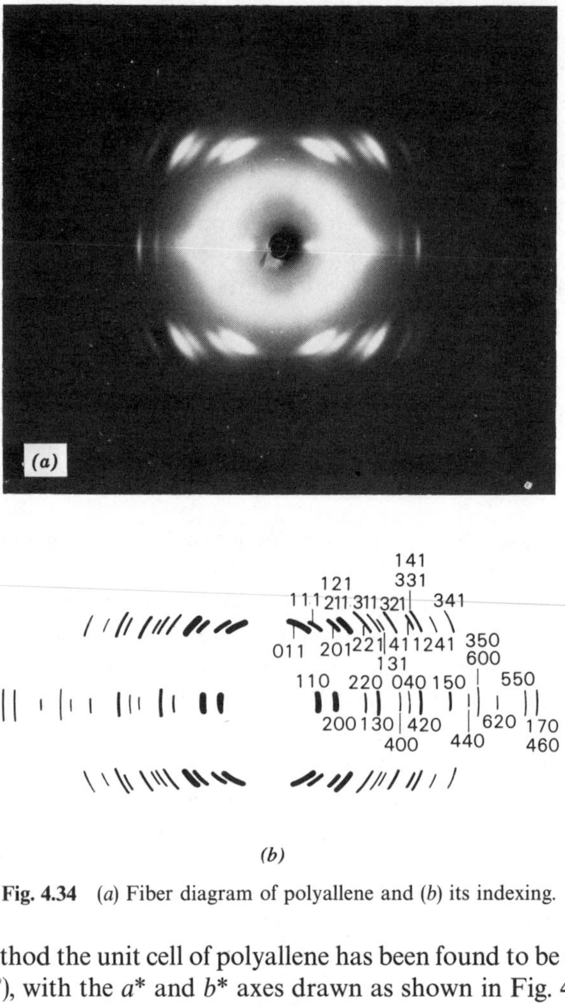

(b)

Fig. 4.34 (*a*) Fiber diagram of polyallene and (*b*) its indexing.

By this method the unit cell of polyallene has been found to be orthorhombic ($\gamma^* = 90°$), with the a^* and b^* axes drawn as shown in Fig. 4.35. On the equator some reflections do not appear because of systematic absences (Section 4.2.7.F); for instance, the $hk0$ reflections with $h + k \neq 2n$ are absent in the case of polyallene. Therefore indexing is not necessarily easy. The lattice of polyallene was found to be rectangular and the c and c^* axes are coincident. In this case the ξ values of the layer reflections hkl are equal to those of the equatorial reflections $hk0$. For example, in polyallene the reflections 111 and 201 have the same ξ as those of 110 and 200, respectively. Hence ξ for the reflections absent on the equator can be determined using the layer reflections having the same hk values. Such information is helpful for indexing.

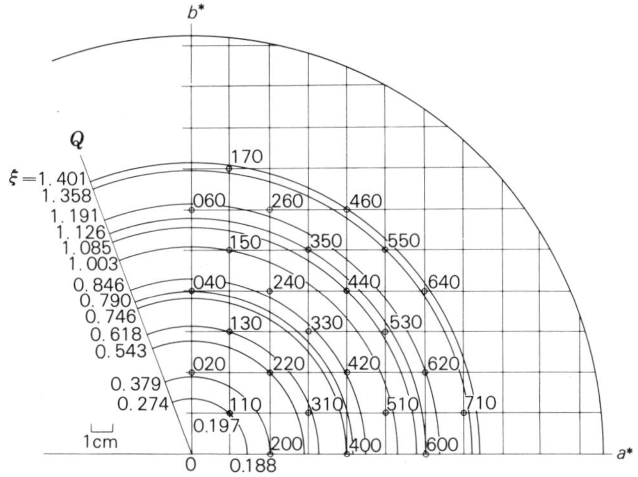

Fig. 4.35 Equatorial plane of the reciprocal lattice of polyallene. $\lambda a^* = 0.188, \lambda b^* = 0.197$.

B. *The Hexagonal and Monoclinic Systems.* In a hexagonal lattice γ is 120° and γ^* is 60° (Table 4.1). In the monoclinic system there are three cases (Section 3.4): (1) a^* is perpendicular to b^*, and c^* is not coincident with the fiber axis c, which is not normal to the ab plane (second setting, Fig. 4.36); (2) a^* and b^* are on the equatorial plane and are not perpendicular to each other, c (fiber axis) being normal to the ab plane (first setting); and (3) a^* and c^* on the equator are not perpendicular to each other and b (fiber axis) is normal to the ac plane (second setting). In cases 2 and 3, trials for determining γ^* ($\neq 90°$) (or $\beta^* \neq 90°$) are not easy, but all reciprocal lattice points having the same h and k (or l) are at the same distance ξ from the c (or b) axis for case 2 [or case 3] as in the case of the orthorhombic system. In case 1 the indexing of the equator is rather easy, but the ξ values of the layer reflections are different from those of equatorial reflections with the same hk; they should be measured by taking the c axis as the origin. The origins of the layer planes do not lie directly above or below the origin on the zero level (the $hk0$ plane), being displaced in the direction of a^* by distances that are multiples of $c^* \cos \beta^*$. Figure 4.36*b* shows ξ for 111 and $\bar{1}11$.

C. *The Triclinic System.* The indexing of a triclinic lattice is difficult because six parameters must be determined. Trials should be made utilizing all the available information. However there are favorable points: the unit cell comprises only one molecular chain in most cases, and the size of the unit cell is rather small. Doubly oriented samples are useful for indexing; nylon 66[26] and poly(*m*-phenylene isophthalamide)[67] are examples. The tilting of crystallites has also been applied successfully to PET, as is explained

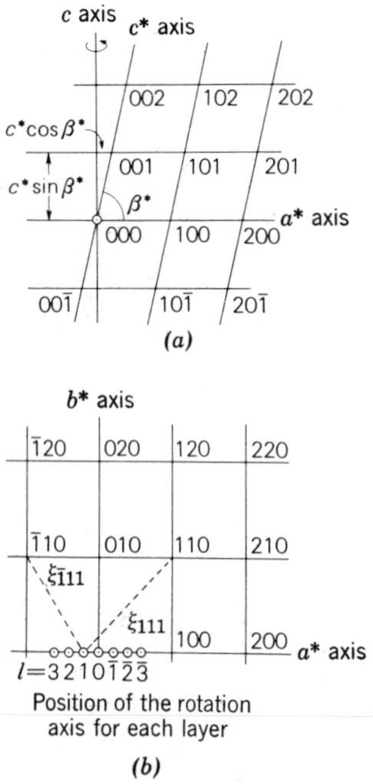

Fig. 4.36 Reciprocal lattice of α-gutta percha. Views along (*a*) the b^* axis and (*b*) the *c* axis. $\lambda a^* = 0.198, \lambda b^* = 0.245, \lambda c^* = 0.180, \alpha^* = \gamma^* = 90^\circ$, and $\beta^* = 78^\circ$.

later (Section 4.12.1).[68] The unit cell of the triclinic form of PEO (Section 4.3.3.A) contains only one molecular chain of the planar zigzag type. The fiber period $c = 7.12$ Å and the projected structure along the *c* axis have been uniquely determined. Although further indexing is difficult because the number of layer reflections is small (five on the second layer and six on the third layer), the method using a comparison of the molecular structure factor (Section 4.2.5) with the observed intensity has overcome this difficulty.[21]

4.5.2 Density and Lattice Constants

If it is assumed that *Z* monomeric units are contained in a unit cell, the density of the crystalline region ρ_{cr} is given by

$$\rho_{cr}\,(\text{g/cm}^3) = \frac{ZM}{N_A V} = \frac{1.66ZM}{V(\text{Å}^3)} \tag{4.52}$$

where N_A is Avogadro's number (6.022×10^{23}), V is the volume of the unit cell (Table 4.1), and M is the molecular weight of the monomeric unit. For polyallene ρ_{cr} is obtained using the lattice constants $a = 8.20$ Å, $b = 7.81$ Å, $c = 3.88$ Å, and $\alpha = \beta = \gamma = 90^\circ$, if $Z = 4$ is assumed. The value of Z is usually chosen by considering the cell dimension, the crystal system, molecular weight, and so on. When the Z value is large, the estimation is difficult and so an integral number is often predicted from the nonintegral Z value calculated according to Eq. 4.52 from the observed density instead of ρ_{cr}.

The density of the sample is usually measured by the flotation method. The flotation medium is prepared by mixing two liquids with higher and lower densities in which the sample is neither dissolved nor swollen. The pycnometer of Lipkin et al. is recommended for density measurement of the flotation medium.[69] The density of polyallene has been found to be 0.97 g/cm^3 at 25°C using water–methanol as the medium.

The observed density of a polymer sample ρ is represented by the equation

$$\frac{1}{\rho} = \frac{x}{\rho_{cr}} + \frac{1-x}{\rho_{am}} \tag{4.53}$$

Here x is the degree of crystallinity and ρ_{am} is the density of the amorphous region, usually lower than ρ_{cr} (90–95%).* Consequently ρ is generally somewhat smaller than ρ_{cr}. The value $\rho = 0.97$ g/cm^3 for polyallene is considered reasonable if $\rho_{cr} = 1.07$ g/cm^3. From this result the lattice constants and also $Z = 4$ are confirmed.

For the lattice constants the fiber period determined from the equator–layer distances on a fiber diagram is not sufficiently accurate. For an accurate determination of the fiber period, the meridional reflections on a Weissenberg photograph (Section 4.2.6.C) are preferable. Another suitable method is to take the photograph with a cylindrical camera in which the sample is set so that the fiber axis is perpendicular to the rotation axis and is oscillated so as to record a suitable meridional reflection.

The relationship between the errors of spacing d and the Bragg angle θ is given by differentiating Eq. 4.1 with respect to θ.

$$\frac{\Delta d}{d} = -\cot\theta\,\Delta\theta \tag{4.54}$$

The relative error of d decreases with increasing θ. Therefore for accurate measurements of spacing, it is preferable to use reflections having higher diffraction angles.

4.5.3 Determination of the Space Group

After indexing the systematic absences given in Table 4.3 can be examined. From the systematic absences for the hkl reflections, the correct lattice type is first selected from P, I, A, B, C, and F. Next the existence of glide planes and screw axes is ascertained on the basis of systematic absences for specific

* *it*-Poly(4-methyl-1-pentene) is an exceptional case (Section 4.3.3.B).[51]

reflections. The space group of α-gutta percha was uniquely determined to be $P2_1/c$-C_{2h}^5, since (1) there is no extinction for the reflections hkl, (2) the lattice is monoclinic, and (3) the systematic absences $k \neq 2n$ for $0k0$ and $l \neq 2n$ for $h0l$ have been found.[39]

The uniaxially oriented sample of polyallene gave only some of the systematic absences, $h + k \neq 2n$ for $hk0$ and $l \neq 2n$ for $00l$. Equatorial and first-layer Weissenberg photographs of a doubly oriented sample revealed more systematic absences, $k + l \neq 2n$ for $0kl$, $h + k \neq 2n$ for $hk0$, and $h \neq 2n$ for $h0l$. Therefore double orientation is very useful in this case. The above systematic absences indicate that $Pnan$-D_{2h}^6 is adequate.[46] $Pnan$ is written as $P\,2/n\,2/a\,2_1/n$ in full symbols (Table 4.4), which is obtained by exchanging the b and c axes for $Pnna$-D_{2h}^6, as given in the *International Tables* Vol. I.[2] See Section 3.4 for an explanation of the exchange of the axes.

There are cases in which sufficient data cannot be obtained because of a reason such as failure of double orientation, or the space group cannot be uniquely determined by systematic absences alone, even after they are fully clarified. In such cases trials should be made for all possible space groups that give the observed systematic absences.

The determination of the systematic absences of poly-1,3-dioxepane was difficult, since many reflections overlap because of the axial ratio $a/b \cong \sqrt{3}$; for example, 110 and 200 overlap. Systematic absences $l \neq 2n$ for $00l$

Table 4.4   Equivalent Points and Systematic Absences of
$P\,2/n\,2/a\,2_1/n$-D_{2h}^6

Number of positions, Wyckoff notation, and point symmetry	Coordinates of equivalent positions	Conditions limiting possible reflections
		General
$8\,e\,1$	$x, y, z; \frac{1}{2} - x, y, \bar{z}; x, \frac{1}{2} - y, \frac{1}{2} - z;$ $\frac{1}{2} - x, \frac{1}{2} - y, \frac{1}{2} + z; \bar{x}, \bar{y}, \bar{z}; \frac{1}{2} + x, \bar{y}, z;$ $\bar{x}, \frac{1}{2} + y, \frac{1}{2} + z; \frac{1}{2} + x, \frac{1}{2} + y, \frac{1}{2} - z$	hkl : no conditions $0kl$: $k + l = 2n$ $hk0$: $h + k = 2n$ $h0l$: $h = 2n$ $h00$: $(h = 2n)$ $0k0$: $(k = 2n)$ $00l$: $(l = 2n)$
		Special: as above plus
$4\,d\,2$	$x, \frac{1}{4}, \frac{1}{4}; \bar{x}, \frac{3}{4}, \frac{3}{4}; \frac{1}{2} + x, \frac{3}{4}, \frac{1}{4}; \frac{1}{2} - x, \frac{1}{4}, \frac{3}{4}$	hkl : $h + k = 2n$
$4\,c\,2$	$\frac{1}{4}, y, 0; \frac{3}{4}, \bar{y}, 0; \frac{3}{4}, \frac{1}{2} + y, \frac{1}{2}; \frac{1}{4}, \frac{1}{2} - y, \frac{1}{2}$	hkl : $h + k + l = 2n$
$4\,b\,\bar{1}$	$0, \frac{1}{2}, 0; \frac{1}{2}, \frac{1}{2}, 0; 0, 0, \frac{1}{2}; \frac{1}{2}, 0, \frac{1}{2}$	hkl : $h = 2n$;
$4\,a\,\bar{1}$	$0, 0, 0; \frac{1}{2}, 0, 0; 0, \frac{1}{2}, \frac{1}{2}; \frac{1}{2}, \frac{1}{2}, \frac{1}{2}$	$k + l = 2n$

have been clarified from the Weissenberg photograph, but it was difficult to determine whether the case is (*a*) $h + k \neq 2n$ for *hk*0, or (*b*) $h \neq 2n$ for *h*00 and $k \neq 2n$ for 0*k*0. In addition, the systematic absence $l \neq 2n$ for *h*0*l* were also expected. Consequently the possible space groups for this polymer were $P2_1cn$-$C_{2v}{}^9$ (systematic absences $h + k \neq 2n$ for *hk*0 and $l \neq 2n$ for *h*0*l*) and $P2_12_12_1$-$D_2{}^4$($h \neq 2n$ for *h*00, $k \neq 2n$ for 0*k*0, and $l \neq 2n$ for 00*l*). Infrared spectra were useful in resolving this case.[70]

For polyoxacyclobutane form II systematic absences – $h + k + l \neq 3n$ for *hkil* and $l \neq 2n$ for $h\bar{h}0l$ have been found, but the determination of space group $R\bar{3}c$-$D_{3d}{}^6$ or $R3c$-$C_{3v}{}^6$ cannot be made from only the above information. From a comparison of the structure factors calculated for the models of the two space groups with the observed systematic absences, the second choice has been found to be correct.[47]

The space groups that give the same systematic absences are listed in Ref. 2 (pp. 349–352).

4.6 SETTING UP MOLECULAR MODELS

Setting up suitable models for molecular conformation is inevitably important in the process of structure analysis, as already stated (Section 1.3).

4.6.1 Bond Distances, Bond Angles, and Internal Rotation Angles

A discussion of standard bond distances and bond angles must be included when steric models are considered. Table 4.5 gives the covalent radii of the atoms involved in polymers, from which rough measures of the bond distances can be obtained. The values in the table are those estimated from the standard bond distances measured on low-molecular-weight compounds mainly by X-ray and electron diffraction, infrared, Raman, and microwave spectroscopy. Figure 4.37 shows the bond lengths and bond angles of a polypeptide chain, proposed by Pauling et al.[72] These values were determined from the results of accurate crystal structure analyses of various amino acids and simple polypeptides. Among these, C_α—N = 1.47 Å agrees with the value calculated using Table 4.5, but C—N = 1.32 Å is much shorter. This shorter distance is considered to be due to the following resonance.

```
—CHR—C—NH—CHR'— ←——→ —CHR—C=N⁺H—CHR'—
     ‖                     |
     O                     O⁻
```

Figure 4.38 shows the molecular structure of polyglycolide[30] as an example of a polymer structure known with high accuracy. This analysis was made

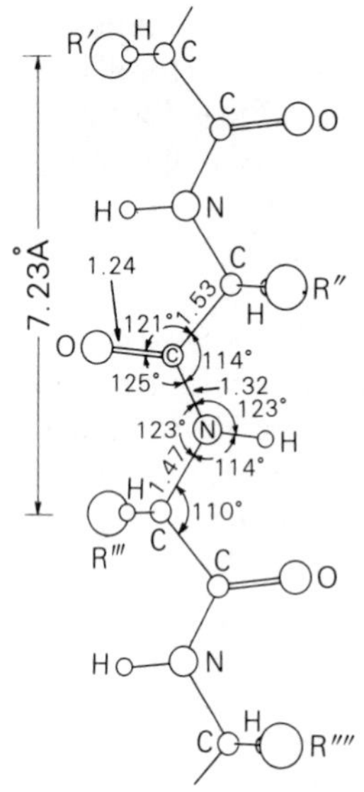

Fig. 4.37 Molecular structure of polypeptide chain. (From Pauling et al.[72])

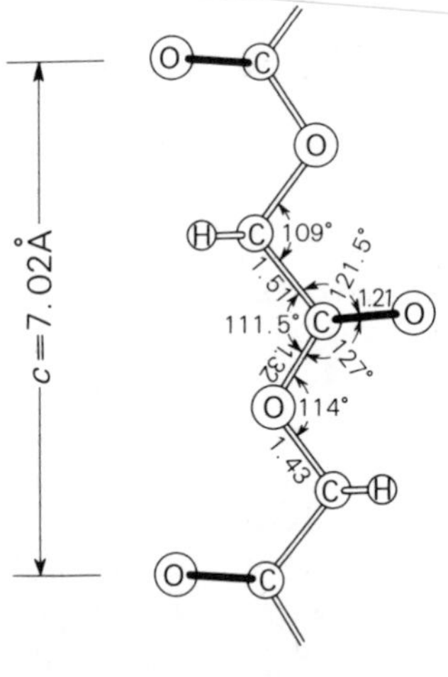

Fig. 4.38 Molecular structure of polyglycolide.[30]

by Fourier synthesis, using integrated intensity data together with anisotropic temperature factors (Section 4.8.1.A). The bond lengths and bond angles of the ester part show values similar to those of low-molecular-weight esters.

Bond angles 109°28′, 120°, and 180° are expected for the sp^3, sp^2, and sp hybridizations, respectively. Bond angles are affected more strongly by the surrounding steric hindrances than are bond lengths. This is reasonable, since the bending force constants are lower by 1 order of magnitude than the stretching force constants, as is described in Section 5.6.2. Examples of appreciable deviations from the tetrahedral angle include 112° for PE,[73] 116° for *it*-polystyrene,[74] and 128° for polyisobutylene.[32] Accordingly reference should be made to the data of related low-molecular-weight compounds. For this purpose, *Tables of Interatomic Distances and Configurations of Molecules and Ions*,[75,76] *Strukturbericht*,[77] *Structure Reports*,[78] *Crystal Structures*,[79] and *Crystal Data Determination Tables*[80] are very useful.

In polymers having the C=C double bond, such as natural rubber, the part of the molecule surrounding the double bond takes a planar form. The amide linkages —CO—NH— are also planar. The —CO—O— bonds in the main chains of polyesters are nearly trans as shown below.

	Angle (°)				
	—C———C—	—C———CO—	—CO———O—	—O———C—	—C———C—
Poly-β-propiolactone[53]	180	180	180	180	
Poly-ε-caprolactone[81]	191	176	173	185	174
Poly(ethylene adipate)[37]	180	—170	162	114	180
Poly(ethylene succinate)[38]	65	176	176	175	—77
Polypivalolactone[82]	—41	—61	—164	178	
Poly-β-hydroxybutyrate[83]	—52	—42	—175	162	
Poly(β-ethyl-β-propiolactone)[31]	—60	—21	179	136	

The internal rotation angle of the bond —CO—O— of polypivalolactone (164°) deviates considerably from *T*.

4.6.2 Equations for Conformational Parameters

The simplest case is when the observed fiber period corresponds to the planar zigzag model. For instance, the observed fiber period of polytetrahydrofuran is 12.07 Å, and analysis has been made on the basis of the symmetry of an essentially planar zigzag, since the calculated fiber period is determined to be 12.12 Å by assuming C—C to be 1.54 Å, C—O to be 1.43 Å, and all the bond angles to be 109°28′.[84]

A. *Helical Polymers.* Equations first derived by Shimanouchi and Mizushima[85] and improved by Miyazawa[86] are used for setting up models

Table 4.5 Covalent Radii of Atoms (A) (From Pauling[71])

Atom	Single bond	Double bond	Triple bond
H	0.30		
C	0.77	0.67	0.60
N	0.70	0.62	0.55
O	0.66	0.57	
F	0.64		
Si	1.17		
P	1.10		
S	1.04	0.94	
Cl	0.99		
Se	1.17		
Br	1.14		
I	1.33		

Reprinted from Linus Pauling, *The Nature of the Chemical Bond*. Copyright © 1939, 1940, 3rd ed. © 1960 by Cornell University. Used by permission of the publisher, Cornell University Press.

for helical polymers. Considered first are the $(\text{—M—})_n$ type helical polymers such as $(\text{—CF}_2\text{—})_n$. Three conformational parameters are defined: the distance ρ of the atom from the helix axis, the angle of rotation θ about the axis on passing from one atom to the next, and the corresponding translation along the axis d. The molecular conformations may be represented by the above three parameters together with the internal coordinates, which include the bond length r, bond angle ϕ, and internal rotation angle τ. For $(\text{—M—})_n$ type polymers, the following equations hold:

$$\cos\left(\frac{\theta}{2}\right) = \cos\left(\frac{\tau}{2}\right)\sin\left(\frac{\phi}{2}\right)$$

$$d\sin\left(\frac{\theta}{2}\right) = r\sin\left(\frac{\tau}{2}\right)\sin\left(\frac{\phi}{2}\right) \tag{4.55}$$

$$2\rho^2(1-\cos\theta) + d^2 = r^2$$

The equations for the $(\text{—M}_1\text{—M}_2\text{—})_n$ type polymers are:

$$d\sin\left(\frac{\theta}{2}\right)=(r_{12}+r_{21})\sin\left(\frac{\tau_{12}+\tau_{21}}{2}\right)\sin\left(\frac{\phi_1}{2}\right)\sin\left(\frac{\phi_2}{2}\right)$$
$$-(r_{12}-r_{21})\sin\left(\frac{\tau_{12}-\tau_{21}}{2}\right)\cos\left(\frac{\phi_1}{2}\right)\cos\left(\frac{\phi_2}{2}\right)$$
$$\cos\left(\frac{\theta}{2}\right)=\cos\left(\frac{\tau_{12}+\tau_{21}}{2}\right)\sin\left(\frac{\phi_1}{2}\right)\sin\left(\frac{\phi_2}{2}\right)$$
$$-\cos\left(\frac{\tau_{12}-\tau_{21}}{2}\right)\cos\left(\frac{\phi_1}{2}\right)\cos\left(\frac{\phi_2}{2}\right) \qquad (4.56)$$

where subscripts 1 and 2 refer to the numbering of the atoms. For a discussion of members $(\text{—M}_1\text{—M}_2\text{—M}_3\text{—})_n$, and so on, see Ref. 86. In practice reasonable combinations of d and θ are usually sought by varying the internal rotation angles, assuming fixed bond lengths and bond angles. Sugeta and Miyazawa[87] have re-formed the equations to matrix formulas that should be suitable for electronic computers.

B. *Glide-type Polymers.* Although the conformational parameter equations for glide-type polymers have been reported by Ganis and Temussi,[88] the method reported by Tai and Tadokoro[89] is explained here. For glide-type conformations, two conditions are necessary: (*a*) the signs of the corresponding internal rotation angles of neighboring asymmetric units* are reversed, and (*b*) the translational unit consists of two asymmetric units. The asymmetric unit consists of m main chain atoms, and the subsequent asymmetric units are distinguished by primes from each other. The numbering of the internal coordinates is defined below.

$$[\text{—}\ M_1\text{—}\ M_2\text{—}\cdots\text{——}M_j\text{—}\cdots\text{—}M_m\text{—}\ M_{1'}\text{—}\ M_{2'}\ \text{—}\cdots\text{——}\ M_{j'}\text{—}\cdots\text{—}M_{m'}\text{—}]_n$$

$$r_{m1}\quad r_{12}\quad r_{j-1,j}\quad r_{m1}\quad r_{12}\quad r_{j-1,j}$$

$$\phi_1\quad \phi_2\quad \phi_j\quad \phi_m\quad \phi_1\quad \phi_2\quad \phi_j$$

$$\tau_{m'1}\quad \tau_{12}\quad \tau_{j-1,j}\quad \tau_{m1'}\quad \tau_{1'2'}\quad \tau_{(j-1)',j'}$$

The jth main chain atom in the nth translational unit is indicated by $M_j^{\,n}$.

* An asymmetric unit or, more strictly, a crystallographic asymmetric unit is the smallest unit related to others by symmety operations. For instance, one monomeric unit of *it*-polystyrene is one asymmetric unit, because the molecular chain is on the threefold screw axis in the lattice. On the other hand, three monomeric units of *it*-PP correspond to one asymmetric unit, since the lattice has no threefold screw axis. However, when one isolated molecular chain of *it*-PP is treated, one monomeric unit may be regarded as one asymmetric unit, because of its approximate threefold screw symmetry.

Condition *a* is given by

$$\tau_{m'1} = -\tau_{m1'}, \qquad \tau_{12} = -\tau_{1'2'}, \cdots, \qquad \tau_{j-1,j} = -\tau_{(j-1)',j'}, \cdots \tag{4.57}$$

Sets of right-hand Cartesian coordinates are chosen as defined in Appendix B. Here $\mathbf{x}_j\,(i)$ denotes the coordinates of the atom M_i represented by the coordinate system (x_j, y_j, z_j) fixed at the atom M_j. Coordinates of M_i can be rewritten using $(j-1)$th coordinates as $\mathbf{x}_{j-1}(i)$ according to Eyring's equation (Appendix B).*

$$\mathbf{x}_{j-1}(i) = \mathbf{A}_{j-1,j}\mathbf{x}_j(i) + \mathbf{B}_{j-1,j} \tag{4.58}$$

where

$$\mathbf{A}_{j-1,j} = \begin{bmatrix} -\cos\phi_j & -\sin\phi_j & 0 \\ \cos\tau_{j-1,j}\ \sin\phi_j & -\cos\tau_{j-1,j}\ \cos\phi_j & -\sin\tau_{j-1,j} \\ \sin\tau_{j-1,j}\ \sin\phi_j & -\sin\tau_{j-1,j}\ \cos\phi_j & \cos\tau_{j-1,j} \end{bmatrix} \tag{4.59}$$

and

$$\mathbf{B}_{j-1,j} = \begin{bmatrix} r_{j-1,j} \\ 0 \\ 0 \end{bmatrix} \tag{4.60}$$

Hereafter *i* is omitted for simplicity.

The transformation of the *j*th coordinates $\mathbf{x}_j^n$ of the *n*th translational unit into the *j*th coordinates $\mathbf{x}_j^{n-1}$ of the $(n-1)$th unit is expressed as

$$\mathbf{x}_j^{n-1} = \mathbf{A}\mathbf{x}_j^n + \mathbf{D}_j \tag{4.61}$$

where

$$\mathbf{A} = \mathbf{A}_{j,j+1'} \cdots \mathbf{A}_{m-1,m}\mathbf{A}_{m,1'} \cdots \mathbf{A}_{(m-1)',m'}\mathbf{A}_{m',1'} \cdots \mathbf{A}_{j-1,j} \tag{4.62}$$

$$\begin{aligned} \mathbf{D}_j = {} & \mathbf{A}_{j,j+1} \cdots \mathbf{A}_{j-2,j-1}\mathbf{B}_{j-1,j}^n + \mathbf{A}_{j,j+1} \cdots \mathbf{A}_{j-3,j-2}\mathbf{B}_{j-2,j-1}^n \\ & + \cdots + \mathbf{A}_{j,j+1}\mathbf{B}_{j+1,j+2}^{n-1} + \mathbf{B}_{j,j+1}^{n-1} \\ & \mathbf{B}_{j-1,j}^n = \mathbf{B}_{j-1,j}^{n-1} \end{aligned} \tag{4.63}$$

Since coordinate systems $\mathbf{x}_j^n$ and $\mathbf{x}_j^{n-1}$ are symmetric with respect to translation, $\mathbf{A}$ equals a unit matrix $\mathbf{E}$. Thus

$$\text{Trace}\ \mathbf{A} = 3 \tag{4.64}$$†

is equivalent to condition *b*. Since $\mathbf{D}_j$ is a vector connecting M_j^{n-1} and M_j^n expressed by the $\mathbf{x}_j^{n-1}$ coordinate system, its absolute value equals the fiber

* See, for example, Refs. 90 and 91 concerning matrix notation.

† The sum of the diagonal elements is called the trace, or *Spur* in German.

period. In practical application, molecular models are chosen from suitable sets of internal rotation angles satisfying the requirements as expressed in Eqs. 4.57 and 4.64 and the fiber period.

C. *New Equations.* The author and his co-workers have derived new equations relating the conformational parameters and internal coordinates.[92] These equations can be applied not only to polymer chains with glide or helical symmetry, but also to those with only translational symmetry. A polymer having k main chain bonds in an asymmetric unit, with fixed bond lengths and bond angles, is now considered. When the fiber period is not fixed, the number of independent internal rotation angles is k for helical polymers ($k - 1$, if the number of monomeric units per turn is given), $k - 1$ for the glide-type polymers, and $k - 3$ for polymers with translational symmetry only. If the fiber period is fixed, the above-mentioned independent variables are further reduced by one. In practical applications of the foregoing methods, the model settings must be made by changing the k variables. However the new method needs only the variation of the independent variables; therefore it reduces the amount of calculation and is suitable for electronic computers. This is especially valuable for polymers having complicated conformations with large k. See also Ref. 93.

D. *Examples of Application.* Poly(*tert*-butylethylene oxide) has a true asymmetric carbon atom in each monomeric unit (Fig. 4.39).[94] Thus there are two optical isomers in the case of the *it*-polymer chain: rectus (R) and sinister (S) (Section 2.2). The racemic sample of *it*-poly(*tert*-butylethylene oxide) gives rise to three crystal modifications, and form I is indexed by a tetragonal lattice with $a = 15.42$ Å and c (fiber axis) $= 24.65$ Å. The unit cell includes four molecular chains, and the chain conformation has been inferred to be a (9/4) helix from the general features of the X-ray pattern (see Section 4.6.4.B).

As shown in Fig. 4.39, there are four kinds of internal rotation angles, τ_1, τ_2, and τ_3 in the main chain and τ_4 in the side chain. The conformations of the main chain are restricted by the helical parameter equations [the equations for $(\text{—M}_1\text{—M}_2\text{—M}_3\text{—})_n$ omitted in Section 4.6.2.A][86] with

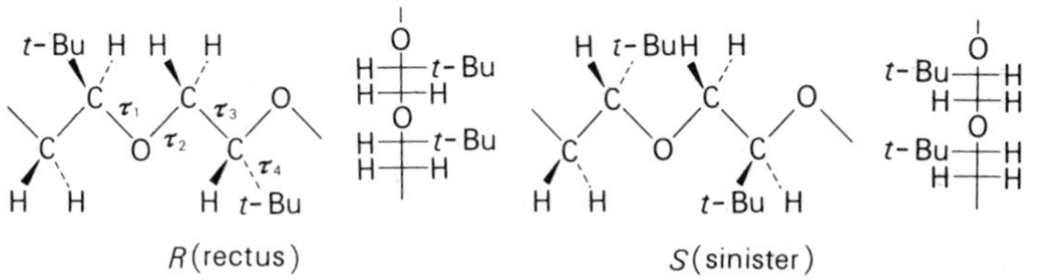

Fig. 4.39 Two optical isomers of *it*-poly (*tert*-butylethylene oxide).

the assumptions of a (9/4) helix; a fiber period of 24.65 Å; the bond lengths C—C = 1.54 Å, C—O = 1.43 Å, and all the bond angles equal to 109°28′. The possible sets of the internal rotation angles are on the two closed curves in a cube defined by three-dimensional Cartesian coordinates and angles τ_1, τ_2, and τ_3 each covering from 0 to 360°. This means that there is only one degree of freedom, for instance, if τ_1 is given, τ_2 and τ_3 are uniquely fixed. Figure 4.40 shows the projection on the $\tau_2\tau_3$ plane, and the two curves correspond to the right- and left-hand helices.

4.6.3 Application of Energy Calculation

The calculation of intramolecular interaction energy can provide suitable molecular models, and so it has already been successfully utilized for the structure analyses of several polymers.[31,32,94–96] In continuation of the preceding section the application to *it*-poly(*tert*-butylethylene oxide) is shown as an example. Details of the energy calculations are described in Chapter 6.

The intramolecular potential energy is calculated along the closed curves of the right-hand helix given in Fig. 4.40. Here it is necessary to perform the calculations for the two configurational isomers *S* and *R* separately. The potential maps around the minima are partially shown in Fig. 4.41, in which the contours are drawn against τ_1 and τ_4. For any τ_1 value, the other two τ values are uniquely fixed. It is sufficient to calculate over the range 0–120° for τ_4, since the *tert*-butyl group has three equivalent methyl groups.

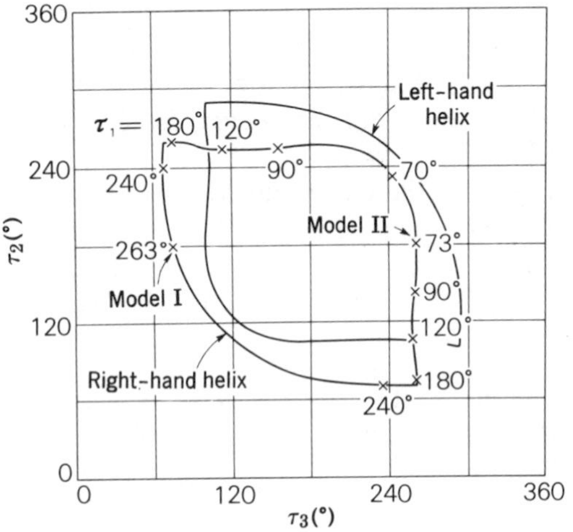

Fig. 4.40 Three-dimensional closed curves for possible conformations of the main chain of *it*-poly (*tert*-butylethylene oxide).[94]

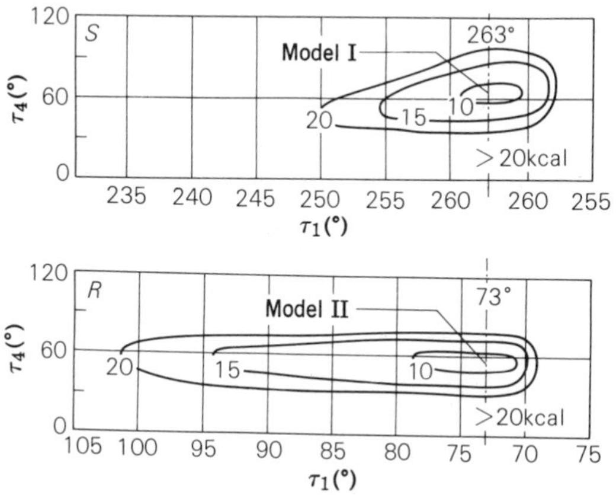

Fig. 4.41 Potential energy maps of *it*-poly (*tert*-butylethylene oxide).[94]

Each optical isomer has a potential minimum, corresponding to a stable conformation of the right-hand helix: model I for an *S* polymer and model II for an *R* polymer. These two minima give the same potential energy value, 8.5 kcal per mole of monomeric unit. Molecular models I and II are shown in Fig. 4.42. The most characteristic difference is that the oxygen atoms (open circles) are at the innermost positions in model I, whereas the methylene groups (filled circles) are innermost in model II. τ_2 is trans and τ_4 is nearly staggered (60°) for both models, while τ_1 and τ_3 are opposite for the two models. Further analysis is described in Section 4.12.3.

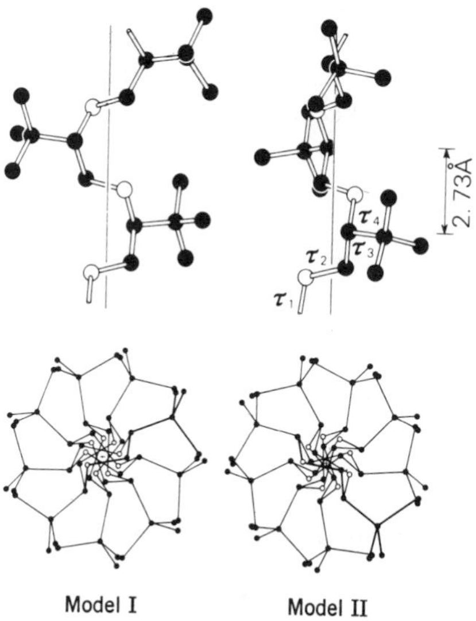

Fig. 4.42. Molecular models of *it*-poly (*tert*-butylethylene oxide): model I [$\tau_1 = 263°$ ($= -97°$), $\tau_2 = 180°$, $\tau_3 = 73°$, $\tau_4 = 64°$] and model II [$\tau_1 = 73°$, $\tau_2 = 183°$, $\tau_3 = -97°$, $\tau_4 = 55°$].[94]

4.6.4 The Cochran-Crick-Vand Formula (the Structure Factor for Helical Polymers)

For X-ray analyses of helical polymers, the theory of Cochran, Crick, and Vand is very important.[97] The diffraction intensity by a helical molecule (per fiber period) is calculated from the square of the following equation $|F|^2$.

$$F\left(R, \psi, \frac{l}{c}\right) = u\sum_n \sum_j f_j J_n(2\pi R r_j) \exp\left[i\left(n\psi - n\varphi_j + \frac{1}{2}n\pi + \frac{2\pi l z_j}{c} \right)\right]$$

$$= \sum_n G_{n,l}(R) \exp\left[in\left(\psi + \frac{1}{2}\pi \right)\right] \tag{4.65}$$

$$G_{n,l}(R) = u\sum_j f_j J_n(2\pi R r_j) \exp\left[i\left(-n\varphi_j + \frac{2\pi l z_j}{c} \right)\right] \tag{4.66}$$

where R, ψ, and l/c are the cylindrical coordinates of a point in reciprocal space, f_j is the atomic structure factor, and r_j, φ_j, and z_j are the cylindrical coordinates of the jth atom (Fig. 4.43).* J_n is a Bessel function, the orders n of which are obtained for reflections on the lth layer line by solving the equation

$$l = tn + um \tag{4.67}$$

where u and t are the numbers of structural units and turns, respectively, in the fiber period and m is an integer. The derivation of these equations is given in Appendix C according to Naya and Nitta.[98]

Although Eq. 4.65 is for one molecular chain, it can be applied to the case of a crystal in which the unit cell includes only one molecular chain. The unit cell structure factor is described in Section 4.7.2 for the case where more than one molecular chain is included in the unit cell.

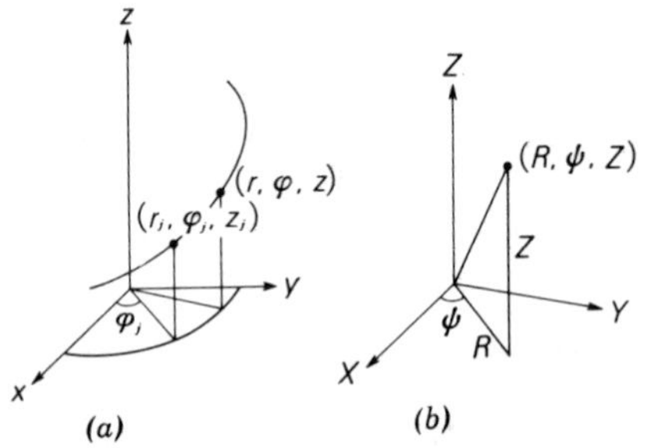

Fig. 4.43 Cylindrical coordinates in (*a*) physical and (*b*) reciprocal spaces.

* z_j is not the fractional coordinate.

Table 4.6 Two Molecular Models of POM[102]

Molecular model	Helix (9/5)	Helix (9/4)
Number of turns in the fiber period	5	4
∠ COC and ∠ OCO	110°53′	123°9′
Internal rotation angle	77°23′	58°50′
Distance of the chain atoms from the helix axis	0.691 Å	0.824 Å
Angle of rotation about the axis between the subsequent chain atoms	100°	80°
Translation along the axis between the subsequent chain atoms	0.961 Å	0.961 Å

A. *Application to Polyoxymethylene.* Structural studies of POM were made in the early years of polymer study by the school of Staudinger.[99a] Hengstenberg[99b] and Sauter[100] found that (*a*) POM has a trigonal unit cell with lattice constants $a = 4.46$ Å and $c = 17.35$ Å, and (*b*) one unit cell contains one molecular chain. Sauter has reported further that the space group for this polymer may be C_3^2 or C_3^3, and that this molecule has a (9/4) helical conformation. On the other hand, Huggins[101] has suggested a (9/5) model, considering that the COC and OCO angles are expected to be close to the tetrahedral angle. Table 4.6 shows the (9/5) and (9/4) helix models calculated according to Eq. 4.55 by assuming that the bond distance C—O = 1.43 Å, ∠ COC = ∠ OCO, and all the internal rotation angles of the main chain are equal.

The author and his co-workers[102] have compared the intensities of the reflections calculated for these two models with the observed X-ray data and concluded that the (9/5) model proposed by Huggins is reasonable. For helices in which the number of chemical units per fiber period u is large, such as in POM, the ψ dependence of $F(R, \psi, l/c)$ is not remarkable. Consequently the average of the square of F with respect to ψ may be used as a good approximation for comparison with the data of the fiber diagram.[103]

$$\left\langle F^2\left(R,\psi,\frac{l}{c}\right)\right\rangle_\psi = \langle FF^*\rangle_\psi = \frac{1}{2\pi}\int_0^{2\pi} FF^*\,d\psi$$

$$= \frac{1}{2\pi}\sum_n\sum_{n'} G_{n,l}(R)G_{n',l}^*(R)\int_0^{2\pi} \exp\left[i(n-n')\left(\psi+\frac{\pi}{2}\right)\right]d\psi \tag{4.68}$$

Since n and n' are integers and F is a periodic function of the period 2π with respect to ψ,

$$\int_0^{2\pi} \exp\left[\mathrm{i}(n-n')\psi\right]\mathrm{d}\psi = \begin{cases} 0 & n \neq n' \\ 2\pi & n = n' \end{cases} \tag{4.69}$$

Therefore

$$\left\langle F^2\left(R, \psi, \frac{l}{c}\right)\right\rangle_\psi = \sum G_{n,l}(R)G_{n,l}{}^*(R) = u^2 \sum_n \left\{\left[\sum_j f_j J_n(2\pi R r_j) \times\cos\left(-n\varphi_j + \frac{2\pi l z_j}{c}\right)\right]^2 + \left[\sum_j f_j J_n(2\pi R r_j)\sin\left(-n\varphi_j + \frac{2\pi l z_j}{c}\right)\right]^2\right\} \tag{4.70*}$$

If the POM molecule is assumed for simplicity to be oriented so that the carbon atom is on the line of $\varphi = 0$ and $z = 0$, then Eq. 4.70 can be written as

$$\left\langle F^2\left(R, \psi, \frac{l}{c}\right)\right\rangle_\psi = u^2 \sum_n \left[f_C J_n(2\pi R r_C) + f_O J_n(2\pi R r_O)\cos \pi m\right]^2 \tag{4.71**}$$

Here the contribution of the hydrogen atoms was omitted. Equation 4.67 is written as Eqs. 4.72 and 4.73 for the (9/5) and (9/4) models, respectively,

$$l = 5n + 9m \tag{4.72}$$

$$l = 4n + 9m \tag{4.73}$$

The first and second solutions of Eq. 4.72 are listed for each layer in Table 4.7. The curves $J_n(2\pi R r_j)$ for $n = 0$–8 are plotted against $2\pi R r_j$ in Fig. 4.44. The broken lines indicate the $2\pi R r_j$ values calculated for the *hkl* reflections by assuming $r_j = 0.691$ Å [r_C and r_O of the (9/5) model of Table 4.6]. For POM only the lower order Bessel functions should make appreciable contributions to the structure factors corresponding to the region of the diffraction angles for which the reflections can be observed. Therefore only the first solutions of Eqs. 4.72 and 4.73 need be taken into account.† Equation 4.71 can be rewritten in a more convenient form by using m for each l in Table 4.7. For example, for the reflections on the fourth and fifth layer lines,

$$\begin{aligned}\langle F^2(R, \psi, 4/c)\rangle_\psi &= u^2[f_C J_{-1}(2\pi R r_C) - f_O J_{-1}(2\pi R r_O)]^2 \\ &= u^2[f_C J_1(2\pi R r_C) - f_O J_1(2\pi R r_O)]^2\end{aligned} \tag{4.74‡}$$

*From $\sum_j f_j J_n(2\pi R r_j)\exp\left[\pm \mathrm{i}(-n\varphi_j + 2\pi l z_j/c)\right] = A_n \pm \mathrm{i}B_n$, $\{\,\} = (A_n + \mathrm{i}B_n)\,(A_n - \mathrm{i}B_n) = A_n^2 + B_n^2$ is obtained.

** From Eqs. 4.67 and 4.70, $-n\varphi_O + 2\pi l z_O/c = -n\pi t/u + \pi l/u = \pi m$.

†This situation does not hold for the cases where r_j is large.

‡$J_{-n}(x) = (-1)^n J_n(x)$.

Table 4.7 The First and Second Solutions of Eq. 4.72 for the (9/5) Helix of POM[a]

l	n	m	n	m
0	0	0	9	−5
			−9	5
1	2	−1	−7	4
2	4	−2	−5	3
3	−3	2	6	−3
4	−1	1	8	−4
5	1	0	−8	5
6	3	−1	−6	4
7	−4	3	5	−2
8	−2	2	7	−3
9	0	1	9	−4
			−9	6
10	2	0	−7	5
11	4	−1	−5	4
12	−3	3	6	−2
13	−1	2	8	−3
14	1	1	−8	6
15	3	0	−6	5
16	−4	4	5	−1
17	−2	3	7	−2
18	0	2	9	−3
			−9	7

[a]For the values of $-l$, sets of n and m with reversed signs are obtained.

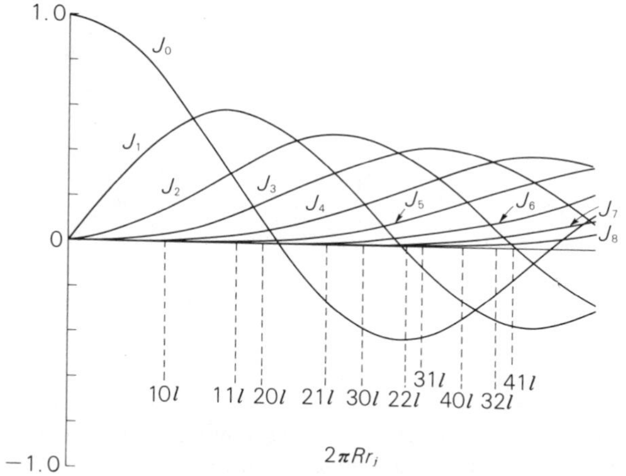

Fig. 4.44 Curves $J_n(2\pi Rr_j)$ for $n = 0–8$.[52]

$$\left\langle F^2\left(R, \psi, \frac{5}{c}\right)\right\rangle_\psi = u^2[f_C J_1(2\pi R r_C) + f_O J_1(2\pi R r_O)]^2 \tag{4.75}$$

for the (9/5) model, and

$$\left\langle F^2\left(R, \psi, \frac{4}{c}\right)\right\rangle_\psi = u^2[f_C J_1(2\pi R r_C) + f_O J_1(2\pi R r_O)]^2 \tag{4.76}$$

$$\left\langle F^2\left(R, \psi, \frac{5}{c}\right)\right\rangle_\psi = u^2[f_C J_1(2\pi R r_C) - f_O J_1(2\pi R r_O)]^2 \tag{4.77}$$

for the (9/4) model. From Eqs. 4.74–4.77, it is clear that: (*a*) in the (9/5) model the X-rays scattered from the CH_2 groups possess phases opposite those from the oxygen atoms on the fourth layer line, while they are in phase on the fifth layer line, and (*b*) the reverse is the case for the (9/4) model. Figure 4.28*a* shows the fiber diagram of Delrin-Acetal Resin, which is characteristic of very strong reflections on the equatorial and fifth layer lines. Table 4.8 gives the average intensities of the reflections on each layer line, the order of Bessel function *n*, and the phase relation between the X-rays scattered from the CH_2 groups and those from the oxygen atoms for the two models. The characteristic feature of the observed average intensities of

Table 4.8 The Average Intensity of the Reflections on Each Layer Line of POM as Compared with the Calculated Results for the Two Models[102]

		(9/5) model		(9/4) model	
l	Estimated average intensity of the layer[a]	*n*	Phase relation	*n*	Phase relation
0	vs	0	+	0	+
1	vw	2	−	−2	−
2	vw	4	+	−4	+
3	w	−3	+	3	−
4	w	−1	−	1	+
5	s	1	+	−1	−
6	vw	3	−	−3	+
7	vw	−4	−	4	−
8	w	−2	+	2	+
9	m	0	−	0	−
18	m	0	+	0	+

[a]The reflections on the first, second, sixth, and seventh layer lines could not be observed for the Derlin sample, but were subsequently observed for the solid-state polymerization product of tetraoxane.[52]

the reflections on the layer lines, especially the low and high intensities of the reflections on the fourth and fifth layer lines, respectively, can be reasonably interpreted in terms of the order of the Bessel functions and the phase relations in the case of the (9/5) model. On the other hand, the expected intensities of the reflections on the third through sixth layer lines for the (9/4) model do not agree at all with the observed ones. Thus the (9/5) model (Fig. 4.45*a*) was found to be adequate.

B. *Considerations for Comparing the Calculated and Observed Molecular Structure Factors.* POM is a suitable example for comparing the calculated molecular structure factors with the observed crystal structure factors. In fact, the molecular structure factors [$G_{n,l}(R)$ of Eq. 4.66] calculated using anisotropic temperature factors show good agreement with the X-ray data obtained from the radiation polymerization product of a tetraoxane single crystal (Fig. 4.28*b*).*

However this is an especially fortunate case in which the following conditions are fulfilled: (*a*) the number of molecular chains in the unit cell N is one; (*b*) the helix radius is of a suitable size, that is, not so large that only the

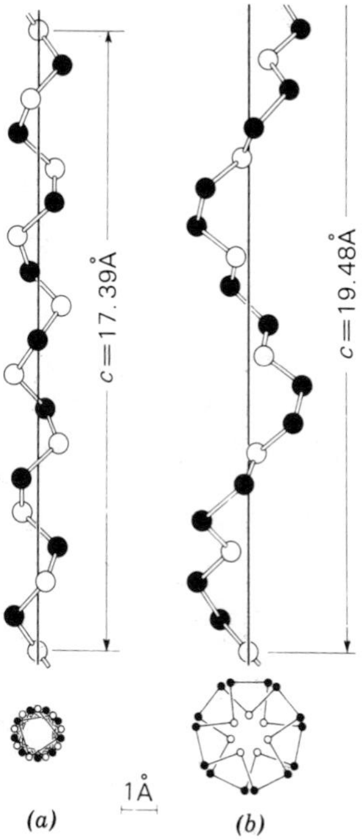

Fig. 4.45 Skeletal models of (*a*) POM and (*b*) PEO: (○) oxygen atom; (●) methylene group.[52,104]

* The temperature factors used here are different from those stated in Section 4.2.7.B. See Ref. 52.

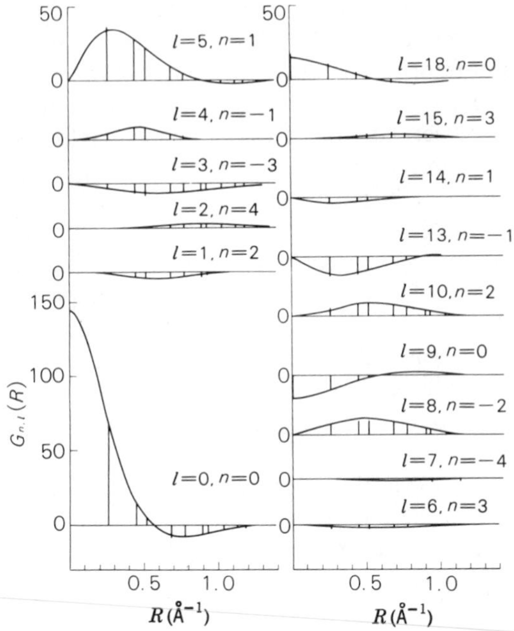

Fig. 4.46 Calculated curves $G_{n,l}(R)$ versus R and the observed structure factors for POM. The signs of the observed values were assumed to be the same as the calculated $G_{n,l}(R)$.[52]

low-order Bessel functions contribute to $G_{n,l}(R)$; (*c*) the number of chemical units per fiber period u is rather large. If these conditions were not fulfilled, such good agreement as shown in Fig. 4.46 would not be obtained.

Even when the above-mentioned conditions are not satisfied, if only condition *b* holds, the values of u and t of the helix can still be estimated from the qualitative consideration of Eq. 4.65, that is, from the comparison of the lowest order n to the average intensity of each layer line. Furthermore, for the purpose of preliminary examination of the molecular conformation, $|F(\text{obs})|^2/N^2$ is often compared with $\langle |F_M|^2 \rangle_\psi$ calculated according to the equation:

$$\left\langle \left| F_M\left(R, \psi, \frac{l}{c}\right) \right|^2 \right\rangle_\psi = u^2 \sum_n \left\{ \left[\sum_j f_j J_n(2\pi R r_j) \cos\left(\frac{2\pi l z_j}{c} - n\varphi_j\right) \right]^2 + \left[\sum_j f_j J_n(2\pi R r_j) \sin\left(\frac{2\pi l z_j}{c} - n\varphi_j\right) \right]^2 \right\} \quad (4.78)$$

Here $|F(\text{obs})|^2$ is the observed intensity $I(\text{obs})$ per unit cell divided by the multiplicity. However, strictly speaking, $|F(\text{obs})|^2/N^2$ should not be compared directly with $\langle |F_M|^2 \rangle_\psi$, as shown below.

Condition c in the case of $N=1$ is considered first. There is no problem concerning the intermolecular interaction in the unit cell if $N=1$. The molecular structure factors have nonzero values only on the planes of $l=$ integer, that is, on the lth layers, where the Laue condition is fulfilled because of the periodicity along the direction of the molecular axis. But molecular structure factors have continuous values on the layer planes. Thus $\langle|F_M|^2\rangle_\psi$ is the averaged value with respect to the continuous value of ψ, which varies from 0 to 2π, and so problems occur for comparison with $|F(\text{obs})|^2/N^2$ when the ψ dependence of the values of F_M is large. The observed intensity $|F(\text{obs})|^2/N^2$ originates from the crystal lattice and so corresponds to the values of the molecular structure factor at the reciprocal lattice points (indicated by h and k) where the Laue conditions are fulfilled. In the case of a uniaxially oriented sample, the average of the values at these discrete points is observed to be $|F(\text{obs})|^2/N^2$. A discrepancy between $\langle|F_M|^2\rangle_\psi$ and $|F(\text{obs})|^2/N^2$ occurs in cases of small u (e.g., $u=2$), such as the planar zigzag and glide-type polymers. POM is an example in which the ψ dependence of F_M is negligibly small because of the large $u(u=9)$.

When $N \geqslant 2$, the intermolecular interaction makes the situation more complex. The unit cell structure factor is represented by Eq. 4.85 (Section 4.7.2). This point should be taken into account when $|F(\text{obs})|^2/N^2$ is compared with $\langle|F_M|^2\rangle_\psi$.

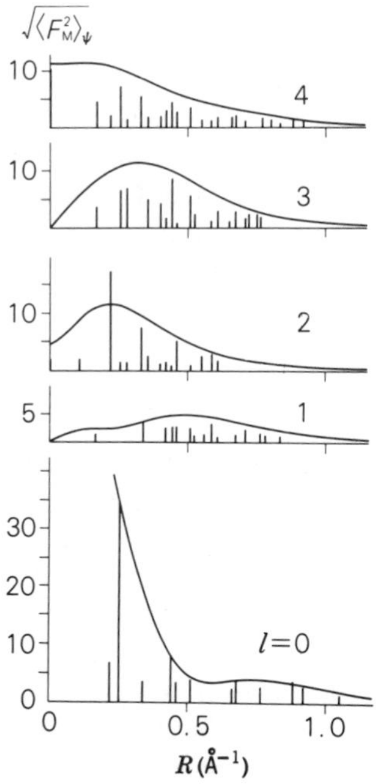

Fig 4.47 $|F(\text{obs})|/N$ (rods) and $[\langle|F_M|^2\rangle_\psi]^{1/2}$ (curves) of poly-1, 3-dioxolane form II.[105]

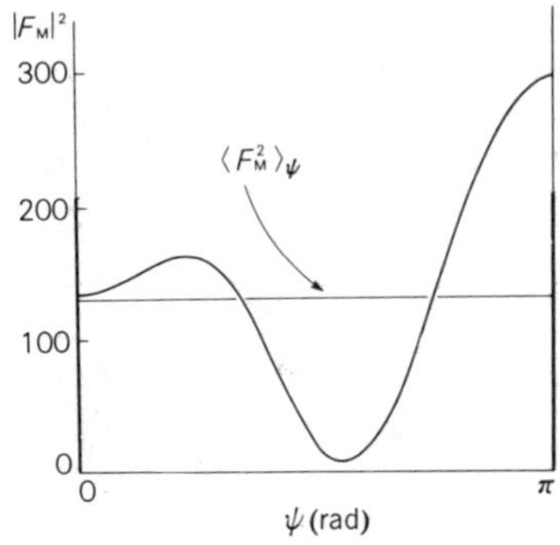

Fig. 4.48 The intensities of the reflections 202 and $\bar{2}02$ of poly-1,3-dioxolane form II (bold vertical rods), the corresponding $|F_M(R,\psi,l/c)|^2$ (curve), and its ψ average (horizontal line).[105]

In Fig. 4.47 the values of $|F(\text{obs})|/N$ of poly-1,3-dioxolane form II are shown by rods, and $[\langle|F_M|^2\rangle_\psi]^{1/2}$ values calculated for the determined crystal structure[20] (glide-type, $N=2$) are denoted by curves. This figure shows how difficult it is to judge the agreement between these two.

In Fig. 4.47 one reflection (202 and $\bar{2}02$ overlapping each other) has a remarkably higher intensity value (rod) than the calculated molecular structure factor (curve) of the second layer. This difference is due to the large ψ dependence of $|F_M|^2$ of the molecular chain. In Fig. 4.48 the calculated intensities of the reflections 202 and $\bar{2}02$ are shown by bold vertical rods (at $\psi=0$ and π, respectively; $\psi=0$ referring to the a^*c^* plane), the corresponding $|F_M(R,\psi,l/c)|^2$ (for $R=0.221$ Å^{-1} and $l=2$) is shown by a curve, and its ψ average by a horizontal line. The comparison that should be made for the reflection 202 (overlapping on $\bar{2}02$) in Fig. 4.47 is between the heights of the bold rods and the horizontal line of Fig. 4.48. There are many reflections for which $|F(\text{obs})|/N$ (rods) show much lower values than the $[\langle F^2{}_M\rangle_\psi]^{1/2}$ curves in Fig. 4.47. These discrepancies may be due to complicated phase relationships resulting from the position and orientation of two molecular chains in the unit cell.

4.6.5 Information from Infrared Spectra

Various models have been proposed by a number of authors for the structure of PEO, but a definite determination has not been achieved. The author and his co-workers proposed a molecular model based on a (7/2) helix[104] and determined the crystal structure after a duration of 10 years.[29] From the general features of the X-ray pattern, a (7/2) helix conformation was inferred, that is, seven monomeric units contained in the fiber period, 19.30 Å, with two turns of the helix. However it was difficult to develop a more detailed structure analysis by the X-ray method alone at that stage of the study, since the unit cell is too large and it contains four molecular chains.

Vibrational analysis played an important role in the selection of the molecular model.

The vibrational modes of an infinitely extended helical molecule of PEO may be treated as a cyclic group or a dihedral group depending on whether or not the helical chain possesses two kinds of twofold axes (one passing through the oxygen atom and the other bisecting the C—C bond). The existence of these twofold axes can be examined by comparing the observed infrared and Raman spectra with the number of normal modes predicted by the selection rules obtained from the factor group analysis (Section 5.5). Table 4.9 gives the results of the factor group analysis, and Fig. 4.49 shows the Raman and polarized infrared spectra.[106,107]* In the case of the dihedral group $[D(4\pi/7)]^{\dagger}$, 10 and 9 normal modes belong to the symmetry species A_1 (Raman active) and A_2 (infrared active, parallel bands), respectively. On the other hand, in the cyclic group $[C(4\pi/7)]$ 19 modes belong to the A species, both infrared (parallel bands) and Raman

Table 4.9 Numbers of Normal Modes and Selection Rules for the Model of the PEO Molecule under the Assumption of Dihedral and Cyclic Groups[104a]

	Number of normal modes	Infrared	Raman
D(4π/7)			
A_1	10	F	A
A_2	$11 - 2(T_{\parallel}, R_{\parallel})$	A	F
E_1	$[21 - 1(T_{\perp})] \times 2$	A	A
E_2	21×2	F	A
E_3	21×2	F	F
C(4π/7)			
A	$21 - 2(T_{\parallel}, R_{\parallel})$	A	A
E_1	$[21 - 1(T_{\perp})] \times 2$	A	A
E_2	21×2	F	A
E_3	21×2	F	F

[a]For explanation of notations, see Section 5.5.

* The Raman spectrum in this figure is reproduced from Ref. 104 and was taken using a mercury lamp as the excitation source. Far better data can now be obtained with a laser source.

† For this notation see Section 5.5.8.

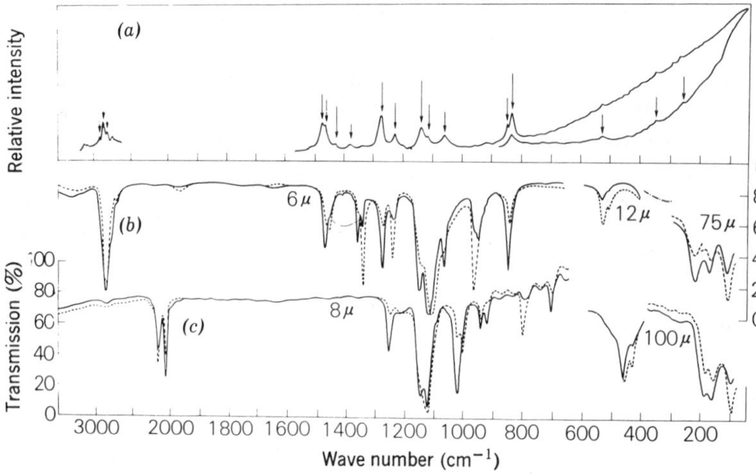

Fig. 4.49 (*a*) Raman spectrum of PEO. Polarized infrared spectra of (*b*) PEO and (*c*) PEO-d_4. In *b* and *c* (———) electric vector perpendicular to elongation; (-------) electric vector parallel to elongation; (-·-·-·-·-·-·-) measured by unpolarized beam.[106,107]

active. In Fig. 4.49 no Raman bands appear at the same frequencies as those of the parallel infrared bands. According to this fact and the number of bands in both spectra, this molecule should have the symmetry of the dihedral group. Using helical parameter equations (Section 4.6.2.A) with the fixed fiber period and $D(4\pi/7)$ symmetry, two types of (7/2) helix models were found to be geometrically possible: model I [internal rotation angles $\tau(\text{CCOC}) = 188°15'$ and $\tau(\text{OCCO}) = 64°58'$] and model II [$\tau(\text{CCOC}) = 111°$ and $\tau(\text{OCCO}) = 202°26'$].

Some bands, especially in the far-infrared region as shown in Fig. 4.49 should be strongly dependent on the molecular conformation. The normal skeletal vibrations were calculated for the two models. The stretching and bending force constants were transferred directly from POM[108] and diethyl ether.[109] The torsional force constants were chosen so that the potential barriers to internal rotation about the C—C and C—O bonds were about 3 kcal/mol. The observed and calculated frequencies for the two models are compared in Table 4.10. It is obvious from the table that the observed frequency and the observed dichroism, 530 (||), 106 (||), 215 ($\perp$), and 165 cm^{-1} ($\perp$), agree far more satisfactorily with the calculated results for model I than with those for model II. The normal vibrations were calculated for a more detailed form of model I, taking into consideration the hydrogen atoms. Good agreement between the observed and calculated frequencies (including the deuterated data) further supported this assumed model.[107]

Since the calculation of structure factors for the crystal was difficult at

Table 4.10 Skeletal Vibrations of PEO[104]

	Observed (cm^{-1})	Calculated (cm^{-1})	
Species	Infrared	Model I	Model II
A_2	1103 (‖) vs	1107	1129
	530 (‖) w	530	412
	106 (‖) m	116	151
E_1	1147 (⊥) s	1137	1129
	1116 (⊥) s	1041	1051
	1060 (⊥) m	954	932
	532 (⊥) w	494	500
	—	350	362
	215 (⊥) w	221	287
	165 (⊥) w	162	112
	—	100	106

that stage of the study, as a first approximation the observed reflection intensities were compared with those calculated for one molecular chain. From the result of the X-ray study, model I was again considered to be more reasonable. Figure 4.45*b* shows the skeletal model of PEO drawn on the same scale as the model of POM in Fig. 4.45*a*. The PEO molecule is a succession of nearly *T*, *T*, and *G* conformations. From Fig. 4.45 it may be well understood that the PEO molecule is a loosely turned helix, while the POM molecule is tightly turned. From these structural features, the marked differences in the physical properties of these two polymers may be understood fairly well. The second stage of the analysis of PEO, resulting in the determination of the crystal structure, is described in Section 4.12.2. Miyazawa et al.[110] and Richards[111] have independently proposed similar molecular structures for PEO using infrared and X-ray diffraction methods, respectively.

The infrared spectroscopic method was utilized for setting up possible molecular models of polyallene,[46] poly-1,3-dioxolane form II,[20] poly-1, 3-dioxepane,[70] and so on.

4.6.6 Cylindrical Patterson Function

The cylindrical Patterson function can be obtained using only the observed reflection intensity data from a fiber diagram with no additional assumptions. See section 4.7.4 concerning the usual Patterson function. One molecular chain in the unit cell as shown in Fig. 4.50*a*,[10] is now considered. Of concern here are the interatomic vectors within the chain itself. If the origins of the

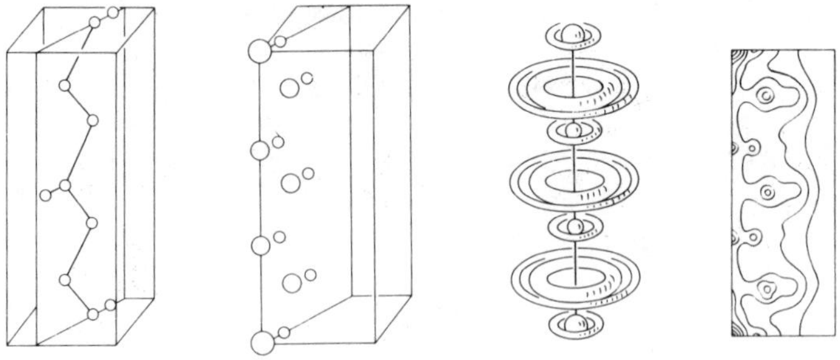

(a) (b) (c) (d)

Fig. 4.50 (*a*) A unit cell including one molecular chain. (*b*) Three-dimensional Patterson function. (*c*) Diagram obtained by rotating *b* around the axis passing through the origin. (*d*) The section of *c* including the axis, which is the cylindrical Patterson function. (From Nyburg.[10])

vectors are all shifted to the origin of the unit cell by parallel translation, then a three-dimensional Patterson function is obtained as shown in Fig. 4.50*b*. Such a figure can be synthesized from the three-dimensional X-ray data using a single crystal or highly doubly oriented sample, but it cannot be obtained from the fiber diagram of a uniaxially oriented sample. When the function shown in Fig. 4.50*b* is rotated around the axis passing through the origin, Fig. 4.50*c* is obtained. The section including the axis (Fig. 4.50*d*), which is the cylindrical Patterson function, is synthesized according to the following equation using the data of the fiber diagram.[112,113]

$$\varphi(r, Z) = \frac{1}{NV}\sum_{l}\sum_{\xi} I(\xi, l) J_0(2\pi\xi r)\cos\frac{2\pi l Z}{c} \qquad (4.79)$$

Here r and Z are, respectively, the cylindrical coordinates expressing the distance from the fiber axis and the distance in the fiber direction in Patterson space. ξ is the Bernal coordinate, $I(\xi, l)$ is the integrated intensity of a reflection of the lth layer line at a distance ξ, V is the volume irradiated, N is the total number of periods in the fiber direction, and $J_0(2\pi\xi r)$ is the zeroth order Bessel function.

it-Poly(methyl methacrylate), whose crystal structure was recently determined after a long roundabout quest,[33,114] is used here to describe the application of the cylindrical Patterson function. For this polymer, models of $(5/2)$[115] and $(5/1)$[116–118] helices have been proposed. After several years of study the author and his co-workers considered the (5/1) helix to be the most reasonable one in 1970.[33] Thereafter they determined the crystal structure, which consists of double strand helices.

It was first considered from the observed X-ray and density data that four helical chains consisting of five monomeric units pass through the unit cell. The molecular conformation can be deduced, provided the four kinds of internal rotation angles shown in Fig. 4.51 are determined. Two skeletal models of (5/1) and (5/2) helices were found to be possible using helical parameter equations with the assumption that five monomeric units are included in the fiber period of 10.50 Å. Figure 4.52*a* is the cylindrical Patterson map calculated using the observed intensities of 48 independent reflections. It is a characteristic feature of this map that a maximum appears at the origin and a minimum at the point $Z = c/2$ and $r = 0$. This feature suggests that the distance between one turn and a succeeding one along the axis corresponds to the fiber period, which excludes the (5/2) helix. The cylindrical Patterson functions synthesized using the molecular structure factors calculated for the (5/1) and (5/2) helices are shown for comparison in Figs. 4.52*c* and 4.52*d*, respectively. The map for the (5/2) helix has a maximum at $Z = c/2$ and $r = 0$ and is not in agreement with the result from the observed data. At this stage of the study, the (5/1) helix model in which the α-CH_3 groups point outward, with $\tau_1 = 180°$, $\tau_2 = -108°$, $\tau_3 = -24°$, and $\tau_4 = 171°$, was considered reasonable.[33]

Thereafter trial-and-error efforts were made assuming a crystal structure model consisting of four (5/1) helices in an orthogonal unit cell of $a = 20.98$ Å, $b = 12.06$ Å, and c (fiber axis) = 10.40 Å, but no reasonable packing was found with this assumption. On the other hand, calculation of the intramolecular interaction energy assuming helical symmetry without fixing the fiber period suggested that helices with larger radii, such as a (12/1) helix, should be more stable than the (5/1) helix (Section 6.3.1.A).[119] Taking these facts into account, the author and his co-workers reexamined the molecular models with larger radii and found that the X-ray data can be explained reasonably by the double strand helix shown in Fig. 4.53.

The double strand helix consists of two chains with the same helix sense and direction (denoted by solid and broken lines) shifted 10.40 Å along the helix axis with respect to each other, each chain having 10 chemical units and 1 turn in the period of 20.80 Å [a (10/1) helix]. Consequently the double strand helix has a translational identity period of 10.40 Å, which is identical to the observed fiber period. As shown in the figure, the α-CH_3 group points outward, and $\tau_1 = 179°$, $\tau_2 = -148°$, $\tau_3 = -30°$, and $\tau_4 = 180°$.

Figure 4.54 shows the crystal structure, in which good packing in the lattice is found. The cylindrical Patterson map synthesized using the calculated structure factors is shown in Fig. 4.52*b*. Agreement with Fig. 4.52*a* is obvious.

The result of the energy calculation indicates that the double helix is more stable by 4.4 kcal per mole of monomer unit than the two isolated (10/1) helices. This suggests that an appreciable stabilization is obtained as a result of the good fit between the two intertwinned chains, although there is no hydrogen bonding between them as in the case of DNA.[120]

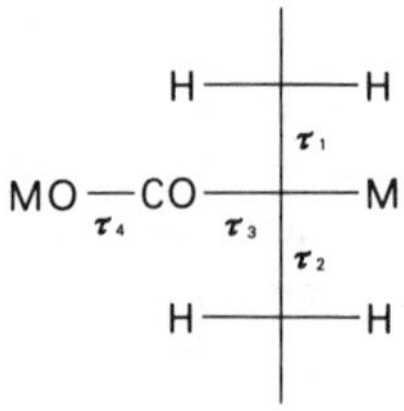

Fig. 4.51 Fischer projection and numbering of the internal rotation angles of *it*-poly (methyl methacrylate). τ_3[MCC(O)O], τ_4[CC(O)OM].[114]

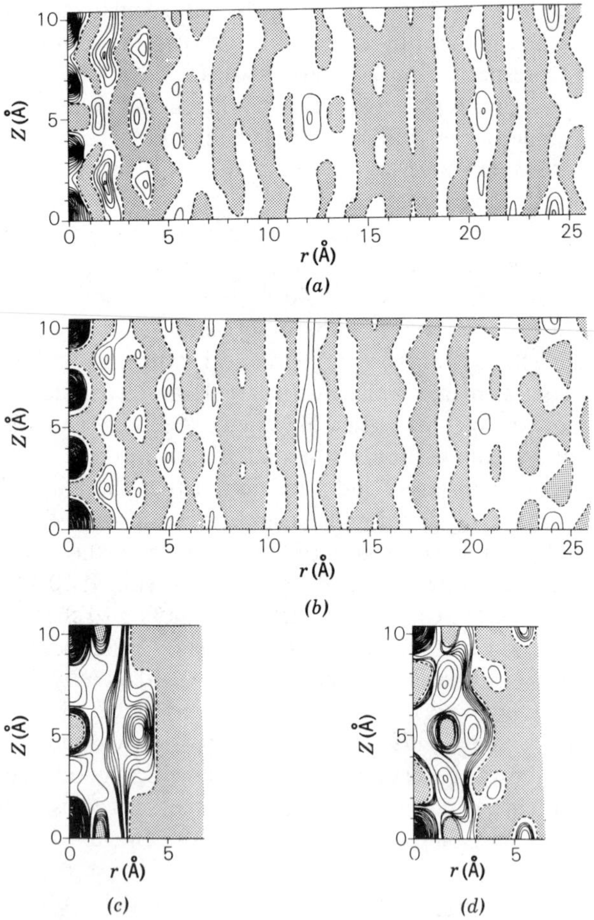

Fig. 4.52 Cylindrical Patterson map of *it*-poly (methyl methacrylate) synthesized from (*a*) the observed intensity, and from the molecular structure factors calculated for (*b*) the double strand helix, (*c*) the (5/1) helix, and (*d*) the (5/2) helix. Halftone indicates the region lower than the zero level.

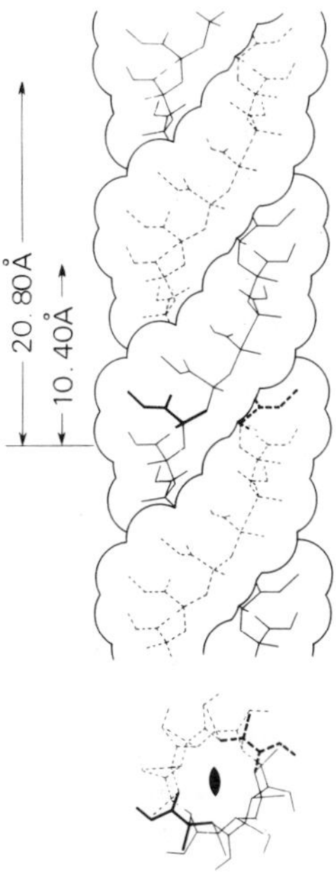

Fig. 4.53 Double strand helix of *it*-poly (methyl methacrylate).[114]

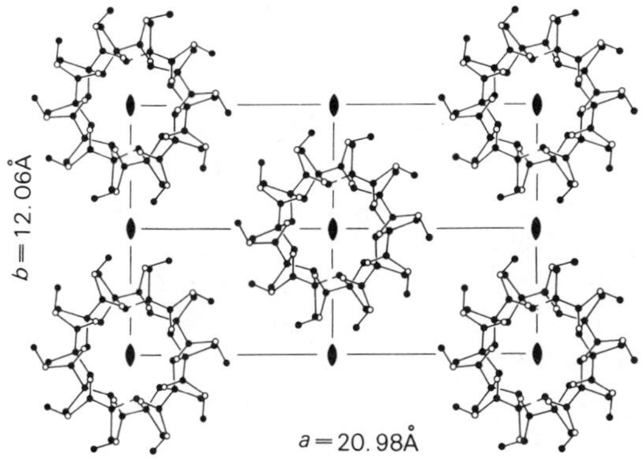

Fig. 4.54 Crystal structure of *it*-poly (methyl methacrylate).[114]

4.6.7 Optical Transforms

The optical transform method is based on the similarity between the scattering of X-rays by an atom and that of visible light by a small pinhole (Fig. 4.55).[121] A figure in which atomic positions in the assumed model of a molecular or crystal structure are replaced by small closed circles is now

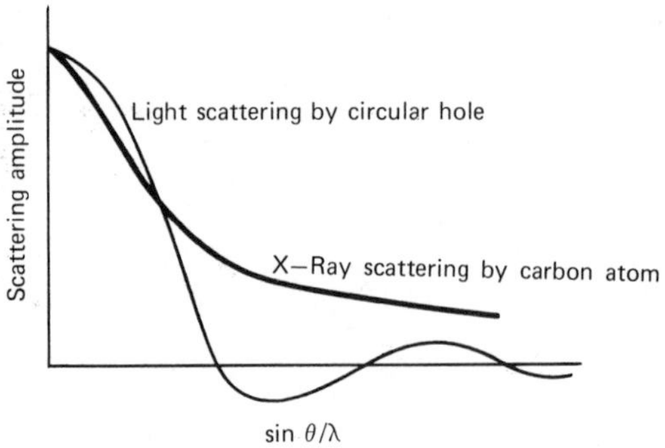

Fig. 4.55 X-Ray atomic scattering factor and visible light scattering by a pinhole. (From Taylor and Lipson.[121])

considered. A parallel beam of monochromatic visible light is incident on a photograph of such a figure taken on a scale small enough to give an optical interference effect. The diffraction patterns obtained in this way are then compared with the X-ray photograph and used for structure analysis. The merit of this method is that molecular structure factors can be obtained conveniently without detailed calculations. Figure 4.56 shows an example of the application of this method to *cis*-1,4-poly-2-*tert*-butylbutadiene $(\text{—}CH_2\text{—}\underset{\displaystyle C(CH_3)_3}{\underset{|}{C}}\text{=}CH\text{—}CH_2\text{—})_n$.[122] The optical transform *b* taken for the molecular model *c* is in good agreement with the sketch of the fiber diagram *a*. This method was used for the structure determination of the β-form of poly(dimethyl ketene) $[\text{—}C(CH_3)_2\text{—}CO\text{—}]_n$,[123] ethylene-*cis*-butene alternating copolymer,[124] and so on. The optical transform method has also been applied to the study of the disordered structure of solid polymers.[125]

4.6.8 Model Molecules

Model molecules made of metal or plastics are very useful for setting up molecular models of high polymers, since figures drawn on paper can not satisfactorily portray a realistic three-dimensional concept. Model molecules are made on a scale, for example, of 20 mm corresponding to 1 Å. It is necessary to reproduce bond lengths and bond angles as accurately as possible. It is especially important in the case of polymers so that the bonds may be fixed at various internal rotation angles to provide the capability of rotation and preserve a conformation after construction. Since considerably long chains must be made, it is essential that the molecules can be handled without unjointing.

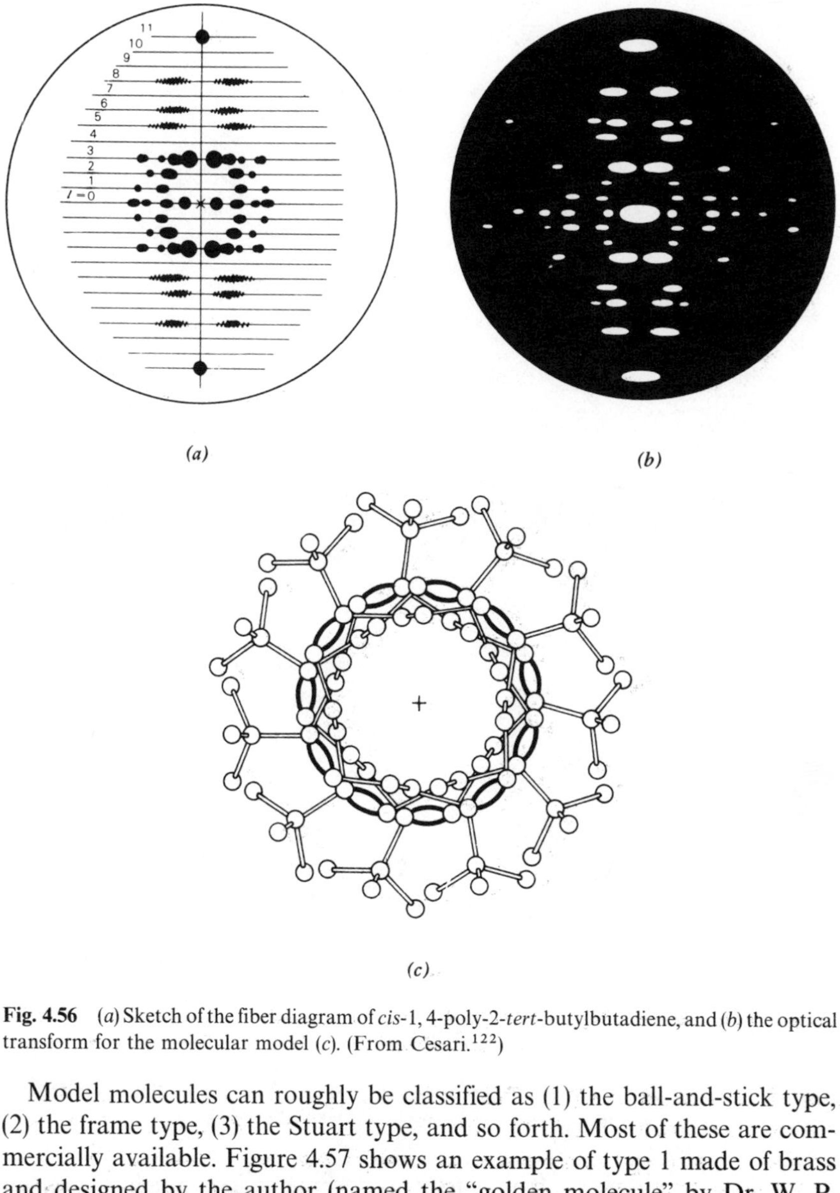

Fig. 4.56 (*a*) Sketch of the fiber diagram of *cis*-1, 4-poly-2-*tert*-butylbutadiene, and (*b*) the optical transform for the molecular model (*c*). (From Cesari.[122])

Model molecules can roughly be classified as (1) the ball-and-stick type, (2) the frame type, (3) the Stuart type, and so forth. Most of these are commercially available. Figure 4.57 shows an example of type 1 made of brass and designed by the author (named the "golden molecule" by Dr. W. P. Slichter of Bell Telephone Lab.). Figure 4.57*a* shows the schematic of the model; each bond can be rotated and fixed at any position. Models illustrating the conformations of a planar zigzag and a $(TG)_3$ helix (e.g., *it*-PP) are shown in Figs. 4.57*b* and 4.57*c*, respectively. The ball-and-stick model is

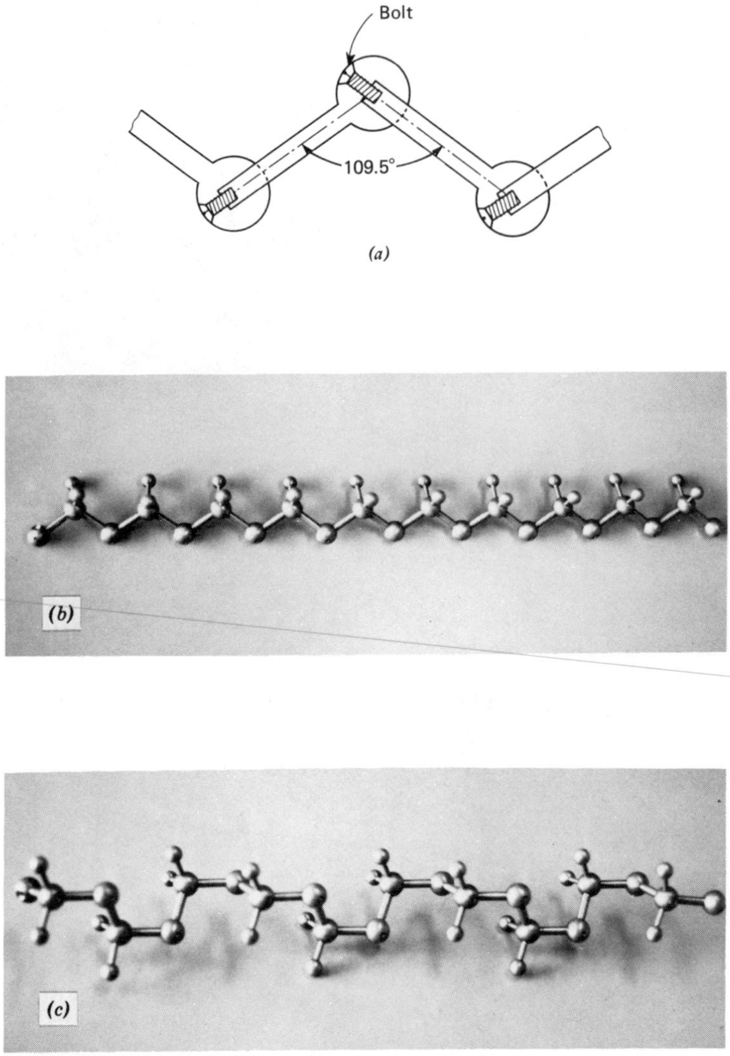

Fig. 4.57 (*a*) Mechanism of the ball-and-stick type model (golden molecule), and the same model assuming the conformations of (*b*) planar zigzag and (*c*) $(TG)_3$ helix.

convenient for observing the conformation directly. The frame model can easily be made of a piece of wire by folding with a plier. It is worthwhile to construct a wire model whenever a good idea comes to the reader even at his home. The Stuart model is made with atomic sizes adjusted to the van der Waals radii; this is useful for examining the effect of steric hindrance.

4.7 CRYSTAL STRUCTURE MODEL AND DETERMINATION OF ATOMIC PARAMETERS

4.7.1 Packing of Molecular Chains in the Crystal Lattice

In the case of high polymers, the molecular axis coincides with the elongation direction, that is, the fiber axis. This feature is helpful for structure analysis.

A. *Polyallene.*[46] The numbering of the carbon atoms is shown below.

$$[\text{—CH}_2\text{—C(=CH}_2\text{)—}]_n$$
$$\text{I} \quad \text{II} \quad \text{III}$$

A reasonable (2/1) helix model (Fig. 4.58) was assumed from the following considerations: (1) good agreement between the observed and calculated fiber periods, (2) the existence of the 2_1 screw axes in the space group *Pnan*, and (3) the infrared spectroscopic evidence (factor group analysis, Section 5.5). This model has three types of symmetry axes: a twofold screw axis coinciding with the molecular axis, twofold axes coinciding with the bisectors of the angles HCH of the main-chain methylene groups, and twofold axes coinciding with the double bonds of the pendant vinylidene groups.

Taking into account the symmetry of the assumed helix model and the fact that the unit cell should contain four monomeric units, equivalent points *c* (on the twofold axis parallel to the *b* axis) and *d* (on the twofold axis parallel to the *a* axis), as given in Table 4.4, can be considered to be suitable positions for carbon atoms. Accordingly, two possible ways of molecular packing may be assumed. One of these is shown in Fig. 4.59, in which C_I atoms of the main-chain methylene groups are on equivalent positions *d*, and C_{II} and C_{III} are on *c*. The hydrogen atoms are placed on general positions *e* in both cases.

Since the center of symmetry exists at the origin of the unit cell in the case of *Pnan*, the structure factor is real (Eq. 4.39). The summation in Eq. 4.39 is made for one of each pair related by the center of symmetry. The structure

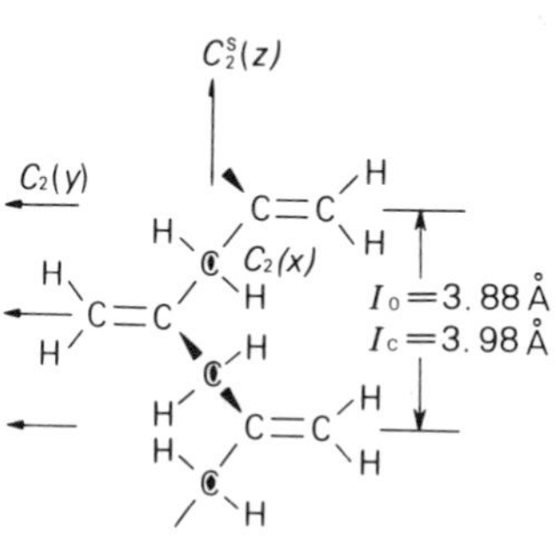

Fig. 4.58 Molecular model of polyallene.[46]

factor for polyallene is derived from the following considerations. The contribution of hydrogen atoms is neglected for simplicity, since the scattering power of hydrogen atoms is very small compared to that of other atoms (see Fig. 4.6). The coordinates $(x, \frac{1}{4}, \frac{1}{4})$ and $(\frac{1}{2}+x, \frac{3}{4}, \frac{1}{4})$ are taken for C_I atoms, and $(\frac{1}{4}, y, 0)$ and $(\frac{1}{4}, \frac{1}{2}-y, \frac{1}{2})$ for C_{II} and C_{III} atoms, corresponding to the first way of packing mentioned above.

$$
\begin{aligned}
F(hkl) = 2f_C\Bigg[\!\Bigg[& \cos 2\pi\left[hx(C_I)+\frac{k+l}{4}\right] + \cos 2\pi\left\{h\left[\frac{1}{2}+x(C_I)\right]+\frac{3k+l}{4}\right\} \\
& + \cos 2\pi\left[\frac{h}{4}+ky(C_{II})\right] + \cos 2\pi\left\{\frac{h}{4}+k\left[\frac{1}{2}-y(C_{II})\right]+\frac{l}{2}\right\} \\
& + \cos 2\pi\left[\frac{h}{4}+ky(C_{III})\right] + \cos 2\pi\left\{\frac{h}{4}+k\left[\frac{1}{2}-y(C_{III})\right]+\frac{l}{2}\right\}\Bigg]\!\Bigg] \\
= 4f_C\Bigg\{ & \cos 2\pi\left[hx(C_I)+\frac{h+2k+l}{4}\right]\cos 2\pi\left(\frac{h+k}{4}\right) \\
& + \cos 2\pi\left[\frac{k+l}{4}-ky(C_{II})\right]\cos 2\pi\left(\frac{h+k+l}{4}\right) \\
& + \cos 2\pi\left[\frac{k+l}{4}-ky(C_{III})\right]\cos 2\pi\left(\frac{h+k+l}{4}\right)\Bigg\} \\
= 4f_C\Bigg\{ & \cos^2 2\pi\left(\frac{h+k}{4}\right)\cos 2\pi\left[hx(C_I)+\frac{k+l}{4}\right] \\
& + \cos^2 2\pi\left(\frac{h+k+l}{4}\right)\cos 2\pi\left[ky(C_{II})+\frac{h}{4}\right] \\
& + \cos^2 2\pi\left(\frac{h+k+l}{4}\right)\cos 2\pi\left[ky(C_{III})+\frac{h}{4}\right]\Bigg\}
\end{aligned}
\tag{4.80}
$$

Here the following formula was used (see Ref. 2, p. 360).

$$
\begin{aligned}
& \cos 2\pi\left(\frac{h+k+l}{4}\right)\cos\left[A \pm 2\pi\left(\frac{h+k+l}{4}\right)\right] \\
& \quad = \cos^2 2\pi\left(\frac{h+k+l}{4}\right)\cos A
\end{aligned}
\tag{4.81}
$$

for any integral values of h, k, and l (including zero).

Equation 4.80 was derived by substituting atomic coordinates into Eq. 4.39, but structure factors for general positions are given in Ref. 2 for all space groups. The formula for $Pnna\text{-}D_{2h}^{6}$ is given on p. 403 of Ref. 2

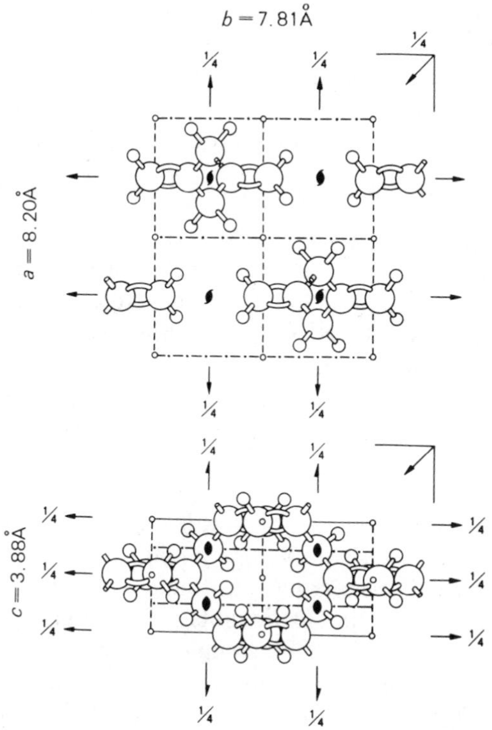

Fig. 4.59 Crystal structure of polyallene.[46]

and is rewritten for *Pnan* as

$$A = 8 \cos 2\pi\left(hx + \frac{k+l}{4}\right)\cos 2\pi\left(lz - \frac{h+k+l}{4}\right)\cos 2\pi\left(ky + \frac{h}{4}\right);$$

$$B = 0 \tag{4.82}$$

If this formula is applied to coordinates $(x, \frac{1}{4}, \frac{1}{4})$ of the special equivalent positions *d* for the C_1 atom, the following equation is obtained:

$$A = 4 \cos^2 2\pi\left(\frac{h+k}{4}\right)\cos 2\pi\left(hx + \frac{k+l}{4}\right) \tag{4.83}$$

where the coefficient 4 is given, since there are four equivalent positions. This equation is the same as the first term of Eq. 4.80. In the same way, by applying coordinates $(\frac{1}{4}, y, 0)$ of the special equivalent positions *c*, the second and third terms of Eq. 4.80 are obtained. Reference 2 (p. 403) shows that Eq. 4.82 can be written more simply as

$$A = 8 \cos 2\pi hx \cos 2\pi lz \cos 2\pi ky \tag{4.84}$$

in the case of $h = 2n$ and $k + l = 2n$. Also given are the equations for the cases $h = 2n$ and $k + l = 2n + 1$; $h = 2n + 1$ and $k + l = 2n$; $h = 2n + 1$ and $k + l = 2n + 1$. Practical calculations are made using these equations.

The calculated structure factors for the first packing model agree quite well with the observed data. Here $F(\text{obs}) = [I(\text{obs})/KpL]^{1/2}$ should be compared with $[m|F(\text{calc})|^2]^{1/2}$, or $\left[\sum_i m_i F_i^2(\text{calc})\right]^{1/2}$ for the case of accidentally overlapped nonequivalent reflections (see Eq. 4.36). Since the C_I atom is at d on the twofold axis and C_{II} and C_{II} are at c on the other twofold axis, only three parameters $x(C_I)$, $y(C_{II})$, and $y(C_{III})$, must be determined. Using 31 independent reflections, the author and his co-workers have determined the following atomic parameters by trial and error: $x(C_I) = 0.358$, $y(C_{II}) = 0.350$, and $y(C_{III}) = 0.524$. The crystal structure is shown in Fig. 4.59. Here an isotropic temperature parameter $B = 8.7\,\text{Å}^2$ was assumed. This value corresponds to $[\langle u^2 \rangle]^{1/2} = 0.33\,\text{Å}$.

B. *Poly(ethylene Succinate)*.[38] The indexing was performed on the basis of an orthorhombic lattice with $a = 7.60$ Å, $b = 10.75$ Å, and c (fiber axis) = 8.33 Å, and the space group was found to be $Pbnb\text{–}D_{2h}^{10}$ from the systematic absences, four monomeric units being contained in the unit cell. As for the molecular conformation having the observed fiber period 8.33 Å, three kinds of possible skeletal conformations were considered: $(TGT\bar{G})_2$, $(TG)_2(T\bar{G})_2$, and $T_3GT_3\bar{G}$. In addition, many possible models were constructed, depending on which skeletal bond takes T and which takes G. On the other hand, since there are eight general equivalent positions in space group $Pbnb$ and there are four molecular chains passing through one unit cell, the molecular chain must have either twofold axes perpendicular to the fiber axis or centers of symmetry. Only two cases are allowed. the molecule itself lies on centers of symmetry as the unit cell, or the molecule has twofold axes perpendicular to the fiber axis. These axes coincide with those of the unit cell, parallel to the b axis. Based on these restrictions, the six molecular models shown in Fig. 4.60 can be proposed. Here symbols C_2 and i are used to distinguish between the models having twofold axes and those having centers of symmetry, respectively. I-C_2 is the model of Erickson and Fuller[126] and III-i is Bunn's model.[127]

The three models having centers of symmetry were excluded, since close agreement between the observed and calculated structure factors of the equatorial $hk0$ reflections could not be obtained at any orientation around the c axis. In such a trial procedure, considerations of a plane group (Section 3.4) is useful. Next the structure factors of $00l$ reflections were examined for three models having twofold axes. I-C_2 and II-C_2 gave the same calculated values and they also agreed well with observed values. However III-C_2

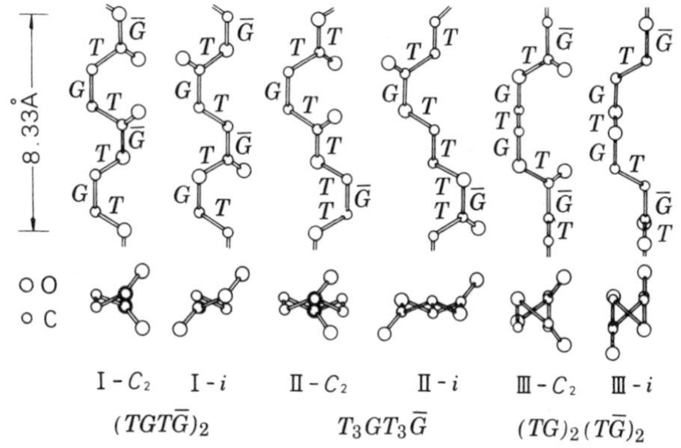

Fig. 4.60 Possible conformations of poly (ethylene succinate). The true conformation is II-C_2.[38]

did not agree; the observed intensity of the 004 reflections was about 10 times that of 002, although the calculated intensities of the two reflections were almost the same. Thus III-C_2 was excluded at this stage. Factor group analyses and normal coordinate treatments for these models were made at the same time. The results suggested the possibility of I-C_2 and II-C_2, but it was not possible to decide which model is the more probable.

In the lattice of space group $Pbnb\text{-}D_{2h}{}^{10}$ the model having C_2 symmetry has translational freedom only along the glide plane parallel to the b axis, since the twofold axes of the molecular model must fit the twofold axes parallel to the b axis of the lattice. By examining the structure factors for all reflections, II-C_2 was found to be reasonable, and the final crystal structure was determined to be that shown in Fig. 4.61.

C. *Poly(ethylene Oxide)–Mercuric Chloride Complex Type I.* Positions of the mercury and chlorine atoms were determined as shown in Fig. 4.62 using the Patterson function (Section 4.7.4). In this case the contribution of the mercuric chloride molecule to the X-ray diffraction intensity is overwhelmingly large compared to that of the PEO molecule (at. nos: Hg, 80; C1, 17; O, 8; C, 6). The structure analysis made at this stage[58] is discussed below.

Spherical models of the mercury and chlorine atoms having radii of 1.5 and 1.8 Å (corresponding to their respective van der Waals radii) were drawn, and then skin layers 2.0 and 1.4 Å thick (corresponding to the van der Waals radii of the methylene group and the oxygen atom, respectively) were added to them.* The center of the carbon or oxygen atom cannot occupy a site

* The van der Waals radii for the methylene group and the oxygen and chlorine atoms were taken from the data of Pauling[71] and that for mercury was taken from the book of Kitaigorodskii.[128]

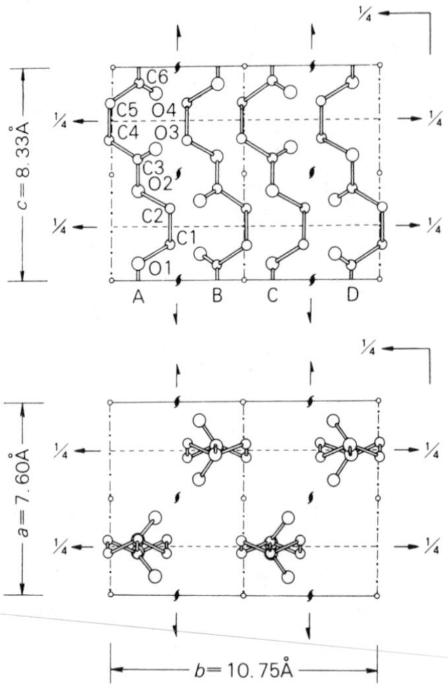

Fig. 4.61 Crystal structure of poly (ethylene succinate).[38]

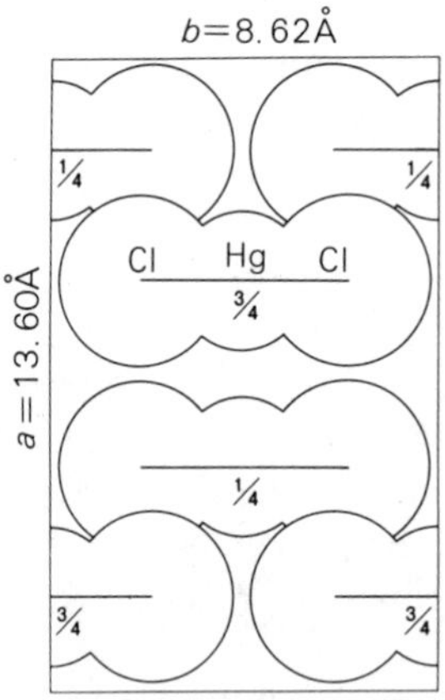

Fig. 4.62 Positions of the mercuric chloride molecules in the crystal lattice of PEO–mercuric chloride complex type I.[58]

inside the 2.0 or 1.4 Å skin layers, respectively. The space that PEO can occupy was investigated by examining the features of several sections drawn on glass plates. The features at the $(x, y, \frac{1}{4})$ and $(x, 0, z)$ sections are shown in Fig. 4.63. In these figures the hatched region can be occupied by both the carbon and oxygen atoms, but the halftone region can be occupied by the oxygen atoms only. The molecular conformation of PEO that can fit the

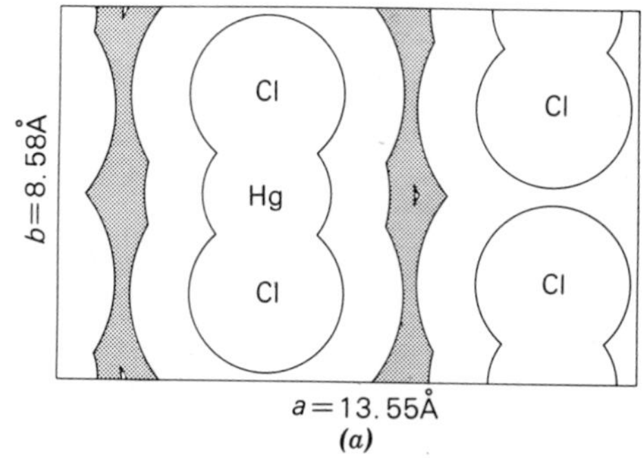

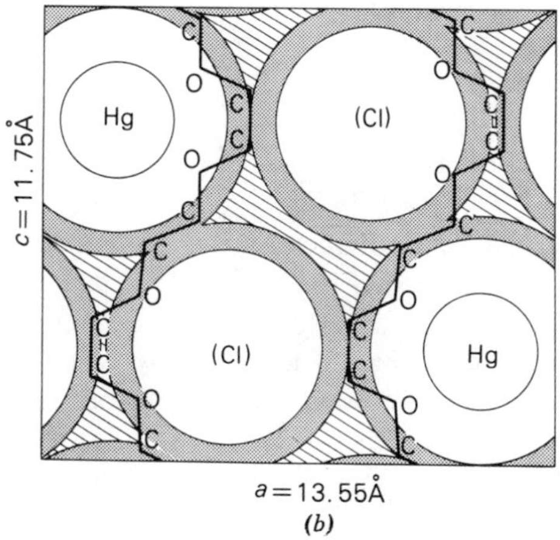

Fig. 4.63 Allowed space for PEO molecules at (*a*) the $(x, y, \frac{1}{4})$ and (*b*) $(x, 0, z)$ sections. The hatched regions are available to both the carbon and oxygen atoms, but the halftone regions are available to the oxygen atoms only. The chlorine atoms indicated with parentheses do not lie on the plane of this section.[58]

allowed space was found to be the $T_5GT_5\bar{G}$ form,

$$\begin{array}{l}—\mathrm{C}—(—\mathrm{O}—\mathrm{C}—\mathrm{C}—\mathrm{O}—\mathrm{C}—\mathrm{C}—\mathrm{O}—\mathrm{C}—\mathrm{C}—\mathrm{O}—\mathrm{C}—\mathrm{C}—)_n \\ \quad\quad T \quad T \quad T \quad T \quad T \quad G \quad T \quad T \quad T \quad T \quad T \quad \bar{G}\end{array}$$

The fiber period calculated for this model is 12.44 Å, which is in fairly good agreement with the observed value, 11.75 Å. Further detailed analysis made by applying Fourier synthesis is described later (Section 4.8.1.B). In Fig. 4.63*b* the oxygen atoms are drawn within the white circles. These are at the final atomic coordinates and the distance Hg···O in this complex is shorter than the usual van der Waals distance. This shortening suggests a special interatomic interaction.

4.7.2 The Unit Cell Structure Factor as Expressed by the Cochran-Crick-Vand Equations

The unit cell structure factor for the lattice including helical molecules can be expressed by using the Cochran-Crick-Vand equations. The position

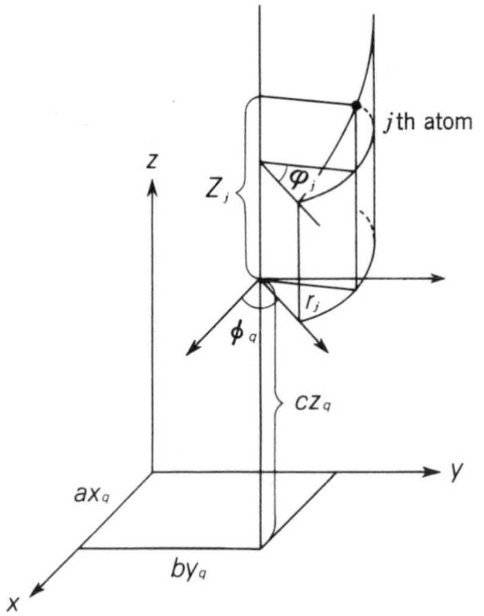

Fig. 4.64 The coordinates of the qth helical molecule in the lattice.

of the qth helix is denoted by fractional coordinates (x_q, y_q, z_q), and the azimuthal angle is denoted by ϕ_q, as shown in Fig. 4.64. Here x_q and y_q indicate the position of the helix axis, and z_q and ϕ_q refer to any suitable atom. Equation 4.65 is multiplied by $\exp[2\pi i(hx_q + ky_q + lz_q)]\exp(-in\phi_q)$ and then summed over all helical molecules in the unit cell. In this process, the right- or left-hand helix sense, and the upward or downward direction of the molecules should be taken into account. Thus the following equation is obtained.

$$F\left(R, \psi, \frac{l}{c}\right) = \sum_{q=1}^{N} \sum_{n} G_{n,l}(R) \exp\left[in\left(\pm\psi + \frac{\pi}{2} \mp \phi_q\right)\right] \times \exp[2\pi i(hx_q + ky_q + lz_q)] \quad (4.85)$$

where

$$G_{n,l}(R) = u\sum_{j} f_j J_n(2\pi R r_j) \exp\left[i\left(\mp n\varphi_j \pm \frac{2\pi l Z_j}{c}\right)\right] \quad (4.86)$$

The upper and lower signs in Eq. 4.85 are used for the right- and left-hand helices, and the upper and lower signs in Eq. 4.86 are for the upward and downward molecules.

Applications of this equation are given later for PEO (Section 4.12.2) and *it*-poly(*tert*-butylethylene oxide) (Section 4.12.3).

4.7.3 Treatment of Intensity Data

The observed reflection data obtained by photographic or counter method are compared to the calculated values of the structure factors (or their squares) after making various corrections (see Sections 4.2.7 and 4.4.2.C). Here it is very important to adjust the observed intensities to the absolute scale; first the intensity scale for the layer lines must be adjusted. In the case of low-molecular-weight single crystals, relationships among the intensities of the layer reflections can be determined by taking their Weissenberg photographs with respect to the different axes or by using other methods. No such methods are applicable to high polymers, since only uniaxially oriented or doubly oriented samples can be obtained instead of three dimensionally oriented samples. Wilson's method for absolute scaling,[129] which is frequently used for low-molecular-weight crystals, is not suitable for high polymers because of the small number of reflections.

Therefore the usual procedure in the case of fiber diagrams is to adjust the scale between the calculated and observed intensity values by trial and error. For adjusting the intensity scale of the layer lines, integrated intensity data on intense reflections measured by a diffractometer on the oriented sample are the most reliable. The following points should be noted for comparison of the layer lines: (1) the differences in incident angle of the diffracted X-rays on the film, (2) the differences in the distance from the sample, (3) the shape of the sample, (4) arc-shaped spreading of the reflection spots on higher layer lines, and so on. Intensity data from a nonoriented sample (Debye-Scherrer rings) may be used for adjusting the scale of layer lines, but caution should be exercised since the temperature factors of oriented and nonoriented samples are not necessarily the same [e.g., for poly(ethylene sulfide): nonoriented sample, 3–5 $Å^2$; oriented sample, 5–7 $Å^2$].

The customary method for adjusting the absolute scale is to first equalize the sum of the observed and calculated structure factors and subsequently employ the least-squares method. Accuracy of the observed intensity data is most important for least-squares refinement.

4.7.4 Patterson's Function

The Patterson function is not usually employed in a polymer study, since polymers rarely contain heavy atoms. However two examples are cited here. The Patterson function was utilized to estimate rough positions of PEO chains in the lattice (Section 4.12.2). The application of the Patterson method and the Fourier method to polymers is suitable only for cases having sufficient reflection data.

The Patterson function is defined by the following equation (Appendix A)[130]:

$$P(uvw) = \int_0^1 \int_0^1 \int_0^1 \rho(xyz)\rho(x+u, y+v, z+w)V \, dx dy dz \qquad (4.87)$$

where V is the volume of the unit cell. Equation 4.87 may be rewritten as

$$P(uvw) = \frac{1}{V}\sum_{-\infty}^{\infty}\sum\sum |F(hkl)|^2 \cos\left[2\pi(hu + kv + lw)\right] \tag{4.88}$$

$P(uvw)$ is synthesized using only the observed values, since $|F(hkl)|$ can be obtained directly from the observed intensity.

The physical meaning of $P(uvw)$ may be understood from Eq. 4.87. $P(uvw)$ assumes a maximum value when $\rho(xyz)$ and $\rho(x+u, y+v, z+w)$ are simultaneously maximized. Therefore $P(uvw)$ shows a maximum value when the atomic positions in the unit cell have the coordinate values (xyz) and $(x+u, y+v, z+w)$ at the same time, in other words, when (uvw) coincides with an interatomic vector. $P(uvw)$ has a large maximum value at the origin. This maximum corresponds to zero vectors from the atoms to themselves. Since the Patterson function is weighted by the electron density, the larger the atomic numbers of the atoms at both ends or at one end of the interatomic vector, the higher the maximum becomes.

A. *Poly(ethylene Oxide)–Mercuric Chloride Complex Type I.* The $P(u, v, 0)$ and $P(u, v, \frac{1}{2})$ maps obtained are shown in Fig. 4.65. The main peaks in the top figures are interpreted as corresponding to the interatomic vectors and are illustrated schematically in the bottom of Fig. 4.65. From these

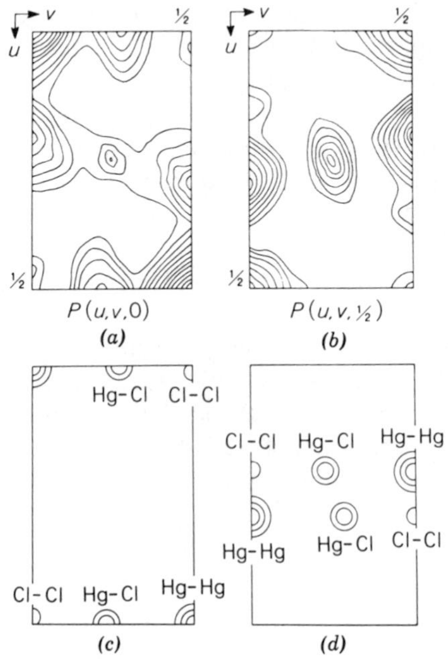

Fig. 4.65 Patterson maps of (*a*) $P(u,v,0)$ and (*b*) $P(u,v,\frac{1}{2})$ of PEO–mercuric chloride complex type I. (*c* and *d*), schematic representations of the interatomic vectors of *a* and *b*, respectively.[58]

figures the fractional coordinates of Hg (0.146, 0, 0.250) and Cl (0.146, 0.275, 0.250) are determined as shown in Fig. 4.62. Contributions of the carbon and oxygen atoms of PEO are not clear because of their low atomic numbers.

B. *Poly-ε-caprolactone.*[81] Poly-ε-caprolactone has an orthorhombic unit cell of space group $P2_12_12_1$-D_2^4 with $a = 7.47$ Å, $b = 4.98$ Å, and c (fiber axis) = 17.05 Å, which includes two nearly planar zigzag chains, each consisting of two monomeric units. At the last stage of analysis the c projection was found as shown in Fig. 4.66, but two unsettled problems remained. One was the relative height of the two chains in the unit cell and the other was the interpretation of the streaks observed on the layer lines. The former is explained herewith, and the latter is discussed in Section 4.10.4.

It may be supposed that the c projection is similar to that of PE, and each atom in one chain has roughly the same height as one of the atoms of the other chain (Fig. 4.67*b*). For the C=O groups, however, four different types of chain arrangement can be considered, as shown in Figs. 4.67*c*–4.67*f*. To examine which is the most reasonable, a one-dimensional Patterson synthesis using the 00*l* reflections ($l = 2$–14) was carried out according to the equation

$$P(w) = \left(\frac{1}{c}\right)\sum |F(00l)|^2 \cos 2\pi lw \qquad (4.89)$$

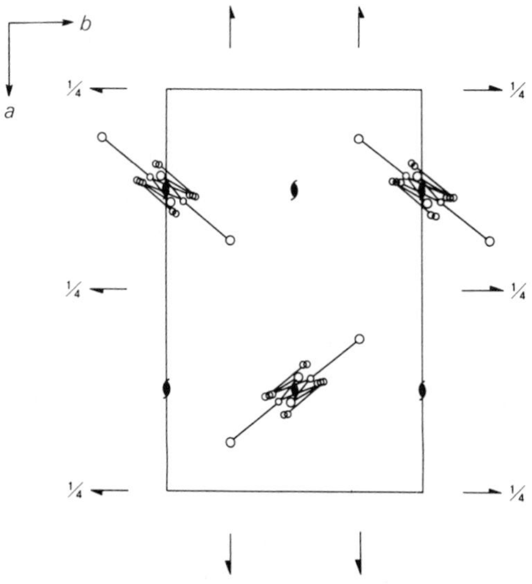

Fig. 4.66 The c projection of poly-ε-caprolactone.[81]

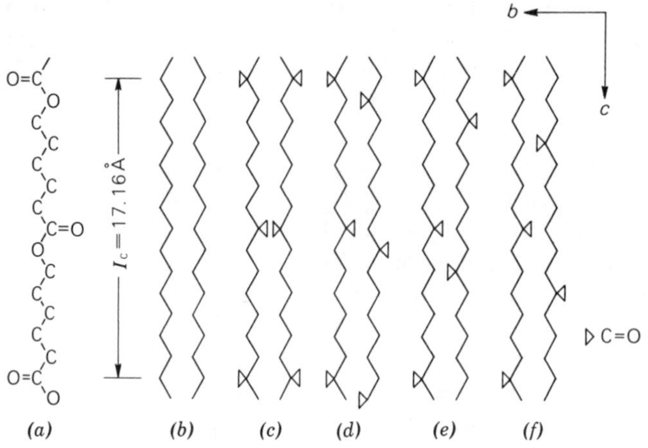

Fig. 4.67 Relative heights of poly-ε-caprolactone molecules in the unit cell.[81]

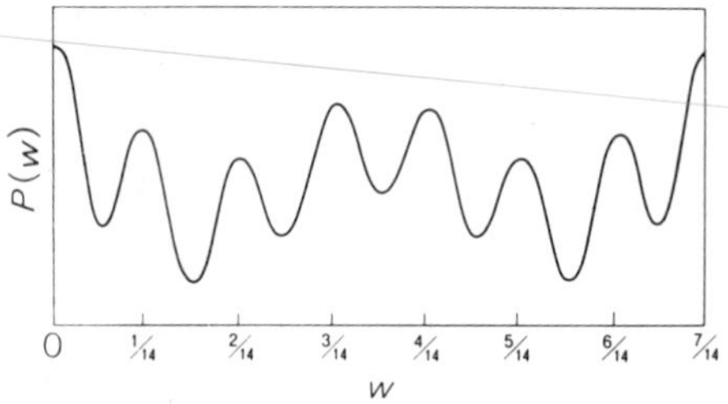

Fig. 4.68 One-dimensional Patterson synthesis along the fiber axis.[81]

The reflection data were taken by Norman's method (Section 4.2.6.C). The result is shown in Fig. 4.68. The strongest peaks at $\frac{3}{14}c$ and $\frac{4}{14}c$ mean that the C=O groups of the two chains in the unit cell are displaced with respect to each other by $\frac{3}{14}c$ or $\frac{4}{14}c$ along the fiber axis. Therefore all but model *f* were eliminated. For example, model *d* must give the strongest peaks at $\frac{1}{14}c$ and $\frac{6}{14}c$. Next, two molecular arrangements are possible from model *f* concerning the relative positions of the methylene groups and the oxygen atoms. These are shown in Fig. 4.69. Further refinement was made using the least-squares method at this stage, and Fig. 4.69*a* was found to be reasonable.

4.7.5 *R*-Factor

After approximate determination of the crystal structure, the following *R*-factor (reliability factor or discrepancy factor) is used as a measure of the

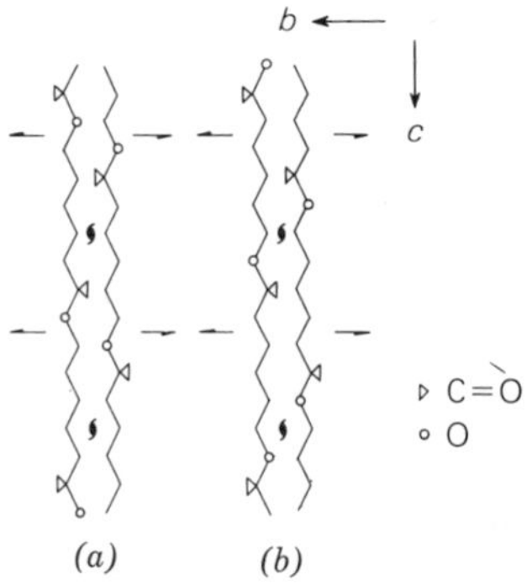

Fig. 4.69 Two possible arrangements of molecules in the unit cell.[81]

agreement between the observed and calculated structure factors.

$$R = \frac{\sum ||F(hkl)_{\text{obs}}| - |F(hkl)_{\text{calc}}||}{\sum |F(hkl)_{\text{obs}}|} \tag{4.90}$$

When reflections overlap accidentally, the following equation is used.

$$R = \frac{\sum \left| \left[\sum_i m_i \left| F_i(\text{obs}) \right|^2 \right]^{1/2} - \left[\sum_i m_i \left| F_i(\text{calc}) \right|^2 \right]^{1/2} \right|}{\sum \left[\sum_i m_i \left| F_i(\text{obs}) \right|^2 \right]^{1/2}} \tag{4.91}$$

where m_i is the multiplicity of the ith reflection. The lower the R-factor, the better the agreement. When the value is less than 20% in the case of high polymers, the model may be considered to be essentially reasonable. It should be noticed that the R-factor is merely a convenient measure and should not be relied on alone. The intensity I, instead of $|F|$, is also used, although not in this book.

4.8 FOURIER SYNTHESES

4.8.1 Fourier Syntheses Using Fractional Coordinates

The electron density $\rho(x, y, z)$ at a point (x, y, z) in a crystal is given by the following equation (see e.g., Ref. 7, p. 11):

$$\rho(xyz) = \frac{1}{V} \sum_{h=-\infty}^{\infty} \sum_{k=-\infty}^{\infty} \sum_{l=-\infty}^{\infty} F(hkl) \exp[-2\pi i(hx + ky + lz)] \tag{4.92}$$

$F(hkl)$ is in general a complex function and is represented by the following form using the phase angle $\alpha(hkl)$.

$$F(hkl) = |F(hkl)| \exp\left[i\alpha(hkl)\right] \tag{4.93}$$

From Eq. 4.93;

$$\rho(xyz) = \frac{1}{V} \sum_{h=-\infty}^{\infty} \sum_{k=-\infty}^{\infty} \sum_{l=-\infty}^{\infty} |F(hkl)| \cos\left[2\pi(hx + ky + lz) - \alpha(hkl)\right] \tag{4.94}$$

The quantity proportional to $|F(hkl)|$ is obtained from the X-ray reflection intensity, but the phase angle $\alpha(hkl)$ cannot be uniquely determined. Accordingly α is determined using the heavy atom method, the isomorphous replacement method, the statistical method, and so forth in the case of low-molecular-weight single crystals. The trial-and-error method is exclusively used in high polymers (except for single crystal forming proteins).

When the lattice has centers of symmetry, $F(hkl)$ is a real function and so Eq. 4.94 can be rewritten as

$$\rho(xyz) = \frac{1}{V}\left\{F(000) + 2\sum_{0}^{\infty} \sum_{-\infty}^{\infty} \sum_{-\infty}^{\infty} F(hkl) \cos\left[2\pi(hx + ky + lz)\right]\right\} \tag{4.95}$$

Here $\alpha(hkl)$ is either 0 or π, and only the sign of $F(hkl)$ should be necessary. There is no high polymer example available in which the analysis was made completely using the Fourier synthesis, since the number of reflections is too small. However there are many cases in which the Fourier synthesis was utilized for refinement after the rough structure had been determined.

First the electron density projected on a plane is calculated, and after enough information on the structure has been obtained, a three-dimensional Fourier synthesis is made. The projection on the *ab* plane, for instance, is given by

$$\rho(xy) = \frac{1}{A} \sum\sum |F(hk0)| \cos\left[2\pi(hx + ky) - \alpha(hk0)\right] \tag{4.96}$$

where A is the area of the projection plane. This calculation is simple, since only the data from the reflection planes parallel to the zone axis* normal to the projection plane are necessary. If the lattice has symmetry axes of even order ($2, 2_1, 4_1$, etc.), the projection perpendicular to these axes has centers of symmetry, even for a lattice having no center of symmetry. Thus the calculation is much simplified by choosing the intersection of the axis and the projection plane as the origin.

A. *Polyglycolide.*[30] Polyglycolide is orthorhombic with space group

* When several planes are parallel to a given axis, these planes are said to belong to a "zone," and the direction is called the "zone axis."

$Pcmn\text{-}D_{2h}^{16}$ and the lattice constants $a = 5.22$ Å, $b = 6.19$ Å, and c (fiber axis) $= 7.02$ Å. A planar zigzag molecular model gives the calculated fiber period 7.16 Å, which is in a fairly good agreement with the value for c. After several trial- and-error procedures, the R-factor was not reduced below 26 %. It was found that reflections from the planes parallel to the bc plane are relatively sharp, while those from the planes normal or closely normal to the ac plane are much broader. Figure 4.70 shows an example of the difference in the breadths of two reflections. The reflection 040 is much broader than the 310 reflection. The visually estimated values of the observed intensities were 187 for the 040 reflection and 411 for the 310 reflection, whereas the integrated intensities were found to be 287 and 247, respectively. Intensities of 21 relatively strong reflections out of 80 visually observed reflections were obtained by a single crystal diffractometer. Using the new data, the R-factor was reduced from 26 to 23%.

Some results of three-dimensional Fourier syntheses are shown in Fig. 4.71.

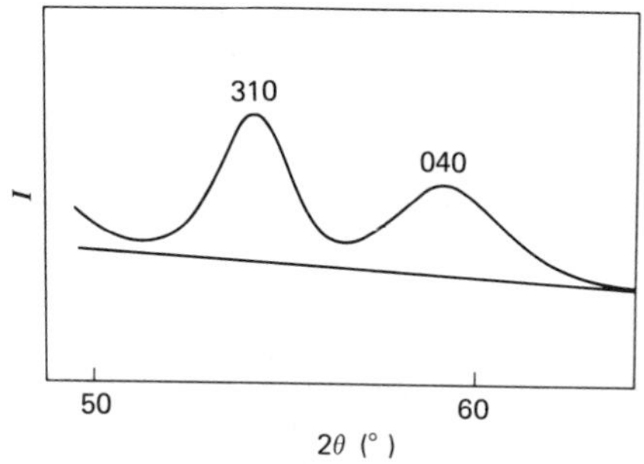

Fig. 4.70 X-Ray diffraction curve on the equator of polyglycolide. The 310 reflection is much sharper than the 040 reflection.[30]

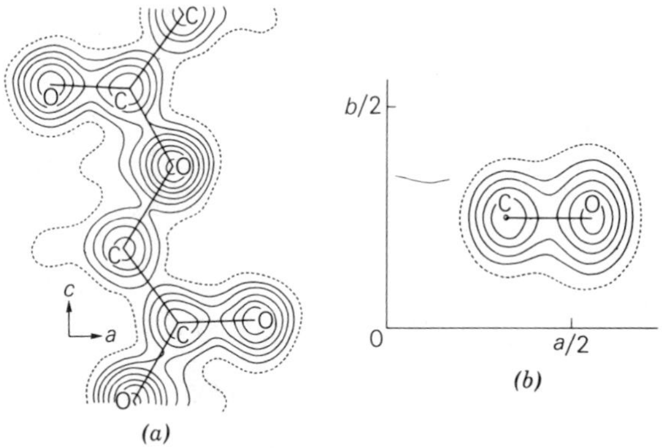

Fig. 4.71 Three-dimensional Fourier map of polyglycolide. (*a*) The section at $y = 0.25$. (*b*) Section through the center of the C=O group, parallel to the *ab* plane. Contours are at intervals of 1 $e/Å^3$, beginning with the 1 $e/Å^3$ contour, which is dotted.[30]

Table 4.11 Temperature Factors of Polyglycolide[30a]

Atom	B_{11}		B_{22}		B_{33}	
C(1)	0.0457	(5 Å^2)	0.0587	(9 Å^2)	0.0230	(4.5 Å^2)
C(2)	0.0366	(4 Å^2)	0.0522	(8 Å^2)	0.0230	(4.5 Å^2)
O(1)	0.0457	(5 Å^2)	0.0587	(9 Å^2)	0.0230	(4.5 Å^2)
O(2)	0.0457	(5 Å^2)	0.0810	(12.4 Å^2)	0.0275	(5.4 Å^2)
H	0.0457	(5 Å^2)	0.0326	(5 Å^2)	0.0254	(5 Å^2)

[a]The values in parentheses are given in the same units as the isotropic temperature factor B. Here, B_{11}, B_{22}, and B_{33} were multiplied by $4a^2$, $4b^2$, and $4c^2$, respectively, since $h^2 B_{11}$ corresponds to $B[(\sin\theta)/\lambda]^2$ (Section 4.2.7.B).

The electron density of the oxygen atom of the carbonyl group was found to be considerably elongated in the *b* direction, implying the presence of an anisotropic vibration. From these Fourier maps all of the atomic coordinates were determined, anisotropic temperature factors being assumed for each atom as shown in Table 4.11. Then the *R*-factor was reduced to 13% by taking the hydrogen atoms into account. The reason for the difference in the breadths of the reflections may be smaller sizes of crystallites along the *b* axis or disorder.

B. *Poly(ethylene Oxide)–Mercuric Chloride Complex Type I.*[58] After the trial-and-error procedure for determining molecular packing in the lattice as described in Section 4.7.1, the intensity data were remeasured

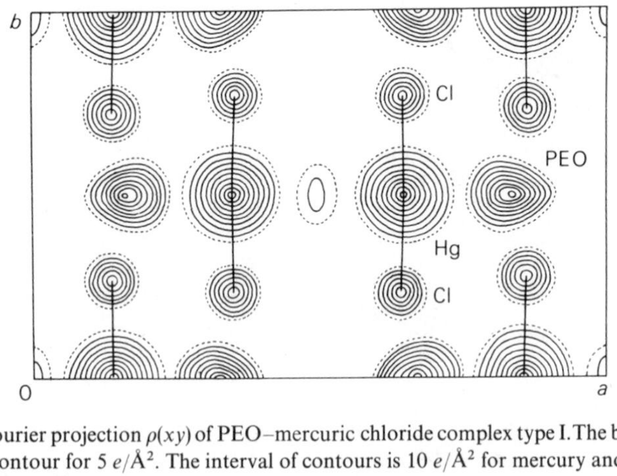

Fig. 4.72 Fourier projection $\rho(xy)$ of PEO–mercuric chloride complex type I. The broken curves indicate the contour for 5 e/Å^2. The interval of contours is 10 e/Å^2 for mercury and 2.5 e/Å^2 for chlorine and PEO.[58]

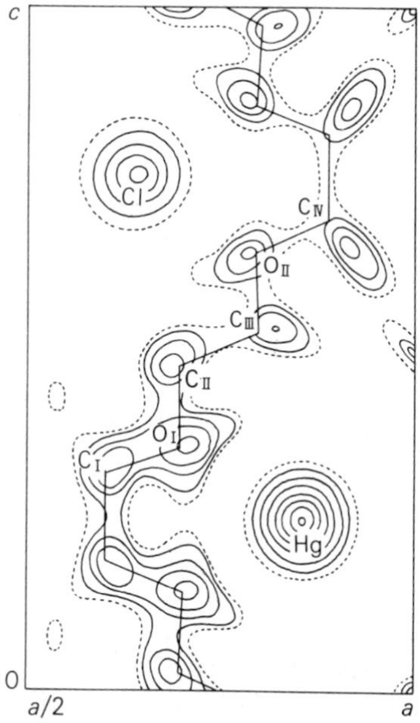

Fig. 4.73 Bounded Fourier projection of PEO–mercuric chloride complex type **I** on the *ac* plane (from $y=-\frac{1}{4}$ to $y=+\frac{1}{4}$). The definition of contours is the same as for Fig. 4.72.[58]

diffractometrically using a powder sample to eliminate the effect of absorption. There were 130 independent reflections. The Fourier synthesis was based on the signs determined from the parameters for the mercury and chlorine atoms, which had been determined at the first stage. The Fourier projection $\rho(xy)$ is shown in Fig. 4.72. In this projection, PEO appears as a single peak, not being resolved into the carbon and oxygen atoms. To examine the distribution of the electron density along the molecular axis, a bounded Fourier projection on the *ac* plane in the range from $y=-\frac{1}{4}$ to $+\frac{1}{4}$ was synthesized as shown in Fig. 4.73. Here the following equation was used (the derivation is given in Ref. 9, p. 387):

$$B\left(x, -\frac{1}{4}\rightarrow\frac{1}{4}, z\right) = \int_{-1/4}^{+1/4} \rho(xyz) b \, \mathrm{d}y$$
$$= \frac{8b}{V}\left(\sum_{h=0}^{\infty}\sum_{l=0}^{\infty} C_{hl}\cos 2\pi hx \cos 2\pi lz - \sum_{h=0}^{\infty}\sum_{l=0}^{\infty} S_{hl}\sin 2\pi hx \right.$$
$$\left. \times \sin 2\pi lz\right) \tag{4.97}$$

where

$$C_{hl} = \left[\frac{1}{2}F(h0l) + \frac{1}{\pi}F(h1l) - \frac{1}{3\pi}F(h3l) + \frac{1}{5\pi}F(h5l) - \cdots\right]$$
$$l = 2n \tag{4.98}$$

$$S_{hl} = \left[\frac{1}{2}F(h0l) + \frac{1}{\pi}F(h1l) - \frac{1}{3\pi}F(h3l) + \frac{1}{5\pi}F(h5l) - \cdots\right]$$
$$l = 2n + 1 \tag{4.99}$$

The figure shows very clearly the electron density distribution corresponding to the $T_5GT_5\bar{G}$ conformation of the PEO molecule, already inferred. The R-factor calculated for mercuric chloride is 21% but is improved remarkably to 17% when the PEO molecule is taken into account.

4.8.2 Fourier Syntheses Using Cylindrical Coordinates

The electron density distribution $\rho(xyz)$ is given by Eq. 4.92 as

$$\rho(xyz) = \frac{1}{V}\sum_h\sum_k\sum_l F(hkl)\exp[-2\pi i(hx + ky + lz)] \tag{4.100}$$

A right-hand Cartesian coordinate system with unit vectors $\mathbf{e}_1$, $\mathbf{e}_2$, and $\mathbf{e}_3$ is now considered. The vector $\mathbf{e}_3$ is taken coincident with the $\mathbf{c}$ axis, and $\mathbf{e}_1$ lies on the **ac** plane, the angle between the positive directions of $\mathbf{e}_1$ and the $\mathbf{a}$ axis being acute. The origins of the direct and reciprocal lattices are taken to be the same. The distance between the reciprocal lattice point hkl and the $\mathbf{c}$ axis is denoted by R_{hkl}, and the azimuthal angle around the $\mathbf{c}$ axis from the $\mathbf{e}_1$ direction is denoted by ψ_{hkl}. The vector $\mathbf{S}$ representing the reciprocal lattice point hkl is given by

$$\begin{aligned}\mathbf{S} &= h\mathbf{a}^* + k\mathbf{b}^* + l\mathbf{c}^* \\ &= R_{hkl}\cos\psi_{hkl}\mathbf{e}_1 + R_{hkl}\sin\psi_{hkl}\mathbf{e}_2 + \left(\frac{l}{c}\right)\mathbf{e}_3\end{aligned} \tag{4.101}$$

The vector $\mathbf{r}$, denoting the point with fractional coordinates x, y, z, is given in terms of the cylindrical coordinates r, φ, and Z by

$$\begin{aligned}\mathbf{r} &= \mathbf{a}x + \mathbf{b}y + \mathbf{c}z \\ &= r\cos\varphi\,\mathbf{e}_1 + r\sin\varphi\,\mathbf{e}_2 + Z\mathbf{e}_3\end{aligned} \tag{4.102}$$

where the Z axis coincides with the $\mathbf{c}$ axis and the $\mathbf{e}_1$ direction is on the plane of $\varphi = 0$. The exponential part of Eq. 4.100 can be written as

$$\begin{aligned}\exp[-2\pi i(hx + ky + lz)] &= \exp(-2\pi i\mathbf{S}\cdot\mathbf{r}) \\ &= \exp\left\{-2\pi i\left[R_{hkl}r\cos\psi_{hkl}\cos\varphi + R_{hkl}r\sin\psi_{hkl}\sin\varphi + \frac{lZ}{c}\right]\right\} \\ &= \exp\left\{-2\pi i\left[R_{hkl}r\cos(\varphi - \psi_{hkl}) + \frac{lz}{c}\right]\right\} \\ &= \exp\left(\frac{-2\pi ilZ}{c}\right)\sum_{n=-\infty}^{\infty} J_n(2\pi R_{hkl}r)\exp\left[in\left(\varphi - \psi_{hkl} - \frac{\pi}{2}\right)\right]\end{aligned} \tag{4.103}$$

Here the following formulas were used:

$$\exp(\mathrm{i}x\cos\phi) = \sum_n J_n(x)\exp\left[\mathrm{i}n\left(\phi + \frac{\pi}{2}\right)\right] \tag{4.104}$$

$$J_n(-x) = (-1)^n J_n(x) = \exp(\mathrm{i}n\pi)J_n(x) \tag{4.105}$$

Substituting Eq. 4.103 in Eq. 4.100, gives the equation

$$\rho(r,\varphi,Z) = \frac{1}{V}\sum_h\sum_k\sum_l\sum_n F(hkl)J_n(2\pi R_{hkl}r)\exp\left[-\mathrm{i}n\left(\psi_{hkl} + \frac{\pi}{2}\right)\right] \times \exp\left[\mathrm{i}\left(n\varphi - \frac{2\pi lZ}{c}\right)\right] \tag{4.106}$$

This is a general formula for the electron density distribution in terms of

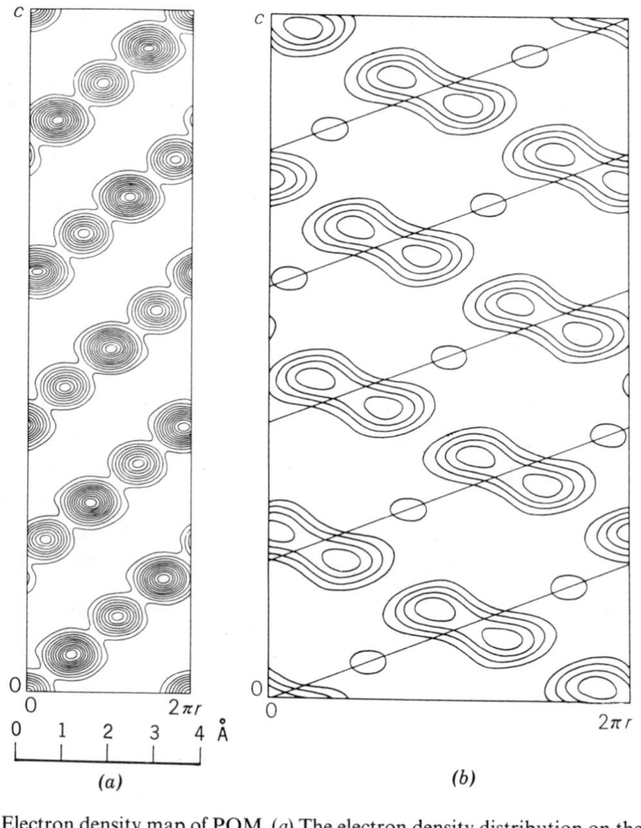

Fig. 4.74 Electron density map of POM. (*a*) The electron density distribution on the cylindrical surface of radius 0.69 Å. (b) [*F*(obs)-*F*(calc)] synthesis on the cylindrical surface of radius 1.52Å.[131]

cylindrical coordinates. If the phase angle of $F(hkl)$ is assumed, $\rho(r, \varphi, Z)$ can be calculated according to this equation.

Figure 4.74*a* is the electron density map of POM on a cylindrical surface of radius 0.69 Å around the helix axis. The cylinder was cut and opened out flat. The alternating sequence of the lower peaks of the carbon atoms and the higher peaks of the oxygen atoms is clearly exhibited. Figure 4.74*b* is the result of the $[F(\text{obs}) - F(\text{calc})]$ synthesis on a cylindrical surface of radius 1.52 Å. In this figure the peaks due to the hydrogen atoms appear, since the contribution of the hydrogen atoms is not taken into account in the calculation of $F(\text{calc})$.*

For the electron density distribution of helical molecules, the following equation was derived by Klug et al.[132]

$$\rho(r, \varphi, Z) = \frac{1}{c}\sum_{l}\sum_{n} \exp\left(\frac{-2\pi i l Z}{c}\right)\int_0^{\infty}\int_0^{2\pi} F_{\mathrm{M}}\left(R, \psi, \frac{l}{c}\right) \times J_n(2\pi R r)\exp\left[in\left(\varphi - \psi - \frac{\pi}{2}\right)\right] R\,\mathrm{d}R\mathrm{d}\psi \qquad (4.107)$$

A knowledge of $F_{\mathrm{M}}\left(R, \psi, \frac{l}{c}\right)$ is necessary for the calculation of this equation. $F_{\mathrm{M}}\left(R, \psi, \frac{l}{c}\right)$ is the molecular structure factor and is different from the unit cell structure factor $F(hkl)$, which is discrete. $F_{\mathrm{M}}(R, \psi, l/c)$ can be estimated by graphical interpolation of the observed unit cell structure factor.

4.9 REFINEMENT BY THE LEAST-SQUARES METHOD

After determination of the rough crystal structure by trial and error, refinement of the analysis can be made by the least-squares method also in the case of polymers. Depending on the parameters to be refined, there are three least-squares methods available: (1) the fractional coordinates of atoms, (2) the cylindrical coordinates of atoms, and (3) the internal coordinates, namely, the bond lengths, bond angles, and internal rotation angles (usually

* Figure 4.74 was drawn using the following equation, which was obtained by carrying out a cylindrical average for all but the central molecule to diminish the contribution of the electron density of neighboring molecules.[131]

$$\rho(r, \varphi, Z) = \frac{1}{V}\sum_{h}\sum_{k}\sum_{l}\sum_{n} G_{n,l}(R_{hkl})\, J_n(2\pi R_{hkl} r)\exp\left[i\left(n\varphi - \frac{2\pi l Z}{c}\right)\right]$$

In the case of a large u, such as POM, Eq. 4.65 can be approximated by $|F| = |G_{n,l}(R)|$ using only the first solution n of Eq. 4.67. Accordingly the observed $|F(hkl)|$ with the sign of $G_{n,l}(R)$ may be substituted for $G_{n,l}(R_{hkl})$ of the equation given above.

the former two being fixed). Method 1 is the least-squares method usually applied to single crystals of low-molecular-weight compounds. At present, method 3 is frequently applied to polymers. Methods 1 and 3 are described below.

Method 2 is performed assuming helical symmetry. Equation 4.85 is used, and its arguments are x_q, y_q, z_q, ϕ_q, r_j, φ_j, Z_j, and so on. This method was applied to *it*-poly(*tert*-butylethylene oxide)[94] and to the early stage of analysis of PEO.[29]

4.9.1 The Least-Squares Method Using Fractional Coordinates as Parameters

Here the observed and calculated structure factors are denoted by F_m(obs) and F_m, respectively. For the least-squares method, the parameters are chosen so as to minimize the value Φ defined in Eq. 4.108; in other words, they are selected to provide the best agreement between F_m(obs) and F_m.

$$\Phi = \sum_{m=1}^{M} w_m[|F_m(\text{obs})| - |F_m|]^2 \tag{4.108}$$

where M is the number of independent observed reflections and w_m is the weight for F_m(obs) and is usually assumed to be one for all reflections in the case of polymers. It is assumed that F_m(obs) has already been approximately corrected to the absolute intensity scale by equalizing the sums of the observed and calculated structure factors. Including the scale factor K for the further adjustment of the observed values to the absolute scale in the process of least-squares refinement gives

$$\begin{aligned} F_m &= KF(hkl) \\ &= K\sum_j f_j \exp\left(-B_j\frac{\sin^2\theta}{\lambda^2}\right)\exp\left[2\pi i(hx_j + ky_j + lz_j)\right] \end{aligned} \tag{4.109}$$

Here isotropic temperature factors are used for simplicity. The parameters u_n $(n = 1, 2, \cdots, N)$ to be determined are the scale factor K, the fractional coordinates (x_j, y_j, z_j), and the temperature parameters B_j of the atoms. The following N equations should be solved to obtain the values of $u_1, u_2, \cdots, u_N$ that minimize Eq. 4.108, since F_m is a function of $u_1, u_2, \cdots, u_N$.

$$\frac{\partial\Phi}{\partial u_n} = -2\sum_m w_m[|F_m(\text{obs})| - |F_m(u_1, u_2, \cdots, u_N)|]\frac{\partial|F_m|}{\partial u_n} = 0$$

$$n = 1, 2, \cdots, N \tag{4.110}$$

If the F_m are linear equations in u_n, as given by

$$F_m(u_1, u_2, \cdots, u_N) = a_{m1}u_1 + a_{m2}u_2 + \cdots + a_{mN}u_N \tag{4.111}$$

and Eq. 4.108 is written as

$$\Phi = \sum_m w_m[|F_m(\text{obs})| - |a_{m1}u_1 + a_{m2}u_2 + \cdots + a_{mN}u_N|]^2 \tag{4.112}$$

then the following N normal equations are obtained by differentiating with respect to u_n and can be solved as simultaneous linear equations.

$$\frac{\partial \Phi}{\partial u_n} = -2\sum_m w_m[|F_m(\text{obs})| - |a_{m1}u_1 + a_{m2}u_2 + \cdots + a_{mN}u_N|](-a_{mn}) = 0$$

$$n = 1, 2, \cdots, N \tag{4.113}$$

However $F_m(u_1, u_2, \ldots, u_N)$ are not linear, but are of the form given in Eq. 4.109. Consequently the small changes of $u_1, u_2, \ldots, u_N$ fixed as the initial values are treated as variables at a stage when the approximate crystal structure has been determined. $|F_m|$ may be expressed as a Taylor's series for small changes of u_n.

$$\begin{aligned} &|F_m(u_1 + \Delta u_1, u_2 + \Delta u_2, \cdots, u_N + \Delta u_N)| \\ &\quad = |F_m(u_1, u_2, \cdots, u_N)| + \sum_{n=1}^{N} \Delta u_n \frac{\partial |F_m(u_1, u_2, \cdots, u_N)|}{\partial u_n} + \cdots \end{aligned} \tag{4.114}$$

By neglecting the higher terms, the following equation is derived from Eq. 4.108.

$$\begin{aligned} \Phi &= \sum_{m=1}^{M} w_m \left[|F_m(\text{obs})| - |F_m(u_1, u_2, \cdots, u_N)| - \sum_{n=1}^{N} \Delta u_n \frac{\partial |F_m|}{\partial u_n} \right]^2 \\ &= \sum_{m=1}^{M} w_m \left[\Delta F_m - \sum_{n=1}^{N} \Delta u_n \frac{\partial |F_m|}{\partial u_n} \right]^2 \end{aligned} \tag{4.115}$$

Since $|F_m|$ can be expressed in the form of Eq. 4.38,

$$|F_m| = K|F(hkl)| = K[A(hkl)^2 + B(hkl)^2]^{1/2} \tag{4.116}$$

the following equation is obtained.

$$\frac{\partial |F_m|}{\partial u_n} = \frac{K}{(A^2 + B^2)^{1/2}} \left(A \frac{\partial A}{\partial u_n} + B \frac{\partial B}{\partial u_n} \right) \tag{4.117}$$

where

$$\frac{\partial |F_m|}{\partial K} = |F(hkl)| \tag{4.118}$$

By differentiating Eq. 4.115 with respect to Δu_q, the following normal equations are obtained, $\partial[\Delta u_n(\partial|F_m|/\partial u_n)]/\partial(\Delta u_q)$ having nonzero values only for $n=q$.

$$\frac{\partial\Phi}{\partial(\Delta u_q)} = -2\sum_{m=1}^{M} w_m\left[\Delta F_m - \sum_{n=1}^{N}\Delta u_n\frac{\partial|F_m|}{\partial u_n}\right]\frac{\partial|F_m|}{\partial u_q} = 0$$

$$q = 1,2,\cdots,N \tag{4.119}$$

It should be noted again that these equations are approximate ones in which the parameters $u_1, u_2, \cdots, u_N$ are chosen as the initial values. Eq. 4.119 is expressed by matrix and vector notations as follows.*

$$\mathbf{FM} - \mathbf{U}\tilde{\mathbf{M}}\mathbf{M} = \mathbf{O} \tag{4.120}$$

Hence

$$\mathbf{U} = \mathbf{FM}(\tilde{\mathbf{M}}\mathbf{M})^{-1} \tag{4.121}$$

where

$$\mathbf{U} = [\Delta u_1, \Delta u_2, \cdots, \Delta u_N] \tag{4.122}$$

$$\mathbf{F} = [\sqrt{w_1}\Delta F_1, \sqrt{w_2}\Delta F_2, \cdots, \sqrt{w_M}\Delta F_M] \tag{4.123}$$

$$\mathbf{M} = \begin{bmatrix} \sqrt{w_1}\dfrac{\partial|F_1|}{\partial u_1}, \cdots, & \sqrt{w_1}\dfrac{\partial|F_1|}{\partial u_N} \\ \cdot & \cdot \\ \cdot & \cdot \\ \cdot & \cdot \\ \sqrt{w_M}\dfrac{\partial|F_M|}{\partial u_1}, \cdots, & \sqrt{w_M}\dfrac{\partial|F_M|}{\partial u_N} \end{bmatrix} \tag{4.124}$$

For high polymers reflections frequently overlap, in which case the roots of the unresolved observed intensities are used for F_m(obs). In this case, the calculated value $|F_m|$ is obtained by

$$|F_m| = \left[\sum_{i=1}^{I} m_i|F_i|^2\right]^{1/2} \tag{4.125}$$

where F_i is the structure factor of the ith reflection among the overlapped reflections, m_i being the multiplicity of the ith reflection and I the number of overlapped reflections. In this case, the partial differential coefficient of $|F_m|$ is given by

$$\frac{\partial|F_m|}{\partial u_n} = \frac{1}{|F_m|}\sum_{i=1}^{I} m_i|F_i|\frac{\partial|F_i|}{\partial u_n} \tag{4.126}$$

* The matrix obtained from a matrix $\mathbf{A}$ by interchanging rows and columns is called the transpose of $\mathbf{A}$ and is denoted by $\tilde{\mathbf{A}}$. The inverse of a matrix $\mathbf{A}$ is denoted by $\mathbf{A}^{-1}$ and is defined by the relation $\mathbf{A}^{-1}\mathbf{A} = \mathbf{A}\mathbf{A}^{-1} = \mathbf{E}$.

Since the off-diagonal elements of $(\widetilde{\mathbf{M}}\mathbf{M})$ in Eq. 4.121 are usually considered to be much smaller than the diagonal elements, the diagonal approximation is often used, in which the off-diagonal elements are neglected except as given below. In most cases the correlation between the scale factor and the temperature factor cannot be neglected. Thus the common temperature parameter B_0 for the whole unit cell is introduced. Usually, the off-diagonal elements correlating K and B_0 are not neglected. For this purpose, Eq. 4.109 is written as

$$F_m = K \exp\left[-B_0 \frac{\sin^2\theta}{\lambda^2}\right] \sum_{j=1}^{J} f_j \exp\left[-B'_j \frac{\sin^2\theta}{\lambda^2}\right] \times \exp\left[2\pi i(hx_j + ky_j + lz_j)\right] \tag{4.127}$$

Frequently used for low-molecular-weight single crystals and polymers in some cases is the block diagonal approximation, in which the off-diagonal elements in the blocks are not neglected; each block associates with the coordinates and temperature factors of the same atom.

4.9.2 The Least-Squares Method Using Internal Coordinates as Parameters (the Constrained Least-Squares Methods)

In the case of high polymers, the least-squares method using fractional coordinates often results in unreasonable bond lengths and bond angles because of insufficient diffraction data. Figure 4.75 shows the results of structure refinements on PET.[133] Figure 4.75*b* shows the bond lengths and bond angles calculated from the fractional coordinates of Bunn et al.[68] These values are not necessarily reasonable, although $R = 19\%$. For example, the C—C distance in the CH_2—CH_2 part is 1.48 Å, differing considerably from the standard value 1.54 Å. Figure 4.75*c* is the result of the least-squares refinement using fractional coordinates. The numbers of parameters to be refined and independent reflections are 23 and 72, respectively. Here the contribution of the hydrogen atoms was omitted, although the hydrogen atoms are usually taken into account in the constrained least-squares method. The R-factor was reduced to 13%, but the bond lengths and bond angles deviate further from the standard values. The method described here is a least-squares method in which the internal rotation angles (the most flexible) can be used as the main parameters by choosing the internal coordinates as parameters and holding the bond lengths and bond angles to the standard values. This method is helpful in reducing the number of parameters to be refined. The insufficient diffraction data are supplemented with reliable bond lengths and bond angles. This is the constrained least-squares method,

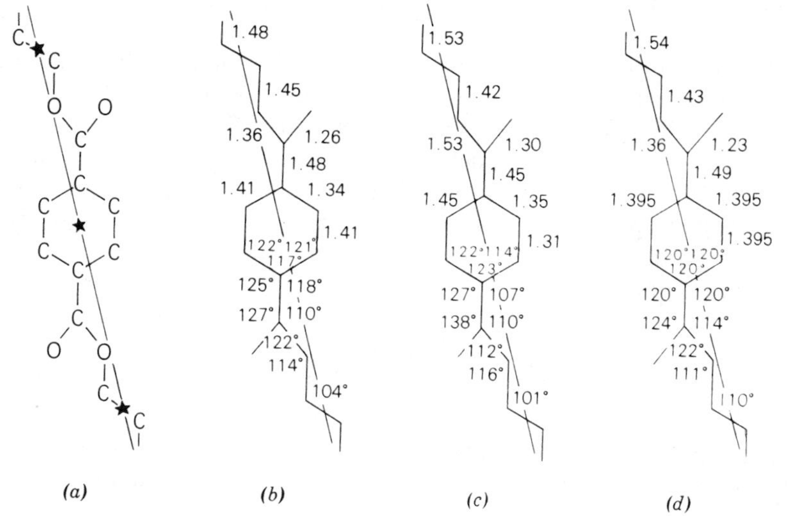

Fig. 4.75 The refinement of the structure analysis of PET. (*a*) Chemical constitution. Stars indicate the centers of symmetry, which coincide with those in the unit cell. The bond lengths and bond angles (*b*) calculated from the coordinates of Daubeny et al.,[68] (*c*) after the least-squares refinement using fractional coordinates, and (*d*) after the constrained least-squares refinement. (From S. Arnott and A. J. Wonacott, *Polymer*, 7,157 (1966) by permission of the publishers, IPC Business Press Ltd. ©.)

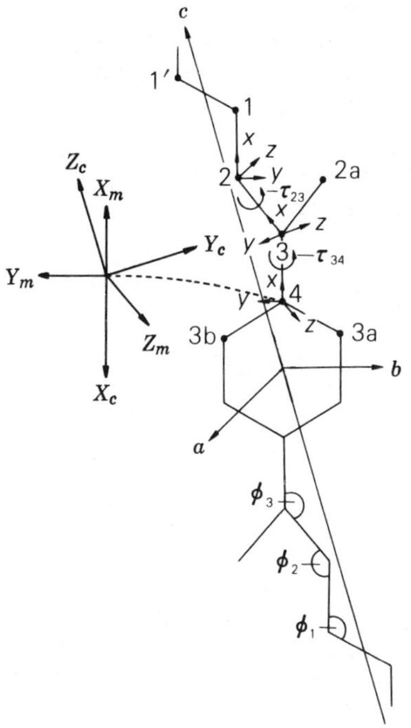

Fig. 4.76 Molecular skeleton of PET. The numbering of the atoms and the coordinate systems used are given. (From S. Arnott and A. J. Wonacott, *Polymer*, 7,157 (1966) by permission of the publishers, IPC Business Press Ltd. ©.)

first applied to high polymers by Arnott and Wonacott.[133] A similar procedure has been used for low-molecular-weight compounds.[134]

Figure 4.76 shows the molecular skeleton of PET with the numbering of the atoms and the coordinate systems used here. The asymmetric unit consists of four main chain atoms (1, 2, 3, 4) and three substituents (2a, 3a, 3b). The hydrogen atoms are ignored for simplicity. If sets of the right-hand Cartesian coordinates as defined in Appendix B are chosen, the coordinates of atom 1 represented by the coordinate system fixed at atom 2 are given by

$$\mathbf{X}_2(1) = \mathbf{B}_{21} = \begin{bmatrix} r_{12} \\ 0 \\ 0 \end{bmatrix} \tag{4.128}$$

where r_{12} is the distance between atoms 1 and 2. As in Eq. 4.58 the coordinates of atom 1 are represented by the coordinate system fixed at atom 4.

$$\mathbf{X}_3(1) = \mathbf{A}_{32}\mathbf{X}_2(1) + \mathbf{B}_{32} = \mathbf{A}_{32}\mathbf{B}_{21} + \mathbf{B}_{32} \tag{4.129}$$

$$\mathbf{X}_4(1) = \mathbf{A}_{43}\mathbf{X}_3(1) + \mathbf{B}_{43} = \mathbf{A}_{43}\mathbf{A}_{32}\mathbf{B}_{21} + \mathbf{A}_{43}\mathbf{B}_{32} + \mathbf{B}_{43} \tag{4.130}$$

Equation 4.130 includes the bond lengths r_{12}, r_{23}, r_{34}, bond angles ϕ_2, ϕ_3, and internal rotation angles τ_{23}, τ_{34}. The coordinate system $\mathbf{X}_4$ is hereafter denoted by $\mathbf{X}_m$, with the sense of the coordinate system fixed at the molecule.

Another right-hand Cartesian coordinate system $\mathbf{X}_c$ is defined as having the same origin as $\mathbf{X}_m$, the Z_c axis being parallel to the c axis and the Y_c axis lying in the bc plane. The coordinates of atom 1 as represented by $\mathbf{X}_c$ are related to $\mathbf{X}_m(1)$ by a matrix $\mathbf{Q}$ including the Eulerian angles θ, ϕ, χ.

$$\mathbf{X}_c(1) = \mathbf{Q}\mathbf{X}_m(1) \tag{4.131}$$

If the Eulerian angles are defined as shown in Fig. 4.77, $\mathbf{Q}$ is given by

$$\mathbf{Q} = \begin{bmatrix} \cos\theta\cos\phi\cos\chi - \sin\phi\sin\chi & -\cos\theta\cos\phi\sin\chi - \sin\phi\cos\chi & \sin\theta\cos\phi \\ \cos\theta\sin\phi\cos\chi + \cos\phi\sin\chi & -\cos\theta\sin\phi\sin\chi + \cos\phi\cos\chi & \sin\theta\sin\phi \\ -\sin\theta\cos\chi & \sin\theta\sin\chi & \cos\theta \end{bmatrix} \tag{4.132}$$

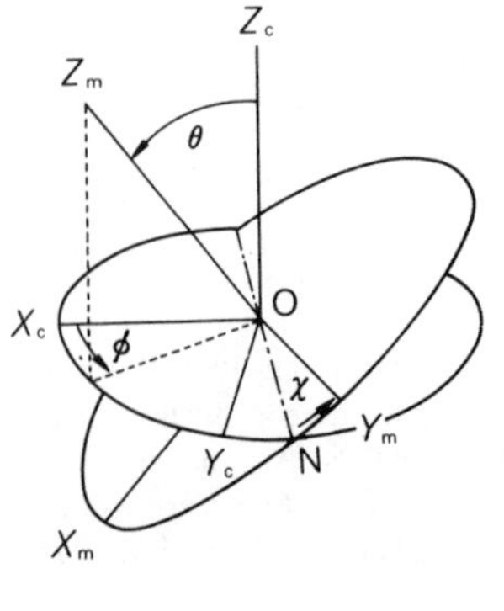

Fig. 4.77 Definition of Eulerian angles. ON is the intersection of the X_cY_c and X_mY_m planes. θ and ϕ are taken according to the direction of the ordinary polar coordinates of Z_m in $\mathbf{X}_c$. χ is an angle in the X_mY_m plane measuring the rotation from ON to OY_m clockwise about the Z_m axis.

$\mathbf{X}_c(1)$ is related with the fractional coordinates $\mathbf{x}(1)$ by

$$\mathbf{x}(1) = \mathbf{C}\mathbf{X}_c(1) + \mathbf{T} = \mathbf{C}\mathbf{Q}\mathbf{X}_m(1) + \mathbf{T} \tag{4.133}$$

where

$$\mathbf{C} = \begin{bmatrix} (a \sin\beta \sin\gamma^*)^{-1} & 0 & 0 \\ (b \sin\alpha \tan\gamma^*)^{-1} & (b \sin\alpha)^{-1} & 0 \\ -\dfrac{\sin\beta\cos\gamma^* + \tan\alpha\cos\beta}{c \tan\alpha \sin\beta \sin\gamma^*} & -(c \tan\alpha)^{-1} & \dfrac{1}{c} \end{bmatrix} \tag{4.134}$$

$$\mathbf{T} = \begin{bmatrix} x_4 \\ y_4 \\ z_4 \end{bmatrix} \tag{4.135}$$

$\mathbf{T}$ comprises the fractional coordinates of atom 4. Thus the fractional coordinates of atom 1 can be represented by $r_{12}, r_{23}, r_{34}, \phi_2, \phi_3, \tau_{23}, \tau_{34}, \theta, \phi, \chi, x_4, y_4$, and z_4. The same procedure can be performed for atoms 3, 2, 2a, 3a, and 3b.

The elements $\partial|F_m|/\partial u_n$ in Eq. 4.124 are now calculated according to the following equation, since $|F_m|$ is not an explicit function of τ_{23}, τ_{34}, θ, ϕ, χ, x_4, y_4, and z_4.

$$\frac{\partial|F_m|}{\partial u_n} = \sum_{p=1}^{P} \left(\frac{\partial|F_m|}{\partial x_p}\frac{\partial x_p}{\partial u_n} + \frac{\partial|F_m|}{\partial y_p}\frac{\partial y_p}{\partial u_n} + \frac{\partial|F_m|}{\partial z_p}\frac{\partial z_p}{\partial u_n} \right) \tag{4.136}$$

where x_p, y_p, z_p are the fractional coordinates of the pth atom and P is the number of atoms involved in the structure factor. The partial differential coefficients in this equation can be easily calculated. For example, $\partial x_1/\partial\tau_{23}$, $\partial y_1/\partial\tau_{23}$, $\partial z_1/\partial\tau_{23}$ are obtained by differentiating $\cos\tau_{23}$ and $\sin\tau_{23}$ in the elements of $\mathbf{x}(1) = \mathbf{CQ}(\mathbf{A}_{43}\mathbf{A}_{32}\mathbf{B}_{21} + \mathbf{A}_{32}\mathbf{B}_{32} + \mathbf{B}_{43}) + \mathbf{T}$ with respect to τ_{23}, since only $\mathbf{A}_{32}$ includes τ_{23}. The overlapped reflections may be treated in the same way as in Eq. 4.125.

Here one important deficiency still remains. The maintenance of the correct stereochemistry at the junction of successive asymmetric units has not been ensured. It is also required in the PET example that the center of the benzene ring coincides with the origin of the unit cell. To introduce these conditions, Lagrange's undetermined multipliers (see, e.g., Ref. 135, p. 324) are employed with the constraining conditions $G_h = 0$ ($h = 1, 2, \cdots, H$; $H < N$). In the case of PET, the following constraining conditions were taken (Fig. 4.76).

$$G_1 = \phi_1 - 109.5^\circ = 0 \tag{4.137}$$

$$G_2 = a^2x_1^2 + b^2y_1^2 + c^2\left(z_1 - \frac{1}{2}\right)^2 + 2bc\cos\alpha\left(z_1 - \frac{1}{2}\right)y_1 + 2ac\cos\beta\left(z_1 - \frac{1}{2}\right)x_1 + 2ab\cos\gamma\, x_1y_1 - \left(\frac{1.54\,\text{Å}}{2}\right)^2 = 0 \tag{4.138}$$

$$G_3 = a^2x_4^2 + b^2y_4^2 + c^2z_4^2 + 2bc\cos\alpha\, z_4y_4 + 2ac\cos\beta\, z_4x_4 + 2ab\cos\gamma\, x_4y_4 - (1.395\,\text{Å})^2 = 0 \tag{4.139}$$

$$G_4 = a^2x_{3a}^2 + b^2y_{3a}^2 + c^2z_{3a}^2 + 2bc\cos\alpha\, z_{3a}y_{3a} + 2ac\cos\beta\, z_{3a}x_{3a} + 2ab\cos\gamma\, x_{3a}y_{3a} - (1.395\,\text{Å})^2 = 0 \tag{4.140}$$

$$G_5 = a^2x_{3b}^2 + b^2y_{3b}^2 + c^2z_{3b}^2 + 2bc\cos\alpha\, z_{3b}y_{3b} + 2ac\cos\beta\, z_{3b}x_{3b} + 2ab\cos\gamma\, x_{3b}y_{3b} - (1.395\,\text{Å})^2 = 0 \tag{4.141}$$

Equation 4.137 ensures that the bond angle $1'12 = 109.5°$, and Eq. 4.138 ensures that the distance $r_{11'} = 1.54$ Å and that the center of the $1-1'$ bond is at $(0, 0, c/2)$. Equations 4.139–4.141 ensure that the benzene ring is centered on the unit cell origin. Equations 4.138–4.141 are derived from the formula for the interatomic distance in a triclinic lattice.

$$r_{12}^2 = a^2(x_1 - x_2)^2 + b^2(y_1 - y_2)^2 + c^2(z_1 - z_2)^2 + 2bc\cos\alpha\,(y_1 - y_2)(z_1 - z_2) + 2ca\cos\beta\,(z_1 - z_2)(x_1 - x_2) + 2ab\cos\gamma\,(x_1 - x_2)(y_1 - y_2) \tag{4.142}$$

Then the best least-squares fit under these constraining conditions is obtained by minimizing

$$\Theta = \sum_{m=1}^{M} w_m \Delta F_m^2 + \sum_{h=1}^{H} \lambda_h G_h \qquad (H < N) \tag{4.143}$$

As in Eq. 4.115, at nearby point $u_n + \Delta u_n$,

$$\Theta = \sum_{m=1}^{M} w_m \left[\Delta F_m - \sum_{n=1}^{N} \Delta u_n \frac{\partial |F_m|}{\partial u_n} \right]^2 + \sum_{h=1}^{H} \lambda_h \left[G_h + \sum_{n=1}^{N} \Delta u_n \frac{\partial G_h}{\partial u_n} \right] \tag{4.144}$$

The values of Δu_n and λ_h that minimize Θ are given by the following $N + H$ linear equations

$$\frac{\partial \Theta}{\partial(\Delta u_q)} = -2 \sum_{m=1}^{M} w_m \left[\Delta F_m - \sum_{n=1}^{N} \Delta u_n \frac{\partial |F_m|}{\partial u_n} \right] \frac{\partial |F_m|}{\partial u_q} + \sum_{h=1}^{H} \lambda_h \frac{\partial G_h}{\partial u_q} = 0 \qquad q = 1, 2, \cdots, N \tag{4.145}$$

$$\frac{\partial \Theta}{\partial \lambda_r} = G_r + \sum_{n=1}^{N} \Delta u_n \frac{\partial G_r}{\partial u_n} = 0 \qquad r = 1, 2, \cdots, H \tag{4.146}$$

Here u_n are the parameters to be refined, that is, the bond lengths, bond angles, internal rotation angles, the fractional coordinates of the atom chosen as the origin of the Cartesian coordinate system fixed on the molecule $\mathbf{X}_m$, the Eulerian angles relating the Cartesian coordinate systems $\mathbf{X}_m$ and $\mathbf{X}_c$, the isotropic temperature factor, and the scale factor. Equations 4.145 and

4.146 may be represented in matrix and vector notation as

$$[U, L]\begin{bmatrix} \hat{M}M & \frac{1}{2}N \\ \frac{1}{2}\tilde{N} & O \end{bmatrix} = [FM, \ \frac{1}{2}G]$$

(4.147)

From this equation, U and L are obtained. Here U, F, and M are the same as in Eqs. 4.122–4.124, and the others are

$$L = [\lambda_1, \lambda_2, \cdots, \lambda_H]$$

(4.148)

$$N = \begin{bmatrix} \dfrac{\partial G_1}{\partial u_1}, \cdots, & \dfrac{\partial G_H}{\partial u_1} \\ \cdot & \cdot \\ \cdot & \cdot \\ \cdot & \cdot \\ \dfrac{\partial G_1}{\partial u_N}, \cdots, & \dfrac{\partial G_H}{\partial u_N} \end{bmatrix}$$

(4.149)

$$G = [-G_1, -G_2, \cdots, -G_H]$$

(4.150)

When the bond lengths and bond angles are to be fixed, these are omitted from the parameters u_n. In the present case, the order of the matrix to be solved is $N + H$, but the number of independent variables is $N - H$.

Figure 4.75d shows the result of the constrained least-squares refinement starting from a model with standard bond lengths and bond angles, the R-factor being reduced in the process from 28 to 23.7%. This result is more reasonable than b and c, since the bond lengths and bond angles have standard values, although the R-factor is higher than those of b and c.

Arnott et al. obtained useful results by applying this method to poly-L-alanine $[-CH(CH_3)-CONH-]_n$ (α- and β-forms),[136,137] poly-L-proline $(-N-CH-CO-)_n$ form II,[138] DNA,[139] RNA,[139] and so on.

$$\begin{array}{ccc} & | & | \\ H_2C & & CH_2 \\ & \diagdown \diagup & \\ & CH_2 & \end{array}$$

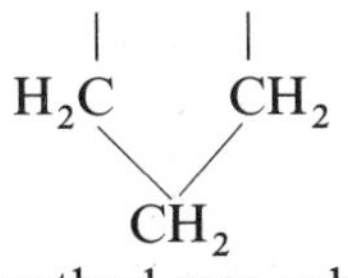

The method was subsequently modified so that the bond lengths, bond angles, and so on between the neighboring asymmetric units can be represented in a general way using the matrices and vectors of the symmetry operations in the unit cell, and the constraining conditions may be easily derived.[39] This method was applied to α-gutta percha,[39] PEO,[29] and so on. The atoms to be taken into the structure factor calculation are limited to an asymmetric unit, and the use of undetermined multipliers is limited only to constraining the distances between pairs of atoms belonging to different neighboring asymmetric units. Consequently the constraining conditions $G_h = 0$ have the simple form

$$G_h = |\mathbf{C}^{-1}(\mathbf{x}_i - \mathbf{x}_{j'})|^2 - d_{ij'}^2 = 0 \qquad h = 1, 2, \cdots, H \tag{4.151}$$

where $\mathbf{C}^{-1}$ is the inverse of the matrix $\mathbf{C}$ which transforms the Cartesian coordinates $\mathbf{X}_c$ into the fractional coordinates $\mathbf{x}$, d_{ij} is the distance to be constrained, the $\mathbf{x}_i$ are the fractional coordinates of the ith atom in one asymmetric unit, and the $\mathbf{x}_{j'}$ are the fractional coordinates of the jth atom in another asymmetric unit. The coordinates $\mathbf{x}_{j'}$ can be related to $\mathbf{x}_j$ in the first asymmetric unit by using the transformation matrix $\mathbf{S}$ and the vector $\mathbf{L}$.

$$\mathbf{x}_{j'} = \mathbf{S}\mathbf{x}_j + \mathbf{L} \tag{4.152}$$

If the bond angle ϕ_i or ϕ_j in Fig. 4.78 is to be taken as the constraining condition, an equation similar to Eq. 4.151 can be written using the distance $h \cdots j'$ or $i \cdots k'$, respectively. Furthermore, the internal rotation angle around the bond $i - j'$ is constrained using the distance $h \cdots k'$.

The case of PEO is now considered. As is described later (Section 4.12.2), the unit cell of $P2_1/a\text{-}C_{2h}^{5}$, $a = 8.05$ Å, $b = 13.04$Å, c (fiber axis) = 19.48Å, and $\beta = 125.4°$, contains four helical chains each consisting of seven monomeric units. At the stage in the calculations where two possible crystal structure models remained, the least-squares method was applied. Had the usual least-squares method been applied to PEO, the number of variables would have exceeded 65 (including one isotropic temperature factor and one scale factor) taking only the skeletal atoms into account. This number is too large compared to the 205 independent observed reflections. Therefore the constrained least-squares method[39,133] was used to refine the crystal structure, with the bond lengths (C—O = 1.43 Å, C—C = 1.54 Å, and C—H = 1.09 Å) and bond angles (∠OCC = 110.0°, ∠COC = 112.0°, and ∠HCH = 109.5°) fixed. The number of adjustable parameters was reduced to 28 (Table 4.12), and the following four constraining conditions were used: (*a*) C(14)—O(1′) = 1.43 Å, (*b*) C(13)... O(1′) = 2.434 Å, (*c*) C(14)

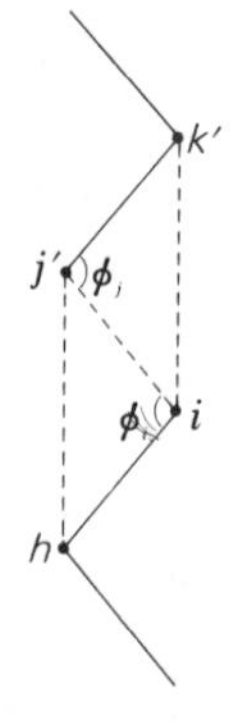

Fig. 4.78 Method for choosing constraining conditions.

Table 4.12 Parameters Obtained by the Constrained Least-Squares Refinement of PEO[29]

Parameter		Final value	Standard deviation[a]
Fractional coordinates of the O(1) atom	x	0.0640	0.003
	y	0.3656	0.003
	z	—0.0424	0.002
Eulerian angles (°) θ		—39.0	2.4
ϕ		—145.4	3.1
χ		2.5	1.9
Internal rotation angles about the skeletal atoms (°)	C(1)—C(2)	57.0	4.9
	C(2)—O(2)	193.8	3.2
	O(2)—C(3)	188.8	3.3
	C(3)—C(4)	67.8	4.5
	C(4)—O(3)	182.9	2.7
	O(3)—C(5)	174.1	3.3
	C(5)—C(6)	74.2	4.3
	C(6)—O(4)	204.3	3.1
	O(4)—C(7)	182.8	3.0
	C(7)—C(8)	49.0	4.2
	C(8)—O(5)	180.3	3.2
	O(5)—C(9)	193.9	2.9
	C(9)—C(10)	91.8	4.5
	C(10)—O(6)	186.2	2.9
	O(6)—C(11)	180.3	3.1
	C(11)—C(12)	60.2	4.0
	C(12)—O(7)	182.0	3.2
	O(7)—C(13)	190.5	3.0
	C(13)—C(14)[b]	199.1	4.5
Overall isotropic temperature parameter (Å^2)		8.36	0.3

[a]See Section 4.11 for standard deviation.
[b]The angle about the bond C(13)—C(14) indicates the dihedral angle O(7)—C(13)—C(14)—H(28). The dihedral angle Y_m—O(1)—C(1)—C(2) was chosen to be 39.0°, where the Y_m axis belongs to the coordinate system $\mathbf{X}_m$ fixed at O(1).

... $C(1') = 2.371$ Å, and (d) $H(28)...O(1') = 2.065$ Å. Condition a constrains the bond distance $C(14)$—$O(1')$, b and c are the conditions $\angle C(13)C(14)O(1') = 110.0°$ and $\angle C(14)O(1')C(1') = 112.0°$, respectively, and d is the condition $\angle H(28)C(14)O(1') = 109.5°$. Here the atoms are numbered as shown below.

<pre>
        H   H   H   H(28)
        | | | |
—[—O(1) – C(1)—C(2)——···——O(7)—C(13)—C(14)—]—O(1')—
        | | | |
        H   H   H   H(27)

        H   H
        | |
     —C(1')—C(2')—
        | |
        H   H
</pre>

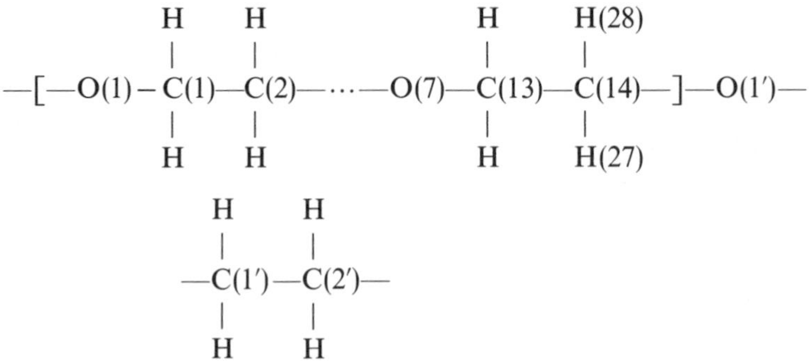

The refinement starting from model A converged at an R-factor of 15.7%, but model B was not improved (34.1%).

In the process of analysis, the packing energy minimization method combined with the constrained least-squares method (described in Section 6.5), are useful.

4.10   CRYSTALLITE SIZE AND DISORDERED STRUCTURE

The crystalline regions of high polymers are not necessarily regular and can include considerable disorder. Some kinds of disorder cause broadening of reflections, which may also originate from the size of the crystallites.

4.10.1   Scherrer's Equation

The mean crystallite size perpendicular to the reflection planes D is related to the broadening of the reflection by Scherrer's equation:

$$D = \frac{K\lambda}{\beta \cos \theta} \tag{4.153}$$

where λ is the wavelength of the X-rays, θ is the diffraction angle, β is the pure X-ray diffraction breadth free of all broadening due to the experimental method employed in observing it, and K is a constant that is commonly assigned a value of 0.9. This equation is derived from Bragg's formula or Laue's condition (see, e.g., Ref. 15, p. 491), and is, strictly speaking, applicable only to systems in which the ordered net planes repeat regularly a certain number of times. It may be useful as a rough measure, but this equation is inadequate when applied to more or less disordered polymer crystallites. From this point of view Hosemann's treatment, described later (Eq. 4.166), is more suitable for polymers.

4.10.2 Distortions of the First and Second Kinds

According to Hosemann,[11] there are two kinds of crystal distortions. In lattices having distortions of the first kind, the distortions are displacements of the structural elements (atoms, monomeric units, etc.) from the equilibrium positions prescribed by the ideal lattice points, and the long range periodicity (order) is preserved with respect to the averaged positions over all the lattice points. The distribution of the displacements may be regarded as a frozen-in thermal vibration at a certain moment (Fig. 4.79*b*). Statistical disorder, such as the configurational disorder of the OH groups in *at*-PVA (Section 4.10.5.A) and the disorder of upward and downward helices in *it*-PP (Fig. 4.79*c*, Section 4.10.5.B), is one case of the distortion of the first kind. Polyketone $[\text{—}(CH_2CH_2)_m\text{—}CO\text{—}]_n$ $(m \geqslant 1)$ obtained by radiation polymerization of ethylene and carbon monoxide shows crystallinity over the range of the nonstoichiometric ratio m.[140] This case also belongs to distortions of the first kind and is called the substitution type (Fig. 4.79*d*). In the frozen-in type distortion *b*, the reflection intensity diminishes with increasing diffraction order, but the broadening of reflection spots does not occur. Diffraction measurements taken at different temperatures are useful for distinguishing between the reflection intensity diminution by distortions of the first kind and those by thermal vibrations. Statistical disorder *c* and substitution type disorder *d* are treated using the structure factors for the average structure (as described later), and therefore it is not necessary to consider the broadening.

In a lattice possessing distortions of the second kind, each lattice point varies in position statistically only in relation to its nearest neighbors instead of in relation to the ideal lattice points, and this situation repeats again for

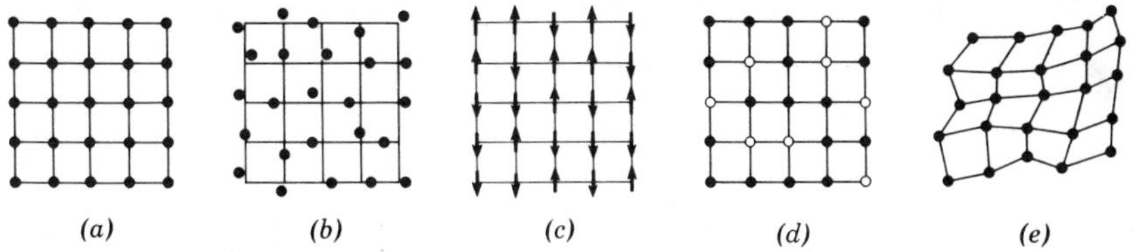

Fig. 4.79 (*a*) Ideal crystal lattice, (*b–d*) distortions of the first kind, and (*e*) distortions of the second kind. (*b*) Frozen-in thermal vibration type, (*c*) statistical disorder of the upward and downward helices, and (*d*) substitution type disorder (open and filled circles indicate different chemical structural units).

subsequent lattice points. As a result long-range order is lost, as shown in Fig. 4.79*e*. This type of distortion is also called paracrystalline; it results in both a diminution of reflection intensity and an increase in reflection breadth with increasing reflection order.

4.10.3 Diffraction and Scattering Intensity of Disordered Structures

A general discussion of X-ray diffraction and scattering by structures ranging from an ideal crystal to a paracrystal is now given. For details, see Refs. 11, 12, and 16–18.

The X-ray scattering intensity $I(\mathbf{S})$, including that from disordered systems, is generally given by

$$\begin{aligned} I(\mathbf{S}) &= F(\mathbf{S})F^*(\mathbf{S}) \\ &= \sum_n f_n \exp(2\pi i \mathbf{S}\cdot\mathbf{R}_n) \sum_{n'} f_{n'}^* \exp(-2\pi i \mathbf{S}\cdot\mathbf{R}_{n'}) \\ &= \sum_n \sum_{n'} f_n f_{n'}^* \exp[-2\pi i \mathbf{S}\cdot(\mathbf{R}_{n'} - \mathbf{R}_n)] \\ &= \sum_m \left(\sum_n f_n f_{n+m}^* \right) \exp(-2\pi i \mathbf{S}\cdot\mathbf{R}_m) \end{aligned} \tag{4.154}$$

Here f_n is the structure factor of the nth scattering unit (atom, molecule, unit cell, etc.) and $\mathbf{R}_m = \mathbf{R}_{n'} - \mathbf{R}_n$ is the vector joining the nth and n'th scattering units. If there is no irregularity of the atomic positions in the lattice and if the crystal is sufficiently large, Eq. 4.154 may be written in the following form, as discussed in Section 4.2.5.

$$I(\mathbf{S}) = |F(\mathbf{S})|^2 |G(\mathbf{S})|^2 \tag{4.155}$$

where $G(\mathbf{S})$ is the Laue function.

In the case of distortions of the first kind, as shown in Figs. 4.79*c* and 4.79 *d*, f_n is usually different from $f_{n'}$. Here $\sum_n f_n f_{n+m}^*$ in Eq. 4.154 may be represented by an average value times the number of terms $N_m \langle f_n f_{n+m}^* \rangle$. For a regular structure in which the structure factor is given by the average value $\langle f \rangle$ of f_n, f_n may be represented in terms of $\langle f \rangle$ and the deviation Δf_n from it,

$$f_n = \langle f \rangle + \Delta f_n \tag{4.156}$$

Since $\langle \Delta f_n \rangle = 0$, the following equation is obtained.

$$\langle f_n f_{n+m}^* \rangle = \langle f_n \rangle^2 + \langle \Delta f_n \Delta f_{n+m}^* \rangle \tag{4.157}$$

If the disorder of each unit cell occurs independently of the surrounding unit cells, $\langle \Delta f_n \Delta f_{n+m}^* \rangle = 0$ holds for all m except $m = 0$. The following equation is derived from Eq. 4.154 by considering only the case of $m=0$ for Eq. 4.157.

$$I(\mathbf{S}) = \sum_m N_m \langle f_n f_{n+m}{}^* \rangle \exp(-2\pi i \mathbf{S}\cdot\mathbf{R}_m)$$
$$= \langle f \rangle^2 |G|^2 + N\langle |\Delta f|^2 \rangle$$
$$= \langle f \rangle^2 |G|^2 + N(\langle f^2 \rangle - \langle f \rangle^2) \qquad (4.158)$$

where N is the number of unit cells scattering the X-rays. $\langle f^2 \rangle$ is equal to $\langle f \rangle^2$ in the regular structure where f_n and $f_{n'}$ are all the same, but they are not equal when disorder exists.* The first term of Eq. 4.158 corresponds to the Bragg reflections and the second term corresponds to diffuse scattering, which shows a continuous and gradually varying intensity distribution.

If distortions of the frozen-in thermal vibration type (or thermal vibrations) (Fig. 4.79*b*) are further considered, Eq. 4.158 is written as

$$I(\mathbf{S}) = \langle f \rangle^2 D^2 G^2 + N(\langle f^2 \rangle - \langle f \rangle^2 D^2)$$
$$= \langle f \rangle^2 D^2 G^2 + N(\langle f^2 \rangle - \langle f \rangle^2) + N\langle f \rangle^2 (1 - D^2) \qquad (4.159)^\dagger$$

where D is the Debye factor, given by

$$D = \exp\left[-8\pi^2 \langle u^2 \rangle \frac{\sin^2\theta}{\lambda^2}\right] = \exp\left[-B\frac{\sin^2\theta}{\lambda^2}\right] \qquad (4.160)$$

D decreases with an increase in the mean square amplitude $\langle u^2 \rangle$ and an increase of the diffraction angle θ (Section 4.2.7.B). Along with the diminution of the Bragg reflections of the first term of Eq. 4.159, the diffuse scattering of the third term appears. $\langle f \rangle^2$ decreases with an increase of θ, while $(1 - D^2)$ is zero at $\theta = 0$, and increases gradually with an increase of θ, being asymptotic to 1. Therefore the third term reveals diffuse scattering with a broad maximum.

If the crystallite has a definite size and shape, the following equation should be considered:

$$I(\mathbf{S}) = \left(\frac{1}{V}\right) \langle f \rangle^2 D^2 \widehat{G^2 | S}(\mathbf{S})|^2 + N(\langle f^2 \rangle - \langle f \rangle^2) + N\langle f \rangle^2 (1 - D^2) \quad (4.161)$$

where V is the volume of the scattering unit and $S(\mathbf{S})$ is the shape factor, which

* The meanings of $\langle |f(\mathbf{S})|^2 \rangle$ and $\langle f(\mathbf{S}) \rangle^2$ may be easily understood by considering the example in which the same numbers of upward and downward molecules are distributed randomly. If the structure factor of each molecular chain is denoted by f_U or f_D, $\langle f^2 \rangle$ and $\langle f \rangle^2$ are given by $\langle f^2 \rangle = (f_U^2 + f_D^2)/2$, and $\langle f \rangle^2 = [(f_U + f_D)/2]^2$. In the case of a disordered structure, $\langle f \rangle$ is used as the structure factor. If there exist upward or downward molecules only, $\langle f^2 \rangle = \langle f \rangle^2$.

† In Eq. 4.159 the Debye factor D is taken into account for the displacements of the scattering unit having the averaged structure factor $\langle f \rangle$ around the equilibrium position, and so the scattering intensity is expressed by $\langle f \rangle^2 D^2$. D is not considered for $\langle f^2 \rangle$.

is the Fourier transform of the shape function representing the actual shape and size of the crystallite. For example, the shape factor for a rectangular parallelepiped with edges $L_1 L_2 L_3$ is given by

$$S_{L_1 L_2 L_3}(X_1 X_2 X_3) = \prod_{i=1}^{3} \frac{\sin(\pi X_i L_i)}{\pi X_i} \tag{4.162}$$

This is obtained by considering Eq. A.6 for the three-dimensional case (Fig. A.1). Consequently if the crystallites are small, $S(\mathbf{S})$ tends to make the reflection spots broader. The symbol $\frown$ in Eq. 4.161 denotes convolution (see Appendix A.3).

If there are distortions of the second kind, the long range order in the lattice is lost, and so $|G(\mathbf{S})|^2$ in Eq. 4.161 should be replaced by the paracrystalline lattice factor $Z(\mathbf{S})$ as shown below.

$$I(\mathbf{S}) = \left(\frac{1}{V}\right)\langle f\rangle^2 D^2 \widehat{Z(\mathbf{S})|S}(\mathbf{S})|^2 + N(\langle f^2\rangle - \langle f\rangle^2) + N\langle f\rangle^2(1 - D^2) \tag{4.163*}$$

Table 4.13 summarizes the above relations.

Table 4.13 X-Ray Scattering by Structures, Including Those with Disorders

System	Diffraction	Diffuse scattering
Ideal crystal without disorder	Sharp	None, Eq. 4.155
Distortions of the first kind		
Statistical disorder and substitution type	Sharp	Second term of Eq. 4.158
Frozen-in thermal vibration type	Sharp; diminish with θ.	Third term of Eq. 4.159
Effect of crystallite size	Broadening of reflections according to the shape factor $\lvert S(\mathbf{S})\rvert^2$	Second and third terms of Eq. 4.161
Distortions of the second kind	Brodening of reflections according to the paracrystalline lattice factor $Z(\mathbf{S})$	Second and third terms of Eq. 4.163

* In Eqs. 4.161 and 4.163, $I(\mathbf{S})$, $\langle f\rangle^2$, D^2, G^2, and $Z(\mathbf{S})$ are dimensionless, and the dimensions of $(1/V)$ and $S(\mathbf{S})$ are $[L^{-3}]$ and $[L^3]$, respectively (see Eq. 4.162). The convolutions in Eqs. 4.161 and 4.163 are performed in reciprocal space, the volume element of which has the dimension $[L^{-3}]$. Therefore the first terms of Eqs. 4.161 and 4.163 are dimensionless.

The paracrystalline lattice factor $Z(\mathbf{S})$ is considered here for a one-dimensional lattice with distortions of the second kind. The lattice points are distributed on a straight line. The distance between two neighboring points fluctuates around the average value $\langle a \rangle$, and its probability distribution is approximated by a Gaussian function. With this approximation, the paracrystalline lattice factor $Z(h)$ for the one-dimensional lattice is given by[11]

$$Z(h) = \frac{1 - \exp(-4\pi^2 g^2 h^2)}{[1 - \exp(-2\pi^2 g^2 h^2)]^2 + 4\sin^2(\pi h)\exp(-2\pi^2 g^2 h^2)} \qquad (4.164)$$

where h is the order of diffraction and has the relation $X = h/\langle a \rangle$ to the X component of the scattering vector $\mathbf{S}$. $g = \sigma/\langle a \rangle$ may be used as a parameter of the fluctuation in the lattice, that is, as a measure of the distortion of the second kind; σ is the standard deviation (Section 4.11) of the Gaussian function. Figure 4.80 shows $Z(h)$ for various g values. If $g \leqslant 0.01$, $Z(h)$ is essentially the usual crystal lattice factor. However, as g increases, the maximum intensity at integral values of h decreases exponentially and the peaks broaden approximately as the square of h. At high values of g more maxima become broadened and decrease in intensity until they merge into the background. The value $g = 0.35$ is the limit where only the first order reflection can be observed.

The shape factor $|S(\mathbf{S})|^2$ for a one-dimensional crystal with the whole length of $N\langle a \rangle$ is given by

$$|S(h)|^2 = \left[\frac{\sin(\pi N h)}{\pi h/\langle a \rangle}\right]^2 \qquad (4.165)$$

This $S(h)$ is the same as the curve shown in Fig. A. 1b, and the peak width is $2/N\langle a \rangle$. If N is sufficiently large, the peak broadening due to the shape factor is negligible. In this case the diffraction intensity is essentially the same as that shown in Fig. 4.80. This figure does not take the shape factor

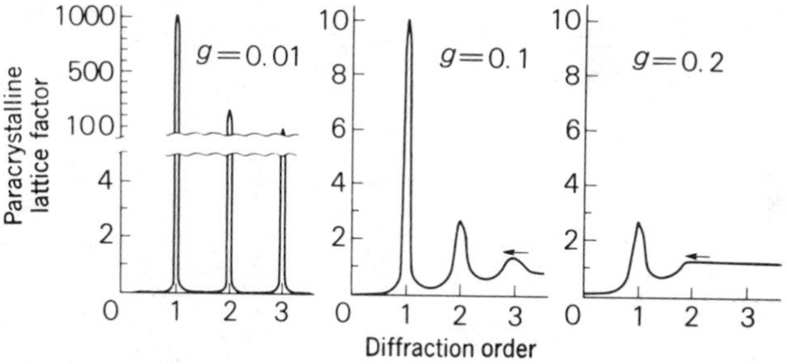

Fig. 4.80 The paracrystalline lattice factor $Z(h)$ for various values of the fluctuation parameter g. The arrows indicate the peak shift to the direction of lower diffraction angles.(From Lindenmeyer and Hosemann.[141])

into account. If N is small, then the peaks broaden as a result of this effect.

Under the Gaussian approximation the value of g can be estimated by measuring the half width $\delta\beta$ for a series of reflections with different diffraction orders, for instance, 110, 220, 330, $\cdots$, and plotting $(\delta\beta)^2$ against h^4, which yields a straight line according to the equation

$$(\delta\beta)^2 = \frac{1}{\langle a\rangle^2}\left(\frac{1}{N^2} + \pi^4 h^4 g^4\right) \tag{4.166}$$

The number of lattice planes N is given by the ordinate intercept and g is given by the slope.[142,143] The crystallite size and distortion parameter can be obtained separately by this method. With the Lorentz approximation, the corresponding equation has the form[144]

$$\delta\beta = \frac{1}{\langle a\rangle}\left(\frac{1}{N} + \pi^2 h^2 g^2\right) \tag{4.167}$$

A new method using only two reflections has been derived recently.[145] Also see Buchanan and Miller's paper.[146]

For X-ray scattering by amorphous samples, see Refs. 12 (Section 11.2), 16 (p. 355), 17 (Section 1–4.4).

4.10.4 Disordered Displacements of the Chains along the Fiber Axis (Poly-β-Propiolactone Form II)

Form II of poly-β-propiolactone shows a unique X-ray fiber pattern (Fig. 4.81): discrete reflections on the equatorial line and continuous scattering

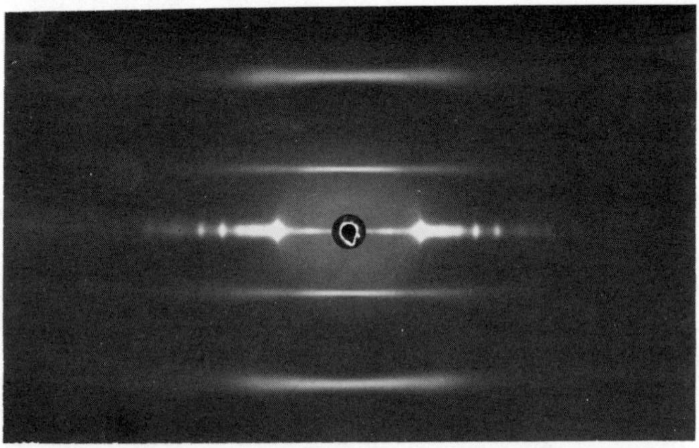

Fig. 4.81 X-Ray fiber photograph of form II of poly-β-propiolactone. *As*-polymerized sample obtained by solid-state polymerization.[53]

("streak") on each layer line.[53] Such streaks have been observed for polyallene form III,[66] poly-ε-caprolactone,[30] polyamides,[26,147] and so on. These features can be explained assuming a disordered crystal structure in which each molecular chain has an essentially regular periodic structure, and also in which the packing of the chains in the lateral direction with respect to the fiber axis is periodic, but the mutual levels of the chains in the fiber direction are irregular. The diffraction spots broaden with an increase of this type of irregularity and ultimately become continuous streaks. The disorder parameter g in this case may be larger than 0.35 according to the paracrystal theory described in the preceding section.

In such a disordered structure the scattered intensity at a position $\mathbf{S}[|\mathbf{S}| = 2\sin\theta/\lambda]$ in reciprocal space is given by[53]

$$I(\mathbf{S}) = \left\{ \left| \sum_m F_m(\mathbf{S}) \right|^2 \frac{\sin^2 \pi N_1 s_1 a}{\sin^2 \pi s_1 a} \frac{\sin^2 \pi N_2 s_2 b}{\sin^2 \pi s_2 b} \frac{\sin^2 \pi s_3 c}{(\pi s_3 c)^2} + N_1 N_2 \sum_m |F_m(\mathbf{S})|^2 \left[1 - \frac{\sin^2 \pi s_3 c}{(\pi s_3 c)^2} \right] \right\} \frac{\sin^2 \pi N_3 s_3 c}{\sin^2 \pi s_3 c} \tag{4.168}$$

where $F_m(\mathbf{S})$ is the structure factor of the chain of the mth state; N_1, N_2, and N_3 are the number of repeated units along the a, b, and c axes, respectively; and s_1, s_2, and s_3 are the components of $\mathbf{S}$ to the three reciprocal axes. The first term in the braces { } corresponds to the crystalline diffraction, and the second term corresponds to the diffuse scattering. $\sin^2 \pi N_3 s_3 c / \sin^2 \pi s_3 c$ on the outside of the braces is the Laue function and is equal to N_3^2 if $s_3 c$ equals l and N_3 is large (l is any integer representing the number of the layer). The function $\sin^2 \pi s_3 c/(\pi s_3 c)^2 = \sin^2 \pi l/(\pi l)^2$ is equal to one if $l = 0$ and it is equal to zero if l is any integer other than zero. (This function is the same as Eq. A.6 with $A = 1$). Consequently, when $l = 0$, only the first term of Eq. 4.168 has a nonzero value. Additionally, if N_1 and N_2 are large, $s_1 a = h$, $s_2 b = k$(h and k: integer), and Eq. 4.168 becomes

$$I(\mathbf{S}) = N_1^2 N_2^2 N_3^2 \left| \sum_m F_m(hk0) \right|^2 \tag{4.169}$$

When l is any integer other than zero, only the second term has a nonzero value:

$$I(\mathbf{S}) = N_1 N_2 N_3^2 \sum_m \left| F_m(s_1 s_2 l) \right|^2 \tag{4.170}$$

Equations 4.169 and 4.170 indicate the appearance of discrete spots on the equator and continuous streaks on the layer lines, respectively.

Next the intensity distribution on the layer lines is compared with the molecular structure factor. Figure 4.82 is the Weissenberg photograph

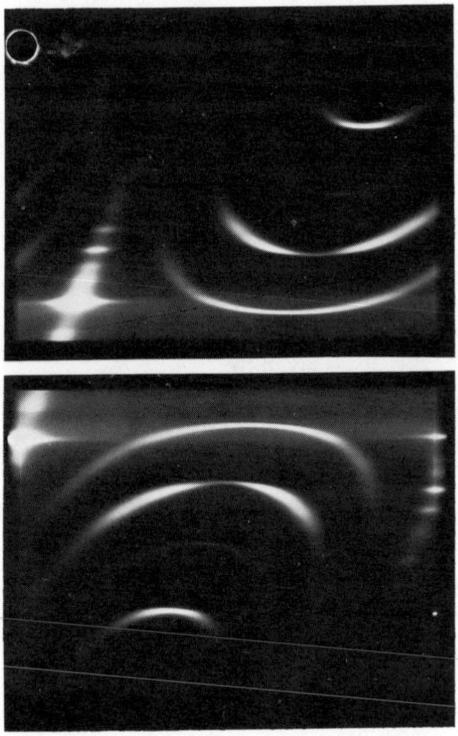

Fig. 4.82 Weissenberg photograph of form II of poly-β-propiolactone.[53]

taken by Norman's method (Section 4.2.6.C). Since the sample is uniaxially oriented, the ψ average of the molecular structure factor should be considered.

$$\left\langle \left| F\left(R,\psi,\frac{l}{c}\right)\right|^2 \right\rangle_\psi = \sum_j \sum_{j'} f_j f_{j'} J_0(2\pi R r_{jj'}) \exp\left[\frac{2\pi i l(z_j - z_{j'})}{c}\right] \quad (4.171)$$

$$r^2_{jj'} = (x_j - x_{j'})^2 + (y_j - y_{j'})^2 \quad (4.172)$$

where J_0 is the zeroth order Bessel function, and x_j, y_j, and z_j are the Cartesian coordinates of the jth atom The derivation of this equation is given in Appendix D. An anisotropic thermal factor $\exp\{-\frac{1}{2}[B_r R^2 + B_l(l/c)^2]\}$, is used, where B_r and B_l are the thermal coefficients in the directions perpendicular and parallel to the c axis, respectively.

The fiber identity period was found to be 4.77 Å from the interval of the layer lines. From the equatorial reflections, an orthogonal cell having the cell dimensions $a' = 7.73$ Å, $b' = 4.48$ A, and $\gamma' = 90°$ was determined by conventional methods. The c dimension of 4.77 Å indicates that the molecular conformation of form II is essentially planar zigzag. In Fig. 4.83 the molecular

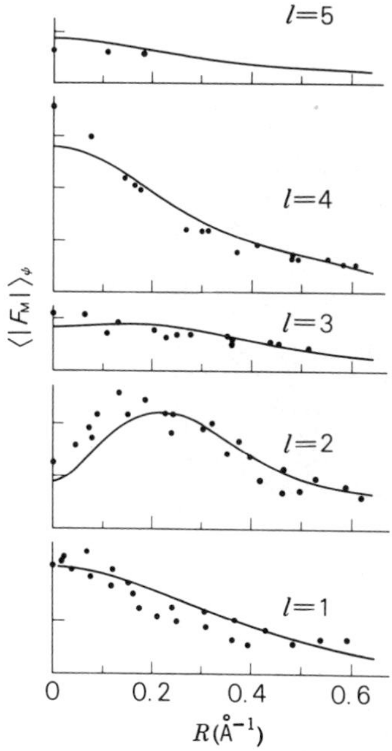

Fig. 4.83 Comparison between the observed (points) and calculated structure factors (curves) for layer lines of poly-β-propiolactone.[53]

structure factor calculated according to Eq. 4.171 is compared with the observed values. The intensities of the first, second, and fourth layer lines are strong, while those of the third and fifth layer lines are weak. The maximum appears at the meridian ($R = 0$) on the first and fourth layer lines, while on the second layer line the intensity is minimum at the meridian. As a whole, these features are well explained by the calculated intensity curves.

4.10.5 Statistically Disordered Structure

A. *Disordered Arrangements of the Side Groups [Atactic Poly(vinyl Alcohol)].* PVA prepared by radical polymerization of vinyl acetate and subsequent hydrolysis gives a well defined fiber diagram (Fig. 4.84), although the configuration is thought to be atactic. Bunn[148] proposed one statistically disordered crystal structure (Fig. 4.85) based on the information that random copolymers of ethylene and vinyl alcohol units are crystalline and the OH groups may replace hydrogen atoms at random on a carbon chain without destroying crystallinity.[149] Although the carbon atoms of the main chain are at regular positions in the lattice, the OH groups are randomly placed in the left- and right-hand positions along the zigzag plane with the same probability. Therefore an averaged structure may be assumed in which the

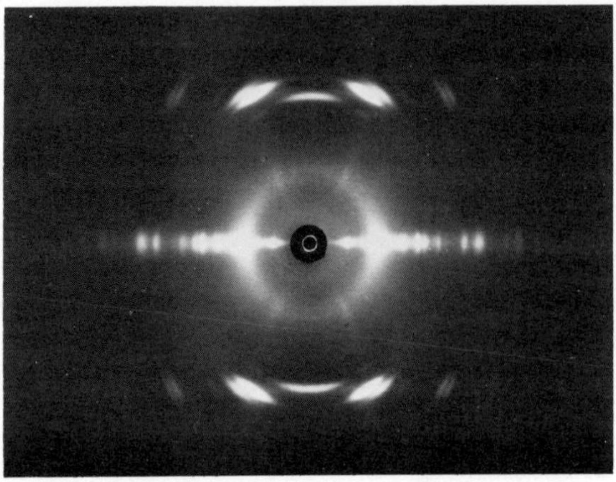

Fig. 4.84 Fiber diagram of *at*-PVA.

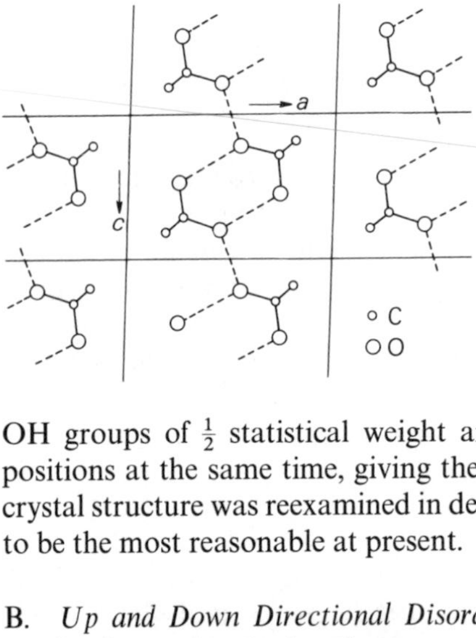

Fig. 4.85 Crystal structure of *at*-PVA. Dashed lines indicate hydrogen bonds. (From Bunn.[148])

OH groups of $\frac{1}{2}$ statistical weight are placed in the left- and right-hand positions at the same time, giving the apparent fiber period of 2.52 Å. This crystal structure was reexamined in detail by Nitta et al.[150] and is considered to be the most reasonable at present.

B. *Up and Down Directional Disorder of Helical Chains* (*Isotactic Polymers*). Isotactic vinyl polymers assume mostly helical structures; for instance, *it*-PP has a $(TG)_3$ (3/1) helix conformation as shown in Fig. 4.86*a*. These can be either right- or left-hand helices pointing either up or down. Thus there are four kinds of conformational settings (Section 7.3.1). From the observed systematic absences the crystal structure of *it*-PP belongs to either space group Cc-C_s^4 or $C2/c$-C_{2h}^6.[151] Figure 4.86*b* shows the crystal structure (two unit cells). In the figure, A, A′, A″, ... are left-hand helices, B, B′, ... are right-hand helices, C, C′, ... are left-hand helices, and D, D′, ... are

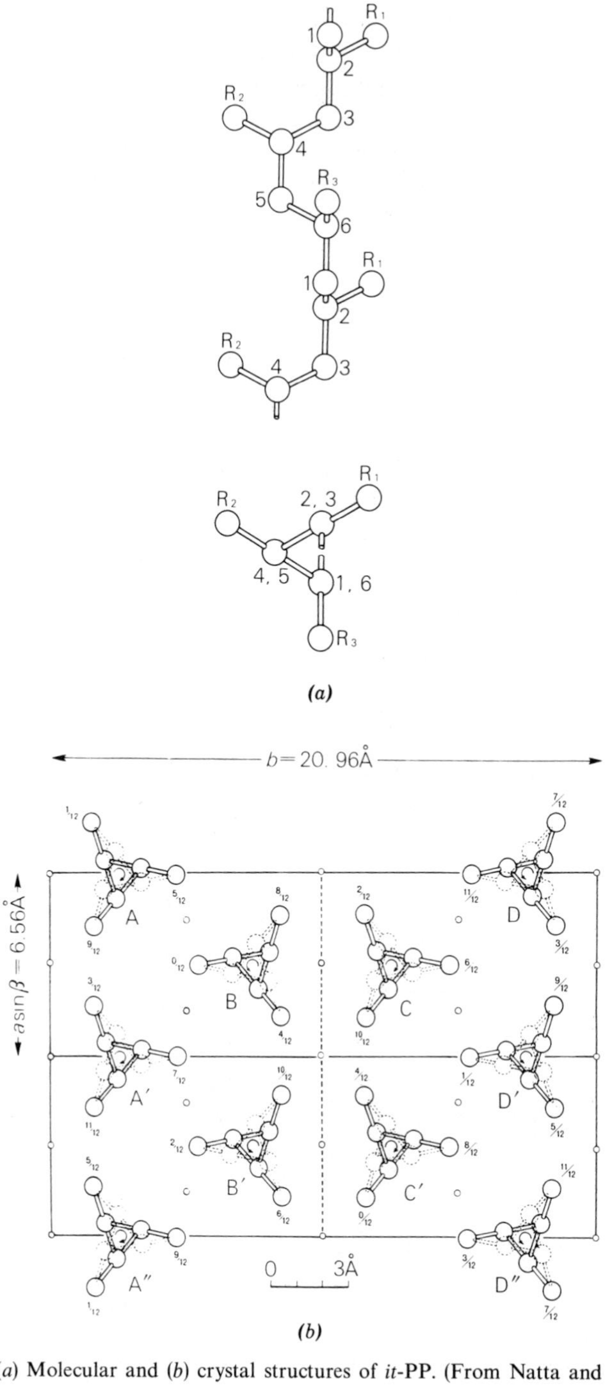

Fig. 4.86 (*a*) Molecular and (*b*) crystal structures of *it*-PP. (From Natta and Corradini.[151])

right-hand helices. The symbol ↻ indicates upward translation with a clockwise rotation, that is, a left-hand helix. Right- or left-hand helices form a sheetlike structure parallel to the *ac* plane. These sheets stack alternately in the direction of the *b* axis. The molecular chain B, indicated by solid lines, is a right-hand helix with up-pointing CH_3 groups. If the space group *Cc* is assumed, there are only two kinds of helices, as indicated by the solid

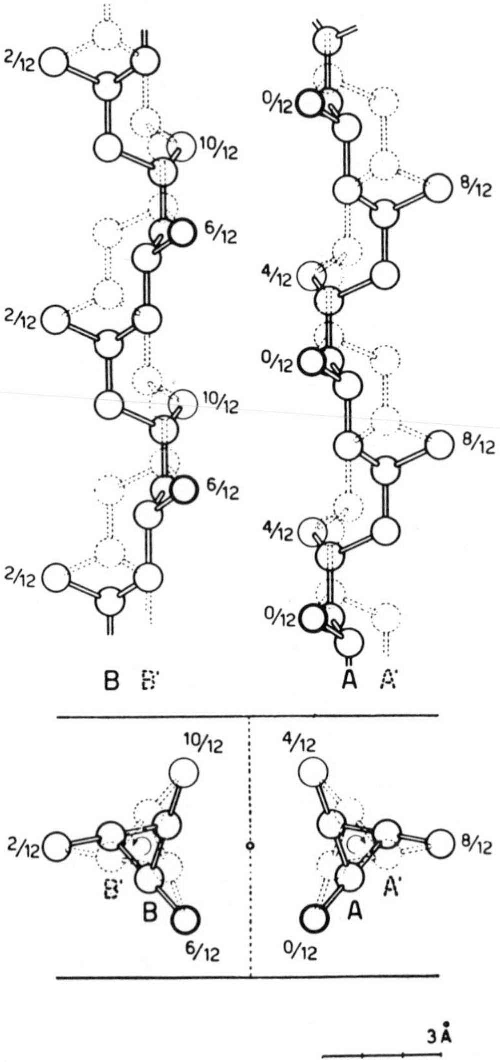

Fig. 4.87 Up- and down-pointing molecules of *it*-PP. (From Natta and Corradini.[153])

lines. The crystallite consists of up- or down-pointing helices only. On the other hand, in the case of $C2/c$, the up- and down-pointing helices (the latter being denoted by dashed lines) coexist at the same lattice positions with the same probability. The choice between the two space groups should be made by comparing the observed structure factors with those calculated for both models. However this problem has not yet been resolved. It may be considered that the structure is locally Cc, but that it is statistically $C2/c$ when it covers a wider range in crystallites. In Fig. 4.87 the up-pointing (solid line) and down-pointing helices (dashed lines) are related by the twofold axes perpendicular to the helix axes.

The crystal structure of *it*-polystyrene presents alternations similar to that of *it*-PP; the space group may be either $R3c\text{-}C_{3v}^{6}$ or $R\bar{3}c\text{-}D_{3d}^{6}$.[74] The crystal structure of a racemic sample of *it*-poly(*tert*-butylethylene oxide) belongs to space group $P\bar{4}n2\text{-}D_{2d}^{8}$ and has a statistically disordered structure comprising up- and down-pointing helices in a ratio of 1 : 1 (Section 4.12.3).[94] On the other hand, *it*-poly(*o*-fluorostyrene) belongs to space group $R3c\text{-}C_{3v}^{6}$ and is a regular structure in which all helices have the same orientation, up- or down-pointing only.[152]

C. *Disorders of Rotation, Translational Displacement, and so forth of Helical Chains* (*Polytetrafluoroethylene*). Figure 4.88 shows the X-ray fiber diagrams of polytetrafluoroethylene taken at 0, 25, and 50°C; each diagram reveals different characteristic features. This polymer has transition points at 19 and 30°C. Below 19°C the crystal structure is triclinic and the molecular conformation is a (13/6) helix.[154] The X-ray pattern taken at 0°C shows strong reflections on the sixth and seventh layer lines, which indicate a subperiod between one-sixth and one-seventh of the true period, namely, 16.88 Å/6.5 = 2.6 Å, which is about the span of one zigzag of a carbon chain. Many sharp reflections appearing on all layer lines indicate a three-dimensional regular crystal structure.

In diagram *b*, taken at 25°C, strong reflections appear on the seventh and eighth layer lines. Above 19°C the fiber period is 19.5 Å and the helix changes from (13/6) to (15/7) by a slight untwisting. The intermolecular distance increases from 5.62 to 5.66 Å. Table 4.14 shows that the diffraction characteristics in the layer lines of the pattern obtained between 19 and 30°C depend on the order of the Bessel function satisfying the selection rule (Eq. 4.67).[155] The fifteenth layer reflections cannot be observed in Fig. 4.88, but they appear in a tilted-fiber photograph. For the interpretation of the transition at 19°C, Klug and Franklin[156] proposed a screw disorder, that is, a combination of chain rotation about the chain axis and translation along the axis. However Clark and Muus[155] pointed out that the screw disorder is not consistent with the observed data, especially the sharp reflections on

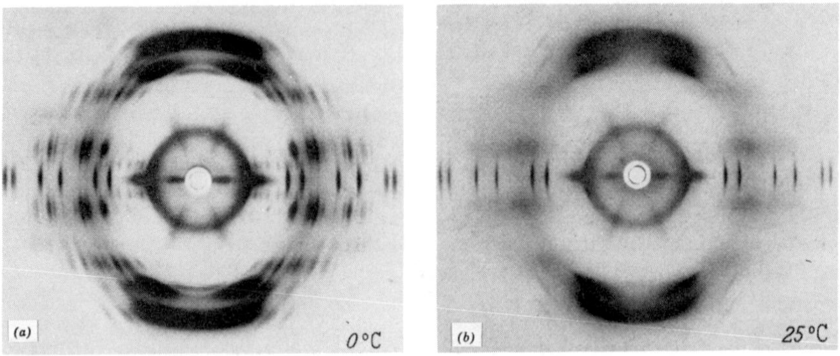

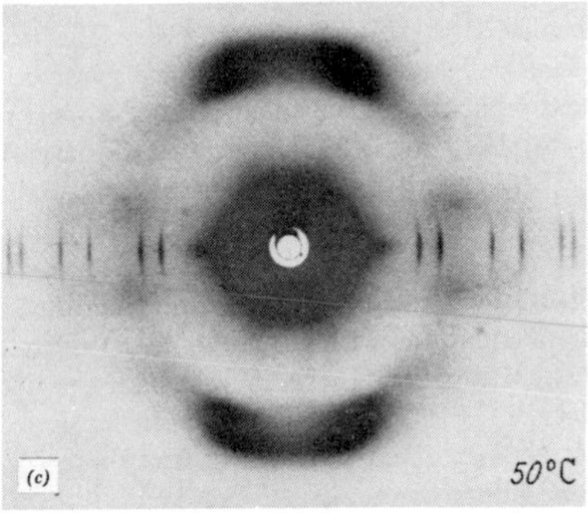

Fig. 4.88 X-Ray fiber diagrams of polytetrafluoroethylene taken at 0, 25, and 50°C. (From Sperati and Starkweather.[154])

Table 4.14 Diffraction Data of Polytetrafluorethylene above the 19°C Transition Point (From Clark and Muus[155])

l	n^a	Characteristic features
0	0	Sharp spots
1	−2	—
2	−4	Continuous distribution
3	−6	Continuous distribution
4	+7, −8	Continuous distribution
5	+5	Continuous distribution
6	+3	Sharp spots and continuous distribution
7	+1	Sharp spots and continuous distribution
8	−1	Sharp spots and continuous distribution
15	0	Sharp spots

[a]Solutions of $l = 7n + 15m$.

the fifteenth layer line. They examined six types of disorder based on the Cochran-Crick-Vand equation (Eq. 4.65): (1) small angular displacements, (2) random angular displacements, (3) small translational displacements, (4) random translational displacements, (5) translational displacements with the distances of integral multiples of the spacing between adjacent residues along the helix axis, and (6) small screw displacements.* From this point of view, the sharp reflections on the equator and the fifteenth layer, where $n = 0$, are interpreted according to case 1, small angular displacements. In case 1, the diffracted intensity I is given by

$$I \cong I_0 \exp(-n^2 \langle \varphi_j^2 \rangle) \tag{4.173}$$

where I_0 is the intensity in the case with no disorder, φ_j is the rotation angle about the chain axis, and j is the numbering of the molecular chain. The reflection intensity is not affected on the layers for which $n = 0$, and it is a function of n^2 and $\langle \varphi_j^2 \rangle$ on the other layers. Continuous streaks appear on some layer lines where the Bragg reflections diminish.

Above the 30°C transition, reflections become diffuse on the layer lines except those for which $n = 0$ (the equator and fifteenth layer), while those on the equator are discrete and sharp up to the melting point at 327°C. These features can be represented by the equation

$$I = \begin{cases} I_0 & n = 0 \\ 0 & n \neq 0 \end{cases} \tag{4.174}$$

This situation corresponds to case 2, where the rotation about the chain axis becomes dominant, but the lateral molecular packing remains in a trigonal form. X-Ray data give a picture of the crystal structure averaged over time and space. Therefore it cannot be determined from the X-ray data alone whether the disorder involves a static misalignment or a special type of molecular motion. For this purpose, broad-line NMR study is useful.[157] For further discussion of the thermal vibrations and diffuse scattering of crystalline polymers with helical molecular conformations see Ref. 158 and for a discussion of a disorder giving rise to streaks along lines of constant ξ in poly(vinylidene fluoride) see Ref. 159.

4.11 ACCURACY OF THE RESULTS OF ANALYSES

As described in Section 4.7.5, the R-factor is used as a rough measure for estimating the agreement between the calculated and observed structure factors. A structure may be considered to be reasonable, if $R \leqslant 20\%$. It has

* These disorders are considered distortions of the first kind, since the displacements are assumed to be symmetrical with respect to the equilibrium positions.

Table 4.15 Examples of *R*-factors in Structure Analyses of Polymers

Polymer	*R*-Factor (%)	Number of independent reflections	Temperature factor B (Å^2)[a]
Polyglycolide[30]	13	118	4–12 (anisotropic)
Polyallene[46]	17.6	31	8.7 (isotropic)
PEO[29]	15.7	205	8.36[b] (isotropic)
it-Poly(propylene sulfide)[161]	13	38	5.0–12.0 (isotropic)
α-Gutta percha[39]	15.9	50	8.56[b] (isotropic)
Poly(pentamethylene sulfide)[162]	8	48	10.0 (isotropic)

[a]A range of values is given for the cases in which atoms have different values of *B*.
[b]A common temperature factor was used for all atoms in the lattice.

been shown[160] that if atoms are distributed randomly in the unit cell, $R =$ 82.2% in the case of a centrosymmetric distribution and 58.6% in the non-centrosymmetric case.

Table 4.15 gives the *R*-factor, the number of independent reflections, and the temperature factor for several polymers. Although the *R*-factor is usually lower than 10% for low-molecular-weight single crystals, it is larger for high polymers. As pointed out earlier in this chapter, compared to the case of low-molecular-weight compounds, it is difficult to obtain data in sufficient quantity and of the required accuracy for polymers. Data of the best quality must be collected using the above-mentioned methods, with special emphasis on preparation of doubly oriented samples, low-temperature measurements, and so on. Nevertheless, a limited amount of X-ray data is a characteristic feature of the study of polymers, and the crystal structure corresponding to the broadened image of the reciprocal lattice is really the object of analyses.

The standard deviation is used for characterizing the accuracy of the parameters obtained by the least-squares method. The standard deviation σ_i for the parameter x_i is given by

$$\sigma_i = \left[\frac{m_{ii} \left(\sum_{m=1}^{M} w_m \Delta F_m^2 \right)}{M - N} \right]^{1/2} \tag{4.175}$$

where m_{ii} is the ith diagonal element of the matrix $(\tilde{\mathbf{M}}\mathbf{M})^{-1}$ in Eq. 4.121 or the inverse matrix of the matrix in Eq. 4.147, $\Delta F_m = |F_m(\text{obs})| - |F_m(\text{calc})|$, w_m is the weight of ΔF_m, M is the number of observed reflections, and N is the number of parameters.[7]

The standard deviation $\sigma(r)$ for the bond distance between atoms A and B,

Table 4.16 Examples of the Accuracy of Analyses of Polymers and Low-Molecular-Weight Single Crystals[a]

Compound	Number of independent reflections	R-factor (%)	Standard deviation
Polyisobutylene[32]	152 (photographic method)	13	Bond lengths: C—C, ~ 0.07 Å, Bond angles: ∠ CCC, ~ 4° Internal rotation angles (skeleton): ~ 6°
PEO[29]	205 (photographic method)	15.7	Internal rotation angles (skeleton): 2.7–4.9°
2,2,4,4-Tetramethylene adipicacid[165]	2005 (counter method)	7.6	Bond lengths (average): 0.005 Å Bond angles (average): 0.2°
1,3,5,7,9-Hexoxane[166] $\overline{\lfloor}$(—CH_2—O—)$_6$$\overline{\rfloor}$	352 (counter method)	5.4	Bond lengths: C—O, 0.002 Å; C—H, 0.02 Å Bond angles: ∠ OCO and ∠ COC, 0.2°; ∠ OCH, 2°; ∠ HCH, 3°

[a]The constrained least-squares method was used for PEO and the usual least-squares method for the others.

whose fractional coordinates are determined independently, is given by

$$\sigma(r) = [\sigma_A^2 + \sigma_B^2]^{1/2} \tag{4.176}$$

where σ_A and σ_B are the standard deviations of the atomic coordinates in the bond direction. If σ is isotropic, there is no selection of direction. For the distance between independent atoms of the same kind,

$$\sigma(r) = 2^{1/2}\sigma_A \tag{4.177}$$

When two atoms are not independent, for instance, when atoms A and B are related by a center of symmetry, then

$$\sigma(r) = 2\sigma_A = 2\sigma_B \tag{4.178}$$

The standard deviation $\sigma(\phi)$ for the bond angle $\phi = \angle \text{ABC}$ is given by

$$\sigma^2(\phi) = \frac{\sigma_A^2}{\text{AB}^2} + \sigma_B^2\left(\frac{1}{\text{AB}^2} - \frac{2\cos\phi}{\text{AB}\cdot\text{BC}} + \frac{1}{\text{BC}^2}\right) + \frac{\sigma_C^2}{\text{BC}^2} \tag{4.179}$$

if atoms A, B, and C are independent.[163] For the standard deviation of the internal rotation angles, see Ref. 164.

The standard deviation σ is so defined that the area under a Gaussian distribution curve in the range of $p_0 \pm \sigma$ is 0.683 times the whole area, p_0 being the true value of the parameter. Since standard deviation in X-ray analysis is based on the formula $\Delta F_m = |F_m(\text{obs})| - |F_m(\text{calc})|$, as shown in Eq. 4.175, it should be noted that in using the results of the structure analysis of polymers the F(obs) are much smaller and less accurate compared to those on low-molecular-weight compounds. In other words, the results of structure analyses reveal that in the crystalline regions, polymers always contain some amount of disorder and are imperfectly arrayed. Table 4.16 compares the accuracy of structure analyses on polymers and on low-molecular-weight compounds.

4.12 EXAMPLES OF ANALYSIS

4.12.1 Poly(ethylene Terephthalate)

The reflection indexing of PET was difficult since this polymer belongs to the triclinic system and also because of tilting crystallites.* Bunn et al.[68] accomplished the indexing using diffraction photographs showing the tilting. In this type of diffraction pattern, obtained from fibers annealed at 210°C

* PET samples free from tilting can be prepared by high speed elongation. The direction and angle of tilting can be varied according to the preparation condition.[167]

and allowed to contract freely, the reflections are displaced, some up and some down, by varying amounts from the normal layer line positions. In the sample the ($\bar{2}30$) plane remained parallel to the elongation axis and the c axis remained inclined to the elongation axis. Prior to this study, Astbury and Brown[168] proposed a triclinic cell of $a = 5.54$ Å, $b = 4.14$ Å, $c = 10.86$ Å, $\alpha = 107°5'$, $\beta = 112°24'$, and $\gamma = 92°23'$ including one molecular chain, but did not make further analysis. Bunn et al. observed 72 independent reflections utilizing the above-mentioned method and indexed them satisfactorily on the basis of the unit cell with $a = 4.56$ Å, $b = 5.94$ Å, $c = 10.75$ Å, $\alpha = 98.5°$, $\beta = 118°$, and $\gamma = 112°$. The length of the c axis, which can be found only approximately from normal fiber diagrams, was determined more accurately from photographs taken by oscillating the fiber axis over an angular range appropriate for the reflection planes nearly normal to the fiber axis, such as ($\bar{1}05$) and ($\bar{2}\bar{1}9$). The tilt displacements of the reflections from normal layer lines are shown schematically in Fig. 4.89. This feature is identical to a rotation photograph of a single crystal whose principal axis is inclined at a few degrees to the rotation axis. For such a case, the method developed previously by Bunn et al.[169] can be applied. Figure 4.90 shows the third layer reciprocal lattice net for tilted PET. The origin of the reciprocal lattice is O_3, which must be imagined as being below the plane of the paper. If the lattice were tilted about a line through O_3 in the basal plane, some of the point on the third layer would move downwards below the paper and others would move upwards. The reflection $0\bar{1}3$ was found to be practically undisplaced from its normal position on the photograph. The reciprocal lattice point $0\bar{1}3$ does not move appreciably up or down. Therefore the lattice must be tilted about a line T_3T_3 almost directly underneath point $0\bar{1}3$ and passing through O_3. Since $\bar{1}\bar{1}3$ is displaced downwards and 003 upwards, the lattice must be tilted over to the right. If the lattice and indexing are correct, all $hk3$ reflections represented by points to the right of T_3T_3 should be displaced downwards and the rest upwards. The reciprocal lattice coordi-

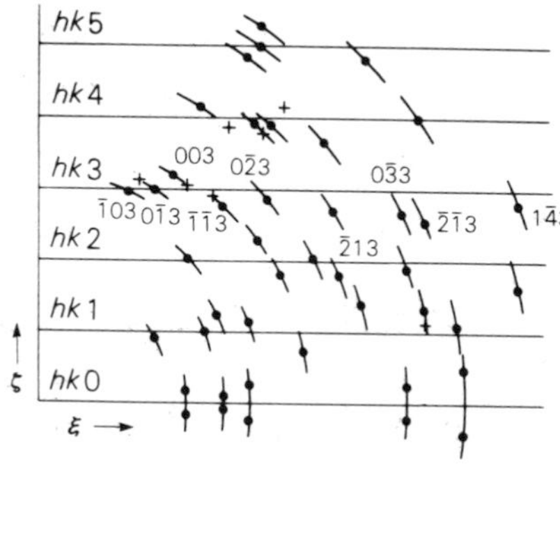

Fig. 4.89 Reciprocal lattice diagram of PET showing tilt displacements. Lines show observed position of reflections, dots and crosses show the calculated positions for Bunn's and Astbury's unit cells, respectively. (From Daubeny et al.[68])

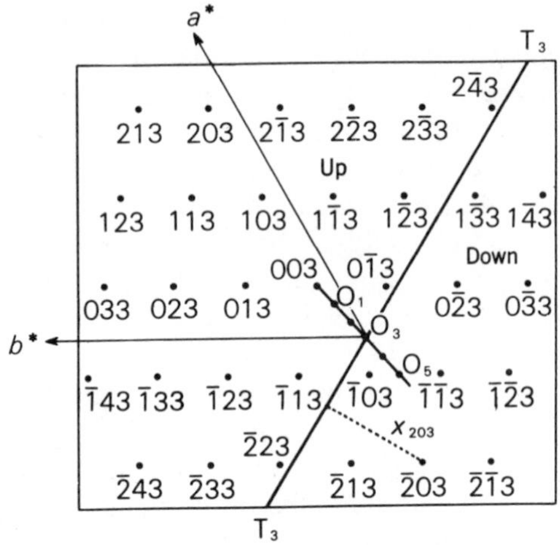

Fig. 4.90 Reciprocal lattice net for *c* projection, illustrating calculation of tilt displacements (third layer). a^* and b^* indicate the axes on the zeroth layer of reciprocal lattice. (From Daubeny et al.[68])

nate ζ (the ζ axis coincides with the elongation direction) is given by

$$\zeta = z \cos \phi - x \sin \phi \tag{4.180}$$

where z is the normal untilted height of the reciprocal lattice layer, ϕ is the angle of tilt, and x is the distance in the diagram of the point from the tilt line $T_3 T_3$ (Fig. 4.91). The same diagram can be used for other layers if alternative origins O_1, O_2, and so on are used. The tilt line for any layer is parallel to $T_3 T_3$ and passes through the origin of that layer. If ϕ is calculated for one reflection by trial, values of ζ can be calculated for all others. Thus the indexing is confirmed. The results of tests of both indexings by this method are shown in Fig. 4.89. The crosses and dots indicate the results of

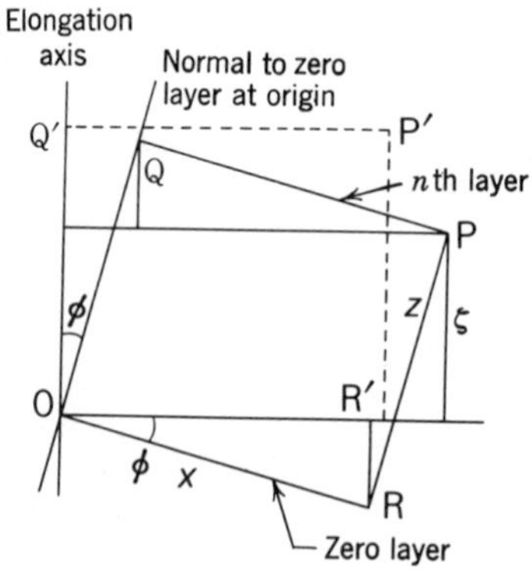

Fig. 4.91 Tilted reciprocal lattice ($OQPR$). $OQ'P'R'$ indicates the untilted reciprocal lattice. (From Bunn et al.[169])

Astbury and Bunn, respectively, and the latter reveal quite reasonable agreement with the observed dıagram. This may be better understood by comparing the indices of the third layer line given in Fig. 4.89 with those in Fig. 4.90. The tilt direction is shown for the real lattice in Fig. 4.92 which is such that the ($\bar{2}30$) plane remains parallel to the elongation direction and the inclination of the (001) plane to the elongation axis increases. The reason for

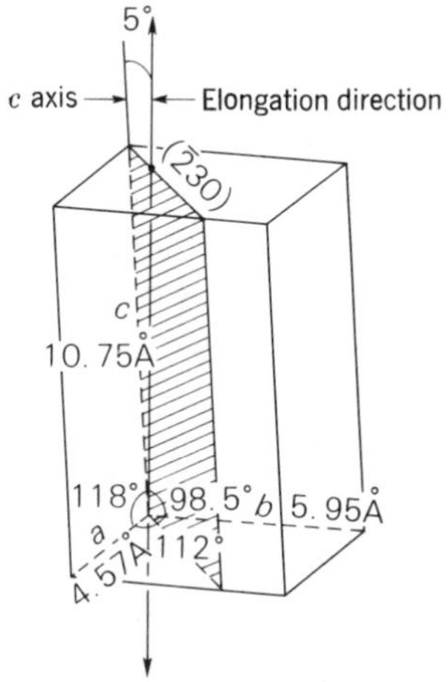

Fig. 4.92 Unit cell of PET and direction of tilt. (From Daubeny et al.[68])

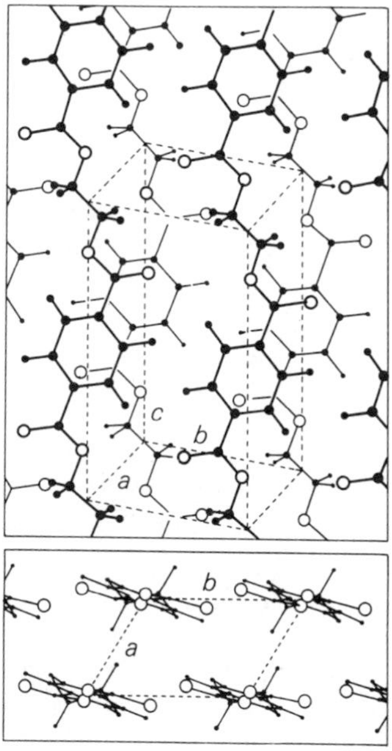

Fig. 4.93 Crystal structure of PET. (From Daubeny et al.[68])

the tilt is not clear at present. Overlapped reflections on the same layer line can be resolved if a tilted sample is prepared. Therefore it would be worthwhile to try to prepare specimens showing this type of orientation. Bunn et al. determined the crystal structure of PET (Fig. 4.93) by carrying out a three-dimensional Fourier synthesis using the intensity data resolved by the tilt method.

4.12.2 Poly(ethylene Oxide)

A uniform (7/2) helical molecular conformation of PEO was proposed from X-ray diffraction data and from infrared and Raman spectroscopic methods, as is mentioned in Section 4.6.5. The study of the second stage[29] was based on more accurate X-ray data obtained with a cylindrical vacuum camera with a diameter 200 mm (Section 4.4.2.A). The spacings were calibrated with respect to aluminum powder. The intensities of 205 independent reflections could be measured, and the integrated intensities of 25 strong reflections were measured by a scintillation counter. The lattice constants were redetermined to be $a = 8.05$ Å, $b = 13.04$ Å, c (fiber axis) $= 19.48$ Å, and $\beta = 125.4°$. The most probable space group was considered to be $P2_1/a\text{-}C_{2h}^5$ from the systematic absences, $h =$ odd for $h0l$ and $k =$ odd for $0k0$. In this space group a molecular chain consisting of seven monomeric units corresponds to an asymmetric unit.

For the space group symmetry $P2_1/a\text{-}C_{2h}^5$ and the uniform helical molecular symmetry isomorphous to point group D_7, Eq. 4.85 is rewritten as follows* for $h + k =$ even:

$$F(hkl) = 4\sum_n G_{n,l}(R)\cos\left[n\left(\mp\phi_q + \frac{\pi}{2}\right) + 2\pi(hx_q + lz_q)\right]\cos(\pm n\psi + 2\pi k y_q) \tag{4.181}$$

and for $h + k =$ odd:

$$F(hkl) = -4\sum_n G_{n,l}(R)\sin\left[n\left(\pm\phi_q + \frac{\pi}{2}\right) + 2\pi(hx_q + lz_q)\right] \times \sin(\pm n\psi + 2\pi k y_q) \tag{4.182}$$

where x_q and y_q are the fractional coordinates of the qth molecular axis, z_q is the fractional coordinate (height) of the origin (the first oxygen atom)

* If the molecular chain at the fractional coordinates (x_q, y_q, z_q) has an azimuthal angle ϕ_q and is a right-hand helix, the azimuthal angle and helix sense should be as follows according to the space group symmetry: (the chain at $-x_q, -y_q, -z_q$), $\phi_q + \pi$ and left hand; (the chain at $\frac{1}{2} + x_q, \frac{1}{2} - y_q, z_q$), $-\phi_q$ and left hand; and (the chain at $\frac{1}{2} - x_q, \frac{1}{2} + y_q, -z_q$), $-\phi_q + \pi$ and right hand. Equations 4.181 and 4.182 are obtained if the summation of Eq. 4.85 is carried out for the four molecular chains taking into account the double signs.

of the qth molecule, and ϕ_q is the rotation angle of the atom at the origin with respect to the ac plane. The upper and lower signs are to be used for right- and left-hand helices, respectively, at the position (x_q, y_q, z_q). $G_{n,l}(R)$ is denoted by the following equation if the hydrogen atoms are not taken into account:

$$G_{n,l}(R) = 7f_O J_n(2\pi R r_O) + 14 f_C J_n(2\pi R r_C) \cos\left(-n\phi_C + \frac{2\pi l Z_C}{c}\right) \quad (4.183)$$

$$l = 2n + 7m \qquad m = 0, \pm 1, \pm 2, \cdots \quad (4.184)$$

If only the lowest order Bessel function obtained from Eq. 4.184 is taken into account for each layer, the summation with respect to n in Eqs. 4.181 and 4.182 can be elminated.

From the molecular parameters reported in the previous paper ($r_O = 0.527$ Å, $\phi_O = 0.0°$, $Z_O = 0.0$Å, $r_C = 1.569$ Å, $\phi_C = 30.6°$, and $Z_C = 0.854$ Å),[104] $G_{n,l}(R)$ values were calculated according to Eq. 4.183. The determination of the molecular position was attempted by using Eqs. 4.181 and 4.182. First the molecular position ($x_q = 0.225$, $y_q = 0.131$ or 0.396, and $z_q = 0.016$ or -0.016) was obtained by calculating the structure factors for the reflections on the equator and the seventh layer, starting from the values given by the Patterson synthesis. Subsequently, by using the reflections on the fourth layer, $y_q = 0.369$ and $z_q = -0.016$ were determined for the right-hand helix molecule. Two values (166 and 346°) for ϕ_q remained possible from the second layer data. These two models are denoted by model A ($x_q = 0.225$, $y_q = 0.369$, $z_q = -0.016$, $\phi_q = 166°$) and model B (x_q, y_q, and z_q are the same as those of model A, and for this case $\phi_q = 346°$). The calculated structure factors for these two models gave poor agreement with the observed ones for the reflections on the first, third, and fifth layers.

Here the least-squares method for helical symmetry (Section 4.9) was applied, but the results gave unsatisfactory agreement; the R-factor is 34.6 and 37.4% for the results based on models A and B, respectively. These unsatisfactory results of the refinement were ascribed to distortion of the molecule from the exact helical symmetry $D(4\pi/7)$. When the usual least-squares method was applied to PEO, the number of variables became greater than 65 if only the skeletal atoms were taken into account. Therefore the constrained least-squares method[39,133] (Section 4.9.2) in which the bond lengths and bond angles were fixed, was applied to refine the crystal structure. Then the number of adjustable parameters was reduced to 28, and refinement starting from model A converged to an R-factor of 15.7%. Model B was not improved significantly (34.1%). The crystal structure is shown in Fig. 4.94, and the molecular structure is shown in Fig. 4.95 along with the starting molecular model having uniform helical symmetry. Examination of the

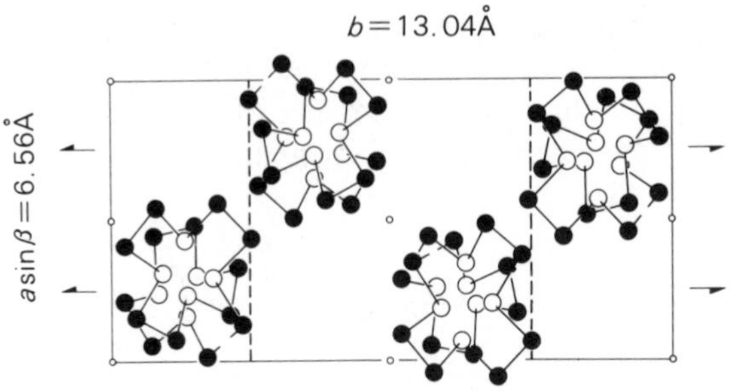

Fig. 4.94 Crystal structure of PEO.[29]

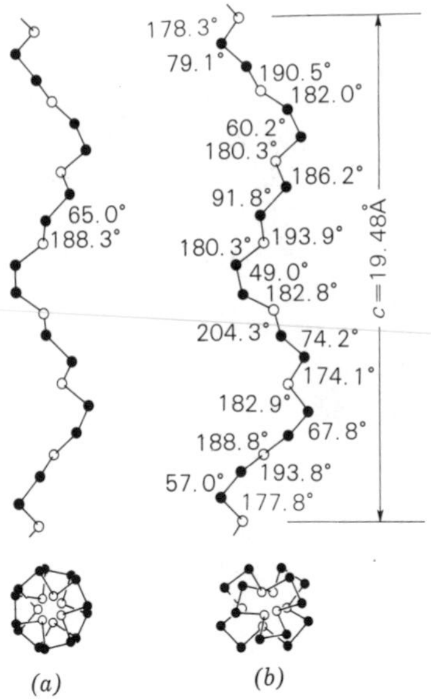

Fig. 4.95 (*a*) Molecular model of PEO with the uniform helical symmetry and (*b*) newly determined molecular structure. Internal rotation angles are given.[29]

various sections of the unit cell normal to the *c* axis confirmed that the molecules are well packed and that they are neither unreasonably close nor too far apart. Therefore the structure shown here is considered to be the most reasonable one at present. The internal rotation angles are considerably distorted from the uniform helix, but the conformation is essentially the (7/2) helix and consists nearly of *TTG* sequences. The average values of the internal rotation angles about the C—O and C—C bonds are 186.0 and 68.4°, respectively. Such a distortion from the uniform helix was subsequently found to be due to intermolecular interactions from a consideration of packing energy minimization method (Section 6.5).*

* The satellite bands appearing in the infrared spectra especially at low temperatures[170] may be due to distortion from the uniform helix.

4.12.3 Isotactic Poly(*tert*-butylethylene Oxide)

As described previously two models were chosen from the result of an intramolecular interaction energy calculation for form I of *it*-poly(*tert*-butylethylene oxide) (Sections 4.6.2.D and 4.6.3).[94] The sample used here was the benzene-insoluble part of the polymerization product of (*RS*)-*tert*-butylethylene oxide with diethylzinc–water (1 :0.7) catalyst. The molecular structure factors were calculated for both models and compared with the observed X-ray reflection data, but it was difficult to determine which model is more reasonable (Fig. 4.96). The results of calculations of normal vibrations (skeletal models) could not provide useful information in this case.

Therefore the unit cell structure factor using the Cochran-Crick-Vand equations (Eq. 4.85) was calculated for both models. Systematic absences, *hk*0 when *h* or *k* is odd, were observed in the equatorial reflections, as shown in Fig. 4.96. This suggests that the lattice with half of the axial length *a* is sufficient for examining the *c* projection and that the origins of the four helical chains can be located in the unit cell having the relation of I, II, III, and IV as shown in Fig. 4.97. Characteristic systematic absences, *h*0*l* when $h + l$ is odd, 0*kl* when $k + l$ is odd, and *hk*9 when *h* or *k* is even, were observed.

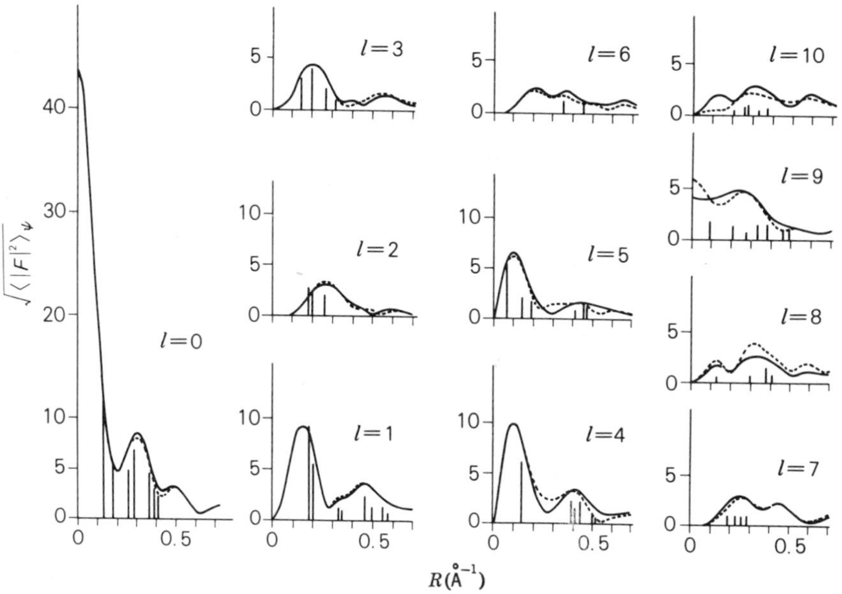

Fig. 4.96 Comparison between the square roots of the observed relative intensities and the calculated molecular structure factors for models I and II: (————) calculated value of model I; (------) calculated value of model II; (vertical rod) observed value.[94]

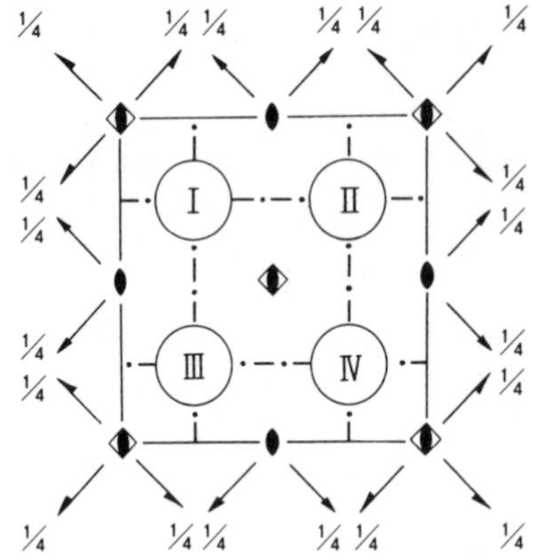

Fig. 4.97 Packing of helices for the space group $P\bar{4}n2 - D_{2d}{}^{8}$.[94]

For example, the 009 reflection was not observed, although the 0018 reflection could be observed clearly. The systematic absences for the reflections $h0l$ and $0kl$ suggest the existence of n glide planes indicated by—·—·—· in Fig. 4.97. Consequently the chains at II and III should be displaced by $c/2$ along the c axis from the chains at I and IV.

Next the unit cell structure factor for the ninth layer was calculated. Only the zeroth order Bessel function was taken into account for the ninth layer line, since the contribution of Bessel functions with the orders $|n| \geqslant 9$ is negligibly small in the present sample. The possible space group satisfying the systematic absences of the equatorial and ninth layer lines is $P4_2\text{-}C_4{}^3$, which belongs to the tetragonal system with four general equivalent positions. In the case of this space group, the intensities of the observable $hk9$ reflections ($h = $ odd and $k = $ odd) can be expressed by

$$I(hk9) = \left[4 \sum_{j=1}^{7} f_j J_0(2\pi R r_j) \cos\left(\frac{18Z_j}{c}\right)\right]^2 + \left[4 \sum_{j=1}^{7} f_j J_0(2\pi R r_j) \sin\left(\frac{18Z_j}{c}\right)\right]^2 \tag{4.185}$$

Accordingly, for this space group the unit cell structure factor per chain of the observable $hk9$ reflections should be almost identical to the molecular structure factor $[\langle |F_M(R, \psi, 9/c)|^2 \rangle_\psi]^{1/2}$. But agreement between the calculated molecular structure factors and the observed values is not so satisfactory, as shown in Fig. 4.96.

Here the tetragonal space groups with more than four general equivalent positions were examined in detail, taking into account the additional systematic absences $h0l$ when $h + l = $ odd and $0kl$ when $k + l = $ odd. In consequence the acceptable space group was found to be $P\bar{4}n2\text{-}D_{2d}{}^8$. The symmetry elements of the space group are shown in Fig. 4.97. This space group has eight general equivalent positions, but the molecular axes should be at positions I–IV given in Fig. 4.97. One problem remains: although the molecular chain has a polar nature, twofold axes perpendicular to the molecular axes exist. Therefore a statistically disordered structure is considered where pairs of the upward and downward isomorphous helices correlated by the twofold axis must exist at definite positions with a probability of $\frac{1}{2}$. For this space group, the intensities of the $hk9$ reflections with $h = \text{odd}$ and $k = \text{odd}$ can be expressed by

$$I(hk9) = \left\{ 4 \sum_{j=1}^{7} f_j J_0(2\pi R r_j) \sin\left[\frac{18\pi(Z_O + Z_j)}{c}\right]\right\}^2 \tag{4.186}$$

where Z_O is the height of the origin oxygen atom from the origin of the unit cell. The value of Z_O was adjusted by trial so as to give good agreement between the calculated and observed intensities of the reflections, especially on the ninth layer line.

At this stage of the study the unit cell structure factors for the space group $P\bar{4}n2\text{-}D_{2d}^{8}$ were calculated by Eq. 4.85 for models I and II. In these calcula-

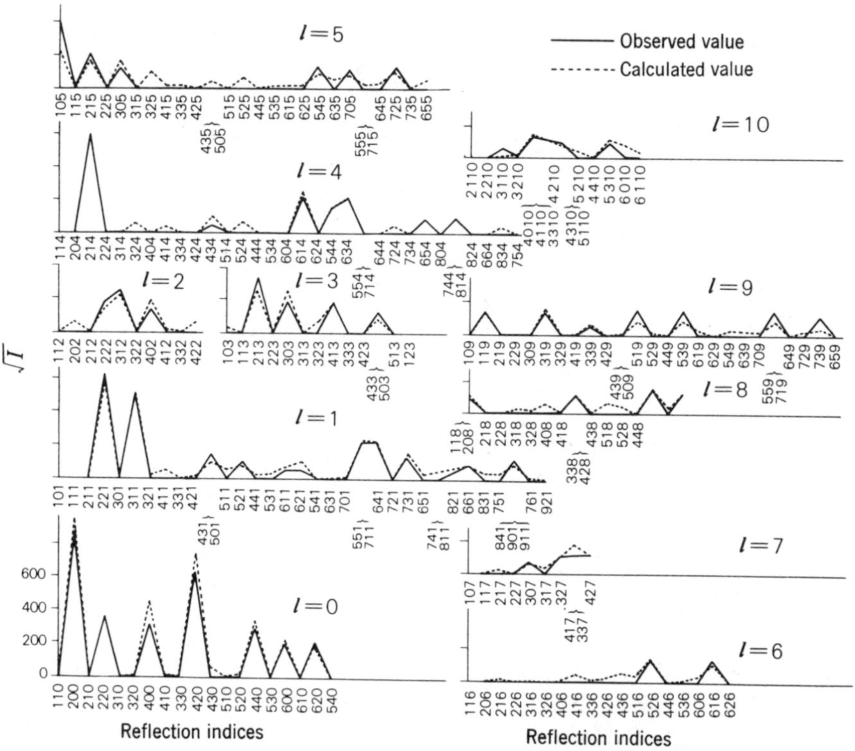

Fig. 4.98 Comparison between the observed (————) and calculated structure factors (------) of *it*-poly (*tert*-butylethyene oxide).[94]

tions the rotation angle ϕ_O, which is defined as the angle between the a axis and the radial direction of the origin oxygen atom, was examined in the vicinity of Z_O already determined. The contribution of the hydrogen atoms was not taken into account.The R-factor of the observed independent 57 reflections was found to be 19% for model I and 27% for model II. Refinement by the least-squares method on the basis of helical symmetry reduced the R-factor to 16% for model I, but no apparent reduction could be found for model II. Accordingly, model I may be considered to be an acceptable molecular model. The comparison between the observed and calculated structure factors is shown in Fig. 4.98.* The isotropic temperature factor was 9.6 Å^2 for all atoms. The crystal structure is shown in Fig. 4.99. The molecular chains at the upper-left and lower-right positions are right-hand S helices. Those in the upper-right and lower-left positions are left-hand R helices. In the figure only the upward helices are shown for simplicity, although the crystal structure is considered to be statistically disordered, consisting

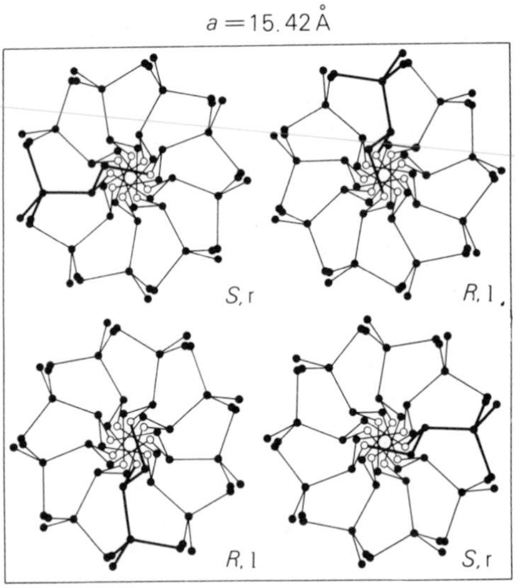

Fig. 4.99 Crystal structure of *it*-poly (*tert*-butylethylene oxide) form I. Only the upward helices are illustrated for simplicity; R, rectus; S, sinister; r, right hand; l, left hand.[94]

* In the space group $P\bar{4}n2\text{-}D_{2d}^{8}$ there are no special systematic absences for the $hk0$ and $hk9$ reflections. But for the present sample the reflections $hk0$ are observed only for the cases in which both h and k are even, and the reflections $hk9$ are observed only when both h and k are odd. The calculated intensities of the reflections $hk0$ and $hk9$ are as follows: zero for $hk0$ with both h and k odd, zero for $hk9$ with both h and k even, and not zero but weak for $hk0$ and $hk9$ with one of h and k odd and the other even. The reflections of the last type may be considered to be nonobservable owing to their weak intensity.

of a random mixture of upward and downward helices in a 1 : 1 ratio. In this crystal structure, although the positions of the main-chain atoms belonging to the upward and downward helices are very different, the side groups occupy nearly the same positions in the lattice, resulting in a good overall packing of molecules. The chain packing appears to be governed essentially by intermolecular interactions among the side groups.

This structure analysis definitely shows the formation of two types of isotatic polymers from the mixture of *R* and *S* monomers in a 1 : 1 ratio, suggesting a stereoselective mechanism of polymerization in the aforementioned catalytic system.

REFERENCES

1. H. Tadokoro, T. Yoshihara, Y. Chatani, and S. Murahashi, *J. Polym. Sci.*, **B2,** 363 (1964).
2. N. F. M. Henry and K. Lonsdale, Eds., *International Tables for X-Ray Crystallography, Vol. I. Symmetry Groups*, 2nd ed., Kynoch Press, Birmingham, 1965.
3. J. S. Kasper and K. Lonsdale, Eds., *International Tables for X-Ray Crystallography, Vol. II. Mathematical Tables*, Kynoch Press, Birmingham, 1959.
4. C. H. MacGillavry, G. D. Rieck, and K. Lonsdale, Eds., *International Tables for X-Ray Crystallography, Vol. III. Physical and Chemical Tables*, Kynoch Press, Birmingham, 1962.
5. J. A. Ibers and W. C. Hamilton, Eds., *International Tables for X-Ray Crystallography, Vol. IV. Revised and Supplementary Tables to Volumes II and III*, Kynoch Press, Birmingham, 1974.
6. R. W. James, *The Optical Principles of the Diffraction of X-Rays, The Crystalline State, II*, L. Bragg, Ed., G. Bell & Sons, London, 1954.
7. H. Lipson and W. Cochran, *The Determination of Crystal Structures, The Crystalline State, III*, 3rd revised ed., L. Bragg, Ed., G. Bell & Sons, London, 1966.
8. C. W. Bunn, *Chemical Crystallography, An Introduction to Optical and X-Ray Methods*, 2nd ed., Oxford University Press, 1961.
9. M. J. Buerger, *Crystal-Structure Analysis*, Wiley, New York, 1960.
10. S. C. Nyburg, *X-Ray Analysis of Organic Structures*, Academic, New York, 1961.
11. R. Hosemann and S. N. Bagchi, *Direct Analysis of Diffraction by Matter*, North-Holland, Amsterdam, 1962.
12. A. Guinier, *Théorie et Technique de la Radiocristallographie*, 3rd ed., Dunod, Paris, 1964.
13. B. E. Warren, *X-Ray Diffraction*, Addison-Wesley, Massachusetts, 1969.
14. M. M. Woolfson, *An Introduction to X-Ray Crystallography*, Cambridge University Press, London, 1970.
15. H. P. Klug and L. E. Alexander, *X-Ray Diffraction Procedures for Polycrystalline and Amorphous Materials*, 2nd ed., Wiley-Interscience, New York, 1974.
16. B. K. Vainshtein, *Diffraction of X-Rays by Chain Molecules*, Elsevier, Amsterdam, 1966.
17. L. E. Alexander, *X-Ray Diffraction Methods in Polymer Science*, Wiley-Interscience, New York, 1969.
18. M. Kakudo and N. Kasai, *X-Ray Diffraction by Polymers*, Kodansha, Tokyo, Elsevier, Amsterdam, 1972.

19. J. D. Bernal, *Proc. Roy. Soc. (Lond.)*, **A113**, 117 (1926).
20. S. Sasaki, Y. Takahashi, and H. Tadokoro, *J. Polym. Sci., Polym. Phys. Ed.*, **10**, 2363 (1972).
21. Y. Takahashi, I. Sumita, and H. Tadokoro, *J. Polym. Sci., Polym. Phys. Ed.*, **11**, 2113 (1973).
22. Y. Chatani, K. Kitahama, H. Tadokoro, T. Yamauchi, and Y. Miyake, *J. Macromol. Sci., Phys.*, **4**, 61 (1970).
23. N. Norman, *Acta Crystallogr.*, **7**, 462 (1954).
24. H. A. Levy and R. D. Ellison, *Acta Crystallogr.*, **13**, 270 (1960).
25. W. Cochran, *J. Sci. Instrum.*, **25**, 253 (1948).
26. C. W. Bunn and E. V. Garner, *Proc. Roy. Soc. (Lond.)*, **A189**, 39 (1947).
27. D. R. Holmes, C. W. Bunn, and D. J. Smith, *J. Polym. Sci.*, **17**, 159 (1955).
28. Y. Kinoshita, *Makromol. Chem.*, **33**, 1 (1959).
29. Y. Takahashi and H. Tadokoro, *Macromolecules*, **6**, 672 (1973).
30. Y.Chatani, K. Suehiro, Y. Okita, H. Tadokoro, and K. Chujo, *Makromol. Chem.*, **113**, 215 (1968).
31. M. Yokouchi, Y. Chatani, H. Tadokoro, and H. Tani, *Polym. J.*, **6**, 248 (1974).
32. T. Tanaka, Y. Chatani, and H. Tadokoro, *J. Polym. Sci., Polym. Phys. Ed.*, **12**, 515 (1974).
33. H. Tadokoro, Y. Chatani, H. Kusanagi, and M. Yokoyama, *Macromolecules*, **3**, 441 (1970).
34. H. Tadokoro, Y. Nishiyama, S. Nozakura, and S. Murahashi, *Bull. Chem. Soc. Jpn.*, **34**, 381 (1961).
35. S. Murahashi, S. Nozakura, and H. Tadokoro, *Bull. Chem. Soc. Jpn.*, **32**, 534 (1959).
36. T. Miyashita, M. Yokouchi, Y. Chatani, and H. Tadokoro, Annual Meeting of the Society of Polymer Science, Japan, Tokyo, Preprint p. 453, 1974.
37. A. Turner-Jones and C. W. Bunn, *Acta Crystallogr.*, **15**, 105 (1962).
38. A. S. Ueda, Y. Chatani, and H. Tadokoro, *Polym. J.*, **2**, 387 (1971).
39. Y. Takahashi, T. Sato, H. Tadokoro, and Y. Tanaka, *J. Polym. Sci., Polym. Phys. Ed.*, **11**, 233 (1973).
40. H. Tadokoro, K. Tatsuka, and S. Murahashi, *J. Polym. Sci.*, **59**, 413 (1962).
41. W. O. Statton and G. M. Godard, *J. Appl. Phys.*, **28**, 1111 (1957).
42. H. Tadokoro, S. Seki, I. Nitta, and R. Yamadera, *J. Polym. Sci.*, **28**, 244 (1958).
43. H. Arimoto, *J. Polym. Sci.*, **A2**, 2283 (1964).
44. Y. Takahashi, H. Tadokoro, and Y. Chatani, *J. Macromol. Sci., Phys.*, **2**, 361 (1968).
45. H. Kakida, D. Makino, Y. Chatani, M. Kobayashi, and H. Tadokoro, *Macromolecules*, **3**, 569 (1970).
46. H. Tadokoro, Y. Takahashi, S. Otsuka, and F. Imaizumi, *J. Polym. Sci. B*, **3**, 697 (1965).
47. H. Tadokoro, Y. Takahashi, Y. Chatani, and H. Kakida, *Makromol. Chem.*, **109**, 96 (1967).
48. Y. Takahashi, Y. Osaki, and H. Tadokoro, *J. Polym. Sci., Polym. Phys. Ed.*, to be published.
49. R. Hasegawa, Y. Tanabe, M. Kobayashi, H. Tadokoro, A. Sawaoka, and N. Kawai, *J. Polym. Sci. A-2*, **8**, 1073 (1970).
50. R. Hasegawa, M. Kobayashi, and H. Tadokoro, *Polym. J.*, **3**, 591 (1972).
51. J. H. Griffith and B. G. Rånby, *J. Polym. Sci.*, **44**, 369 (1960).
52. T. Uchida and H. Tadokoro, *J. Polym. Sci. A-2*, **5**, 63 (1967).
53. K. Suehiro, Y. Chatani, and H. Tadokoro, *Polym. J.*, **7**, 352 (1975).

54. Y. Chatani, S. Nakatani, and H. Tadokoro, *Macromolecules*, **3,** 481 (1970).
55. Y. Chatani, T. Takahagi, T. Kusumoto, and H. Tadokoro, *J. Polym. Sci., Polym. Phys. Ed.*, to be published.
56. Shōdō Kobayashi, H. Tadokoro, and Y. Chatani, *Makromol. Chem.*, **112,** 225 (1968).
57. G. Carazzolo and M. Mammi, *J. Polym. Sci. A*, **1,** 965 (1963).
58. R. Iwamoto, Y. Saito, H. Ishihara, and H. Tadokoro, *J. Polym. Sci. A-2*, **6,** 1509 (1968).
59. M. Yokoyama, H. Ishihara, R. Iwamoto, and H. Tadokoro, *Macromolecules*, **2,** 184 (1969).
60. Y. Chatani, Y. Ueda, and H. Tadokoro, *Rep. Prog. Polym. Phys. Jpn.*, **20,** 179 (1977).
61. H. J. Goldschmidt, *High Temperature X-Ray Diffraction Techniques*, International Union of Crystallography, N. V. A. Oosthoek's Uitgevers Mij, Utrecht, 1964.
62. B. Post, *Low-Temperature X-Ray Diffraction*, International Union of Crystallography, N. V. A. Oosthoek's Uitgevers Mij, Utrecht, 1964.
63. T. Ito and H. Marui, *Polym. J.*, **2,** 768 (1971).
64. C. Nakafuku and T. Takemura, *Jpn. J. Appl. Phys.*, **14,** 599 (1975).
65. H. D. Flack, *J. Polym. Sci. A-2*, **10,** 1799 (1972).
66. H. Tadokoro, M. Kobayashi, Koh Mori, Y. Takahashi, and S. Taniyama, *J. Polym. Sci. C*, **22,** 1031 (1969).
67. H. Kakida, Y. Chatani, and H. Tadokoro, *J. Polym. Sci., Polym. Phys. Ed.*, **14,** 427 (1976).
68. R. de P. Daubeny, C. W. Bunn, and C. J. Brown, *Proc. Roy. Soc. (Lond.)*, **A226,** 531 (1954).
69. M. R. Lipkin, J. A. Davison, W. T. Harvey, and S. S. Kurtz, Jr., *Ind. Eng. Chem. Anal. Ed.*, **16,** 55 (1944).
70. S. Sasaki, Y. Takahashi, and H. Tadokoro, *Polym. J.*, **4,** 172 (1973).
71. L. Pauling, *The Nature of the Chemical Bond*, 3rd ed., Cornell University Press, Ithaca, 1960.
72. L. Pauling, R. B. Corey, and H. R. Branson, *Proc. Natl. Acad. Sci. U.S.A.*, **37,** 205 (1951).
73. C. W. Bunn, *Trans. Faraday Soc.*, **35,** 482 (1939).
74. G. Natta, P. Corradini, and I. W. Bassi, *Nuovo Cimento, Suppl.*, **15,** 68 (1960).
75. *Tables of Interatomic Distances and Configuration in Molecules and Ions*, Special Publication No. 11 (1958); *Supplement*, Special Publication No. 18, The Chemical Society, London, 1965.
76. O. Kennard, D. G. Watson, F. H. Allen, N. W. Isaacs, W. D. S. Motherwell, R. C. Pettersen, and W. G. Town, Eds., *Molecular Structure and Dimensions*, Vol. A1, *Interatomic Distances 1960–1965, Organic and Organometallic Crystal Structures*, Crystallographic Data Centre Cambridge, International Union of Crystallography, N. V. A. Oosthoek's Uitgevers Mij, Utrecht, 1972.
77. P. P. Ewald and C. Herman, Eds., *Strukturbericht*, Bd. 1–7, Akademische Verlagsgesellschaft, Leipzig, 1913–1939.
78. International Union of Crystallography, Ed., *Structure Reports*, Vols. 8–27, N. V. A. Oosthoek's Uitgevers Mij, Utrecht, 1940–1962.
79. R. W. G. Wyckoff, Ed., *Crystal Structures*, Vols. I–V, Interscience, New York, 1948–1966.
80. J. D. H. Donnay and H. M. Ondik, Eds., *Crystal Data Determination Tables*, Vol. I. *Organic Compounds*, 3rd ed., American Crystallographic Association, 1972.
81. Y. Chatani, Y. Okita, H. Tadokoro, and Y. Yamashita, *Polym. J.*, **1,** 555 (1970).

82. G. Perego, A. Melis, and M. Cesari, *Makromol. Chem.*, **157,** 269 (1972).
83. M. Yokouchi, Y. Chatani, H. Tadokoro, K. Teranishi, and H. Tani, *Polymer*, **14,** 267 (1973).
84. K. Imada, T. Miyakawa, Y. Chatani, H. Tadokoro, and S. Murahashi, *Makromol. Chem.*, **83,** 113 (1965).
85. T. Shimanouchi and S. Mizushima, *J. Chem. Phys.*, **23,** 707 (1955).
86. T. Miyazawa, *J. Polym. Sci.*, **55,** 215 (1961).
87. H. Sugeta and T. Miyazawa, *Biopolymers*, **5,** 673 (1967).
88. P. Ganis and P. A. Temussi, *Makromol. Chem.*, **89,** 1 (1965).
89. K. Tai and H. Tadokoro, *Macromolecules*, **7,** 507 (1974).
90. H. Margenau and G. M. Murphy, *The Mathematics of Physics and Chemistry*, Van Nostrand, New York, 1943.
91. E. B. Wilson, Jr., J. C. Decius, and P. C. Cross, *Molecular Vibrations*, McGraw-Hill, New York, 1955.
92. M. Yokouchi, H. Tadokoro, and Y. Chatani, *Macromolecules*, **7,** 769 (1974).
93. N. Go and K. Okuyama, *Macromolecules*, **9,** 867 (1976).
94. H. Sakakihara, Y. Takahashi, H. Tadokoro, N. Oguni, and H. Tani, *Macromolecules*, **6,** 205 (1973).
95. H. Kusanagi, H. Tadokoro, Y. Chatani, and K. Suehiro, *Macromolecules*, **10,** 405 (1977).
96. M. Yokouchi, Y. Chatani, and H. Tadokoro, *J. Polym. Sci., Polym. Phys. Ed.*, **41,** 81 (1976).
97. W. Cochran, F. H. C. Crick, and V. Vand, *Acta Crystallogr.*, **5,** 581 (1952).
98. S. Naya and I. Nitta, *Kwansei Gakuin Univ. Annu. Studies*, **15,** 1 (1966); S. Tanaka and S. Naya, *J. Phys. Soc. Jpn.*, **26,** 982 (1969).
99. (a) H. Staudinger, H. Johner, R. Signer, G. Mie, and J. Hengstenberg, *Z. Physik. Chem.*, **126,** 425 (1927); (b) J. Hengstenberg, *Ann. Phys.*, **84,** 245 (1927).
100. E. Sauter, *Z. Phys. Chem.*, **B18,** 417 (1932); **B21,** 186 (1933).
101. M. L. Huggins, *J. Chem. Phys.*, **13,** 37 (1945).
102. H. Tadokoro, T. Yasumoto, S. Murahashi, and I. Nitta, *J. Polym. Sci.*, **44,** 266 (1960).
103. D. R. Davies and A. Rich, *Acta Crystallogr.*, **12,** 97 (1959).
104. H. Tadokoro, Y. Chatani, T. Yoshihara, S. Tahara, and S. Murahashi, *Makromol. Chem.*, **73,** 109 (1964).
105. S. Sasaki, Y. Takahashi, and H. Tadokoro, Annual Meeting of the Society of Polymer Science, Japan, Tokyo, Preprint p. 88, 1972.
106. Y. Matsui, T. Kubota, H. Tadokoro, and T. Yoshihara, *J. Polym. Sci. A*, **3,** 2275 (1965).
107. T. Yoshihara, H. Tadokoro, and S. Murahashi, *J. Chem. Phys.*, **41,** 2902 (1964).
108. H. Tadokoro, *J. Chem. Phys.*, **33,** 1558 (1960).
109. M. Hayashi, *Nippon Kagaku Zasshi*, **78,** 222 (1957).
110. T. Miyazawa, K. Fukushima, and Y. Ideguchi, *J. Chem. Phys.*, **37,** 2764 (1962).
111. J. R. Richards, Ph. D. Thesis, University of Pennsylvania, 1961; *Dissertation Abstr.*, **22,** 1029 (1961).
112. C. H. MacGillavry and E. M. Bruins, *Acta Crystallogr.*, **1,** 156 (1948).
113. R. E. Franklin and R. G. Gosling, *Acta Crystallogr.*, **6,** 678 (1953).
114. H. Kusanagi, H. Tadokoro, and Y. Chatani, *Macromolecules*, **9,** 531 (1976).

115. J. D. Stroupe and R. E. Hughes, *J. Am. Chem. Soc.*, **80,** 2341 (1958).

116. M. D'Alagni, P. De Santis, A. M. Liquori, and M. Savino, *J. Polym. Sci.*, **B2,** 925 (1964).

117. A. M. Liquori, Q. Anzuino, V. M. Coiro, M. D'Alagni, P. De Santis, and M. Savino, *Nature*, **206,** 358 (1965).

118. V. M. Coiro, P. De Santis, A. M. Liquori, and L. Mazzarella, *J. Polym. Sci. C*, **16,** 4591 (1969).

119. H. Tadokoro, K. Tai, M. Yokoyama, and M. Kobayashi, *J. Polym. Sci., Polym. Phys. Ed.*, **11,** 825 (1973).

120. J. D. Watson and F. H. C. Crick, *Nature*, **171,** 737 (1953).

121. C. A. Taylor and H. Lipson, *Optical Transforms*, G. Bell & Sons, London, 1964.

122. M. Cesari, *J. Polym. Sci. B*, **2,** 453 (1964).

123. P. Ganis and P. A. Temussi, *Eur. Polym. J.*, **2,** 401 (1966).

124. P. Corradini and P. Ganis, *Makromol. Chem.*, **62,** 97 (1963).

125. R. Hosemann, *J. Polym. Sci. C*, **20,** 1 (1967).

126. C. S. Fuller and C. L. Erickson, *J. Am. Chem. Soc.*, **59,** 344 (1937).

127. C. W. Bunn, *Proc. Roy. Soc. (Lond.)*, **A180,** 67 (1942).

128. A. I. Kitaigorodskii, *Organic Chemical Crystallography*, Consultants Bureau, New York, 1961.

129. A. J. C. Wilson, *Acta Crystallogr.* **2,** 318 (1949).

130. A. L. Patterson, *Z. Krist.*, **90,** 517 (1935).

131. Y. Takahashi and H. Tadokoro, *J. Polym. Sci., Polym. Phys. Ed.*, **16,** 1219 (1978).

132. A. Klug, F. H. C. Crick, and H. W. Wyckoff, *Acta Crystallogr.*, **11,** 199 (1958).

133. S. Arnott and A. J. Wonacott, *Polymer*, **7,** 157 (1966).

134. C. Scheringer, *Acta Crystallogr.*, **16,** 546 (1963).

135. G. S. Rushbrooke, *Introduction to Statistical Mechanics*, Oxford at the Clarendon Press, London, 1949.

136. S. Arnott and A. J. Wonacott, *J. Mol. Biol.*, **21,** 371 (1966).

137. S. Arnott, S. D. Dover, and A. Elliott, *J. Mol. Biol.*, **30,** 201 (1967).

138. S. Arnott and S. D. Dover, *Acta Crystallogr.*, **B24,** 599 (1968).

139. S. Arnott, S. D. Dover, and A. J. Wonacott, *Acta Crystallogr.*, **B25,** 2192 (1969).

140. Y. Chatani, T. Takizawa, S. Murahashi, Y. Sakata, and Y. Nishimura, *J. Polym. Sci.*, **55,** 811 (1961).

141. P. H. Lindenmeyer and R. Hosemann, *J. Appl. Phys.*, **34,** 42 (1963).

142. R. Bonart, R. Hosemann, and R. L. McCullough, *Polymer*, **4,** 199 (1963).

143. R. Hosemann and W. Wilke, *Faserforsch. Textiltech.*, **15,** 521 (1964).

144. R. Hosemann and W. Wilke, *Makromol. Chem.*, **118,** 230 (1968).

145. W. Vogel, J. Haase, and R. Hosemann, *Z. Naturforsch.*, **29a,** 1152 (1974).

146. D. R. Buchanan and R. L. Miller, *J. Appl. Phys.*, **37,** 4003 (1966).

147. R. Bonart, *Kolloid-Z. Z. Polym.*, **231,** 438 (1969).

148. C. W. Bunn, *Nature*, **161,** 929 (1948).

149. C. W. Bunn and H. S. Peiser, *Nature*, **159,** 161 (1947).

150. I. Nitta, I. Taguchi, and Y. Chatani, *Annu. Rep. Inst. Fiber Res. Osaka Univ.*, **10,** 1 (1957).

151. G. Natta and P. Corradini, *Nuovo Cimento, Suppl.*, **15,** 40 (1960).

152. G. Natta, P. Corradini, and I. W. Bassi, *Nuovo Cimento, Suppl.*, **15,** 83 (1960).
153. G. Natta and P. Corradini, *Nuovo Cimento, Suppl.*, **15,** 9 (1960).
154. C. A. Sperati and H. W. Starkweather, Jr., *Adv. Polym. Sci.*, **2,** 465 (1961).
155. E. S. Clark and L. T. Muss, *Z. Krist.* **117,** 108, 119 (1962).
156. A. Klug and R. E. Franklin, *Discuss. Faraday Soc.*, **25,** 104 (1958).
157. D. Hyndman and G. F. Orglio, *J. Appl. Phys.*, **31,** 1849 (1960).
158. S. Tanaka, *J. Phys. Soc. Jpn.*, **36,** 1096, 1396 (1974).
159. Y. Takahashi, M. Kohyama, and H. Tadokoro, *Macromolecules*, **9,** 870 (1976).
160. A. J. C. Wilson, *Acta Crystallogr.*, **3,** 397 (1950).
161. H. Sakakihara, Y. Takahashi, H. Tadokoro, P. Sigwalt, and N. Spassky, *Macromolecules*, **2,** 315 (1969).
162. Y. Gotoh, H. Sakakihara, and H. Tadokoro, *Polym. J.*, **4,** 68 (1973).
163. D. W. J. Cruickshank and A. P. Robertson, *Acta Crystallogr.*, **6,** 698 (1953).
164. R. H. Stanford, Jr. and J. Waser, *Acta Crystallogr.*, **A28,** 213 (1972).
165. E. Benedetti, C. Pedone, and G. Allegra, *Macromolecules*, **3,** 16 (1970).
166. Y. Chatani, T. Ohno, T. Yamauchi, and Y. Miyake, *J. Polym. Sci., Polym. Phys. Ed.*, **11,** 369 (1973).
167. T. Asano, *Polym. J.*, **5,** 72 (1973).
168. W. T. Astbury and C. J. Brown, *Nature*, **158,** 871 (1946).
169. C. W. Bunn, H. S. Peiser, and A. Turner-Jones, *J. Sci. Instrum.*, **21,** 10 (1944).
170. T. Yoshihara, H. Tadokoro, and S. Murahashi, *J. Chem. Phys.*, **41,** 2902 (1964).

Chapter 5

Infrared Absorption and Raman Spectra

5.1 CHARACTERISTIC FEATURES OF INFRARED AND RAMAN SPECTRA OF HIGH POLYMERS

As is described later (Sections 5.2.3 and 5.2.4), infrared absorption and Raman spectra originate from molecular vibrations (and rotations in the case of gases) that cause changes in the dipole moment and polarizability of the molecule, respectively. These spectra are unique to each molecule and therefore reflect the structure of a molecule, especially the masses of the constituent atoms and the intramolecular forces acting among them. In some cases the intermolecular forces operating in the crystalline state are also important.

For details the reader is referred to the following suitable literature: textbooks on infrared absorption and Raman spectra,[1–18] references on infrared absorption spectra of high polymers,[18–28] references on Raman spectra of high polymers,[11,12,18,28–33] standard spectra including those of high polymers,[34–36] and an index of publications on infrared spectroscopy.[37]

5.1.1 Infrared Absorption Spectra of High Polymers

1. For most polymers, film samples of suitable thickness for infrared measurements can be prepared easily. In general, infrared measurement requires samples about 5 mm × 10 mm, with a thickness between several microns and several tens of microns. For far-infrared measurements, thicker and larger films are necessary.

2. Crystallization effects appear in the spectra. The intensities of so-called "crystallization-sensitive bands" vary with the degree of crystallinity. Indeed, these bands have been used as measure of crystallinity.

3. Chain molecules and crystallites can be oriented by stretching or rolling. By measuring the polarized infrared spectra of such oriented polymer

samples, useful information about the molecular or crystal structure can be obtained. Since oriented samples of high polymers can be easily prepared, the polarization measurements are readily applied. Such measurements are not necessarily possible for low-molecular-weight compounds. Dichroism and pleochroism are measured for the uniaxially and doubly oriented samples, respectively, and give very useful information for vibrational analysis.

4. Although most synthetic polymers consist of sequences of more or less simple structural units, their overall structures, at first glance, appear to be highly complicated, and the application of detailed theoretical analysis has been considered difficult. However many polymer molecules are arranged regularly in the crystalline regions, and so can be treated theoretically by utilizing the symmetry properties of the chain or crystal. Factor group analysis gives the number of normal modes and selection rules for infrared and Raman spectra. The use of electronic computers has made it possible to calculate the normal modes of vibrations of many crystalline polymers.

5. Comparatively few solvents are applicable to polymers, but the data obtained from solutions give useful information on the isolated polymer molecule. The infrared spectra of carbon tetrachloride solutions of PE provide an estimate of the degree of branching without crystallization effects.[38] The solution spectra of stereoregular poly(methyl methacrylate),[39] *it*-polystyrene,[40,41] and so on indicate the presence of special conformations in solution, especially at low temperatures.

5.1.2 Raman Spectra of High Polymers

1. Although only a few Raman studies on high polymers were reported before the appearance of the laser source, excellent Raman data have been obtained in the last several years since the development of these efficient laser sources.

2. Raman spectra give information complementary to infrared spectra, since selection rules are different for each spectrum. The stretching vibrations of the S—S linkages in rubber, the C=C bonds, and so on are strong in Raman spectra, but very weak or unobservable in infrared spectra.

3. Comparatively small bulk samples (film shape not required as in infrared measurements) or powder samples can be measured with the Raman technique. Turbid samples can be used without much trouble. Another great merit is that samples containing water or glass can be measured.

4. Low-frequency measurements (even lower than 10 cm^{-1}) can be made on suitable samples and provide information on lattice vibrations. The range from nearly 0 to 4000 cm^{-1} can be covered continuously for

one sample by a single spectrometer. The spectrum is recorded on a linear scale with scattering intensity.

5. Polarization measurements are applicable to oriented samples. Only a few studies have been reported so far, but development of this technique may be anticipated.

6. Liquid samples, especially aqueous solutions, may be conveniently measured. Depolarization ratios (Section 5.4.1.B) give useful information on the symmetry properties of molecular vibrations. Solvents can be selected from a wider range in Raman spectroscopy than in infrared studies.

7. Strong fluorescence occasionally interferes with the measurements of Raman scattering, depending on samples. Such fluorescence often disappears after irradiation with an Ar^+ ion laser.

8. Unusually strong Raman scattering appears occasionally when the frequency of the light source is close to that of the electronic transitions in the sample. Such a phenomenon is called resonance Raman scattering and gives information about the electronic states and the vibrations at the excited electronic energy levels.[8] The resonance Raman scattering technique will become important for polymers when laser sources of various wavelengths become available.

5.2 BASIC PRINCIPLES OF INFRARED ABSORPTION AND RAMAN SPECTRA

Here the basic principles are surveyed to the extent necessary to understand the descriptions of the infrared and Raman experiments.

5.2.1 Normal Vibrations

Since the observed infrared and Raman spectra reflect essentially the normal vibrations, the normal vibrations of an N atomic molecule are now described. The molecular vibrations may be treated by classical mechanics as explained below.

The potential energy is conveniently written using internal displacement coordinates; Δr, $\Delta\phi$, and $\Delta\tau$ are, respectively, the displacements from the equilibrium values of the bond length r, bond angle ϕ, and internal rotation angle τ. The internal displacement coordinates are generally denoted by $R_i (i = 1, 2, \cdots)$. For small displacements, the potential energy V may be expressed as a Taylor power series in terms of the displacement R_i:

$$V = V_0 + \frac{1}{1!}\sum_i \left(\frac{\partial V}{\partial R_i}\right)_0 R_i + \frac{1}{2!}\sum_{i,j}\left(\frac{\partial^2 V}{\partial R_i \partial R_j}\right)_0 R_i R_j + \cdots \tag{5.1}$$

By assuming that the energy of the equilibrium state is zero, V_0 may be eliminated. Furthermore, $(\partial V/\partial R_i)_0 = 0$, since the potential energy for each R_i should be a minimum at the equilibrium position. For small amplitudes of vibration, the cubic and higher terms in the R's can be neglected. Therefore the potential energy of the N atomic molecule is expressed by

$$V = \tfrac{1}{2}(f_{11}R_1{}^2 + f_{22}R_2{}^2 + \cdots + f_{nn}R_n{}^2 + 2f_{12}R_1R_2 + \cdots + 2f_{n-1,n}R_{n-1}R_n) \tag{5.2}$$

with

$$f_{ij} = f_{ji} = \left(\frac{\partial^2 V}{\partial R_i \partial R_j}\right)_0 \tag{5.3}$$

where n is the number of vibrational degrees of freedom, which is $3N - 6$ for nonlinear N atomic molecules, and $3N - 5$ for linear molecules.*

The general potential energy curve for a bond stretching motion as a function of the interatomic distance has the form shown by the solid line in Fig. 5.1. The harmonic approximation, which neglects the third and higher terms in Eq. 5.1, corresponds to the parabolic curve indicated by the dashed line and is a satisfactory approximation if the amplitude of vibration is very small.

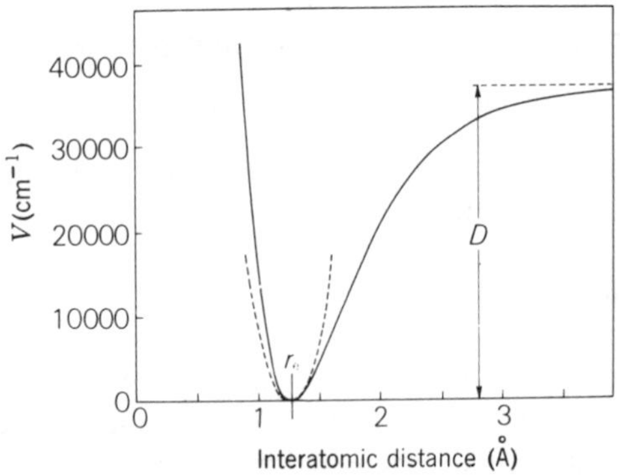

Fig. 5.1 The potential energy curve (————) for the HCl molecule. The parabola (-------) is the harmonic approximation. r_e is the equilibrium interatomic distance and D is the spectroscopic heat of dissociation (From G. Herzberg, *Spectra of Diatomic Molecules*, Copyright 1950 by D. Van Nostrand Company, published by Litton Educational Publishing, Inc.)

* These numbers refer to isolated gas molecules. In the case of isolated polymer chains including p atoms in a unit cell, the number of factor group vibrations (Section 5.5.5) is $3p - 4$ (four corresponds to three translations and a rotation around the chain axis). For three-dimensional crystals, the number is $3p - 3$ (three corresponds to pure translations). The number of internal displacement coordinates is larger than n in some cases, but the redundant coordinates can be eliminated using some suitable transformations (Section 5.6.1).

The kinetic energy T is expressed by using the Cartesian displacement coordinates:

$$T = \frac{1}{2} \sum_{\alpha=1}^{N} m_\alpha [(\Delta \dot{x}_\alpha)^2 + (\Delta \dot{y}_\alpha)^2 + (\Delta \dot{z}_\alpha)^2] \tag{5.4}$$

where m_α is the mass of the αth atom and $\dot{x} = dx/dt$. The Cartesian displacement coordinates are transformed into the internal displacement coordinates using the **B** matrix as described later (Eq. 5.144). The pure translations and rotations of the molecule included in the Cartesian displacement coordinates can be separated from the internal vibrations as is shown later (Eq. 5.140). In terms of the internal displacement coordinates the kinetic energy can be expressed as

$$T = \frac{1}{2}(a_{11}\dot{R}_1{}^2 + a_{22}\dot{R}_2{}^2 + \cdots + a_{nn}\dot{R}_n{}^2 + 2a_{12}\dot{R}_1\dot{R}_2 + \cdots + 2a_{n-1,n}\dot{R}_{n-1}\dot{R}_n) \tag{5.5}$$

If $l_{11}, \cdots, l_{nn}$ involved in the transformation equations

$$\begin{aligned} R_1 &= l_{11}Q_1 + l_{12}Q_2 + \cdots + l_{1n}Q_n \\ R_2 &= l_{21}Q_1 + l_{22}Q_2 + \cdots + l_{2n}Q_n \\ &\cdots \\ R_n &= l_{n1}Q_1 + l_{n2}Q_2 + \cdots + l_{nn}Q_n \end{aligned} \tag{5.6}$$

are chosen suitably, Eqs. 5.2 and 5.5 can be transformed into simple forms involving only the square terms of Q's and $\dot{Q}$'s.

$$V = \frac{1}{2}(\lambda_1 Q_1{}^2 + \lambda_2 Q_2{}^2 + \cdots + \lambda_n Q_n{}^2) \tag{5.7}$$

$$T = \frac{1}{2}(\dot{Q}_1{}^2 + \dot{Q}_2{}^2 + \cdots + \dot{Q}_n{}^2) \tag{5.8}$$

Such coordinates ($Q_1, Q_2, \cdots, Q_n$) are called normal coordinates. If normal coordinates are used, the equations of motion are separated into n equations, each of which includes only one Q_k:

$$\frac{d}{dt}\left(\frac{\partial T}{\partial \dot{Q}_k}\right) + \frac{\partial V}{\partial Q_k} = \ddot{Q}_k + \lambda_k Q_k = 0 \qquad k = 1, 2, \cdots, n \tag{5.9}$$

The solution to this equation is

$$Q_k = K_k \cos(\lambda_k^{1/2} t + \varepsilon_k) \tag{5.10}$$

where K_k and ε_k are the amplitude and phase of the kth normal vibration, respectively. The frequency of the kth vibration has the relation

$$4\pi^2 \nu_k{}^2 = \lambda_k \tag{5.11}$$

with the characteristic value λ_k. In vibrational spectroscopy, the wave number $\tilde{\nu}_k$ (cm^{-1}) is frequently used instead of the frequency ν_k. The wave number is defined as the number of waves involved in 1 cm and is related to the frequency ν (sec^{-1}) by the velocity of light, $c = 2.99793 \times 10^{10}$ cm/sec, and to the wavelength λ (cm) by

$$\tilde{\nu} = \frac{\nu}{c} = \frac{1}{\lambda} \tag{5.12}$$

In one normal vibration, say the kth, all the atomic nuclei vibrate with the same frequency ν_k and the same phase ε_k. Any given molecular vibration can be resolved into a sum involving normal vibrations with suitable amplitudes and phases.

5.2.2 Vibrations of Diatomic and Triatomic Molecules

A diatomic molecule consisting of two atoms of masses m_1 and m_2 with the bond length r is now considered. A force constant k is used under the harmonic approximation. Since the vibrational mode is only the bond stretching, the potential and kinetic energies corresponding to Eqs. 5.7 and 5.8 are given by

$$V = \frac{1}{2}k(\Delta r)^2 = \frac{1}{2}\left(\frac{k}{\mu}\right)Q^2 = \frac{1}{2}\lambda Q^2 \tag{5.13*}$$

and

$$T = \tfrac{1}{2}\mu(\Delta\dot{r})^2 = \tfrac{1}{2}\dot{Q}^2 \tag{5.14}$$

with the reduced mass $\mu = m_1 m_2/(m_1 + m_2)$. Therefore

$$\tilde{\nu} = \frac{\nu}{c} = \left(\frac{1}{2\pi c}\right)\left(\frac{k}{\mu}\right)^{1/2} \tag{5.15}$$

From this equation the frequency of the diatomic molecule can be calculated.

Figure 5.2 is a schematic representation of the normal vibrations of the water-type triatomic molecule. The three kinds of vibrations are symmetric stretching ν_1, bending ν_2, and antisymmetric stretching ν_3. If the bond lengths and bond angle are denoted as shown in the figure, the internal displacement coordinates consist of Δr_1, Δr_2, and $\Delta\phi$. Then the normal coordinates are given by

$$Q_1 \cong \frac{N_1(\Delta r_1 + \Delta r_2)}{2^{1/2}}, \qquad Q_2 \cong N_2 \Delta\phi, \qquad Q_3 = \frac{N_3(\Delta r_1 - \Delta r_2)}{2^{1/2}} \tag{5.16}$$

* It is necessary to distinguish between λ (characteristic value) in Eq. 5.13 and λ (wavelength) in Eq. 5.12.

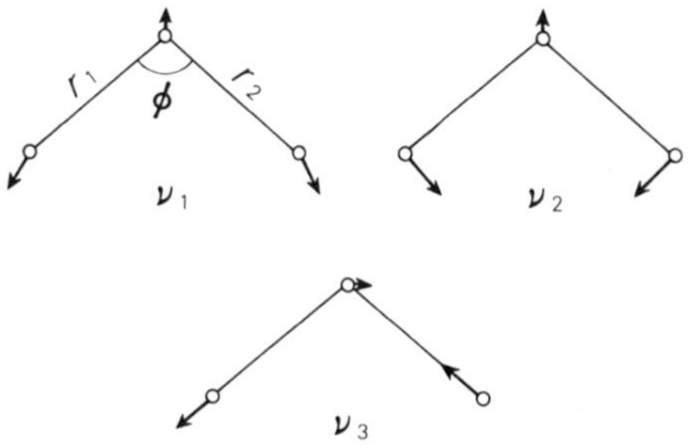

Fig. 5.2 Schematic drawing of the normal vibrations of the H_2O type triatomic molecule.

where the N's are normalization factors. Here the symbol $\cong$ is used for Q_1 and Q_2. This means that Q_1 includes a small fraction of $\Delta\phi$, and Q_2 includes a small fraction of $(\Delta r_1 + \Delta r_2)/2^{1/2}$ as is discussed in Section 5.6.2. For frequencies see p. 276.

5.2.3 Quantum Mechanical Foundation of Infrared Spectra

Calculations of molecular vibrations have been performed using classical mechanics as an approximation. However, in principle, the occurrence of infrared and Raman spectra should be discussed in terms of quantum mechanics. If the potential and kinetic energies are given by Eqs. 5.7 and 5.8, Schroedinger's equation can be written as

$$\frac{d^2\psi(Q_k)}{dQ_k{}^2} + \frac{8\pi^2}{h^2}\left[E(k) - \frac{1}{2}\lambda_k Q_k{}^2\right]\psi(Q_k) = 0 \qquad k = 1, 2, \cdots, n \tag{5.17}$$

where h is Planck's constant, ψ is the wave function, and E is the characteristic value. It is clear (see, e.g., Ref. 42, p. 267) that ψ satisfies the condition of the wave function; it is continuous, one-valued, and finite. The energy levels take discrete and equidistant values:

$$E_v = h\nu(v + \tfrac{1}{2}) \qquad v = 0, 1, 2, \cdots \tag{5.18}$$

where ν is the same as the frequency given in Eq. 5.11 and v is the vibrational quantum number. The quantity k is hereafter omitted for simplicity. The wave function satisfying Eq. 5.17 is given by

$$\psi_v(Q) = N_v \exp\left(\frac{-\gamma Q^2}{2}\right) H_v(\gamma^{1/2}Q) \tag{5.19}$$

where $N_v = (\gamma/\pi)^{1/4}(2^v v!)^{-1/2}$, $\gamma = 4\pi^2\nu/h$, and $H_v(\gamma^{1/2}Q)$ is a Hermite polynomial of degree v, the first three of which are

$$H_0(x) = 1 \qquad H_1(x) = 2x \qquad H_2(x) = 4x^2 - 2 \tag{5.20}$$

The values of Eq. 5.19 for $v = 0$ and 1 are

$$\psi_0(Q) = \left(\frac{\gamma}{\pi}\right)^{1/4} \exp\left(\frac{-\gamma Q^2}{2}\right) \tag{5.21}$$

$$\psi_1(Q) = \left(\frac{\gamma}{\pi}\right)^{1/4} (2\gamma)^{1/2} Q \exp\left(\frac{-\gamma Q^2}{2}\right) \tag{5.22}$$

Absorption or emission of light of frequency ν occurs by a transition that satisfies the Bohr frequency condition.

$$h\nu = E_{v'} - E_v \tag{5.23}$$

The intensity of infrared absorption is proportional to the square of the integral (Ref. 42, p. 302)

$$[\boldsymbol{\mu}_{v'v}] = \int_{-\infty}^{+\infty} \psi_{v'}(Q_k)\boldsymbol{\mu}\psi_v(Q_k)\mathrm{d}Q_k \tag{5.24}$$

This integral is called the transition moment. $\boldsymbol{\mu}$ is an electric dipole moment whose component is denoted by (μ_x, μ_y, μ_z), assuming a Cartesian coordinate system (x, y, z) fixed to the center of gravity of the molecule. For a vibrational infrared activity at least one of the components of $[\boldsymbol{\mu}]_{v'v}$, that is, $[\mu_x]_{v'v}$, $[\mu_y]_{v'v}$, or $[\mu_z]_{v'v}$, must be different from zero. In this way, the infrared activity of a molecular vibration depends on the nonvanishing value of the transition moment. Rules indicating which transitions may occur are called selection rules.

One of the components of $\boldsymbol{\mu}$, say, μ_x can be expanded as a power series in Q_k, and the third and higher terms are usually neglected assuming small amplitudes,

$$\mu_x = (\mu_x)_0 + \sum_k \left(\frac{\partial \mu_x}{\partial Q_k}\right)_0 Q_k \tag{5.25}$$

Integration of Eq. 5.24 after substitution of Eq. 5.25 shows that $[\mu_x]_{v'v} = 0$ unless $v' = v \pm 1$ (Ref. 1, p. 80). This indicates that only the transition of $\Delta v = \pm 1$ is permitted under the harmonic approximation and that it is infrared active. Using the orthogonality property of the functions $\psi_v(Q)$, the transition moment becomes

$$[\mu_x]_{v+1,v} = \left[(\mu_x)_0 + \sum_{k' \neq k} \left(\frac{\partial \mu_x}{\partial Q_{k'}}\right)_0 Q_{k'}\right] \int_{-\infty}^{+\infty} \psi_{v+1}(Q_k)\psi_v(Q_k)\mathrm{d}Q_k$$

$$+ \left(\frac{\partial \mu_x}{\partial Q_k}\right)_0 \int_{-\infty}^{+\infty} \psi_{v+1}(Q_k) Q_k \psi_v(Q_k)\,\mathrm{d}Q_k$$

$$= \left(\frac{\partial \mu_x}{\partial Q_k}\right)_0 \left[\frac{(v+1)h}{8\pi^2 \nu_k}\right]^{1/2} \tag{5.26}$$

which is proportional to $(\partial\mu_x/\partial Q_k)_0$ (Ref. 4, p. 289). Therefore a normal vibration is infrared active only when the dipole moment of the molecule is changed during the vibration. The transition in which the quantum number changes by one is called the fundamental transition, and the resulting absorption is referred to as the fundamental band. In addition to the fundamental bands, there occur overtone and combination bands, which are due to anharmonicity (Section 5.2.6). Of these the most important vibration corresponds to the transition $v = 0 \rightarrow v = 1$. The fundamental vibrations of the water molecule are observed at 3657 (ν_1), 1599 (ν_2), and 3756 cm^{-1} (ν_3).

5.2.4 Raman Effect

When a light wave with frequency ν_0 falls on a molecule, the electrons in the molecule are polarized by the oscillating field.

$$E = E_0 \cos 2\pi\nu_0 t \tag{5.27}$$

If the dipole moment induced in the molecule is given by

$$P = \alpha E = \alpha E_0 \cos 2\pi\nu_0 t \tag{5.28}$$

where α is the polarizability, then such an oscillating dipole emits light at the same frequency as the incident light. This phenomenon is called Rayleigh scattering.

However, if the polarizability α is modulated by vibrations, α can be expanded as a power series in Q_k. Here the third and higher terms can be neglected assuming small vibration amplitudes,

$$\alpha = \alpha_0 + \sum_k \left(\frac{\partial\alpha}{\partial Q_k}\right)_0 Q_k \tag{5.29}$$

Since the vibration is given by

$$Q_k = Q_k^0 \cos 2\pi\nu_k t \tag{5.30}$$

substituting Eqs. 5.29 and 5.30 into Eq. 5.28 gives

$$P = \left[\alpha_0 + \sum_k \left(\frac{\partial\alpha}{\partial Q_k}\right)_0 Q_k^0 \cos 2\pi\nu_k t\right] E_0 \cos 2\pi\nu_0 t$$

$$= E_0\left\{\alpha_0 \cos 2\pi\nu_0 t + \frac{1}{2}\sum_k \left(\frac{\partial\alpha}{\partial Q_k}\right)_0 Q_k^0 \left[\cos 2\pi(\nu_0 + \nu_k)t + \cos 2\pi(\nu_0 - \nu_k)t\right]\right\} \tag{5.31}$$

This equation indicates that the light waves at frequencies $(\nu_0 \pm \nu_k)$ are scattered in addition to the Rayleigh scattering at frequency ν_0. The scattering with frequencies $(\nu_0 \pm \nu_k)$ is referred to as the Raman scattering. The lines with the frequency $\nu_0 + \nu_k$ are called the anti-Stokes lines, and those with $\nu_0 - \nu_k$ are called the Stokes lines. The condition for a vibration to be Raman active is that the value of $(\partial\alpha/\partial Q_k)_0$ in Eq. 5.31 is not zero; in other words, the molecular polarizability is changed by such a vibration.

The induced dipole and the electric field are both vectors and are thus connected by

$$\begin{bmatrix} P_x \\ P_y \\ P_z \end{bmatrix} = \begin{bmatrix} \alpha_{xx} & \alpha_{xy} & \alpha_{xz} \\ \alpha_{xy} & \alpha_{yy} & \alpha_{yz} \\ \alpha_{xz} & \alpha_{yz} & \alpha_{zz} \end{bmatrix} \begin{bmatrix} E_x \\ E_y \\ E_z \end{bmatrix} \tag{5.32}$$

or, in a short-hand notation, by

$$\mathbf{P} = \boldsymbol{\alpha}\mathbf{E} \tag{5.33}$$

where x, y, and z are a Cartesian coordinate system fixed to the molecule. The polarizability $\boldsymbol{\alpha}$ is a symmetric tensor of the second rank and consists of six independent components α_{xx}, α_{yy}, α_{zz}, α_{xy}, α_{yz}, and α_{zx}. This equation indicates, for instance, that the y component of the induced dipole moment by the x component of the electric field E_x is given by $\alpha_{xy}E_x$.

According to quantum mechanics, the intensity of Raman scattering is proportional to the square of the transition moment (Ref. 4, p. 50)

$$[\boldsymbol{\alpha}]_{v'v} = \int \psi_{v'}(Q_k)\boldsymbol{\alpha}\psi_v(Q_k)\,\mathrm{d}Q_k \tag{5.34}$$

Under the harmonic approximation, the selection rule is $\Delta v = \pm 1$.

The equations for the transition moment as given below are obtained in the same way as those resulting from infrared absorption studies.

$$[\alpha_{xx}]_{v+1,v} = \left(\frac{\partial\alpha_{xx}}{\partial Q_k}\right)_0 \int \psi_{v+1}(Q_k)Q_k\psi_v(Q_k)\,\mathrm{d}Q_k \tag{5.35}$$

The intensities of the Stokes and anti-Stokes lines are proportional to the numbers of molecules on the energy levels v and $v + 1$, respectively. According to the Maxwell-Boltzmann distribution law, the ratio of these numbers is equal to $\exp(-h\nu/kT)$, and so the Stokes lines are stronger than the corresponding anti-Stokes lines.

5.2.5 Symmetry and Selection Rules

Selection rules can be determined from the molecular symmetry. For example, the water-type triatomic molecule has the symmetry of point group C_{2v}. If the x, y, and z axes are taken as shown in Fig. 5.3, the symmetry operations are E, C_2, $\sigma_v(xz)$, and $\sigma_v(yz)$.* To classify normal coordinates in terms of symmetry properties, the term "symmetry species" is used. In the case of C_{2v}, there are four symmetry species designated by A_1, A_2, B_1, and B_2. The notations A and B are symmetric (the normal coordinates do not change sign) and antisymmetric (the normal coordinates change sign), respectively, with respect to the symmetry operation C_2. The notations 1 and 2 indicate species symmetric and antisymmetric to $\sigma_v(xz)$, respectively. Such relations are summarized in Table 5.1 for the case of the water molecule, $+1$ representing symmetric and -1 representing antisymmetric. The normal coordinates Q_1 and Q_2 in Eq. 5.16 are symmetric to all the symmetry operations and so belong to the symmetry species A_1. Q_3 is antisymmetric to C_2 and $\sigma_v(yz)$ and belongs to B_1. Table 5.1 also indicates symmetry species associated with translations of the molecule along the x, y, and z axes (denoted by T_x, T_y, and T_z) and rotation around the three axes (R_x, R_y, R_z).

As is described earlier, the necessary condition for the normal vibration ν_k to be infrared active is that at least one of the three components of the transition moment be nonzero. To satisfy this condition, at least one of the quantities

$$\psi_{v'}(Q_k)\mu_x\psi_v(Q_k), \qquad \psi_{v'}(Q_k)\mu_y\psi_v(Q_k), \qquad \psi_{v'}(Q_k)\mu_z\psi_v(Q_k)$$

should remain unchanged for any of the symmetry operations; in other words, it is totally symmetric.† This condition holds irrespective of the harmonic approximation. Under the harmonic approximation, only the transition from $v = 0$ to $v' = 1$ is usually taken into account as the fundamental vibration. $\psi_0(Q_k)$ is not affected by the change of the sign of Q_k, and $\psi_1(Q_k)$ changes in the same way as Q_k (Eqs. 5.21 and 5.22). On the other hand, μ_x, μ_y, and μ_z belong to the same symmetry species as T_x, T_y, and T_z, respectively. Therefore, when Q_k belongs to the same symmetry species as any one of T_x, T_y, and T_z, the mode is infrared active. The direction of the transition

* The choice of the x, y, and z axes differs from author to author. Here it is taken in the manner of Herzberg[2] and Wilson et al.[4]

† One of the components of Eq. 5.24 $[\mu_x]_{v'v}$ is considered here. Since this is a definite integral, the value should not be affected by the choice of coordinates. $Q_3 = N_3(\Delta r_1 - \Delta r_2)/2^{1/2}$ has only the component in the x direction. Therefore if the symmetry operation C_2 or $\sigma_v(yz)$ is applied, the sign of Q_3 is reversed. Since $\psi_0(Q_3)$ is not affected by the change of the sign of Q_3, and $\psi_1(Q_3)$ and μ_x change their sign by these symmetry operations, the integrand of $[\mu_x]_{1,0}$ for Q_3 is totally symmetric with respect to all symmetry operations; namely, it is an even function, and therefore $[\mu_x]_{1,0}$ is not zero.

On the other hand, for $[\mu_y]_{1,0}$ and $[\mu_z]_{1,0}$, the sign of $\psi_1(Q_3)$ is changed by the above two symmetry operations, but μ_y and μ_z can be placed outside the integral. Therefore the integrands are not totally symmetric; that is, they are odd functions, and so $[\mu_y]_{1,0}$ and $[\mu_z]_{1,0}$ for Q_3 are zero.

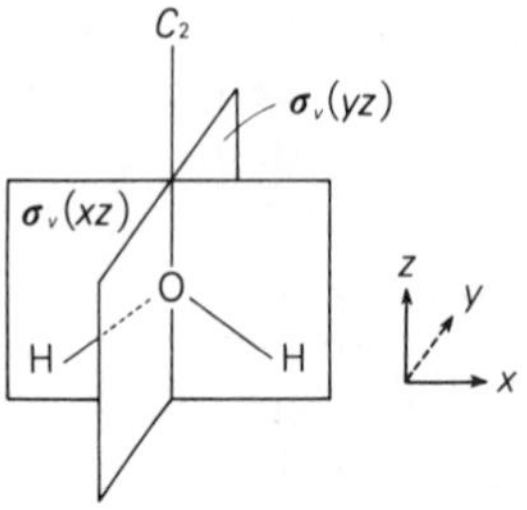

Fig. 5.3 Symmetry of H_2O molecule (From E. B. Wilson, Jr., J. C. Decius, and P. C. Cross, *Molecular Vibrations*, Copyright McGraw-Hill Book Co., New York, 1945. Used with permission of McGraw-Hill Book Co.).

Table 5.1 Normal Vibrations and Symmetry Properties of a Water-type Triatomic Molecule

	E	C_2	$\sigma_v(xz)$	$\sigma_v(yz)$	Normal vibrations	Translations and rotations of the molecule	Components of the polarizability tensor
A_1	$+1$	$+1$	$+1$	$+1$	ν_1, ν_2	T_z	$\alpha_{xx}, \alpha_{yy}, \alpha_{zz}$
A_2	$+1$	$+1$	-1	-1	—	R_z	α_{xy}
B_1	$+1$	-1	$+1$	-1	ν_3	T_x, R_y	α_{xz}
B_2	$+1$	-1	-1	$+1$	—	T_y, R_x	α_{yz}

moment of a mode coincides with that of T_x, T_y, or T_z belonging to the same symmetry species. It is understood from this result that ν_1, ν_2, and ν_3 are all infrared active and that the transition moment is in the z direction for ν_1 and ν_2 and in the x direction for ν_3.

In the same way, for the ν_k vibration to be Raman active Q_k must belong to the same symmetry species as at least one of α_{xx}, α_{yy}, α_{zz}, α_{xy}, α_{yz}, and α_{zx}. The symmetry species to which α_{xx}, and so on belong in the case of C_{2v} are shown in Table 5.1. Therefore the ν_1, ν_2, and ν_3 vibrations of the water molecule are all Raman active. The method of their determination is described later (Eq. 5.102).

In the cases of gas, liquid, and nonoriented samples, the direction of the infrared transition moment should be averaged over the whole steric angle, since the molecules assume all orientations. However, since crystals and oriented polymer samples have inherent specific orientations, polarization measurements can be made. The infrared absorption intensity is proportional to the square of the scalar product of the transition moment and the oscillating electric vector $\{[\boldsymbol{\mu}]_{v'v} \cdot \mathbf{E}\}^2$.

In the case of Raman spectra, the depolarization ratio can be measured for nonoriented samples; the polarization measurements can also be made on crystals and oriented polymer samples. These are described in detail in Sections 5.4.1.B and 5.4.2.

5.2.6 Vibrations and Anharmonicity

Up to this point only the harmonic approximation has been treated; the fourth and higher terms in Eq. 5.1 have been neglected, and the third and higher terms in Eqs. 5.25 and 5.29 have been omitted. The main features of molecular vibrations can be understood using the harmonic approximation, but there are some problems. For instance, in the case of the hydrogen chloride molecule, weak absorption bands are observed at 5668, 8347, 10,923, and 13,397 cm^{-1} in addition to the fundamental vibrational band at 2886 cm^{-1}. These bands correspond to the transitions $v = 0 \rightarrow v = 2$, 3, 4, and 5 and are called overtones. In the spectrum of water vapor, weak bands also appear at 5332 and 10,613 cm^{-1}, corresponding to the wave numbers $\nu_2 + \nu_3$ and $2\nu_1 + \nu_3$, respectively. They are called combination tones. From the wave numbers of the overtones of hydrogen chloride it can be seen that the separation between energy levels decreases slowly with increasing v. This phenomenon cannot be explained with the harmonic approximation and can be interpreted only in terms of anharmonicity. In short, for real molecules the higher terms should not be neglected if all bands are to be taken into account.

To a good approximation the solid curve in Fig. 5.1 can be expressed by the Morse function,

$$V = D[1 - \exp(-aq)]^2 \tag{5.36}$$

with

$$q = r - r_e \tag{5.37}$$

Equation 5.36 is now expanded according to Eq. 5.1, in powers of q:

$$V(q) = D[a^2q^2 - a^3q^3 + \tfrac{7}{12}a^4q^4 - \cdots] \tag{5.38}$$

If the Schroedinger wave equation is solved using the first three terms of this expansion, the energy levels become

$$E_v = hc\tilde{\nu}_e[(v + \tfrac{1}{2}) - x_e(v + \tfrac{1}{2})^2] \tag{5.39}$$

with

$$\tilde{\nu}_e = \left(\frac{a}{2\pi c}\right)\left(\frac{2D}{\mu}\right)^{1/2} \tag{5.40}$$

$$x_e = \frac{hc\tilde{\nu}_e}{4D} \tag{5.41}$$

where $\tilde{\nu}_e$ is the wave number corrected for anharmonicity. In other words, it is the wave number for the harmonic potential energy curve having the same curvature as that of the true potential energy curve with an infinitesimal amplitude. For anharmonic vibrations the wave functions are different from the Hermite polynomial in Eq. 5.19, and the selection rule $\Delta v = \pm 1$ does not hold. This selection rule is modified if the third and higher terms are added in Eqs. 5.25 and 5.29. The anharmonicity is responsible for the appearance of overtones and combination tones which are forbidden for the harmonic oscillator. The wave numbers of the overtones of hydrogen chloride can also be interpreted using Eq. 5.39. The measure of the extent of anharmonicity is given by $\tilde{\nu}_e x_e$ (cm^{-1}) and is proportional to $\tilde{\nu}_e^2$. Accordingly the higher the frequency, the greater is the anharmonicity. The anharmonicity correction has not been made for most polyatomic molecules because of the complexity of the calculation.

5.2.7 Band Assignments

Band assignments are not the ultimate objective of a polymer study, but it is one of the most important steps in the process of spectral interpretation. Although the methods of assignment are essentially the same as those used for nonpolymeric compounds, the following methods are particularly applicable to high polymers. (1) Information about the characteristic group frequencies: The concept of characteristic group frequencies based on empirically accumulated data is most useful in the preliminary interpretation of infrared spectra. Many books[5,7,10] and charts, for example, Colthup's,[43] provide such information. For Raman characteristic frequencies, see Ref. 44. However, it should be noted that the vibrational frequencies not only depend on the structure of the specific atomic groups, but are affected to a greater or less extent by the structure of the other parts of the molecule. A term such as the "C=O frequency" may be used only to emphasize the large vibrational amplitude of this bond; more accurately, this frequency is due to a normal vibration characteristic of the molecule as a whole. (2) Comparison with the spectra of similar substances. (3) Factor group analysis. (4) Dichroic and pleochroic measurements and comparison with crystal structure. (5) Spectral measurements with samples in different states: influence of temperature. (6) Normal coordinate treatments. (7) Use of deuterated derivatives.

5.3 MEASUREMENTS OF INFRARED ABSORPTION SPECTRA

Dispersion spectrophotometers using either prisms or gratings have been widely employed in the past. However the Fourier transform spectrophotometers, which give spectra in terms of the Fourier transform of the measured interferograms by computer, are now also available commercially. Using such an apparatus, the sensitivity is increased by 2 orders of magnitude and thus the time needed to make measurements become correspondingly shorter (e.g., 4 sec for covering the whole region) compared with the dispersion spectrophotometer. Consequently the following measurements are facilitated: (1) spectra with very high resolving power, (2) time-dependent spectra with fast rate (elongation, crystallization, etc.), (3) spectra of unstable samples, (4) spectra of samples with very low transmission and small samples, and so forth. These have been impossible to measure until now. Moreover, the absorption due to solvents or adsorbents can be subtracted from spectra by using the computers.

5.3.1 Sample Preparation

It is important to adjust the thickness (or concentration) of the sample to obtain a suitable intensity of each band in the spectrum. In qualitative treatments, a preferable criterion is to adjust the percent transmission of the most intense band to be about 10%. In quantitative treatments, it is desirable to choose the specimen thickness so that the percent transmission of the band to be used is between 20 and 60% (see Fig. 5.14).

A. *Films.* Interference fringes often appear in the spectra of film samples of uniform thickness as shown in the 4000–2000 cm^{-1} region in Fig. 5.4*a*. The interference fringes arise at the wave numbers corresponding to the constructive and destructive interferences owing to the phase difference among the rays reflected $2N$ times ($N = 0, 1, 2, \cdots$) in the film, as shown in Fig. 5.5. When the film is very thick compared to the wavelengths, the spacing of the interference fringes is too small to be observed according to the equation

$$nd = \frac{m}{2(\tilde{\nu}_1 - \tilde{\nu}_2)} \tag{5.42}$$

where n is the refractive index of the sample, d is the thickness of the film (cm), $\tilde{\nu}_1$ and $\tilde{\nu}_2$ are the two positions in the spectrum (cm^{-1}), and m is the number of interference fringes between the two positions. If the film thickness is nonuniform to some extent, the interference is averaged over the irradiated area of the film giving no fringes. Figure 5. 4*b* is the spectrum of a polystyrene film cast from a benzene solution on a slightly uneven surface, and no interference fringes can be observed. Sandwiching the film between a pair of sodium chloride or potassium bromide plates wetted with either Nujol or hexachlorobutadiene also removes the interference fringes. Alternatively,

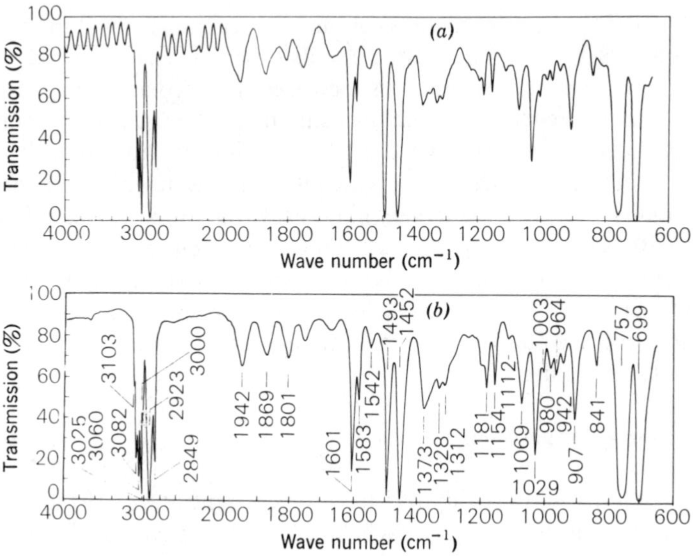

Fig. 5.4 Infrared absorption spectra of polystyrene films of (*a*) uniform thickness and (*b*) slight nonuniformity of thickness. Interference fringes appear in *a*. Standard wave numbers of bands (Plyler and Peters[45]) are indicated in *b*.

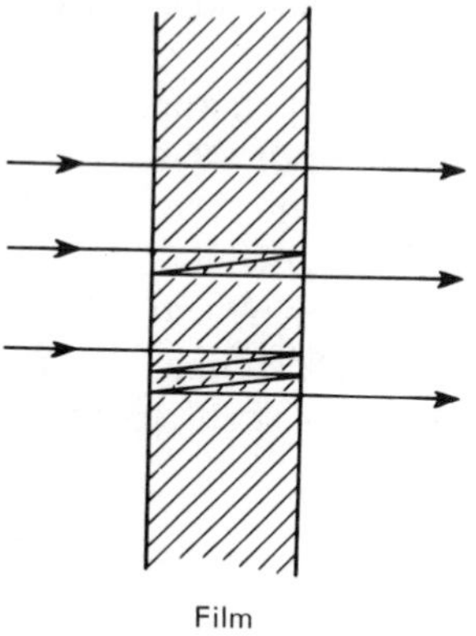

Fig. 5.5 Origin of interference fringes.

crumpling or sandpapering the film removes the interference fringes, but surface roughness tends to disturb the base line at shorter wavelengths. The film thickness can be estimated from Eq. 5.42 if the refractive index is known, and the refractive index can be obtained if the thickness can be measured by other methods.

1. Casting from Solution. Two examples suffice to explain the method. A film of *it*-polystyrene may be conveniently cast from a toluene solution

(about 4–8% polymer) at room temperature on a smooth surface (glass plate, a metal box, aluminum foil, or water surface). Very thin films have been prepared on a water surface. Films cast on an aluminum foil can be easily stripped off because of the flexibility of the foil. Since films prepared by casting are thin, solvent removal is almost complete after the film is heated at 80°C for 10 min. For most studies suitable film thickness is 20–80 μ. *it*-Polystyrene film cast from a toluene solution is partially crystalline. An amorphous sample can be obtained by melting a cast film wrapped with an aluminum foil between two metal plates in Wood's alloy at about 240°C for 15–30 sec and subsequent quenching in ice–water. An oriented sample is prepared by stretching a low crystalline film very slowly in boiling water to about five times the original length using an elongation device (Section 4.3.1). Heat treatment is carried out in Wood's alloy or in glycerol at about 180°C for 30 min.

at-PVA is dissolved in water at about 80°C with stirring. The most suitable thickness of the PVA film for infrared measurements is about 7–10 μ. Since it is difficult to strip off such a thin film of this polymer from a glass plate, films are cast in a polystyrene box with a flat inner bottom surface placed horizontally in an electric oven at about 50°C. A small level is used for setting the box horizontally. The film prepared in this way is of low crystallinity; the density is about 1.291 g/cm^3, corresponding to the degree of crystallinity equal to 0.29.[46] If the film put between a folded wire gauze of fine mesh stainless steel in silicone oil is heated at 180°C for 30 min, the degree of crystallinity increases to about 0.55 (density: 1.307 g/cm^3) with almost no thermal decomposition. An oriented film is prepared by stretching the slightly moistened original film (breath on the sample) by hand over a small flame.

The solvents should not react with the polymers chemically. If the evaporation of solvent is incomplete, confusing absorption bands due to the solvent appear in the spectra. It is desirable to choose solvents having no intense bands in the spectral region considered. Moreover, as much solvent as possible should be removed from the specimen. For this purpose, heating under evacuation is useful. It is necessary to confirm the absence of the solvents in the film by comparing the polymer spectra with the spectra of the solvents used. Suitable solvents include benzene for polyisobutylene, poly(vinyl isobutylether), polybutadiene, polyisoprene, and polychloroprene; methanol for poly(vinyl acetate); formic acid aqueous solution (30–80%) for nylon 6; tetrahydrofuran for poly(vinyl chloride); pyridine–water or acetic acid for poly(vinyl formal); dimethyl formamide or rhodanate aqueous solution for polyacrylonitrile; and dichloroethylene for polycarbonates. See also Ref. 47.

2. Preparation from Melt. For illustration, the formation of a PET film

(melting point is about 267°C) is described. A pellet is placed between two flat steel plates, fixed by a pair of metal frames with bolts, and heated in Wood's alloy at about 240°C. After it is pressed between the plates to obtain an appropriate thickness, the sample is then quenched in cold water. This procedure forms an amorphous specimen about 20–50 μ thick. The sample is stretched in hot water at 80°C followed by heat treatment in air at 190°C for 20 min. A heating press is also useful, but rapid quenching is difficult to carry out.

Amorphous thin films of PET can be prepared by the following method.[48] A molten sample is blown into a bubble of thin film on the trumpet-shaped end of a glass tube just as a soap bubble is blown (Fig. 5.6). Before the polymer is melted it should be dried by heating, since the moist polymer decomposes easily.

3. Rolling. Films of suitable thickness can be prepared from most thermoplastic polymers by rolling at a suitable temperature (Section 4.3.1.B). In this case caution is needed for the resulting orientation of the film which can be uniaxial or double.

4. Heating-Press. Film specimens are prepared by pressing a polymer sample sandwiched with two flat metal sheets between a pair of suitably heated metal blocks. This procedure gives no uniaxial or double orientation, but it tends to give a planar orientation in which molecular axes are parallel to the film surface, with a random direction of the axes in the plane.

Another useful technique is lamination or impregnation. The polymer under study is placed between or melted into sheets of other materials that have transparent regions in the infrared. Mica, silver chloride, and PE have been used.[49] A small pellet of the polymer is placed between two sheets of mica (or silver chloride). This assembly is then inserted between stainless steel plates and placed in a hydraulic laboratory press. The sample is then heated into the plastic temperature region at sufficient pressure to sheet the

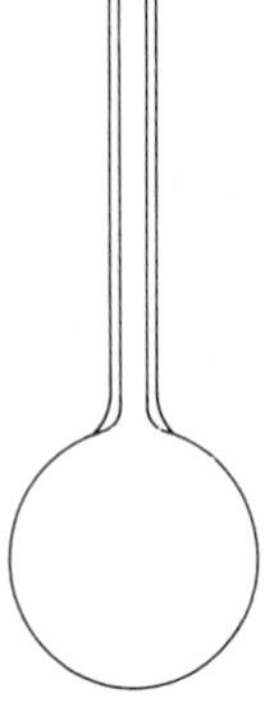

Fig. 5.6 Trumpet-shaped glass tube and a spherically blown polymer film produced with the tube.

sample into a transparent sandwich film. After it is removed from the press, this laminated film is used for the infrared measurement. Thin cleavage crystals of mica produce interference fringes that disappear on sandwiching. Mica is relatively free of absorption bands in the region from 3300 to 1600 cm^{-1}. Silver chloride has no absorption bands in the infrared region higher than 400 cm^{-1}, and it can be rolled into thin sheets. This technique was used for butadiene–acrylonitrile copolymers and others.[49] When lamination is not feasible, an impregnation technique using a polyethylene film is applied. This procedure consists of successive pressings and rollings.

When commercial film samples are used, special caution is necessary to avoid errors due to residual solvents and plasticizers. In general, suitable film thickness for infrared measurements is 3–20 μ for polar polymers containing atoms such as oxygen and nitrogen and 30–100 μ for nonpolar polymers. For measuring the film thickness, a dial gauge or micrometer is convenient. Since absorption bands of various intensities are usually present in a spectrum, at least three different film thicknesses for each polymer should be taken.

B. *Other Preparation Techniques of Specimens.* Should it be impossible to form a satisfactory film for measurements by the usual techniques, the following methods have been found useful:

1. Paste Method. A pair of spectra, taken using Nujol or hexachlorobutadiene as mulling agent, can cover the whole infrared region. For polymers, hexachlorobutadiene seems to give better spectra than perfluorokerosene. It is most important to grind the sample as finely as possible to reduce excessive scattering of the beam.
2. Potassium Bromide Disk Method. Caution is needed for fine grinding of both the sample and potassium bromide, purification of potassium bromide, drying of the materials, uniform pressing, and good evacuation of the disk press. Finely chopped fiber can be also used in this technique. A disk with a diameter of 12 mm and a thickness of about 1 mm requires about 200 mg of potassium bromide powder and 8–0.6 mg of polymer powder. If only a small quantity of sample is available, a disk may be prepared with the sample only in the area on which the beam is incident (see the hatched part in Fig. 5.7*a*). Also spectra on samples of a few tens of micrograms can be obtained with the micro-disk technique (Fig. 5.7*b*). Apparatus using a disk of only 5–2.5 mm in diameter is commercially available.

Grinding the polymer sample with a small quantity of potassium bromide powder (and Dry Ice if necessary) simplifies the preparation of a finely powdered sample, although most polymer samples are elastic to some degree. In this case, addition of a suitable solvent to the grinding medium to promote swelling and/or partial dissolution is helpful.

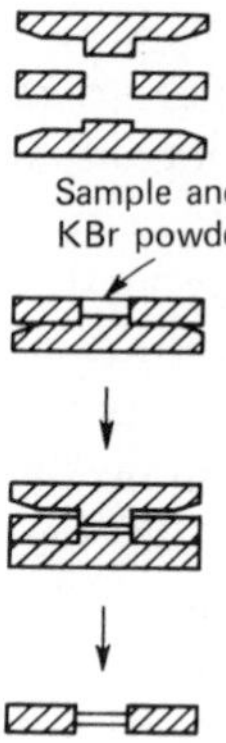

Fig. 5.7 Potassium bromide disk method. (*a*) For powdered sample of small quantity, only the hatched part of disk may be filled by the sample, and the remainder by potassium bromide only. (*b*) Microdisk method.

3. Arranging Single Filaments. The infrared spectra of filaments can be measured by arranging them compactly side by side. The filament is wound round a rectangular U-shaped metal holder, put on its side, and fixed to the holder with a binding agent.[50] Wetting the filaments with hexafluorobutadiene or Nujol reduces the excessive scattering of radiation and helps to keep the filaments in position.[51]

4. Microtomed Sections.[18] When the above-mentioned techniques cannot be applied for any reason, such as crosslinking, a section of the required thickness may be cut from a large sample with a microtome and mounted between two potassium bromide plates. The temperature may be varied so as to adjust the hardness of the sample to make it suitable for microtoming.

5. Solution. Only a few infrared spectra of polymer solutions have been reported.[38–41] Carbon tetrachloride (transmitting at least 25% incident energy in the regions 4000–1610, 1500–1270, 1200–1020, and 960–860 cm^{-1}), carbon disulfide (the same in 4000–2350, 2100–1640, 1385–875, and 845–650 cm^{-1}), and so on are used as solvents.[5] Carbon disulfide was used for the quantitative measurements of specific bonds in polybutadiene.[52]

6. Other Methods. The surface of the polymer can be scraped mechanically using very fine steel wool impregnated with potassium bromide powder.[53] The resulting powder sample is pressed into a microdisk. Surface

components of only about 50 Å in depth can be examined by this method.

The attenuated total reflection (ATR) method[54] can be applied to samples such as vulcanized rubber. Such samples are difficult to examine by transmission methods.

The infrared spectra of pyrolizates of polymers (liquid or gas) are useful for chemical analysis of polymers. For this purpose it is important to obtain only the specific components for the polymer under examination by removing common components such as methane and ethylene with a Dry Ice–methanol trap.[55]

5.3.2 Deuterated Samples

Spectral changes on isotopic substitution, especially on replacement of hydrogen by deuterium, give information about the extent of the contribution of a particular group to some specific spectral bands; the greater the contribution of the group to the vibrational mode, the larger is the frequency shift. However the spectral changes observed on deuteration, especially in the "fingerprint region," are in most cases so complicated that the data can be utilized satisfactorily only in combination with normal mode calculations.

Nevertheless, infrared and Raman spectra of deuterated samples are useful for (1) assignments of bands, (2) elucidation of mechanisms of polymerization reactions (e.g., Ref. 56), (3) separation of the bands due to the crystalline and amorphous regions in the case where the reaction rates of deuterium exchange are much different in these two regions, for instance, cellulose,[57] and (4) estimation of crystallinity in relation to case 3. Various deuterated compounds are now commercially available (see e.g., Isotope Catalogue of Merck, Sharp & Dohme).

A. *Deuteration of OH and NH Groups.* It is well known that OH and NH groups are easily converted into OD and ND, respectively, by an exchange reaction with deuterium oxide, but CH groups are not easily converted except under special conditions.* Thus, when PVA is brought into contact with deuterium oxide, the following reaction takes place and change is observed in the infrared spectra (Fig. 5.8).

$$(\text{—}CH_2\text{—}\underset{\displaystyle OH}{\underset{|}{CH}}\text{—})_n \xrightarrow{D_2O} (\text{—}CH_2\text{—}\underset{\displaystyle OD}{\underset{|}{CH}}\text{—})_n$$

* In the case of compounds having keto–enol tautomerism, such as ketones, the CH_2 or CH_3 group adjacent to the carbonyl group undergoes exchange reaction with deuterium oxide. Acrylonitrile is another exception (see next section).

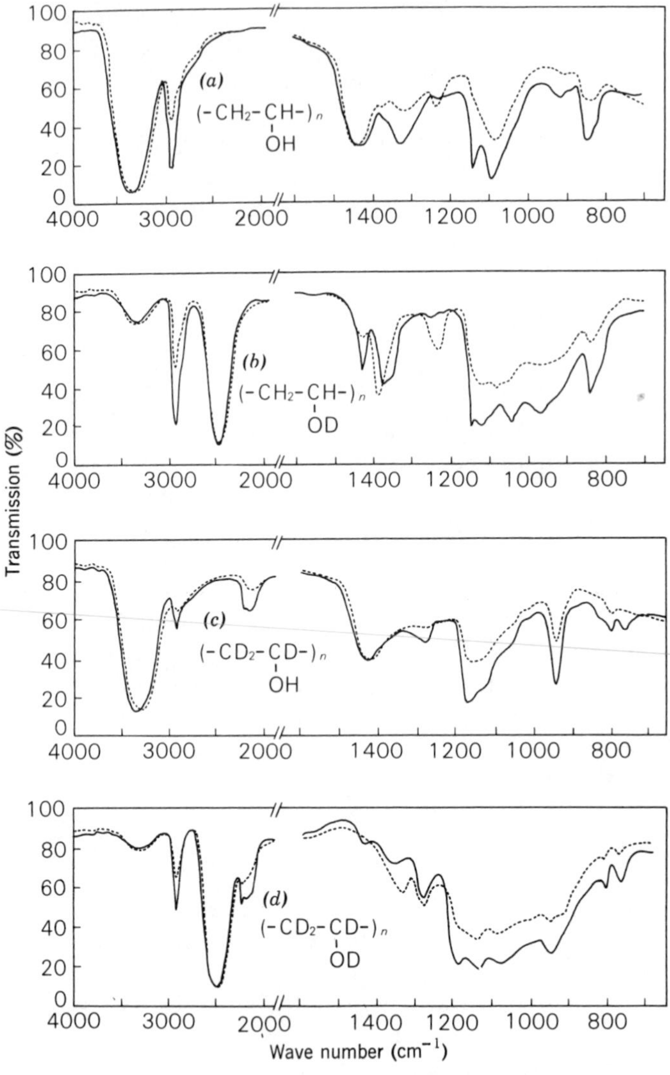

Fig. 5.8 Polarized infrared spectra of PVA and three kinds of deuterated PVAs. (————) Electric vector perpendicular to elongation. (– – – – –) Electric vector parallel to elongation.[58]

One example of deuteration is briefly described.[59] A PVA film (about 7 μ in thickness, and 20 mm × 30 mm in size), placed between folded sheets of fine mesh stainless steel gauze, is inserted into a glass tube A (Fig. 5.9) and dried at 60°C for about 48 hr under high vacuum (10^{-4} torr or below) using a liquid air trap (stopcock C_1 being open and C_2 being closed). After it is

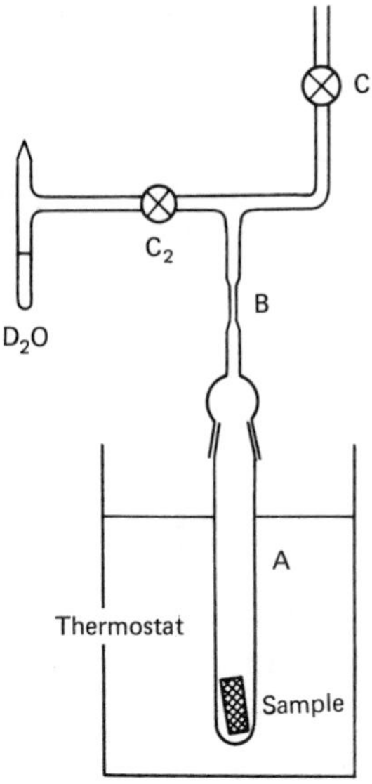

Fig. 5.9 Apparatus for deuteration.[59]

dried, C_1 is closed, the temperature of the sample is lowered to room temperature, and then by opening C_2, the sample is equilibrated with saturated deuterium oxide vapor. After it stands for about 24 hr, the sample is dried again, and the above procedures are repeated several times. After it is sealed off at *B*, the vessel is removed. Measurements are made by putting the films between NaCl plates with or without hexachlorobutadiene. Using this method the author has obtained samples deuterated up to 90–92% for crystalline PVA. However the same procedure gives a much lower extent of deuteration for nylon 6. The crystalline regions of PVA can be easily deuterated in contrast to those of nylon 6 and cellulose, possibly because of continuously formed hydrogen bonds. This is an interesting area for research. Because of the easily deuterated crystalline regions, an oriented PVA film deuterated about 90% can be prepared starting from an oriented crystallized film using the procedure mentioned above. Films of PVA, nylon, cellulose, and so on of suitable thickness for infrared measurements rehydrogenate rapidly in the atmosphere. For instance, a cellulose film less than 5 μ thick has been shown to rehydrogenate almost completely in a few minutes.[57]

B. *Deuteration of CH Groups.* It is necessary to polymerize a deuterated monomer to obtain a high polymer in which CH groups are deuterated. The preparation of such polymers is often laborious, but the resulting polymers give invaluable information. Leitch et al.[60] first measured the infrared spectrum of a deuterated PE prepared by the following method.

$$CH_2N_2 \xrightarrow{\ D_2O\ } CD_2N_2 \xrightarrow[\text{polymerization}]{\text{decomposition}} (-CD_2-)_n$$

Two examples of the preparation of deuterated polymers are given below. For PVA[58] :

$$CaC_2 \xrightarrow{\ D_2O\ } C_2D_2$$

$$(CH_3CO)_2O \xrightarrow{\ D_2O\ } CH_3COOD$$

$$\xrightarrow{\text{addition}} \begin{array}{c} CD_2{=}CD \\ | \\ OCOCH_3 \end{array}$$

$$\xrightarrow{\text{polymerization}} \begin{array}{c} (-CD_2-CD-)_n \\ | \\ OCOCH_3 \end{array} \xrightarrow{\text{hydrolysis}} \begin{array}{c} (-CD_2-CD-)_n \\ | \\ OH \end{array}$$

$$\xrightarrow{\ D_2O\ } \begin{array}{c} (-CD_2-CD-)_n \\ | \\ OD \end{array}$$

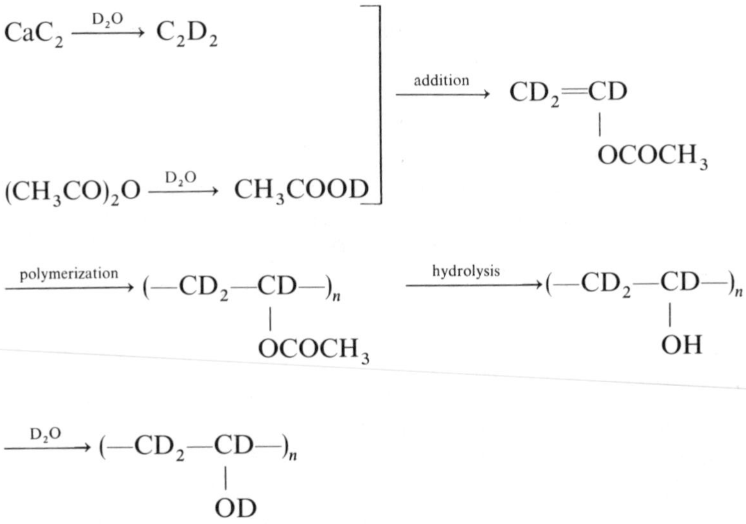

For Polyacrylonitrile[61-63] :

$$C_2D_2 + DCN \longrightarrow CD_2{=}CD{-}CN \longrightarrow \begin{array}{c} (-CD_2-CD-)_n \\ | \\ CN \end{array}$$

$$\uparrow$$

$$KCN + D_2SO_4$$

$$\xrightarrow{CaO,H_2O} CD_2{=}CH{-}CN \longrightarrow \begin{array}{c} (-CD_2-CH-)_n \\ | \\ CN \end{array}$$

$$CH_2{=}CH{-}CN \xrightarrow{D_2O,CaO} CH_2{=}CD{-}CN \longrightarrow \begin{array}{c} (-CH_2-CD-)_n \\ | \\ CN \end{array}$$

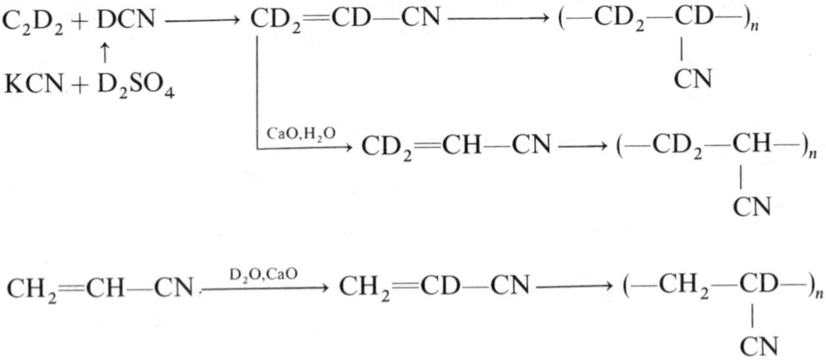

C. *Examples of Deuterated Polymers.* Further examples of deuterated polymers are given below:

PE polymerized from *cis*- and *trans*-dideuteroethylenes

$(\text{—CHD—CHD—})_n$[64]; *it*-PP $[\text{—CH}_2\text{CH(CH}_2\text{D)—}]_n$,[65] $[\text{—CD}_2\text{CH(CH}_3\text{)—}]_n$,[66,67] $[\text{—CH}_2\text{CD(CH}_3\text{)—}]_n$,[66,67] $[\text{—CH}_2\text{CH(CD}_3\text{)—}]_n$,[66,67] $[\text{—CD}_2\text{CD(CD}_3\text{)—}]_n$,[68] $[\text{—CD}_2\text{CD(CH}_3\text{)—}]_n$[68]; threo and erythro *dit*-PP $[\text{—CHD—CH(CH}_3\text{)—}]_n$[67,69]; *it*-polystyrene $(\text{—CH}_2\text{—CH—})_n$,[70,71] $(\text{—CD}_2\text{—CD—})_n$,[70,71] $(\text{—CD}_2\text{—CH—})_n$,[72]

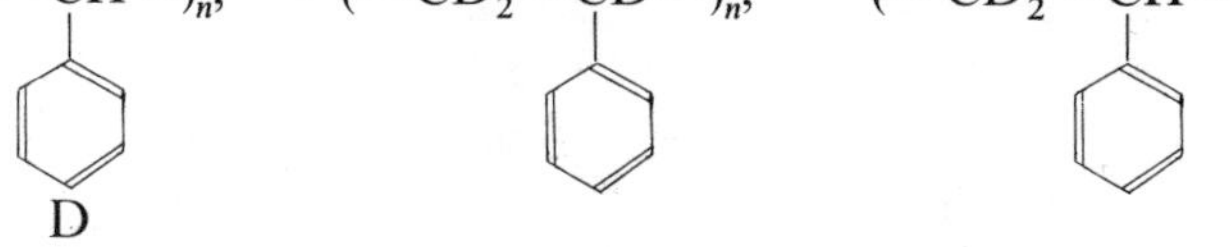

$(\text{—CH}_2\text{—CD—})_n$,[73] $(\text{—CH}_2\text{—CH—})_n$[74]; poly(vinyl chloride) $(\text{—CD}_2\text{-}$

$\text{CDCl—})_n$,[75,76] $(\text{—CH}_2\text{CDCl—})_n$,[76] $(\text{—CHD—CHCl—})_n$,[76] $(\text{—CHD—CDCl—})_n$[76]; poly(vinylidene chloride)$(\text{—CD}_2\text{CCl}_2\text{—})_n$,[77] poly(methyl methacrylate) $[\text{—CH}_2\text{C(CH}_3\text{)(COOCD}_3\text{)—}]_n$,[78] $[\text{—CD}_2\text{C(CD}_3\text{)(COOCH}_3\text{)—}]_n$,[78] $[\text{—CD}_2\text{C(CD}_3\text{)(COOCD}_3\text{)—}]_n$[78]; polyallene $[\text{—CD}_2\text{-C(=CD}_2\text{)—}]_n$[79]; polybutadiene $(\text{—CD}_2\text{—CH=CH—CD}_2\text{—})_n$,[80,81] $(\text{—CDH—CH=CH—CH}_2\text{—})_n$,[81] $(\text{—CH}_2\text{—CD=CH—CH}_2\text{—})_n$,[82] $(\text{—CH}_2\text{—CD=CD—CH}_2\text{—})_n$,[82] $(\text{—CD}_2\text{—CD=CD—CD}_2\text{—})_n$[82]; polyisoprene $[\text{—CD}_2\text{—C(CH}_3\text{)=CH—CD}_2\text{—}]_n$[80]; POM$(\text{—CD}_2\text{O—})_n$[83,84]; PEO$(\text{—CD}_2\text{CD}_2\text{O—})_n$,[85,86] (polymerized from *cis*- and *trans*-dideuterated monomers) $(\text{—CHD—CHD—O—})_n$[56,87]; polytetrahydrofuran $[\text{—(CD}_2)_4\text{—O—}]_n$,[88] $(\text{—CD}_2\text{CH}_2\text{CH}_2\text{CD}_2\text{—O—})_n$[88]; poly(ethylene sulfide) $(\text{—CD}_2\text{CD}_2\text{S—})_n$[89]; PET $(\text{—CH}_2\text{CH}_2\text{OCO—}$ 

$\text{—COO—})_n$,[90–92] $(\text{—CD}_2\text{CD}_2\text{OCO—}$  $\text{—COO—})_n$,[91–93] $(\text{—CD}_2\text{-}$
$\text{CD}_2\text{OCO—}$  $\text{—COO—})_n$.[92]

For syntheses of deuterated compounds, see Refs. 94–96.

5.3.3 Wave Number Calibration

The approximate positions of measured absorption bands can be obtained from the counter of the spectrophotometer as well as the chart scale. However these values represent rough measures, and calibration should be made by

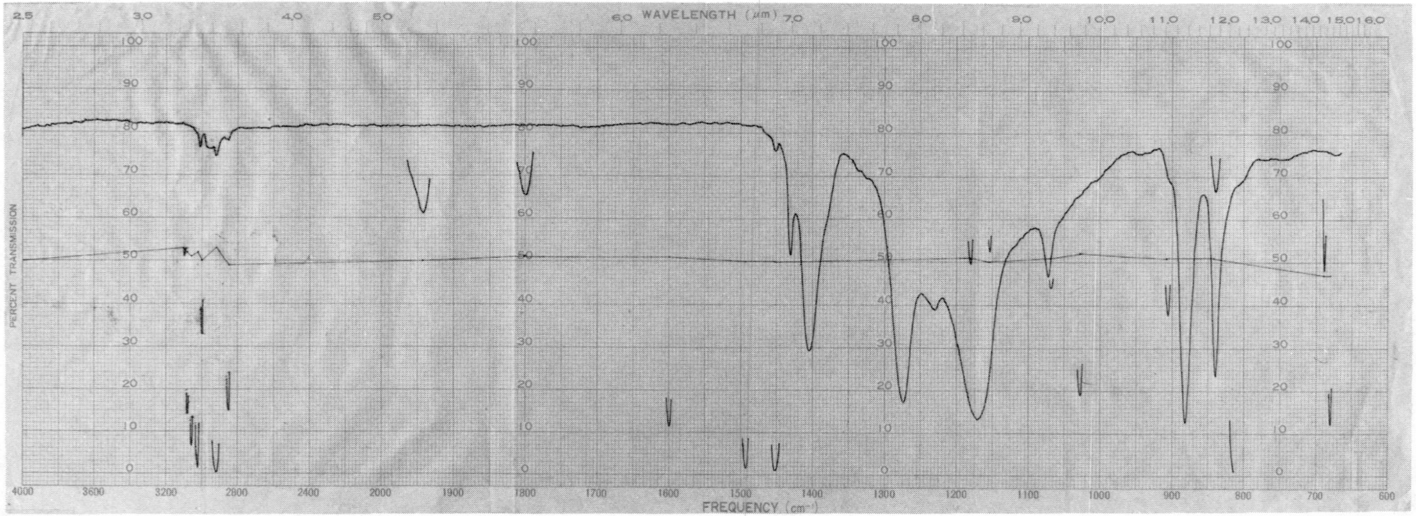

Fig. 5.10 Wave number calibration curve using polystyrene and 1, 2, 4-trichlorobenzene (two bands of the longer wavelength side). The main spectrum is of poly (vinylidene fluoride) form I.

superimposing spectra of standard materials in which the wave numbers of the bands have already been determined accurately.

Since most polymers do not have sharp absorption bands compared to gases, the spectrum of polystyrene is frequently used as the standard in the rock salt region (4000–650 cm^{-1}). Accurate wave numbers of the spectral bands of polystyrene are indicated in Fig. 5.4*b*. For the purpose of calibration, only the curves around the peaks of polystyrene with known wave numbers are recorded on the chart immediately after scanning the sample under study. Figure 5.10 shows an example of such a calibration for the case of poly(vinylidene fluoride) form I using polystyrene. This procedure is recommended for the measurement of new materials or important samples. The calibration curve is constructed so that the correct wave number of the band under examination is obtained by adding a correction to the observed wave number with a plus or minus sign depending on whether the calibration curve is above or below the 50% transmission line, respectively.

The bands in the potassium bromide region (700–400 cm^{-1}) are usually calibrated by using 1,2,4-trichlorobenzene, and those in the far-infrared region (lower than 400 cm^{-1}) are calibrated by the rotation spectrum of water. In addition, indene, methylcyclohexane, ammonia, water, and so on are sometimes used as standards. For further standard wave numbers see Refs. 97 and 98.

5.3.4 Treatment of Absorption Intensity

Infrared spectra are usually given in terms of percent transmission, defined by $(I/I_0) \times 100$, where I_0 is the intensity of the incident radiation and I is the intensity of the transmitted radiation. Where the Lambert-Beer law,

$$I = I_0 \exp(-kct) \tag{5.43}$$

holds, optical density or absorbance D is expressed by

$$\begin{aligned} D &= \log_{10}\left(\frac{I_0}{I}\right) = \log_{10} \mathrm{e} \ln\left(\frac{I_0}{I}\right) \\ &= 0.4343 \ln\left(\frac{I_0}{I}\right) = 0.4343kct \end{aligned} \tag{5.44}$$

where k is the absorption coefficient, t is the thickness of the sample, and c is the concentration. Accordingly, D is a quantity directly proportional to the amount of the absorbing material present in the beam path. This law should hold, accurately speaking, only for dilute gases and dilute solutions, but it has often been assumed to apply to other systems. In discussing absorption intensity, optical density D should be used. In most spectrophotometers

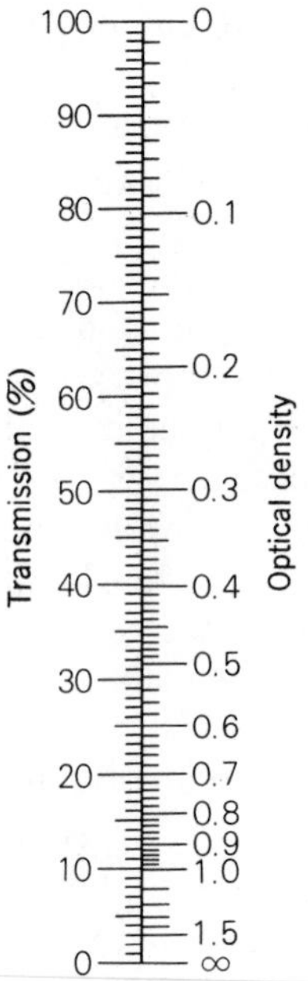

Fig. 5.11 Relationship between transmission (%) and optical density.

used to date, the scale of the recorder chart is often provided in terms of transmission. A conversion scale is given in Fig. 5.11 for convenience.

In practice the problem is the estimation of I_0. Although accurate treatments have been developed for gases and solutions, the so-called base line method, which is the most commonly used technique, especially for polymers, is now described. As shown in Fig. 5.12*a*, a tangential line is drawn between two points of minimum absorption *A* and *B* chosen on the both sides of the band under examination. This line is assumed to be the base line (100% transmission). Taking I_0 and I as shown in the figure, the optical

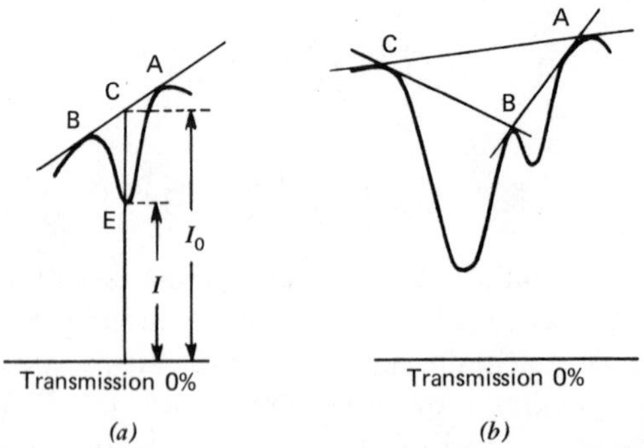

Fig. 5.12 Base line method.

density $D = \log_{10}(I_0/I)$ of the band is obtained by subtracting the optical density at point C from that at E on the optical density scale shown in Fig. 5.11. The choice of the base line is arbitrary. There are alternative choices for joining $A-C$ as shown in Fig. 5.12*a* and for joining $A-B$ and $B-C$ as shown in Fig. 5.12*b*. In such cases, the base line that gives good linearity on the calibration curve after comparing data from various samples should be chosen.

In the examination of the intensity change of a given absorption band, for instance, a crystalline band, the correction for the different thicknesses of the films is made using bands for which the intensity depends only on the thickness of the film, namely, an internal standard. The most recommended procedure is to prepare films having similar thicknesses.

When two or more bands overlap in the spectrum, the optical density curve is often resolved into individual bands for the purpose of accurate treatment of absorption intensity. If a Lorentzian function is assumed for each band, the optical density at the wave number $\tilde{\nu}$ is given by

$$D(\tilde{\nu}) = \sum_{i=1}^{m} \frac{a_i}{b_i^2 + (\tilde{\nu} - \tilde{\nu}_{0i})^2} \tag{5.45}$$

where $2b_i = (\Delta\tilde{\nu})_{1/2}$ is the half width, a_i/b_i^2 is the optical density at the peak, and $\tilde{\nu}_{0i}$ is the wave number. Subscript i indicates the ith band. Summation is over the bands in the given wave number range. A set of values is assumed first for parameters b_i and $\tilde{\nu}_{0i}$. Substituting an arbitrarily chosen m observed values of $D(\tilde{\nu})$ into Eq. 5.45 gives a set of linear equations with m unknown parameters a_i. Substitution of the parameters a_i into Eq. 5.45 and solution

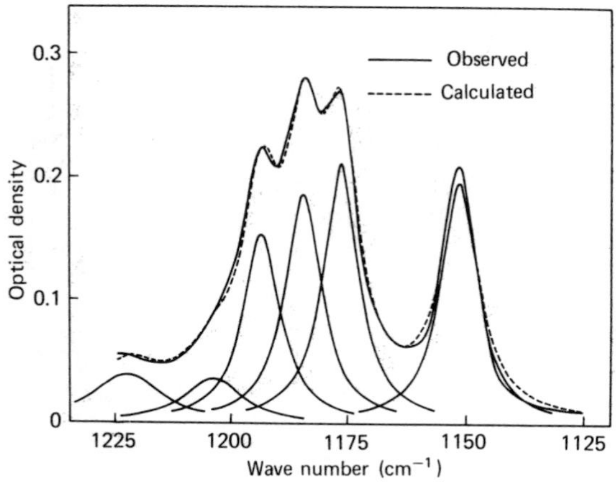

Fig. 5.13 Separation of the absorption spectrum of styrene–styrene-α-d_1 copolymer into Lorentzian curves.[99]

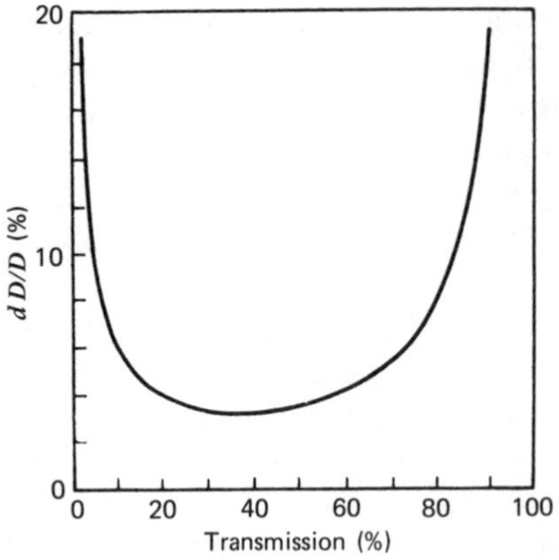

Fig. 5.14 The percentage error in optical density calculated assuming 1% errors for the 100 and 0% transmission lines [Reprinted with permission from *Anal. Chem.* **23**, 273 (1951). Copyright by the American Chemical Society.]

of the linear equations yield the calculated absorption curve. Trials are repeated by adjusting the parameters b_i and $\tilde{\nu}_{0i}$ until the calculated curve fits the observed one, or the least-squares method is applied. An example of the band resolution is shown in Fig. 5.13.

Figure 5.14 shows the percent error in optical density as calculated by assuming 1% errors for the 100 and 0% transmission lines. Since the most suitable absorption of the specimen lies in the range of 20–60% transmission, the proper thickness should be chosen.

5.3.5 Polarization Measurements

As already stated, polymer films can be uniaxially oriented by stretching, and in several polymers they can be doubly oriented by rolling. Figure 5.15 is a schematic representation of the relation between a plane-polarized beam and an oriented C=O bond. If the electric vector of the radiation is oriented parallel to the transition moment vector of the C=O stretching vibration as shown in Fig. 5.15*a*, the intensity of the C=O stretching band is maximum; on the other hand, if the electric vector is perpendicular to the C=O direction as in Fig. 5.15*b*, it disappears. In other words, a nonvanishing

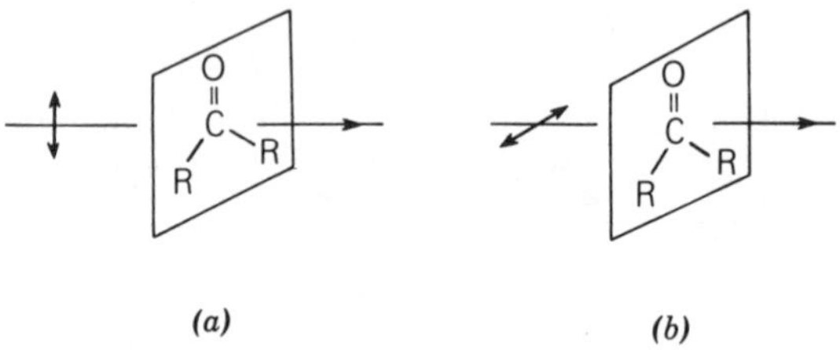

Fig. 5.15 Polarized infrared radiation and oriented C=O group. (From Jones and Sandorfy[5]).

scalar product of the transition moment vector $[\boldsymbol{\mu}]$ and the electric vector of the incident beam $\mathbf{E}$ is the necessary condition for infrared absorption.

$$|[\boldsymbol{\mu}]|\,|\mathbf{E}|\cos\theta \neq 0 \tag{5.46}$$

where θ is the angle between $[\boldsymbol{\mu}]$ and $\mathbf{E}$.

A. *Polarizers.* The most commonly used polarizer consists of a pile of thin silver chloride plates (about six) inclined at 26.5° to the beam as shown in Fig. 5.16 so that the angle of incidence α is Brewster's angle

$$\tan\alpha = n \tag{5.47}$$

where n is the refractive index of silver chloride and is about 2.1. At each reflection (internal and external) a fraction of the radiation whose electric vector is perpendicular to the plane of incidence is reflected away. The transmitted beam from six sheets then consists mostly of vibrations in the plane of incidence, about 95%.[101] It should be noted that in the quantitative interpretation of polarization measurements, a deficiency in pure polarization of about 5% leads to errors in dichroic ratios. Silver chloride should be stored in the dark, since its transparency is decreased by exposure to ultraviolet radiation. Silver chloride should not be mounted in contact with metals whose ionization tendency is larger than that of silver. In the frequency region lower than 500 cm^{-1}, PE films are used as the polarizer material instead of silver chloride. The refractive index of PE in the far-infrared region is 1.46, and Brewster's angle is about 55°. A fractional polarization higher than 95% is obtained.

A new type of infrared polarizer, namely, a wire-grid polarizer, has recently become available commercially. It consists of a grating formed on a silver chloride, silver bromide, KRS-5, or PE plate by uniformly spaced parallel strips of evaporated gold or aluminum (0.3 μ for infrared and 0.7 μ for far-infrared). Only the component of radiation with an electric vector perpendicular to the lines can pass through the polarizer. The fractional polarization for this polarizer is higher than 97%.[102–104]

Special care should be taken regarding polarization in the spectrophotometer itself. Figure 5.17 shows the transmission curves of a double beam

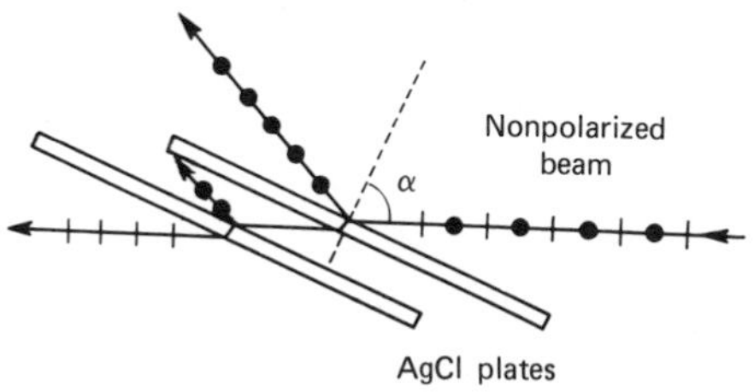

Fig. 5.16 Principle of the silver chloride polarizer; only two plates are shown. The vertical dash and dot indicate that the vibrational direction of the electric vector is parallel and perpendicular to the page, respectively.

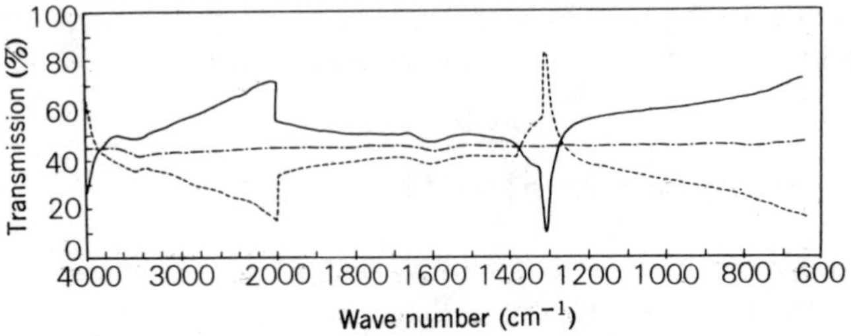

Fig. 5.17 Polarization character in a grating spectrophotometer (Japan Spectroscopic Co., DS-402G). (———) polarization component **E** vertical, (– – –) **E** horizontal, and (—·—·—) **E** at a 45° angle to the vertical direction. The discontinuity at 2000 cm^{-1} is due to the automatic exchange of gratings and the peak at 1300 cm^{-1} is due to the character of the specific grating.

grating spectrophotometer measured by placing a silver chloride polarizer in the sample beam. The solid line denotes the curve measured with the polarizer set in such a way that the electric vector of the polarized radiation is horizontal; the dashed line is obtained after rotation of the polarizer by 90° with the electric vector being vertical. The dash–dot line indicates the case when the electric vector is 45° to the vertical direction. A remarkable polarization property is noticed. In the present example, for instance, the polarization component parallel to the vertical direction has a transmission of only about 20% in the 2000 and 700 cm^{-1} regions. Therefore a suitable position for the polarizer would be behind the exit slit or before the entrance slit. If, due to the design of the apparatus, the only available space is the sample chamber, two polarizers, one placed in the sample beam and one in the reference beam, should be employed to obtain good spectra.

B. *Uniaxially Oriented Samples.* As seen in Fig. 5.17, it is desirable to make measurements with the polarizer placed at the 45° position. This arrangement is recommended if the sample is large enough to cover the whole cross section of the incident beam even when it is placed at an angle of 45° to the slit as shown in Fig. 5.18. After the spectrum is measured for one

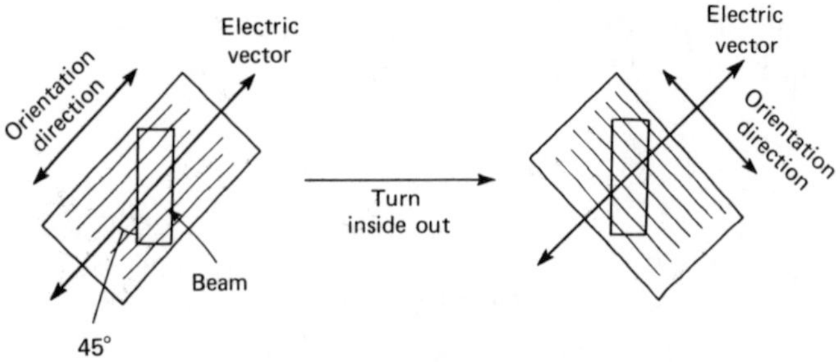

Fig. 5.18 Polarization measurements of an oriented film of sufficiently large size.

orientation, another measurement is made by turning the film as indicated. However, if the film is not large enough for this method, it should be placed in the vertical orientation. In this case the measurement is made by setting the polarizer vertical and then horizontal.

Figure 5.8 shows the polarized infrared spectra of uniaxially oriented PVA and deuterated PVAs. The solid and dashed lines denote the spectra measured so that the electric vector is perpendicular or parallel to the elongation direction of the sample (perpendicular or parallel radiation), respectively. The band at about 2930 cm^{-1} in Fig. 5.8*a* is assigned to the CH_2 stretching vibrations. This band is strong in perpendicular radiation and weak in parallel radiation, since the plane of the CH_2 group is normal to the molecular axis as shown in Fig. 4.85. Figure 5.8*a* shows this result. The bands strong in perpendicular radiation are called the perpendicular bands and are indicated by σ or $\perp$. The bands strong in parallel radiation are called the parallel bands and are indicated by π or $\|$.

The dichroic ratio R is used for the quantitative expression of infrared dichroism.

$$R = \frac{\log (I_0/I_{\|})}{\log (I_0/I_{\perp})} = \frac{D_{\|}}{D_{\perp}} \tag{5.48}$$

where I_0 is the intensity of the incident beam, $I_{\|}$ and $D_{\|}$ are the intensity of the transmitted beam and optical density in parallel radiation, and $I_{\perp}$ and $D_{\perp}$ are those in perpendicular radiation. The dichroic ratio is greater than one for parallel bands and smaller than one for perpendicular bands. To treat the dichroic ratio accurately, corrections are necessary for the polarization characteristics of the spectrophotometer, slit width, incomplete orientation of the sample, and so on. For a further discussion, see Refs. 21 and 27.

In the case of a uniaxially oriented sample, if the direction of the transition moment for one absorbing group makes an angel θ with the elongation direction, the dichroic ratio is given by[105]

$$R = 2 \cot^2 \theta \tag{5.49}$$

When $\theta = 54°44'$, $R = 1$, giving rise to no dichroism. Good examples for this case are the CH_2 symmetric stretching bands of *it*-polystyrene and *it*-PP appearing at 2849 and 2844 cm^{-1}, respectively (Figs. 5.19 and 5.20).[65] In these polymer molecules, which have the $(TG)_3$ conformation, the direction of the transition moment of the CH_2 symmetric stretching vibration, namely, the bisector of the angle HCH of the CH_2 group, is considered to make an angle of about 55° with the molecular axis (Fig. 5.21). If each CH_2 group is treated independently, R may be close to one. Rigorously speaking, this problem should be treated under the factor group $C(2\pi/3)$ (see Section 5.5.8). The parallel band belonging to the A symmetry species and the per-

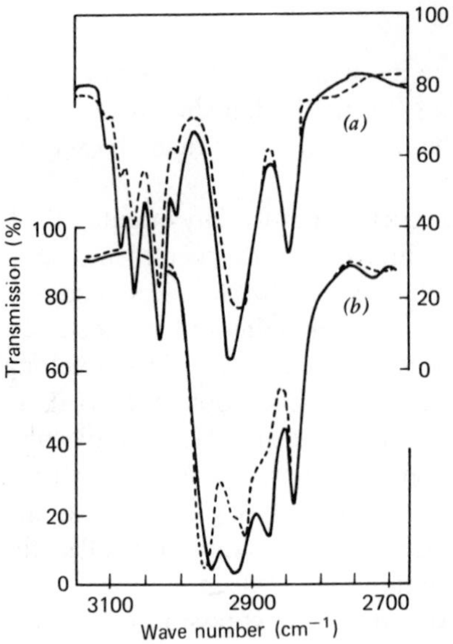

Fig. 5.19 Polarized infrared spectra of (*a*) *it*-polystyrene and (*b*) *it*-PP. (———) Perpendicular radiation. (-----) Parallel radiation.[65]

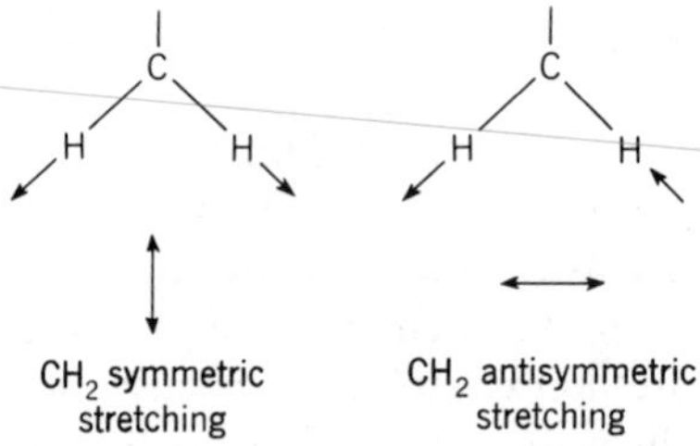

Fig. 5.20 CH_2 symmetric and antisymmetric stretching vibrations. Large arrows indicate the directions of the transition moment.

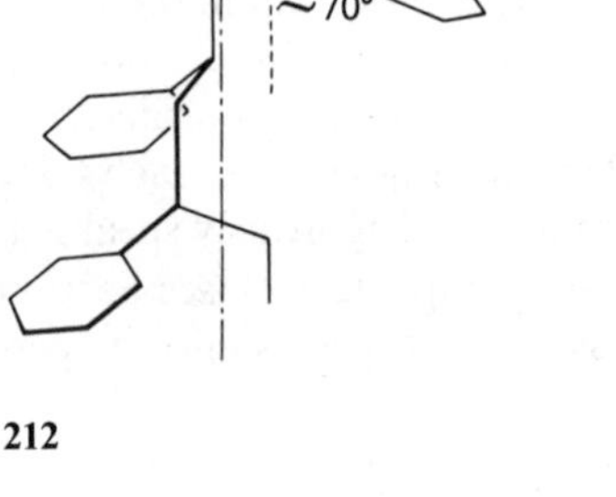

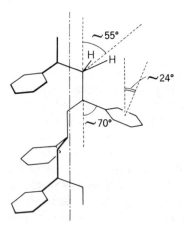

Fig. 5.21 Molecular conformation of *it*-polystyrene in the crystalline state. (From Natta and Corradini.[215])

pendicular band belonging to the *E* species may appear with similar intensity. The CH_2 antisymmetric stretching vibration is nearly perpendicular to the molecular axis. Thus no dichroic nature is associated with the CH_2 symmetric stretching bands and distinct perpendicular dichroism is associated with the CH_2 antisymmetric stretching bands (about 2924 cm^{-1}) as shown in Fig. 5.19.

If an unpolarized spectrum of an oriented sample is compared to that of an unoriented sample of the same polymer, bands that appear as the perpendicular bands in the polarized spectra are found with considerably reduced intensity compared to the bands appearing as the parallel bands in the polarized spectra. This fact is important for a quantitative treatment of the band intensity, especially in association with the effect of orientation, and can be interpreted as follows. The unpolarized light is considered to be incident in the y direction to the film in the plane xz as shown in Fig. 5.22. In the unoriented sample, the directions of the transition moments of the absorbing units distribute randomly; each component in the x, y, z directions has a probability of $\frac{1}{3}$. In a perfect uniaxially oriented sample with the z axis as the orientation direction, the components of the transition moment of the parallel bands are given by (0, 0, 1) and those of the perpendicular bands are given by ($\frac{1}{2}$, $\frac{1}{2}$, 0). Therefore, in the case of an unoriented sample, two thirds of all the absorbing units contribute to absorption, since only the x and z components can contribute. On the other hand, in oriented samples, all units contribute to the parallel bands, while one half of the units contribute to the perpendicular bands.

C. *Doubly Oriented Samples.* In the case of doubly oriented samples, pleochroism, that is, absorption anisotropy in the plane normal to the elongation direction, can be observed for perpendicular bands. The infrared studies of doubly oriented polymer samples were first carried out by Elliott et al. on nylon 66, PVA, and so on in the 3 μ region.[106]

Pleochroic measurements have been made on several bands in *at*-PVA.[107] Figure 5.23 is a schematic representation of the relation between the cross section perpendicular to the direction of rolling (the direction of the molecular chain axis) of the doubly oriented film and the incident polarized infrared

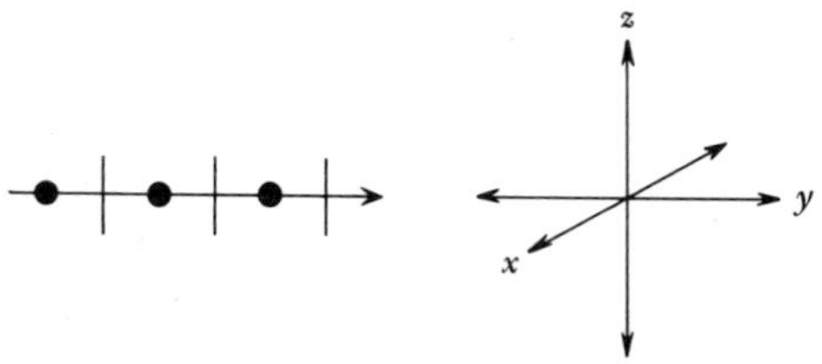

Fig. 5.22 Incidence of unpolarized light on a sample.

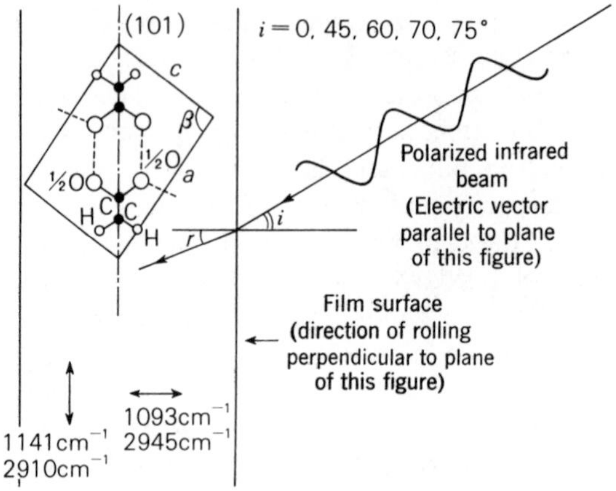

Fig. 5.23 The relation between the cross section of the doubly oriented PVA film and the incident polarized beam.[107]

radiation. As shown in this figure, both the electric vector and the direction of propagation of the polarized infrared beam are perpendicular to the direction of rolling. The spectra were observed at various angles of incidence i (0, 45, 60, and 70°) by rotating the sample around the vertical axis (the direction of rolling). The (101) plane in PVA coincides with the zigzag plane of the molecule and orients itself parallel to the rolled plane.

The absorption intensity of the perpendicular band having the transition moment parallel to the rolled plane should be strongest when i is 0° and decreases as i increases. The intensity of the band with the transition moment normal to the rolled plane should exhibit the reverse behavior. In view of this fact, it is possible to determine the approximate direction of the transition moment from the results of absorption measurements. The CH_2 stretching bands of PVA are observed at 2910 (symmetric stretching) and 2945 cm^{-1} (antisymmetric stretching). The direction of the transition moment coincides with the bisector of the HCH angle in the symmetric stretching and is perpendicular to the bisector in the antisymmetric stretching as shown in Fig. 5.20. Figure 5.24 shows the spectra taken with different angles of incidence i of the light whose electric vector is perpendicular to the elongation direction.[107] When i is 0°, that is, the incident beam is normal to the film plane, the symmetric stretching band at 2910 cm^{-1} is strong and becomes weaker as the beam is inclined. The antisymmetric stretching band at 2945 cm^{-1} shows the reverse behavior. This result indicates that the zigzag plane is nearly parallel to the rolled plane, and the perpendicular bands at 1141 and 1093 cm^{-1} have transition moments parallel and perpendicular to the rolled plane, respectively.

Pleochroic measurements can also be made with the following method.

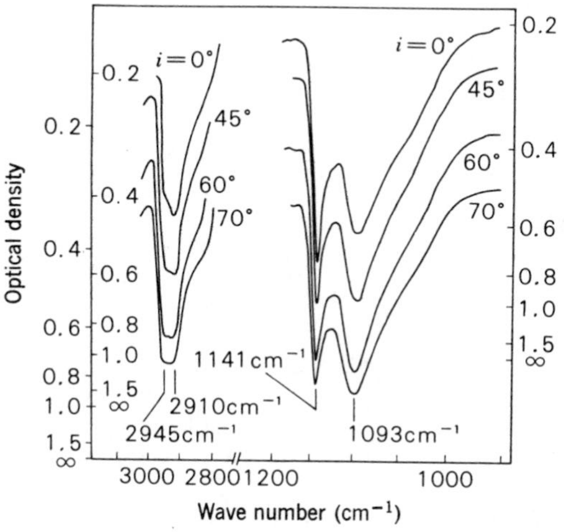

Fig. 5.24 Infrared spectra of doubly oriented PVA film taken at different inclinations (i denotes the angle of incidence).[107]

Polarized infrared spectra are measured using perpendicular radiation on both uniaxially and doubly oriented samples of similar thickness. The bands that are more intense in the doubly oriented sample than in the uniaxially oriented sample may be considered to have the transition moments nearly parallel to the rolled plane. The transition moments of the bands showing the reverse behavior are nearly perpendicular to the rolled plane.[108]

Further references on pleochroic studies are given for PET,[48,109] PVA,[110,111] nylon 6,[112] nylon 66,[106] and polytetrahydrofuran.[88,113]

D. *Preferentially Oriented Samples.* It has been confirmed by X-ray diffraction that a PE sample in which the a axis (the molecular axis is the c axis) turns into the stretching direction can be obtained by heat relaxation of a stretched sample.[114] The same type of preferred orientation is also found in some machine-extruded films. The spectra of such a PE film are shown in Fig. 5.25.[115] The solid line denotes the spectrum taken using the polarized radiation with the electric vector parallel to the machine direction, and the dashed line indicates the electric vector perpendicular to the machine direction. This is interpreted in terms of factor group analysis in which the transition moment of the 730 cm^{-1} band is considered to be parallel to the a axis, and that of the 720 cm^{-1} band is considered to be parallel to the b axis, as is described in Section 5.5.7.

In addition, preferred orientations have been found for various polymers in the early stages of stretching. For instance, in poly(vinyl chloride) the 603 cm^{-1} band shows parallel dichroism at an elongation ratio of 2.5 and

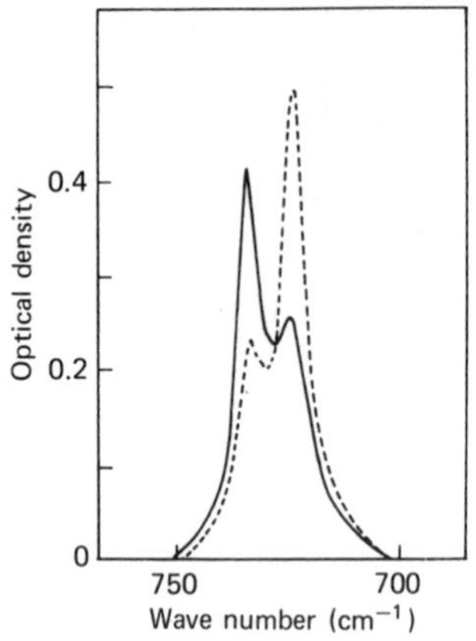

Fig. 5.25 Polarized spectra of machine-extruded PE film. (———) Electric vector parallel to machine direction. (-------) Electric vector perpendicular to machine direction. (From Krimm.[115])

turns into perpendicular dichroism at a sevenfold elongation.[116] This band is assigned to the CCl stretching vibration and has been interpreted as follows. The *a* axis preferentially orients parallel to the elongation direction at a 2.5-fold elongation [in poly(vinyl chloride) the *a* axis is perpendicular to the molecular axis], and then the *c* axis (molecular axis) orients parallel to the elongation direction at a sevenfold elongation. Such a preferred orientation seems to arise when crystallites grow predominantly in a direction perpendicular to the molecular axis, as has been found in polymers such as PE,[117–119] PET,[120] and nylon 66.[117]

5.3.6 Measurements at High and Low Temperatures or under High Pressure or Stress

Conduction heating cells[121] are easily constructed and are also available commercially. Blowing apparatus has also been used for crystallization measurements of PET.[122] In high-temperature measurements, special care must be taken to eliminate the radiation from the sample. There is no problem when the chopper is located between the radiation source and the sample. However the chopper is placed between the sample and the detector in most double beam spectrophotometers. Thus radiant energy from the sample is amplified and reduces the apparent optical density. A double chopping system is used in the Hitachi FIS-III far-infrared spectrophotmeter, Perkin-Elmer Type 180, and so on to avoid this difficulty.

Spectral measurements of crystals at low temperatures have the following merits: (1) bands become sharper and have better separation as a result of the diminution of the effect of thermal motions; (2) the spectra reveal variation of the intermolecular interaction; (3) spectra of low-temperature crystal modifications, as well as unstable crystal modifications, can be observed.

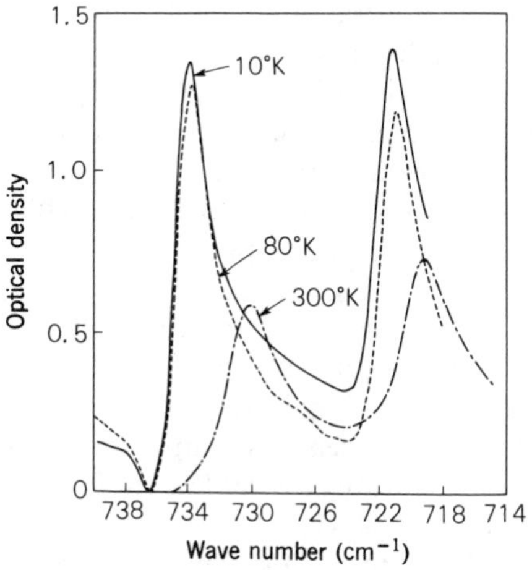

Fig. 5.26 Infrared spectra of PE single crystals taken at 10, 80, and 300°K. (From Shen et al.[123])

Cryostats for low-temperature infrared measurements down to the liquid helium temperature are commercially available (for instance, from Oxford Instrument Co. Ltd.).

As an example of infrared spectra taken at low temperatures, Fig. 5.26 shows the temperature dependence of the 730 and 720 cm^{-1} bands of PE.[123] These bands shift to higher frequency on lowering the temperature, revealing an increase of intermolecular interaction.

Using an anvil-type high-pressure cell,[124] Brown observed the spectra of polytetrafluoroethylene at 60–80 kbar (for conversion of units see Appendix H). The remarkable decrease of the intensities of the 1210 and 640 cm^{-1} bands led him to suggest that a transition into the planar zigzag modification had occurred. Since the transmitting area of the anvil is small, a beam condenser was used in the experiment (see next section). Wu[125] measured the spectra of PE at 40 kbar using a different type of high-pressure cell[126] with NaCl windows and found that the splitting width of the CH_2 rocking bands (see Section 5.5.7) increases up to 30 cm^{-1} at 35 kbar.

Wool and Statton[127,128] measured changes of the polarized spectra of several polymers under stress relaxation and creep. For instance, in the spectra of perpendicular radiation of *it*-PP the 975 cm^{-1} band is shifted to higher frequencies by about 1–2 cm^{-1} when the stress is relaxed from 22.7×10^3 to 8.3×10^3 psi.

5.3.7 Measurement of Small Samples

For most spectrophotometers the effective cross-sectional area of the radiation beam at the sample position is about 12 mm × 6 mm. If the sample size is less than this, only part of the radiation beam is effectively used at the sample. In this case the portion of the beam outside the sample should be masked by aluminum foil, and the reference beam should correspondingly be balanced by using an attenuater. For samples of smaller size special illuminating and condensing systems are used. These usually consist of potassium bromide lenses or aluminized curved mirrors. Such beam condensers (condensing ratio is about 1:4–1:5) are commercially available. For samples of even smaller size, a microspectroscopic apparatus is useful.

A. *Infrared Microspectroscopic Apparatus.* A microspectroscopic apparatus can be used to measure infrared spectra on small filaments, oriented samples, single crystals, and small amounts of liquid. Measurements are also possible on small selected parts of an inhomogeneous sample, such as a part of a spherulite. The apparatus are commercially available from Perkin-Elmer, Hitachi, Japan Spectroscopic Co., and others.

Figure 5.27 shows the optical layout of the Hitachi IM-3 infrared microscopic apparatus. It is used by setting in the sample beam passage of a double beam spectrophotometer. Since the air path length of the apparatus is about 1 m, a 1 m gas cell must be placed in the reference beam for compensation. The radiation beam from the source proceeds upwards by reflection at mirror *A* and forms a reduced image of the source (× 1/8.3) on the sample

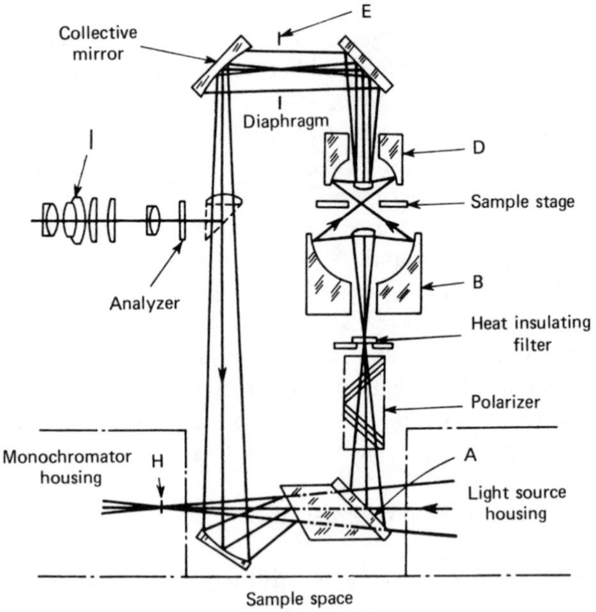

Fig. 5.27 Optical arrangement of the Hitachi IM-3 infrared microspectroscopic apparatus.

stage after passing through the Schwalzschild condenser B (combination of a small concave mirror and a large convex mirror with a centered hole). The beam again forms an enlarged image (× 30) at diaphragm E after passing through an objective mirror system D. The beam is reflected from two ellipsoidal and one plane mirror and forms an image at H, the normal point of entry with the apparatus not in position. The sample can be observed by using a viewer I (× 160). A polarizer is used if necessary. To avoid heating of the sample due to the condensed beam, an interference filter is used for samples with low melting points. The effective size at the sample is 0.87 mm × 1.45 mm when the diaphragm is fully opened. Under this condition the intensity of the beam after it has passed through the apparatus is reduced to 20–25% of the original intensity. To compensate for this effect, an attenuator is inserted into the reference beam. According to the author's experience, the spectrum taken on a sample of the size 18 μ × 1.0 mm is still useful, although the accuracy is reduced.

Figure 5.28 shows the optical arrangement of the Japan Spectroscopic Co. IRM-1 infrared microspectroscopic apparatus. This is for a double-beam spectrophotometer. The sample beam side is designed symmetrically with respect to the reference beam side. The condenser reduction ratio is 1 : 8. The beam condenser system consists of two ellipsoidal mirrors (M_2, M_3) and two plane mirrors, each having a central hole (M_1, M_4). The sample is inserted between M_1 and M_4, about 10 mm apart. The sample can be observ-

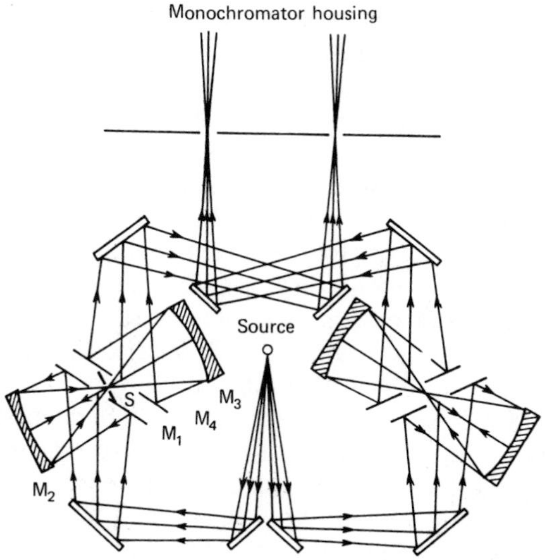

Fig. 5.28 Optical arrangement of the Japan Spectroscopic Co. IRM-1 microspectroscopic apparatus.

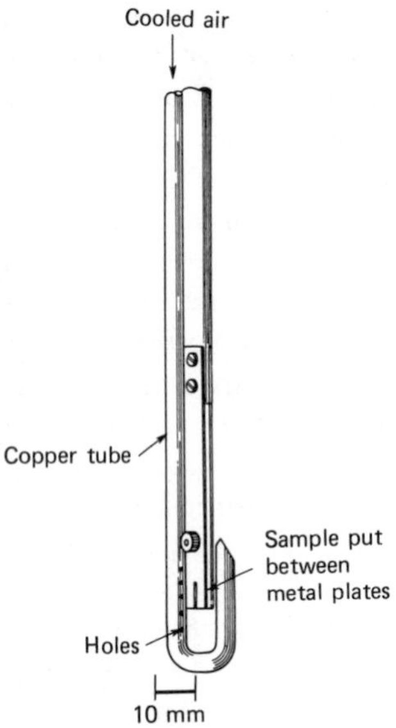

Fig. 5.29 Sample holder and cooling device for the Japan Spectroscopic Co. IRM-1 infrared microspectroscopic apparatus.

ed by magnifying with a viewer through a hole at the center of M_2. Because of the simplicity of the optical system, efficiency of the incident radiation is about 75%. Figure 5.29 shows a sample holder. To reduce heating of the sample, a copper tube with small holes is attached through which cooled air can be supplied. Polarization measurements can also be made using the polarizer in the spectrophotometer.

B. *Examples of Measurements with Microspectroscopic Apparatus.* Spectral measurements of single filaments of several polymers including nylon 66 using the Perkin-Elmer apparatus have been reported.[120,129,130] Pleochroic measurements on a highly doubly oriented PVA[111] and PET[48] are now considered. The X-ray patterns of the samples used in the present measurements are shown earlier in Figs. 4.22 and 4.23. As shown in Fig. 5.30, the polarized infrared beam is incident to the sample in the end direction defined in Fig. 4.22. The spectra taken with the electric vector parallel and perpendicular to the rolled plane are compared.

The doubly oriented sample was embedded in *n*-butyl methacrylate resin and then cut into a thin section, about 10 μ thick, perpendicular to the rolling direction with a microtome. The following procedure was used. A rectangular shaped piece (1.5 mm × 20 mm) was cut out from the central part of the doubly oriented sample (the thickness is 0.08 mm for PVA and 0.2 mm for PET) and was then fixed on a U-shaped wire holder with a thread as shown in Fig. 5.31. The sample was put in a gelatin capsule, which was then filled

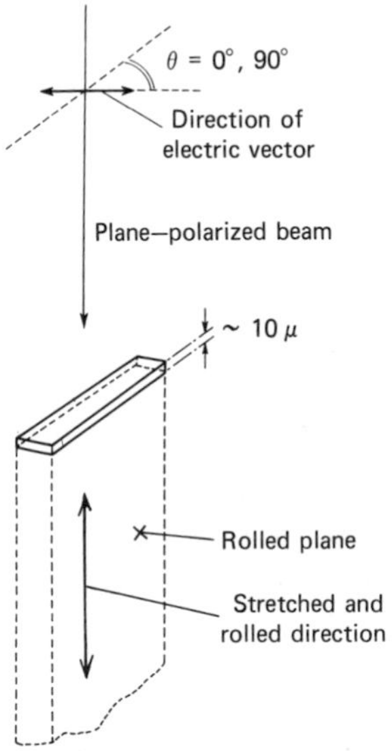

Fig. 5.30 Microspectroscopic measurements of a small section of a doubly oriented sample.[48]

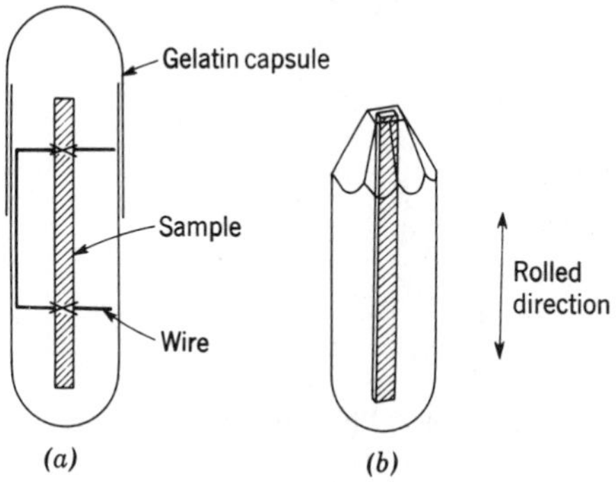

Fig. 5.31 Sample embedded in acrylic resin.

with *n*-butyl methacrylate containing 0.5–1% benzoyl peroxide as a polymerization initiator. The polymerization was accomplished by letting the solution stand for one day at about 55°C. After dissolving the capsule with hot water, the resin was cut into the shape shown in Fig. 5.31*b*, and sections were made using a microtome.

Figure 5.32 shows the results obtained for PVA,[111] and Fig. 5.33*a* for PET.[48] Solid and dashed lines indicate the spectra taken with the electric vector parallel and perpendicular to the rolled plane, respectively. Although the pleochroic measurements of a doubly oriented PVA film by the tilting method are described in Section 5.3.5.C, a far more distinct result can be

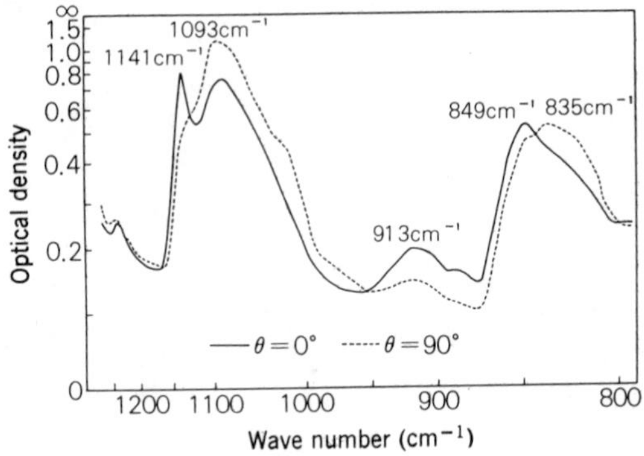

Fig. 5.32 Infrared spectra of a highly doubly oriented PVA sample obtained by the microspectroscopic method. (————) Electric vector parallel to the rolled plane. (-------) Electric vector perpendicular to the rolled plane.[111]

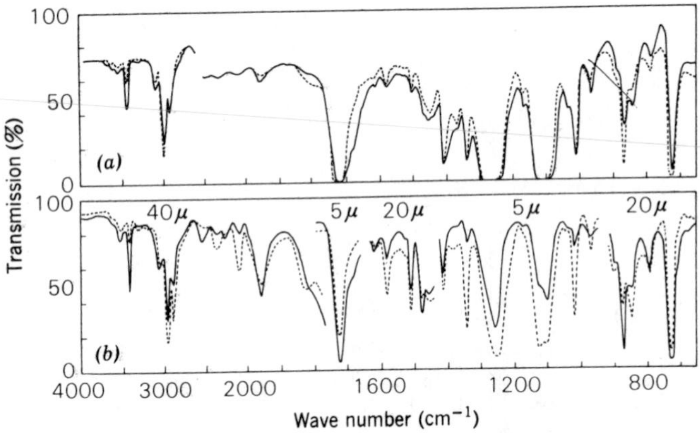

Fig. 5.33 (*a*) Infrared spectra of a highly doubly oriented PET sample obtained by the microspectroscopic method: (————) electric vector parallel to the rolled plane; (– – – – – –) Electric vector perpendicular to the rolled plane. (*b*) Polarized infrared spectra of a uniaxially oriented PET film: (————) perpendicular radiation; (– – – – – –) parallel radiation. The film thickness is indicated in the spectra.[48]

obtained using the technique shown in Fig. 5.32. For PET useful information about band assignments have also been obtained in relation to the orientation of the benzene rings (parallel to the rolled plane).

If an oriented sample of sufficient size cannot be obtained for the usual spectral measurements, the microspectroscopic technique is very useful. The spectra of POM-d_2 shown in Fig. 5.47 were measured in this way.[84]

5.4 MEASUREMENT OF RAMAN SPECTRA

The 632.8 nm line (red) of the He–Ne laser (power: 50–80 mW) and 488.0 (blue) and 514.5 nm (green) lines of the Ar^+ ion laser (power: 100–3000 mW) have been used most often as the exciting source.

5.4.1 Measurement of Polymer Samples

A. *Setting of Samples.* The laser beam is usually condensed to a diameter of about 0.3 mm with a convex lens and is focused on the sample. Therefore, in this case a small pellet of sample is suitable rather than the film-shaped samples used in the case of infrared absorption measurements. Figure 5.34 shows methods of sample arrangement that are often used for polymers. For samples of fairly good transparency, *a* or *c* is suitable. To obtain strong Raman scattering, the excitation beam should be incident as close as possible to the surface facing the observation direction. For nontransparent or colored samples, *b* is suitable. Powdered samples can be measured in pellet form using the potassium bromide disk technique. Of these methods, *b* is the simplest and has been used most frequently. The side and back surfaces of the sample are covered with an aluminum foil. To prevent decomposition or melting of the sample due to the laser irradiation, a rotating cell *f* may be

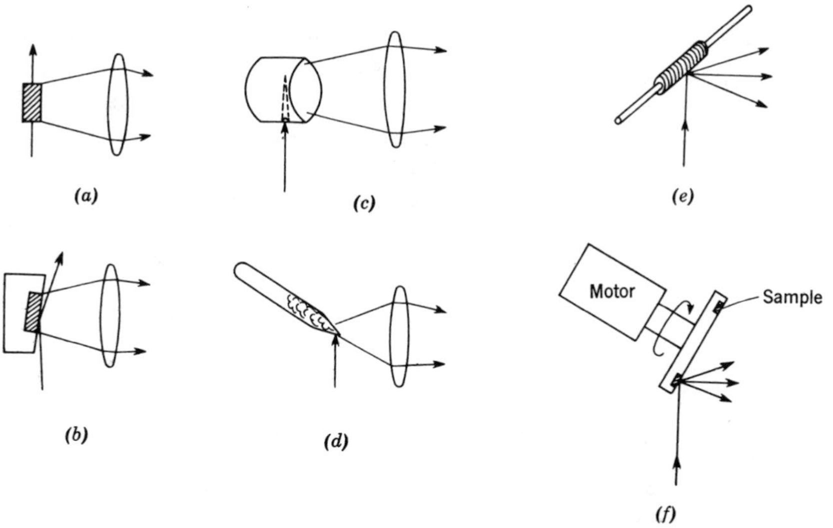

Fig. 5.34 Methods of sample arrangement: (*a*) vertical illumination; (*b*) oblique front illumination; (*c*) vertical illumination of a disk having a conical cavity; (*d*) powder sample; (*e*) fiber sample wound round a glass tube; (*f*) rotating cell. Sample is packed in a circular groove.

used. A stream of cooled nitrogen or air is also useful in reducing temperature.

Liquid samples are easily measured using an ampule (for an injection) after careful purification and then setting it on the sample stage. For solution studies, water may be used as solvent. This is a great advantage over the case of infrared measurements.

Elimination of the strong fluorescence frequently appearing in the Raman spectra of polymer samples is difficult. The usual purification by repeated reprecipitation does not always help. Active carbon is useful in some cases. In addition, preirradiating the sample with the laser beam for some time before recording the spectrum often reduces the fluorescence. Different exciting frequencies may also help.

B. *Depolarization Ratio.* The depolarization ratio can be measured for transparent liquid samples and optically clear solids. As shown in Fig. 5.35, the laser light polarized with the electric vector parallel to the vertical direction (the z axis) is incident to a liquid sample or a solution from the left-hand side in the y axis. Then the scattering in the x direction is observed using an analyzer. The observed intensity of the component polarized in the y direction is denoted by $I_{\perp}$ (Fig. 5.35*a*), and the component polarized in the z direction is denoted by $I_{\parallel}$ (Fig. 5.35*b*). The depolarization ratio is defined by

$$\rho = \frac{I_{\perp}}{I_{\parallel}} \tag{5.50}$$

The value of ρ is lower than 0.75 for totally symmetric vibrations (polarized lines; p) and is 0.75 for all other vibrations (depolarized lines; dp) (Ref. 4, p. 48). Figure 5.36 shows the Raman spectra of carbon tetrachloride; the solid line denotes $I_{\parallel}$, and the dashed line denotes $I_{\perp}$. The value of ρ is 0.75 for the 218 cm^{-1} band (doubly degenerated bending vibration, dp) and is less than 0.01 for the 459 cm^{-1} band (totally symmetric stretching, p).

For transparent solid samples, the depolarization ratio can also be observed. Vasko and Koenig[131] measured the depolarization ratio of the

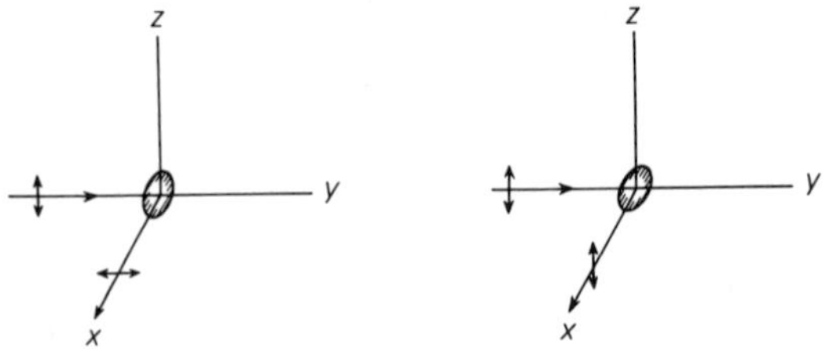

(*a*) $I_{\perp}$ (*b*) $I_{\parallel}$

Fig. 5.35 Measurement of depolarization ratio.

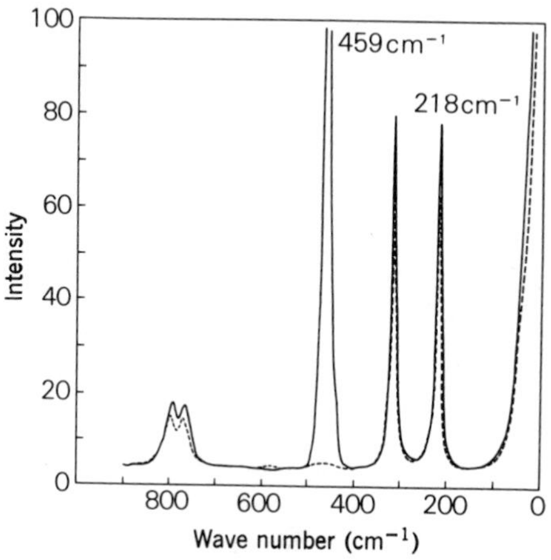

Fig. 5.36 Raman spectra of carbon tetrachloride indicating depolarization ratio. (———) $I_{\parallel}$, (---------) $I_{\perp}$.

lines for a smectic *it*-PP sample prepared by quenching the melt in ice–water. They found the similarity between the values for the smectic sample and the corresponding values measured for a molten sample, and deduced that *it*-PP takes the (3/1) helix conformation in the molten state, essentially similar to the conformation in the smectic sample. Measurements of the depolarization ratio for other polymers are also possible. However the measurements may be difficult for samples in which refractive indices are discrete over a small range. Furthermore birefringent samples cannot yield meaningful depolarization ratios because of the disturbance of the polarization direction.

5.4.2 Polarized Raman Spectra of Oriented Samples

Characteristic Raman spectra of an oriented sample can be observed using the combination of the incident direction, the direction of the electric vector of the exciting light, the observing direction, and the direction of the electric vector of the scattered light.

A. *Single Crystals.* It is now supposed that a rectangular single crystal is placed in position as shown in Fig. 5.37 and that the incident radiation is coming along the y axis with the electric vector along the z axis and observation is along the x axis with the electric vector along the y axis. The polarized Raman spectrum measured in this arrangement is indicated by the symbol $y(zy)x$ according to the notation proposed by Porto et al.[132] The letters on the left- and right-hand sides outside the parentheses indicate the directions of the incident light and the observation direction, respectively. The letters in the parentheses on the left- and right-hand sides are the directions of the electric vector of the incident and scattered lights, respectively. In the spec-

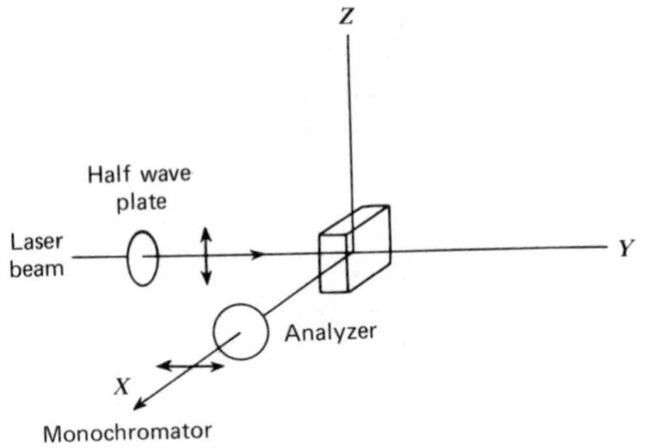

Fig. 5.37 Measurement of the polarized Raman spectrum of a single crystal. $y(zy)x$.

trum denoted by $y(zy)x$, the vibrations that change only the α_{yz} component of the polarizability tensor are observed. The polarization directions of the incident and scattered lights can be changed by inserting a half-wave plate and rotating an analyzer, respectively. As a sample for polarization measurements, a cube of several millimeters is suitable; the preparation of the sample is easy compared with the preparation in the case of infrared absorption.

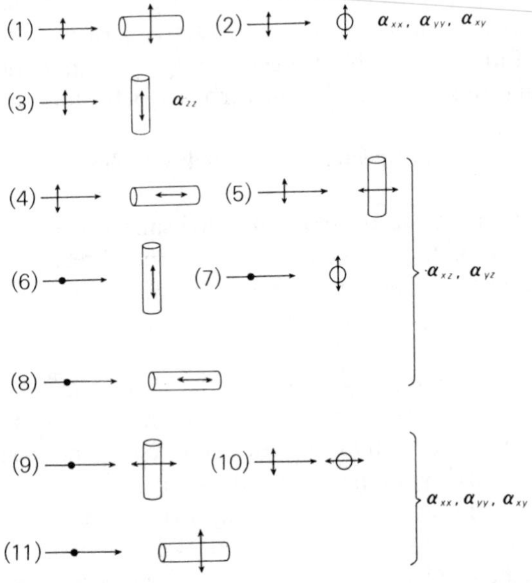

Fig. 5.38 Arrangements of a uniaxially oriented sample for polarization measurements and the observed components of polarizability tensor. ↕ and • indicate the direction of the vibrating electric vector, parallel and perpendicular to the page, respectively. The orientation direction (cylindrical direction) is along the z axis.

B. *Uniaxially Oriented Polymer Samples.* For a uniaxially oriented sample all components of the polarizability tensor cannot be observed independently since some of them overlap in a spectrum. Figure 5.38 shows the tensor components observed for a uniaxially oriented sample when the exciting light is incident from left-hand side and the scattering is observed normal to the page. Here the orientation direction (cylindrical axis) is along the z axis.

Figure 5.39 shows the polarized Raman spectra of a uniaxially oriented PE observed by Carter.[133] The three spectra from the top in the figure are the ones observed with the arrangements (2), (3), and (5) of Fig. 5.38, respectively. The axes a, b, and c in Fig. 5.39 correspond to x, y, and z in Fig. 5.38. Interpretation of these spectra indicates that the 1370 cm^{-1} line should be assigned to the CH_2 twisting mode of the B_{1g} species (see Sections 5.5.7 and 5.7.1). In this book the relationship between a, b, and c axes and the symmetry species is chosen in the same way as that of the factor goup analysis discussed in the subsequent section. Therefore the choice of these axes in Fig. 5.39 is different from Carter's original paper.[133] Polarized Raman spectra also have been measured for *it*-PP,[134] PET,[135-137] poly(methyl methacrylate),[138] and so on.

In addition, polarized Raman spectra of doubly oriented specimens of *at*-PVA and PET have been measured by a method similar to that used for single crystals and have given useful information.[139]

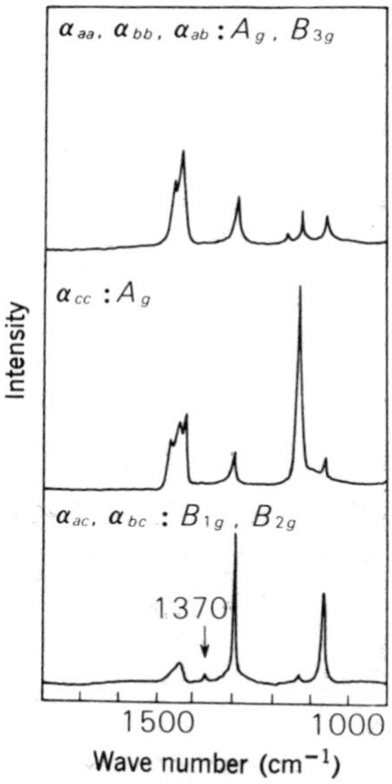

Fig. 5.39 Polarized Raman spectra of a uniaxially oriented PE. The axes are taken differently from in the original paper (see text). (From Carter.[133])

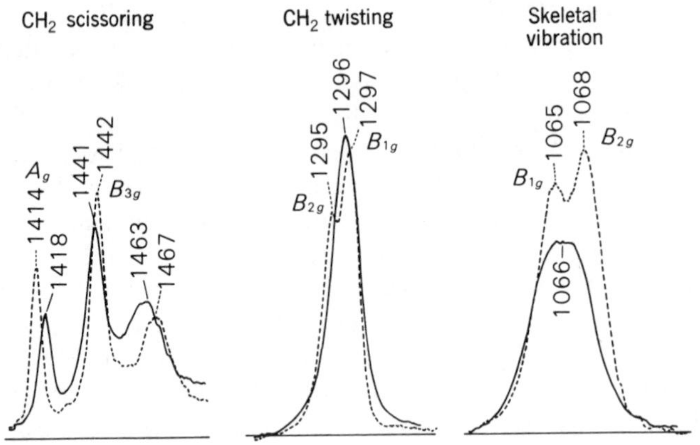

Fig. 5.40 Raman spectra of PE at 25°C (———) and at − 160°C (------). A_g, B_{1g}, and so on denote the symmetry species of the crystal (see Fig. 5.45). (From Boerio and Koenig.[140])

5.4.3 Measurements at Various Temperatures

The Raman measurements at high and low temperatures and under high pressures are carried out in the same way as described for X-ray and infrared measurements. Glass can be used as windows in Raman measurements.

Figure 5.40 shows the Raman spectra of PE taken at room and liquid nitrogen temperatures.[140] The Raman active vibrations in a single molecular chain of PE should split into two Raman active vibrational modes in the crystal as a result of intermolecular interaction (see Section 5.5.7). Such a splitting of Raman lines is observed only for the CH_2 scissoring vibration modes at a room temperature; these are 1441 (the B_{3g} mode of crystal) and 1418 cm^{-1} lines (the A_g mode of crystal). At liquid nitrogen temperature, the CH_2 twisting lines at about 1296 cm^{-1} and the C—C stretching bands at about 1066 cm^{-1} show splittings of 2–3 cm^{-1}.

5.5 FACTOR GROUP ANALYSIS

Since crystalline polymers consist of mostly simple sequence units, theoretical analysis of vibrational spectra can be made utilizing the symmetry of the molecular chain and crystal. Factor group analysis is one of the most important processes for the theoretical analysis of vibrational spectra of high polymers. This method can be used to determine the number of normal vibrations belonging to each symmetry species and also selection rules for infrared absorption and Raman spectra.

In this section, the factor group analysis of a one-dimensional space group

(Section 3.3) for a single chain of PE[141,142] is given as an example. The treatment is then extended to the PE crystal,[142] for which a three-dimensional space group analysis is required (Section 3.4). In addition, a single chain of POM is discussed as an example of helical molecules. The fundamental concepts, definitions, and theorems of group theory are briefly described. For a more detailed study of group theory, many good textbooks are available, such as Refs. 4, 21, 143–146.

5.5.1 Assumption of Infinitely Long Chains (Using Born's Cyclic Boundary Condition)

The translational symmetry operations, which are infinite in number and denoted by $T_n (n = -\infty, \cdots, -1, 0, +1, \cdots, +\infty)$, T_0 being the identity operation, are introduced in Section 3.3. For mathematical convenience, Born's cyclic boundary condition in which the translation T_N is considered to be equivalent to the identity T_0 is now introduced:

$$T_N = T_0 \tag{5.51}$$

Therefore there are a finite number of translations T_n $(n = 1, 2, \cdots, N)$. The cyclic boundary condition means that the infinitely long straight chain is replaced by a closed circle with finite repeat units, as shown in Fig. 5.41.

Similar considerations apply to a three-dimensional crystal. A translation may, in this case, be denoted by $\mathbf{T}_{n_1 n_2 n_3}$, with the three subscripts corresponding to translations along the three crystal axes. Born's cyclic boundary condition is then described as

$$\mathbf{T}_{N_1 N_2 N_3} = \mathbf{T}_{N_1 00} = \mathbf{T}_{0 N_2 0} = \mathbf{T}_{00 N_3} = \mathbf{T}_{000} \tag{5.52}$$

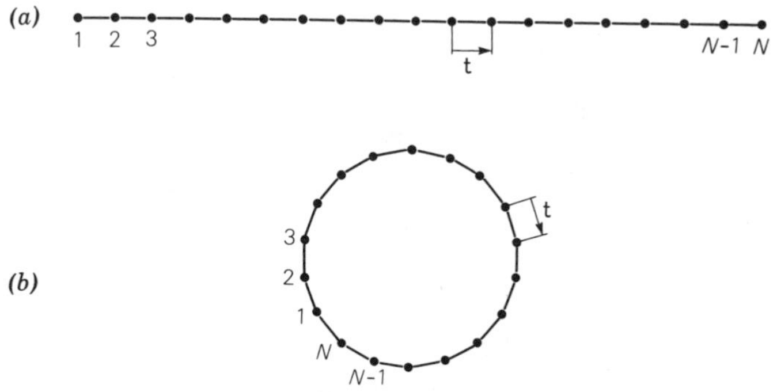

Fig. 5.41 Born's cyclic boundary condition. (*a*) One-dimensional lattice of N repeat units. (*b*) Same as *a*, but closed to form a circle. (From Zbinden.[21])

5.5.2 Outline of Group Theory

The elements A, B, C,$\cdots$, in the set **G** are now considered. These elements may be numbers, functions, symmetry operations, and so on. In the set of symmetry operations (not the symmetry elements) of a single molecular chain of PE, a product $T_1 i$ defined by translation T_1 operated after inversion i is also a symmetry operation.* The set **G** forms a group if **G** satisfies the following four conditions.

1. The product of any two elements, A and B, in **G** or the square of any element, say $AA = A^2$, is again an element of **G**. As shown in Fig. 3.3, a product of two operations $C_2(y)i$ produces the following transformations in sequence:

$$\mathrm{H_I}(j) \xrightarrow{i} \mathrm{H_{IV}}(j') \xrightarrow{C_2(y)} \mathrm{H_{III}}(j'')$$

Here the correlation between the integers j, j', and j'' depends on the positions of the symmetry elements i and $C_2(y)$ chosen. By comparing the result with the transformation produced by the operation $\sigma_g(xz)$:

$$\mathrm{H_I}(j) \xrightarrow{\sigma_g(xz)} \mathrm{H_{III}}(j)$$

it can be shown that a combined operation $C_2(y)i$ at any position is identical to the operation $\sigma_g(xz)$ combined with one of the translational operations T_n.

$$C_2(y)i = T_n\sigma_g(xz) \tag{5.53}$$

More generally, it may be written as

$$\mathbf{T}C_2(y)\mathbf{T}i = \mathbf{T}\sigma_g(xz) \tag{5.54}$$

where $\mathbf{T}C_2(y)$ denotes the collections of all the elements $T_nC_2(y)$. For

* There are an infinite number of symmetry elements and symmetry operations along the PE chain for $C_2(y)$, $C_2(z)$, i, and $\sigma(yz)$ as shown in Fig. 3.3. On the other hand, there is only one symmetry element for each of $\sigma(xy)$, $\sigma_g(xz)$, and $C_2^{\,s}(x)$. The symmetry operations involving any two symmetry elements of the former type differ from each other only by a pure translation. The result of the operations are compared with respect to centers of symmetry $i_\mathrm{I}(j)$ and $i_\mathrm{II}(j)$ at the centers of the bonds $\mathrm{C_I}(j)$—$\mathrm{C_{II}}(j)$ and $\mathrm{C_{II}}(j)$—$\mathrm{C_I}(j+1)$, respectively. For example,

$$\mathrm{H_I}(j) \xrightarrow{i_\mathrm{I}(j)} \mathrm{H_{IV}}(j); \qquad \mathrm{H_I}(j) \xrightarrow{i_\mathrm{II}(j)} \mathrm{H_{IV}}(j+1)$$

Thus inversion with respect to an element is identical to the operation of another inversion followed by a translation T_n; that is, $i_\mathrm{II}(j) = T_1 i_\mathrm{I}(j)$, $i_\mathrm{I}(j+1) = T_2 i_\mathrm{I}(j)$, and so on. Therefore it may be concluded that the collection of all the operations $T_n i$, which is written simply as $\mathbf{T}i$, forms an identical set independently of the position of the element i chosen. The same applies to $C_2(y)$, $C_2(z)$, and $\sigma(yz)$.

simplicity hereafter, the following form is used.

$$C_2(y)i = \sigma_g(xz) \tag{5.55}$$

Similarly, any product of two symmetry operations will again be a symmetry operation of the group.

2. There is one and only one element in **G** that satisfies the relation

$$EA = AE = A \tag{5.56}$$

Element E is called the identity element. In the above example, T_0 or T_0E is the identity element.

3. The associative law of multiplication must hold.

$$(AB)C = A(BC) \tag{5.57}$$

In the above case it is easy to verify, for example, that

$$[\sigma(xy)C_2(y)]i = \sigma(xy)[C_2(y)i] \tag{5.58}$$

$$[\sigma(xy)C_2(y)]i\colon \mathrm{H_I}(j) \xrightarrow{\;i\;} \mathrm{H_{IV}}(j) \xrightarrow{\;\sigma(xy)C_2(y) = \sigma(yz)\;} \mathrm{H_{IV}}(j)$$

$$\sigma(xy)[C_2(y)i]\colon \mathrm{H_I}(j) \xrightarrow{\;C_2(y)i = \sigma_g(xz)\;} \mathrm{H_{III}}(j) \xrightarrow{\;\sigma(xy)\;} \mathrm{H_{IV}}(j)$$

4. For every element A there is only one element X that satisfies the relation

$$AX = XA = E \tag{5.59}$$

Element X is the inverse of A and is represented by A^{-1}. In the present example, $(T_n)^{-1} = T_{-n}$ and $(i)^{-1} = i$.

The collection of all the symmetry operations of a molecule satisfies the four rules and so forms a group. The number of elements in a group is called the order of the group. This group for PE, or that for any infinitely extended polymer chain, contains an infinite number of elements and is called a one-dimensional space group.

The product AB is not necessarily the same as BA. When AB and BA are the same, A and B are said to commute. Groups in which combination is commutative are said to be Abelian.*

5.5.3 Space Group and Factor Group

Symmetry operations in the PE molecule can be obtained by multiplication

* The point groups such as C_2, C_3, and D_{2h} are Abelian, but D_3 obtained by combining C_2 with C_3, is non-Abelian. Although the one-dimensional space group of the PE chain is non-Abelian, its factor group is Abelian. See Refs. 143 and 144.

of a translation T_n ($n = -\infty, \cdots, 0, \cdots, +\infty$) and an operation in the set of E, $C_2(y)$, $C_2(z)$, $C_2^s(x)$, i, $\sigma(xy)$, $\sigma(yz)$, and $\sigma_g(xz)$. If the collection of all translations in the space group **G** is represented by the symbol **T**, **T** forms a group by itself; that is, **T** is a subgroup of **G**. The subgroup **T** has the important property that every element of the form $X^{-1}T_nX$ is contained in **T**, where T_n and X are arbitrarily chosen elements of **T** and **G**, respectively. In other words, every conjugate element of **T** is again an element of **T**. This can be written as

$$X^{-1}\mathbf{T}X = \mathbf{T} \tag{5.60}$$

Here the equality sign means that both sides represent the identical collection of elements. Such a subgroup is called a self-conjugate subgroup or an invariant subgroup.

A group can be subdivided into cosets by using a subgroup. A coset is defined as a set of elements obtained by multiplying one element of the group by all elements of the subgroup to its left or its right. Using **T** as the subgroup, the space group **G** may be subdivided into right or left cosets:

$$\begin{aligned}\mathbf{G} = \mathbf{T}E + \mathbf{T}C_2(y) + \mathbf{T}C_2(z) + \mathbf{T}C_2^s(x) + \mathbf{T}i + \mathbf{T}\sigma(xy) \\ + \mathbf{T}\sigma(yz) + \mathbf{T}\sigma_g(xz)\end{aligned} \tag{5.61}$$

or

$$\begin{aligned}\mathbf{G} = E\mathbf{T} + C_2(y)\mathbf{T} + C_2(z)\mathbf{T} + C_2^s(x)\mathbf{T} + i\mathbf{T} + \sigma(xy)\mathbf{T} \\ + \sigma(yz)\mathbf{T} + \sigma_g(xz)\mathbf{T}\end{aligned} \tag{5.62}$$

Here the plus signs mean that **G** is composed of these sets of elements, and equality sign has the same meaning as in Eq. 5.60.

If **T** is an invariant subgroup, any left coset is identical to the corresponding right coset. If the cosets of **T** are regarded as elements, the collection of **T** and its cosets forms a group of order 8. This group is called the factor group and is designated by **G**/**T**. Here the elements of the factor group $\mathbf{T}E$, $\mathbf{T}C_2(y), \cdots$ are represented in an abbreviated form by E, $C_2(y), \cdots$.

Two groups are said to be isomorphous if there is a one-to-one correspondence between the elements in one group A, B, $C, \cdots$ and those of the other group A', B', $C', \cdots$, that is, a relation such as

$$A'B' = C' \quad \text{if} \quad AB = C \tag{5.63}$$

holds for every combination of elements. Thus two isomorphous groups give the same multiplication table. The factor group of a space group (one-dimensional as well as three-dimensional) is known to be isomorphous to one of point groups. The factor group of the one-dimensional space group of the PE molecule is isomorphous to the point group D_{2h}, giving the multiplication table shown in Table 5.2.

Table 5.2 Multiplication Table for the Factor Group (Isomorphous to the Point Group D_{2h}) of the One-Dimensional Space Group of PE Molecule (From Tobin[141])[a,b]

	E	$C_2(y)$	$C_2(z)$	$C_2^s(x)$	i	$\sigma(xy)$	$\sigma(yz)$	$\sigma_g(xz)$
E	E	$C_2(y)$	$C_2(z)$	$C_2^s(x)$	i	$\sigma(xy)$	$\sigma(yz)$	$\sigma_g(xz)$
$C_2(y)$	$C_2(y)$	E	$C_2^s(x)$	$C_2(z)$	$\sigma_g(xz)$	$\sigma(yz)$	$\sigma(xy)$	i
$C_2(z)$	$C_2(z)$	$C_2^s(x)$	E	$C_2(y)$	$\sigma(xy)$	i	$\sigma_g(xz)$	$\sigma(yz)$
$C_2^s(x)$	$C_2^s(x)$	$C_2(z)$	$C_2(y)$	E	$\sigma(yz)$	$\sigma_g(xz)$	i	$\sigma(xy)$
i	i	$\sigma_g(xz)$	$\sigma(xy)$	$\sigma(yz)$	E	$C_2(z)$	$C_2^s(x)$	$C_2(y)$
$\sigma(xy)$	$\sigma(xy)$	$\sigma(yz)$	i	$\sigma_g(xz)$	$C_2(z)$	E	$C_2(y)$	$C_2^s(x)$
$\sigma(yz)$	$\sigma(yz)$	$\sigma(xy)$	$\sigma_g(xz)$	i	$C_2^s(x)$	$C_2(y)$	E	$C_2(z)$
$\sigma_g(xz)$	$\sigma_g(xz)$	i	$\sigma(yz)$	$\sigma(xy)$	$C_2(y)$	$C_2^s(x)$	$C_2(z)$	E

[a] $\sigma(xy)C_2(y) = \sigma(yz)$, for example, is found at the intersection of row $\sigma(xy)$ and column $C_2(y)$. This convention is important for non-Abelian groups. See Ref. 144, p. 527.
[b] The factor **T** of the invariant subgroup is omitted for simplicity.

5.5.4 Matrix Representation of a Group

It is possible to associate every element R of a group with a square matrix $\mathbf{D}(R)$. If such a set of matrices $\mathbf{D}(E)$, $\mathbf{D}(A)$, $\mathbf{D}(B),\cdots$ forms a group isomorphous to the group $E, A, B,\cdots$, it is called a representation of the group.

$$\mathbf{D}(A)\,\mathbf{D}(B) = \mathbf{D}(C) \quad \text{if} \quad AB = C \tag{5.64}$$

The representation of the factor group for the PE molecule is isomorphous to the point group D_{2h}. In the factor group analysis translation **T** is treated as identity E, and so it is sufficient to treat the point group D_{2h} instead of the factor group. Such a mathematical treatment deals only with the vibrational modes (called factor group vibrations) with the corresponding atoms* in all repeating units moving in phase; only the vibrations of this type are infrared and Raman active as is explained later (Section 5.5.9).

To construct a representation of a given symmetry group, a basis coordinate system must be defined first. In vibrational problems it is convenient to describe molecular motions by using a displacement coordinate system. Using a Cartesian coordinate system, the displacement coordinates $\mathbf{q}$ for the whole molecule of PE are represented by the following column vector using Born's cyclic boundary condition.

* The atoms symmetric with respect to translations, for instance, $H_I(j)$, $H_I(j \pm 1)$, $H_I(j \pm 2),\cdots$ in Fig. 3.3.

$$\mathbf{q} = \begin{bmatrix} \mathbf{q}(1) \\ \vdots \\ \mathbf{q}(j) \\ \vdots \\ \mathbf{q}(N) \end{bmatrix} \tag{5.65}$$

where $\mathbf{q}(j)$ is a column vector consisting of the coordinates associated with the atoms in the jth repeating unit; namely,

$$\mathbf{q}(j) = \begin{bmatrix} \Delta x_{C_I}(j) \\ \Delta y_{C_I}(j) \\ \Delta z_{C_I}(j) \\ \Delta x_{H_I}(j) \\ \Delta y_{H_I}(j) \\ \Delta z_{H_I}(j) \\ \Delta x_{H_{II}}(j) \\ \Delta y_{H_{II}}(j) \\ \Delta z_{H_{II}}(j) \\ \Delta x_{C_{II}}(j) \\ \Delta y_{C_{II}}(j) \\ \Delta z_{C_{II}}(j) \\ \Delta x_{H_{III}}(j) \\ \Delta y_{H_{III}}(j) \\ \Delta z_{H_{III}}(j) \\ \Delta x_{H_{IV}}(j) \\ \Delta y_{H_{IV}}(j) \\ \Delta z_{H_{IV}}(j) \end{bmatrix} \tag{5.66}$$

where $\Delta x_{C_I}(j)$, $\Delta y_{C_I}(j)$, and $\Delta z_{C_I}(j)$ are the coordinates representing displacements from the equilibrium position of the atom $C_I(j)$ along the x, y, and z directions, respectively.

Instead of $\mathbf{q}$, the above-mentioned factor group vibrations can be represented by a column vector $\mathbf{q}^{op}$ of a finite order,

$$\mathbf{q}^{op} = \left(\frac{1}{N}\right)^{1/2} \sum_{j=1}^{N} \mathbf{q}(j) \tag{5.67}$$

where $(1/N)^{1/2}$ is the normalization factor and the superscript op means optically active. Hereafter the elements of $\mathbf{q}^{\text{op}}$ are abbreviated for simplicity as

$$\Delta x_{\text{C}_\text{I}} = \left(\frac{1}{N}\right)^{1/2} \sum_{j=1}^{N} \Delta x_{\text{C}_\text{I}}(j) \tag{5.68}$$

A symmetry operation, say, $C_2(y)$, transforms $\mathbf{q}^{\text{op}}$ to a new vector $\mathbf{q}_1^{\text{op}}$:

$$\mathbf{q}^{\text{op}} = \begin{bmatrix} \Delta x_{\text{C}_\text{I}} \\ \Delta y_{\text{C}_\text{I}} \\ \Delta z_{\text{C}_\text{I}} \\ \cdot \\ \cdot \\ \cdot \\ \Delta x_{\text{H}_\text{IV}} \\ \Delta y_{\text{H}_\text{IV}} \\ \Delta z_{\text{H}_\text{IV}} \end{bmatrix} \xrightarrow{C_2(y)} \mathbf{q}_1^{\text{op}} = \begin{bmatrix} -\Delta x_{\text{C}_\text{I}} \\ \Delta y_{\text{C}_\text{I}} \\ -\Delta z_{\text{C}_\text{I}} \\ \cdot \\ \cdot \\ \cdot \\ -\Delta x_{\text{H}_\text{III}} \\ \Delta y_{\text{H}_\text{III}} \\ -\Delta z_{\text{H}_\text{III}} \end{bmatrix} \tag{5.69}$$

In terms of a matrix this transformation is expressed by

$$\mathbf{D}[C_2(y)]\mathbf{q}^{\text{op}} = \mathbf{q}_1^{\text{op}} \tag{5.70}$$

where $\mathbf{D}[C_2(y)]$ is the representation matrix corresponding to the element $C_2(y)$ of the factor group and is given by

$$\mathbf{D}[C_2(y)] = \begin{bmatrix} \mathbf{A} & \mathbf{0} & \mathbf{0} & \mathbf{0} & \mathbf{0} & \mathbf{0} \\ \mathbf{0} & \mathbf{0} & \mathbf{A} & \mathbf{0} & \mathbf{0} & \mathbf{0} \\ \mathbf{0} & \mathbf{A} & \mathbf{0} & \mathbf{0} & \mathbf{0} & \mathbf{0} \\ \mathbf{0} & \mathbf{0} & \mathbf{0} & \mathbf{A} & \mathbf{0} & \mathbf{0} \\ \mathbf{0} & \mathbf{0} & \mathbf{0} & \mathbf{0} & \mathbf{0} & \mathbf{A} \\ \mathbf{0} & \mathbf{0} & \mathbf{0} & \mathbf{0} & \mathbf{A} & \mathbf{0} \end{bmatrix} \tag{5.71}$$

with

$$\mathbf{A} = \begin{bmatrix} -1 & 0 & 0 \\ 0 & 1 & 0 \\ 0 & 0 & -1 \end{bmatrix} \tag{5.72}$$

Similarly, the representation matrices for every element in the factor group

G/**T** can be constructed.

$$
\mathbf{D}(E) = \begin{bmatrix} \mathbf{A} & \mathbf{0} & \mathbf{0} & & & \\ \mathbf{0} & \mathbf{A} & \mathbf{0} & & \mathbf{0} & \\ \mathbf{0} & \mathbf{0} & \mathbf{A} & & & \\ & & & \mathbf{A} & \mathbf{0} & \mathbf{0} \\ & \mathbf{0} & & \mathbf{0} & \mathbf{A} & \mathbf{0} \\ & & & \mathbf{0} & \mathbf{0} & \mathbf{A} \end{bmatrix} \qquad \mathbf{A} = \begin{bmatrix} 1 & 0 & 0 \\ 0 & 1 & 0 \\ 0 & 0 & 1 \end{bmatrix}
$$

$$
\mathbf{D}[C_2(z)] = \begin{bmatrix} & & & \mathbf{A} & \mathbf{0} & \mathbf{0} \\ & \mathbf{0} & & \mathbf{0} & \mathbf{A} & \mathbf{0} \\ & & & \mathbf{0} & \mathbf{0} & \mathbf{A} \\ \mathbf{A} & \mathbf{0} & \mathbf{0} & & & \\ \mathbf{0} & \mathbf{A} & \mathbf{0} & & \mathbf{0} & \\ \mathbf{0} & \mathbf{0} & \mathbf{A} & & & \end{bmatrix} \qquad \mathbf{A} = \begin{bmatrix} -1 & 0 & 0 \\ 0 & -1 & 0 \\ 0 & 0 & 1 \end{bmatrix}
$$

$$
\mathbf{D}[C_2{}^s(x)] = \begin{bmatrix} & & & \mathbf{A} & \mathbf{0} & \mathbf{0} \\ & \mathbf{0} & & \mathbf{0} & \mathbf{0} & \mathbf{A} \\ & & & \mathbf{0} & \mathbf{A} & \mathbf{0} \\ \mathbf{A} & \mathbf{0} & \mathbf{0} & & & \\ \mathbf{0} & \mathbf{0} & \mathbf{A} & & \mathbf{0} & \\ \mathbf{0} & \mathbf{A} & \mathbf{0} & & & \end{bmatrix} \qquad \mathbf{A} = \begin{bmatrix} 1 & 0 & 0 \\ 0 & -1 & 0 \\ 0 & 0 & -1 \end{bmatrix}
$$

$$
\mathbf{D}(i) = \begin{bmatrix} & & & \mathbf{A} & \mathbf{0} & \mathbf{0} \\ & \mathbf{0} & & \mathbf{0} & \mathbf{0} & \mathbf{A} \\ & & & \mathbf{0} & \mathbf{A} & \mathbf{0} \\ \mathbf{A} & \mathbf{0} & \mathbf{0} & & & \\ \mathbf{0} & \mathbf{0} & \mathbf{A} & & \mathbf{0} & \\ \mathbf{0} & \mathbf{A} & \mathbf{0} & & & \end{bmatrix} \qquad \mathbf{A} = \begin{bmatrix} -1 & 0 & 0 \\ 0 & -1 & 0 \\ 0 & 0 & -1 \end{bmatrix}
$$

$$
\mathbf{D}[\sigma(xy)] = \begin{bmatrix} \mathbf{A} & \mathbf{0} & \mathbf{0} & & & \\ \mathbf{0} & \mathbf{0} & \mathbf{A} & & \mathbf{0} & \\ \mathbf{0} & \mathbf{A} & \mathbf{0} & & & \\ & & & \mathbf{A} & \mathbf{0} & \mathbf{0} \\ & \mathbf{0} & & \mathbf{0} & \mathbf{0} & \mathbf{A} \\ & & & \mathbf{0} & \mathbf{A} & \mathbf{0} \end{bmatrix} \qquad \mathbf{A} = \begin{bmatrix} 1 & 0 & 0 \\ 0 & 1 & 0 \\ 0 & 0 & -1 \end{bmatrix}
$$

$$\mathbf{D}[\sigma(yz)] = \begin{bmatrix} \mathbf{A} & \mathbf{0} & \mathbf{0} & & & \\ \mathbf{0} & \mathbf{A} & \mathbf{0} & & \mathbf{0} & \\ \mathbf{0} & \mathbf{0} & \mathbf{A} & & & \\ & & & \mathbf{A} & \mathbf{0} & \mathbf{0} \\ & \mathbf{0} & & \mathbf{0} & \mathbf{A} & \mathbf{0} \\ & & & \mathbf{0} & \mathbf{0} & \mathbf{A} \end{bmatrix} \qquad \mathbf{A} = \begin{bmatrix} -1 & 0 & 0 \\ 0 & 1 & 0 \\ 0 & 0 & 1 \end{bmatrix}$$

$$\mathbf{D}[\sigma_g(xz)] = \begin{bmatrix} & & & \mathbf{A} & \mathbf{0} & \mathbf{0} \\ & \mathbf{0} & & \mathbf{0} & \mathbf{A} & \mathbf{0} \\ & & & \mathbf{0} & \mathbf{0} & \mathbf{A} \\ \mathbf{A} & \mathbf{0} & \mathbf{0} & & & \\ \mathbf{0} & \mathbf{A} & \mathbf{0} & & \mathbf{0} & \\ \mathbf{0} & \mathbf{0} & \mathbf{A} & & & \end{bmatrix} \qquad \mathbf{A} = \begin{bmatrix} 1 & 0 & 0 \\ 0 & -1 & 0 \\ 0 & 0 & 1 \end{bmatrix} \tag{5.73}$$

The representation matrices thus constructed depend on the coordinates adopted as the basis. Let $\mathbf{p}$ be another coordinate vector obtained from $\mathbf{q}$ by transformation with a matrix $\mathbf{U}$

$$\mathbf{p} = \mathbf{U}\mathbf{q} \tag{5.74}$$

If a symmetry operation R transforms $\mathbf{q}$ and $\mathbf{p}$ to $\mathbf{q}_1$ and $\mathbf{p}_1$, respectively, then

$$\mathbf{p}_1 = \mathbf{U}\mathbf{q}_1 \tag{5.75}$$

Using the representation matrix with respect to $\mathbf{q}$ gives

$$\mathbf{D}(R)\mathbf{q} = \mathbf{q}_1 \tag{5.76}$$

From these three equations,

$$\mathbf{U}\,\mathbf{D}(R)\mathbf{U}^{-1}\mathbf{p} = \mathbf{p}_1 \tag{5.77}$$

Therefore if $\mathbf{p}$ is used as the basis, the representation matrices take the form $\mathbf{U}\mathbf{D}(R)\mathbf{U}^{-1}$, which is called the similarity transformation of $\mathbf{D}(R)$ by $\mathbf{U}$. If $\mathbf{U}$ is orthogonal, then $\mathbf{U}^{-1} = \tilde{\mathbf{U}}$, and, for this, the transformation becomes an orthogonal transformation $\mathbf{U}\,\mathbf{D}(R)\tilde{\mathbf{U}}$, $\tilde{\mathbf{U}}$ being the transpose of $\mathbf{U}$.

If a transformation matrix $\mathbf{U}$ is found so that the transformed representation matrix $\mathbf{U}\mathbf{D}(R)\tilde{\mathbf{U}}$ has the block-out form

$$\mathbf{U}\mathbf{D}(R)\tilde{\mathbf{U}} = \begin{bmatrix} \mathbf{P}_1(R) & & \mathbf{0} \\ & \mathbf{P}_2(R) & \\ \mathbf{0} & & \ddots \end{bmatrix} \tag{5.78}$$

for every symmetry operation R, then $\mathbf{D}(R)$ is said to be reducible; otherwise it is irreducible. Each of the factors $\mathbf{P}_1(R)$, $\mathbf{P}_2(R)$,$\cdots$ is again a representation matrix as shown below. Since the equation

$$\mathbf{D}(A)\mathbf{D}(B) = \mathbf{D}(AB) \tag{5.79}$$

can be written as

$$\begin{bmatrix} \mathbf{P}_1(A) & & \mathbf{0} \\ & \mathbf{P}_2(A) & \\ \mathbf{0} & & \ddots \end{bmatrix} \begin{bmatrix} \mathbf{P}_1(B) & & \mathbf{0} \\ & \mathbf{P}_2(B) & \\ \mathbf{0} & & \ddots \end{bmatrix} = \begin{bmatrix} \mathbf{P}_1(A)\mathbf{P}_1(B) & & \mathbf{0} \\ & \mathbf{P}_2(A)\mathbf{P}_2(B) & \\ \mathbf{0} & & \ddots \end{bmatrix}$$
$$= \begin{bmatrix} \mathbf{P}_1(AB) & & \mathbf{0} \\ & \mathbf{P}_2(AB) & \\ \mathbf{0} & & \ddots \end{bmatrix} \tag{5.80}$$

then

$$\mathbf{P}_1(A)\mathbf{P}_1(B) = \mathbf{P}_1(AB),\ \mathbf{P}_2(A)\mathbf{P}_2(B) = \mathbf{P}_2(AB), \cdots \tag{5.81}$$

If $\mathbf{P}_1(R)$, $\mathbf{P}_2(R)$,$\cdots$cannot be reduced further, they are called irreducible representations. The order of the irreducible representation appearing in the symmetry groups is less than 4.

Summation of the diagonal elements of $\mathbf{D}(R)$, that is, the trace of $\mathbf{D}(R)$, gives the character of the representation and is usually designated by the symbol $\chi(R)$. Since the trace of a matrix is invariant with respect to a similarity transformation, the character provides a convenient identification of an irreducible representation. See Ref. 144, p. 303.

Two elements A and B in a group are said to be conjugate with each other if

$$B = X^{-1}AX \tag{5.82}$$

for any element X in the group. A set of mutually conjugate elements forms a class of the group. It is apparent from Eq. 5.82 that the character of every element in a single class is identical. For the case of the factor group isomorphous to the point group D_{2h} all elements form respective classes, and there are, therefore, eight classes in the group.*

All irreducible representations of D_{2h} are one dimensional, and they are classified into eight symmetry species: A_g, B_{1g}, B_{2g}, B_{3g}, A_u, B_{1u}, B_{2u}, and B_{3u}; their characters are given in Table 5.3.† The letters A and B distinguish the species that are, respectively, symmetric and antisymmetric with respect to a rotation axis, and the subscripts g and u indicate those symmetric (*gerade* in the German) and antisymmetric (*ungerade*) with respect to the center of

* The number of the symmetry species is always equal to that of the classes for each group. In any Abelian group, for instance, D_{2h}, each element forms a class by itself. The situation is different for non-Abelian groups; for example, in the point group D_9, which is explained later for POM, 18 elements are divided into six classes.

† The choice of x, y, and z in Table 5.3 is the same as those of Refs. 4 and 142.

Table 5.3 Factor Group Analysis of PE Molecule (from Krimm et al.[142])

D_{2h}	E	$C_2(y)$	$C_2(z)$	$C_2^s(x)$	i	$\sigma(xy)$	$\sigma(yz)$	$\sigma_g(xz)$	n_i	$n_i(T)$	$n_i(R)$	n'_i	Infrared	Raman
A_g	1	1	1	1	1	1	1	1	3	0	0	3	Inactive	$\alpha_{xx}, \alpha_{yy}, \alpha_{zz}$
B_{1g}	1	−1	1	−1	1	1	−1	−1	2	0	0	2	Inactive	α_{xy}
B_{2g}	1	1	−1	−1	1	−1	−1	1	1	0	0	1	Inactive	α_{zx}
B_{3g}	1	−1	−1	1	1	−1	1	−1	3	0	R_x	2	Inactive	α_{yz}
A_u	1	1	1	1	−1	−1	−1	−1	1	0	0	1	Inactive	Inactive
B_{1u}	1	−1	1	−1	−1	−1	1	1	3	T_z	0	2	μ_z	Inactive
B_{2u}	1	1	−1	−1	−1	1	1	−1	3	T_y	0	2	μ_y	Inactive
B_{3u}	1	−1	−1	1	−1	1	−1	1	2	T_x	0	1	μ_x	Inactive
N_R	6	2	0	0	0	2	6	0						
$(\pm 1 + 2\cos\phi)$	3	−1	−1	−1	−3	1	1	1						
$\chi(R)$	18	−2	0	0	0	2	6	0						

symmetry. For D_{2h}, 1, 2, and 3 designate the species symmetric with respect to $C_2(z)$, $C_2(y)$, and $C_2^s(x)$, respectively. The character tables for point groups are given in many books, for instance, Refs. 2 and 4.

The representation matrices $\mathbf{D}(R)$ for the PE molecule can be reduced by using a transformation matrix

$$\mathbf{U}=\begin{bmatrix} (\frac{1}{2})^{1/2}\mathbf{E} & \mathbf{0} & \mathbf{0} & (\frac{1}{2})^{1/2}\mathbf{E} & \mathbf{0} & \mathbf{0} \\ (\frac{1}{2})^{1/2}\mathbf{E} & \mathbf{0} & \mathbf{0} & -(\frac{1}{2})^{1/2}\mathbf{E} & \mathbf{0} & \mathbf{0} \\ \mathbf{0} & \frac{1}{2}\mathbf{E} & \frac{1}{2}\mathbf{E} & \mathbf{0} & \frac{1}{2}\mathbf{E} & \frac{1}{2}\mathbf{E} \\ \mathbf{0} & \frac{1}{2}\mathbf{E} & -\frac{1}{2}\mathbf{E} & \mathbf{0} & \frac{1}{2}\mathbf{E} & -\frac{1}{2}\mathbf{E} \\ \mathbf{0} & \frac{1}{2}\mathbf{E} & \frac{1}{2}\mathbf{E} & \mathbf{0} & -\frac{1}{2}\mathbf{E} & -\frac{1}{2}\mathbf{E} \\ \mathbf{0} & \frac{1}{2}\mathbf{E} & -\frac{1}{2}\mathbf{E} & \mathbf{0} & -\frac{1}{2}\mathbf{E} & \frac{1}{2}\mathbf{E} \end{bmatrix} \tag{5.83}$$

where

$$\mathbf{E}=\begin{bmatrix} 1 & 0 & 0 \\ 0 & 1 & 0 \\ 0 & 0 & 1 \end{bmatrix}$$

By orthogonal transformation, a diagonal matrix of order 18 is obtained for each symmetry operation.

$$\mathbf{UD}(E)\tilde{\mathbf{U}}=\begin{bmatrix} \mathbf{A} & & & & & \\ & \mathbf{A} & & & \mathbf{0} & \\ & & \mathbf{A} & & & \\ & & & \mathbf{A} & & \\ & \mathbf{0} & & & \mathbf{A} & \\ & & & & & \mathbf{A} \end{bmatrix} \qquad \mathbf{A}=\begin{bmatrix} 1 & 0 & 0 \\ 0 & 1 & 0 \\ 0 & 0 & 1 \end{bmatrix}$$

$$\mathbf{UD}[C_2(y)]\tilde{\mathbf{U}}=\begin{bmatrix} \mathbf{A} & & & & & \\ & \mathbf{A} & & & \mathbf{0} & \\ & & \mathbf{A} & & & \\ & & & -\mathbf{A} & & \\ & \mathbf{0} & & & \mathbf{A} & \\ & & & & & -\mathbf{A} \end{bmatrix} \qquad \mathbf{A}=\begin{bmatrix} -1 & 0 & 0 \\ 0 & 1 & 0 \\ 0 & 0 & -1 \end{bmatrix}$$

$$\mathbf{UD}[C_2(z)]\tilde{\mathbf{U}}=\begin{bmatrix} \mathbf{A} & & & & & \\ & -\mathbf{A} & & & \mathbf{0} & \\ & & \mathbf{A} & & & \\ & & & \mathbf{A} & & \\ & \mathbf{0} & & & -\mathbf{A} & \\ & & & & & -\mathbf{A} \end{bmatrix} \qquad \mathbf{A}=\begin{bmatrix} -1 & 0 & 0 \\ 0 & -1 & 0 \\ 0 & 0 & 1 \end{bmatrix}$$

$$\mathbf{UD}[C_2^s(x)]\tilde{\mathbf{U}} = \begin{bmatrix} \mathbf{A} & & & & & \\ & -\mathbf{A} & & & \mathbf{0} & \\ & & \mathbf{A} & & & \\ & & & -\mathbf{A} & & \\ & \mathbf{0} & & & -\mathbf{A} & \\ & & & & & \mathbf{A} \end{bmatrix} \qquad \mathbf{A} = \begin{bmatrix} 1 & 0 & 0 \\ 0 & -1 & 0 \\ 0 & 0 & -1 \end{bmatrix}$$

$$\mathbf{U}[\mathbf{D}(i)]\tilde{\mathbf{U}} = \begin{bmatrix} \mathbf{A} & & & & & \\ & -\mathbf{A} & & & \mathbf{0} & \\ & & \mathbf{A} & & & \\ & & & -\mathbf{A} & & \\ & \mathbf{0} & & & -\mathbf{A} & \\ & & & & & \mathbf{A} \end{bmatrix} \qquad \mathbf{A} = \begin{bmatrix} -1 & 0 & 0 \\ 0 & -1 & 0 \\ 0 & 0 & -1 \end{bmatrix}$$

$$\mathbf{UD}[\sigma(xy)]\tilde{\mathbf{U}} = \begin{bmatrix} \mathbf{A} & & & & & \\ & \mathbf{A} & & & & \\ & & \mathbf{A} & & \mathbf{0} & \\ & & & -\mathbf{A} & & \\ & \mathbf{0} & & & \mathbf{A} & \\ & & & & & -\mathbf{A} \end{bmatrix} \qquad \mathbf{A} = \begin{bmatrix} 1 & 0 & 0 \\ 0 & 1 & 0 \\ 0 & 0 & -1 \end{bmatrix}$$

$$\mathbf{UD}[\sigma(yz)]\tilde{\mathbf{U}} = \begin{bmatrix} \mathbf{A} & & & & & \\ & \mathbf{A} & & & \mathbf{0} & \\ & & \mathbf{A} & & & \\ & & & \mathbf{A} & & \\ & \mathbf{0} & & & \mathbf{A} & \\ & & & & & \mathbf{A} \end{bmatrix} \qquad \mathbf{A} = \begin{bmatrix} -1 & 0 & 0 \\ 0 & 1 & 0 \\ 0 & 0 & 1 \end{bmatrix}$$

$$\mathbf{UD}[\sigma_g(xz)]\tilde{\mathbf{U}} = \begin{bmatrix} \mathbf{A} & & & & & \\ & -\mathbf{A} & & & \mathbf{0} & \\ & & \mathbf{A} & & & \\ & & & \mathbf{A} & & \\ & \mathbf{0} & & & -\mathbf{A} & \\ & & & & & -\mathbf{A} \end{bmatrix} \qquad \mathbf{A} = \begin{bmatrix} 1 & 0 & 0 \\ 0 & -1 & 0 \\ 0 & 0 & 1 \end{bmatrix} \tag{5.84}$$

Each diagonal element, that is, each one-dimensional irreducible representation matrix, belongs to one of the eight species of D_{2h}. The first and second irreducible representations, for example, have the characters shown below for the symmetry operations and therefore belong to the B_{3u} and B_{2u} species, respectively.

R	E	$C_2(y)$	$C_2(z)$	$C_2^s(x)$	i	$\sigma(xy)$	$\sigma(yz)$	$\sigma_g(xz)$
First representation	1	−1	−1	1	−1	1	−1	1
Second representation	1	1	−1	−1	−1	1	1	−1

$$\mathbf{UD}(R)\widetilde{\mathbf{U}} = \begin{bmatrix} B_{3u} \\ & B_{2u} \\ & & B_{1u} \\ & & & B_{1g} & & & & & & & & & & \mathbf{0} \\ & & & & A_g \\ & & & & & B_{3g} \\ & & & & & & B_{3u} \\ & & & & & & & B_{2u} \\ & & & & & & & & B_{1u} \\ & & & & & & & & & B_{2g} \\ & & & & & & & & & & B_{3g} \\ & & & & & & & & & & & A_g \\ & & & & & & & & & & & & B_{1g} \\ & & & & & & & & & & & & & A_g \\ & & & & & & & & & & & & & & B_{3g} \\ & & & & & & & & & & & & & & & A_u \\ & & \mathbf{0} & & & & & & & & & & & & & & B_{1u} \\ & & & & & & & & & & & & & & & & & B_{2u} \end{bmatrix} \quad (5.85)$$

This is represented simply as

$$\Gamma = 3A_g + 2B_{1g} + B_{2g} + 3B_{3g} + A_u + 3B_{1u} + 3B_{2u} + 2B_{3u} \quad (5.86)$$

signifying that the representation $\mathbf{UD}(R)\widetilde{\mathbf{U}}$ consists of numbers of different irreducible representations given on the right-hand side of the equation.

The basis coordinate vector associated with the reduced representation is given by $\mathbf{Uq}^{\text{op}}$. The elements of $\mathbf{Uq}^{\text{op}}$ can be classified according to the species to which they belong and are called symmetry coordinates.

$$\begin{aligned} A_g: S_1 &= (\tfrac{1}{2})^{1/2}(\Delta y_{\mathrm{C_I}} - \Delta y_{\mathrm{C_{II}}}) \\ S_2 &= \tfrac{1}{2}(\Delta z_{\mathrm{H_I}} - \Delta z_{\mathrm{H_{II}}} + \Delta z_{\mathrm{H_{III}}} - \Delta z_{\mathrm{H_{IV}}}) \\ S_3 &= \tfrac{1}{2}(\Delta y_{\mathrm{H_I}} + \Delta y_{\mathrm{H_{II}}} - \Delta y_{\mathrm{H_{III}}} - \Delta y_{\mathrm{H_{IV}}}) \end{aligned}$$

$$
\begin{aligned}
B_{1g}: S_4 &= (\tfrac{1}{2})^{1/2}(\Delta x_{\mathrm{C_I}} - \Delta x_{\mathrm{C_{II}}}) \\
S_5 &= \tfrac{1}{2}(\Delta x_{\mathrm{H_I}} + \Delta x_{\mathrm{H_{II}}} - \Delta x_{\mathrm{H_{III}}} - \Delta x_{\mathrm{H_{IV}}}) \\
B_{2g}: S_6 &= \tfrac{1}{2}(\Delta x_{\mathrm{H_I}} - \Delta x_{\mathrm{H_{II}}} + \Delta x_{\mathrm{H_{III}}} - \Delta x_{\mathrm{H_{IV}}}) \\
B_{3g}: S_7 &= (\tfrac{1}{2})^{1/2}(\Delta z_{\mathrm{C_I}} - \Delta z_{\mathrm{C_{II}}}) \\
S_8 &= \tfrac{1}{2}(\Delta y_{\mathrm{H_I}} - \Delta y_{\mathrm{H_{II}}} + \Delta y_{\mathrm{H_{III}}} - \Delta y_{\mathrm{H_{IV}}}) \\
S_9 &= \tfrac{1}{2}(\Delta z_{\mathrm{H_I}} + \Delta z_{\mathrm{H_{II}}} - \Delta z_{\mathrm{H_{III}}} - \Delta z_{\mathrm{H_{IV}}}) \\
A_u: S_{10} &= \tfrac{1}{2}(\Delta x_{\mathrm{H_I}} - \Delta x_{\mathrm{H_{II}}} - \Delta x_{\mathrm{H_{III}}} + \Delta x_{\mathrm{H_{IV}}}) \\
B_{1u}: S_{11} &= (\tfrac{1}{2})^{1/2}(\Delta z_{\mathrm{C_I}} + \Delta z_{\mathrm{C_{II}}}) \\
S_{12} &= \tfrac{1}{2}(\Delta z_{\mathrm{H_I}} + \Delta z_{\mathrm{H_{II}}} + \Delta z_{\mathrm{H_{III}}} + \Delta z_{\mathrm{H_{IV}}}) \\
S_{13} &= \tfrac{1}{2}(\Delta y_{\mathrm{H_I}} - \Delta y_{\mathrm{H_{II}}} - \Delta y_{\mathrm{H_{III}}} + \Delta y_{\mathrm{H_{IV}}}) \\
B_{2u}: S_{14} &= (\tfrac{1}{2})^{1/2}(\Delta y_{\mathrm{C_I}} + \Delta y_{\mathrm{C_{II}}}) \\
S_{15} &= \tfrac{1}{2}(\Delta y_{\mathrm{H_I}} + \Delta y_{\mathrm{H_{II}}} + \Delta y_{\mathrm{H_{III}}} + \Delta y_{\mathrm{H_{IV}}}) \\
S_{16} &= \tfrac{1}{2}(\Delta z_{\mathrm{H_I}} - \Delta z_{\mathrm{H_{II}}} - \Delta z_{\mathrm{H_{III}}} + \Delta z_{\mathrm{H_{IV}}}) \\
B_{3u}: S_{17} &= (\tfrac{1}{2})^{1/2}(\Delta x_{\mathrm{C_I}} + \Delta x_{\mathrm{C_{II}}}) \\
S_{18} &= \tfrac{1}{2}(\Delta x_{\mathrm{H_I}} + \Delta x_{\mathrm{H_{II}}} + \Delta x_{\mathrm{H_{III}}} + \Delta x_{\mathrm{H_{IV}}})
\end{aligned}
\tag{5.87}
$$

Each symmetry coordinate has the symmetry properties defined by the symmetry species. The normal coordinates can be expressed by the linear combinations of the symmetry coordinates belonging to the same symmetry species. When the Cartesian displacement coordinates are used for the vibrations of an infinitely extended chain molecule, as in the present case, four nongenuine vibrations, that is, the pure translations along the x, y, and z axes and the pure rotation around the x axis, are included. They can be expressed by a linear combination of the symmetry coordinates in the respective species.*

For treating vibrations of a single molecular chain, internal displacement

* The translation along the z axis, for example, is expressed by a linear combination of S_{11} and S_{12} and hence belongs to the B_{1u} species.

$$S(T_z) = N_1(2^{1/2} m_{\mathrm{C}} S_{11} + 2 m_{\mathrm{H}} S_{12})$$

where N_1 is the normalization factor, and m_{C} and m_{H} represent the masses of the carbon and hydrogen atoms, respectively. The pure rotation around the x axis is represented by the following linear combination and clearly belongs to the B_{3g} species.

$$S(R_x) = N_2(2^{1/2} m_{\mathrm{C}} |y_{\mathrm{C}}{}^0| S + 2 m_{\mathrm{H}} |y_{\mathrm{H}}{}^0| S_9 - 2 m_{\mathrm{H}} |z_{\mathrm{H}}{}^0| S_8)$$

where N_2 is the normalization factor, $y_{\mathrm{C}}{}^0$ and $y_{\mathrm{H}}{}^0$ are the equilibrium values of the y coordinates of the carbon and hydrogen atoms, respectively, and $z_{\mathrm{H}}{}^0$ is the equilibrium value of the z coordinate of the hydrogen atoms.

coordinates are more frequently used than Cartesian displacement coordinates as is described in Section 5.6.2.

5.5.5 Number of Normal Modes

Based on the results obtained above, a general equation giving the number of normal modes involved in each symmetry species can be derived. Using Cartesian displacement coordinates, a set of representation matrices of order $3N$ is obtained, N being the number of atoms per unit cell. Since the character of $\mathbf{D}(R)$, denoted by $\chi(R)$, is invariant under a similarity transformation,

$$\chi(R) = \sum_i n_i \chi_i(R) \tag{5.88}$$

where n_i represents the number of the ith irreducible representation, whose character is designated by $\chi_i(R)$, appearing in the reduced representation $\mathbf{UD}(R)\tilde{\mathbf{U}}$. Therefore n_i gives the number of symmetry coordinates, that is, the number of normal modes including nongenuine vibrations (pure translations and pure rotation) belonging to the ith symmetry species. Multiplying both sides of Eq. 5.88 by $\chi_j(R)$ and summing over all the operations R gives

$$\sum_R \chi(R)\chi_j(R) = \sum_R \sum_i n_i \chi_i(R)\chi_j(R) \tag{5.89}$$

The orthogonality of the character (see Ref. 4, p. 341)

$$\sum_R \chi_i(R)\chi_j(R) = \begin{cases} 0 & i \neq j \\ g & i = j \end{cases} \tag{5.90}$$

where g represents the order of the group gives

$$\sum_R \chi(R)\chi_i(R) = n_i g \tag{5.91}$$

or

$$n_i = \frac{1}{g}\sum_R \chi(R)\chi_i(R) \tag{5.92}$$

The reducible representation matrix $\mathbf{D}(R)$ based on a Cartesian displacement coordinate system is divided into three-dimensional factors $\mathbf{A}$, each associated with the transformation of three coordinates of one atom, as shown in Eqs. 5.71 and 5.72. The transformation matrix $\mathbf{A}$ for a proper or improper rotation* through an angle ϕ ($0 \leqslant \phi < 2\pi$) about, say, the z axis

* $\sigma(xy)$, $\sigma(yz)$, and $\sigma_g(xz)$ are improper rotations ($\phi = 0$) about the z, x, and y axes, respectively, and i is an improper rotation about any axis ($\phi = \pi$).

is generally given by

$$\mathbf{A} = \begin{bmatrix} \cos\phi & \sin\phi & 0 \\ -\sin\phi & \cos\phi & 0 \\ 0 & 0 & \pm 1 \end{bmatrix} \tag{5.93}$$

where the plus and minus signs correspond to proper and improper rotations, respectively. From Eqs. 5.71 and 5.73 it is seen that the factor **A** appears on the diagonal only if it is associated with the atom that does not move, or moves to the position of the corresponding atom in the other unit cell by the symmetry operation R. Therefore the character $\chi(R)$ can be obtained by summing the traces of **A**'s for such atoms. If the number of such atoms per unit cell is expressed by N_R, then

$$\chi(R) = \begin{cases} N_R(1 + 2\cos\phi) & \text{proper rotation} \\ N_R(-1 + 2\cos\phi) & \text{improper rotation} \end{cases} \tag{5.94}$$

The values of $(\pm 1 + 2\cos\phi)$ for various symmetry operations are summarized in Table 5.4. For the case of PE, the values of N_R and $(\pm 1 + 2\cos\phi)$ are given in Table 5.3. As Fig. 3.3 shows, for example, by the symmetry operation with respect to $C_2(y)$ passing through the $C_{II}(j)$ atom, $C_{II}(j)$ does not move, while $C_I(j)$ and $C_I(j+1)$ change position with each other; the latter two are the corresponding atoms and are counted as nonmoving atoms. All the hydrogen atoms move to noncorresponding positions. Therefore $N_R = 2$ for $C_2(y)$. Thus the number of normal modes for each species is calculated according to

$$A_g: n_i = \tfrac{1}{8}[(6)(3)(1) + (2)(-1)(1) + (0)(-1)(1) + (0)(-1)(1) + (0)(-3)(1) + (2)(1)(1) + (6)(1)(1) + (0)(1)(1)] = \tfrac{24}{8} = 3$$

$$B_{1g}: n_i = \tfrac{1}{8}[(6)(3)(1) + (2)(-1)(-1) + (0)(-1)(1) + (0)(-1)(-1) + (0)(-3)(1) + (2)(1)(1) + (6)(1)(-1) + (0)(1)(-1)] = \tfrac{16}{8} = 2$$

Table 5.4 Character of Symmetry Operation

Proper rotation		Improper rotation	
$R(=C_n^k)$	$1 + 2\cos\phi$	$R(=S_n^k)$	$-1 + 2\cos\phi$
$C_1^k = E$	3	$S_1^1 = \sigma$	1
C_2^1	-1	$S_2^1 = i$	-3
C_3^1, C_3^2	0	S_3^1, S_3^2	-2
C_4^1, C_4^3	1	S_4^1, S_4^3	-1
C_6^1, C_6^5	2	S_6^1, S_6^5	0

For the other species, the results are given in the column headed n_i in Table 5.3.

Pure translations of the molecule as a whole can be expressed in terms of three Cartesian coordinates. Using the coordinate vector as the basis gives the representation matrices of the form expressed by Eq. 5.93, the character being given as $(\pm 1 + 2\cos\phi)$. Consequently the number of the pure translations involved in the ith species is given by

$$n_i(T) = \frac{1}{g}\sum_R (\pm 1 + 2\cos\phi)\chi_i(R) \tag{5.95}$$

For the PE molecule, this gives one translation for each of the B_{1u}, B_{2u}, and B_{3u} species, corresponding to those along the x, y, and z axes, respectively.

Pure rotation of a molecule as a whole can be represented by the angular momentum vector, which is defined as the sum of the vector products of two three-dimensional vectors $\sum_\alpha \mathbf{r}_\alpha \times \mathbf{p}_\alpha$. Here $\mathbf{r}_\alpha$ and $\mathbf{p}_\alpha$ represent the coordinate and momentum vectors of atom α, respectively, and the summation covers all atoms of the molecule. If the Cartesian coordinate system describing $\mathbf{r}$ and $\mathbf{p}$ is transformed by a matrix $\mathbf{A}$ of Eq. 5.93, the vector product $\mathbf{r} \times \mathbf{p}$ is transformed*:

$$\mathbf{r} \times \mathbf{p} = \begin{bmatrix} r_y p_z - r_z p_y \\ r_z p_x - r_x p_z \\ r_x p_y - r_y p_x \end{bmatrix} \longrightarrow \begin{bmatrix} \pm(r_y p_z - r_z p_y)\cos\phi \pm (r_z p_x - r_x p_z)\sin\phi \\ \mp(r_y p_z - r_z p_y)\sin\phi \pm (r_z p_x - r_x p_z)\cos\phi \\ r_x p_y - r_y p_x \end{bmatrix} \tag{5.96}$$

Hence the representation matrix for a pure rotation may be written as

$$\mathbf{D}(R)_{\text{rot}} = \begin{bmatrix} \pm\cos\phi & \pm\sin\phi & 0 \\ \mp\sin\phi & \pm\cos\phi & 0 \\ 0 & 0 & 1 \end{bmatrix} \tag{5.97}$$

and the character is given by

$$\chi(R)_{\text{rot}} = 1 \pm 2\cos\phi \tag{5.98}$$

Then the number of pure rotations for each species is given by

* $\mathbf{r}$ is transformed as

$$\begin{bmatrix} \cos\phi & \sin\phi & 0 \\ -\sin\phi & \cos\phi & 0 \\ 0 & 0 & \pm 1 \end{bmatrix} \begin{bmatrix} r_x \\ r_y \\ r_z \end{bmatrix} = \begin{bmatrix} r_x\cos\phi + r_y\sin\phi \\ -r_x\sin\phi + r_y\cos\phi \\ \pm r_z \end{bmatrix}$$

Substituting this and the corresponding result for $\mathbf{p}$ into the left-hand side of Eq. 5.96, gives the right-hand-side vector.

$$n_i(R) = \frac{1}{g}\sum_R (1 \pm 2\cos\phi)\chi_i(R) \tag{5.99}$$

Application of this equation to the PE molecule gives one pure rotation in each of the B_{1g}, B_{2g}, and B_{3g} species, corresponding to the rotation about the z, y, and x axes, respectively. The pure rotations about the z and y axes, however, should be omitted since they are not factor group vibrations. Thus there is only one pure rotation about the molecular axis.

Subtracting the number of the nongenuine vibrations, $n_i(T) + n_i(R)$, from the total number of normal modes n_i gives the number n_i' of the internal vibrations for each species.

By using Eqs. 5.92, 5.95, and 5.99, the group theoretical analysis can be also made for simple molecules, for instance, a water molecule for which the result is given in Table 5.1.

5.5.6 Selection Rules for Infrared and Raman Spectra

As is described in Section 5.2.5, for the normal vibration v_k to be infrared active, the normal coordinate Q_k must belong to the same symmetry species as one of the components μ_x, μ_y, and μ_z of the dipole moment vector. The components μ_x, μ_y, and μ_z or their linear combinations belong to the same symmetry species as those of the corresponding components of the pure translation of the molecule. For the case of PE, μ_x, μ_y, and μ_z belong to the B_{3u}, B_{2u}, and B_{1u} species, respectively, and these three species are infrared active. This result gives important information about the polarization properties of the infrared absorption bands. The B_{3u}, B_{2u}, and B_{1u} vibrations are excited only by radiation having the electric vector parallel to the x, y, and z directions, respectively. Infrared dichroism or pleochroism offers one of the most informative techniques for determining the symmetry species of the absorption bands.

The v_k vibration is Raman active only when the normal coordinate Q_k belongs to the same symmetry species as at least one of the components α_{xx}, α_{yy}, α_{zz}, α_{xy}, α_{yz}, and α_{xz} of the polarizability tensor. In Eq. 5.33, **P** and **E** are, respectively, three-dimensional electric polarization and electric field vectors defined in a Cartesian coordinate system. Since $\mathbf{E}_1 = \mathbf{D}(R)\mathbf{E}$, $\mathbf{P}_1 = \mathbf{D}(R)\mathbf{P} = \mathbf{D}(R)\boldsymbol{\alpha}\mathbf{E} = \mathbf{D}(R)\boldsymbol{\alpha}\mathbf{D}(R)^{-1}\mathbf{E}_1$. Hence the polarizability tensor in the new coordinate system is transformed according to

$$\mathbf{D}(R)\boldsymbol{\alpha}\mathbf{D}(R)^{-1} = \begin{bmatrix} \cos\phi & \sin\phi & 0 \\ -\sin\phi & \cos\phi & 0 \\ 0 & 0 & \pm 1 \end{bmatrix} \begin{bmatrix} \alpha_{xx} & \alpha_{xy} & \alpha_{xz} \\ \alpha_{xy} & \alpha_{yy} & \alpha_{yz} \\ \alpha_{xz} & \alpha_{yz} & \alpha_{zz} \end{bmatrix} \begin{bmatrix} \cos\phi & -\sin\phi & 0 \\ \sin\phi & \cos\phi & 0 \\ 0 & 0 & \pm 1 \end{bmatrix} \tag{5.100}$$

Therefore six components of the polarizability tensor are transformed according to

$$\begin{bmatrix} \alpha_{xx} \\ \alpha_{yy} \\ \alpha_{zz} \\ \alpha_{xy} \\ \alpha_{yz} \\ \alpha_{xz} \end{bmatrix} \xrightarrow{R} \begin{bmatrix} \cos^2\phi & \sin^2\phi & 0 & 2\cos\phi\sin\phi & 0 & 0 \\ \sin^2\phi & \cos^2\phi & 0 & -2\cos\phi\sin\phi & 0 & 0 \\ 0 & 0 & 1 & 0 & 0 & 0 \\ -\sin\phi\cos\phi & \sin\phi\cos\phi & 0 & 2\cos^2\phi - 1 & 0 & 0 \\ 0 & 0 & 0 & 0 & \pm\cos\phi & \mp\sin\phi \\ 0 & 0 & 0 & 0 & \pm\sin\phi & \pm\cos\phi \end{bmatrix} \begin{bmatrix} \alpha_{xx} \\ \alpha_{yy} \\ \alpha_{zz} \\ \alpha_{xy} \\ \alpha_{yz} \\ \alpha_{xz} \end{bmatrix} \tag{5.101}$$

The transformation matrix on the right-hand side is the representation matrix of the polarizability tensor. Hence the number of α's (or their combinations) belonging to each symmetry species is given by

$$n_i(\alpha) = \frac{1}{g}\sum_R 2\cos\phi\,(2\cos\phi \pm 1)\chi_i(R) \tag{5.102}$$

The species with nonvanishing $n_i(\alpha)$ is Raman active. For the case of PE, the three diagonal elements of the polarizability tensor (α_{xx}, α_{yy}, and α_{zz}) belong to the Raman-active species A_g with $n_i(\alpha) = 3$. The off-diagonal elements, α_{xy}, α_{xz}, and α_{yz} belonging, respectively, to B_{1g}, B_{2g}, and B_{3g} species have $n_i = 1$. As is described in Section 5.4.2.A, if an oriented sample is excited by a laser beam whose electric vector is parallel to, say, the x direction, and the y component of the scattered light is measured, then a Raman spectrum associated with the vibrations that belong to the species containing the component α_{xy} is obtained. Recent polarized Raman data obtained by using a laser spectrophotometer have been reported for various kinds of single crystals and also oriented polymers.

The number of normal modes and selection rules for the infrared and Raman spectra under each symmetry species for the PE molecule are given in Table 5.3. Of the 14 normal modes, 5 are infrared-active fundamentals, 8 are Raman active, and 1 is inactive in both infrared and Raman spectra. The modes active in infrared are inactive in Raman spectra and vice versa; that is, the mutual exclusion rule holds in this case.

5.5.7 Polyethylene

The polarized infrared and Raman spectra of PE are reproduced in Figs. 5.42 and 5.43. From the factor group analysis described in the preceding sections, the B_{1u} and B_{2u} modes are expected to give rise to perpendicular bands, and the B_{3u} modes to parallel bands. The bands at 2919, at 2851,

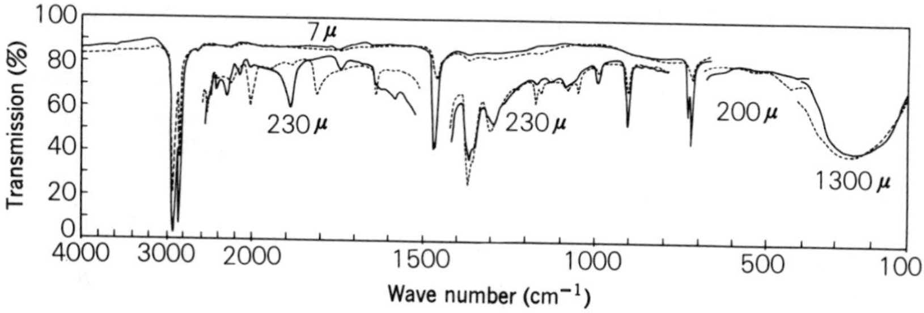

Fig. 5.42 Polarized infrared spectra of a uniaxially oriented PE film. (———) Perpendicular radiation. (------) Parallel radiation. The film thickness is indicated in the spectra.

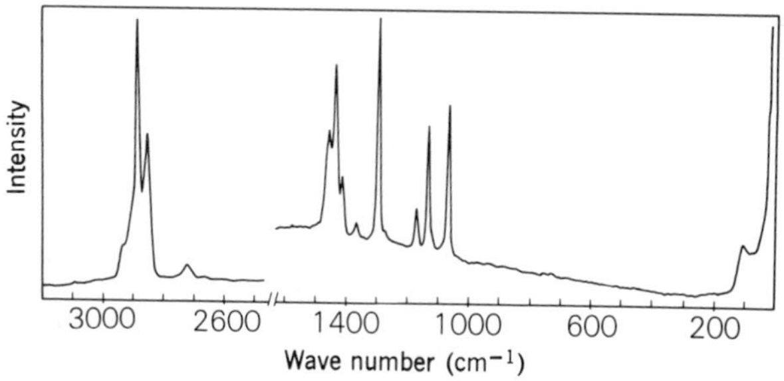

Fig. 5.43 Raman spectra of PE.

around 1460, and around 725 cm^{-1} show strong perpendicular polarization and are assigned to B_{1u} or B_{2u}. The weak parallel band at 1176 cm^{-1} is assigned to B_{3u}. Based on the factor group analysis and the knowledge of the vibrations of a CH_2 group from the detailed interpretation of small molecules, the vibrational modes of the PE single chain are classified as shown in Table 5.5. The vibrational forms approximated in terms of the symmetry coordinates are shown in Fig. 5.44.

Although the main features of the PE spectra can be predicted from consideration of a single chain, a complete interpretation must take account of the crystal structure. The crystal structure belongs to the space group $D_{2h}{}^{16}$, which has a factor group isomorphous to the point group D_{2h}. The procedure for analysis is similar to that discussed for the case of a single chain. The results of analysis are given in Table 5.6.*

Attention is now focused on the influence of intermolecular interaction in the crystal on the vibrational spectra. Since there are two molecular chains

* *a*, *b*, and *c* axes in Table 5.6 are chosen in the same way as in Wilson et al. (Ref. 4, p. 327), where $x = c$, $y = b$, and $z = a$ were assumed. Care must be taken in the choice of the axes, since different authors frequently do not agree.

Table 5.5 Classification of Vibrational Modes of PE Chain (From Krimm et al.[142])

Type of motion	Group mode	Phase between groups[a]	Symmetry species
Internal hydrogen vibrations of CH_2 group	CH_2 symmetric stretching, $\nu_s(CH_2)$	0	A_g
		π	$B_{2u}(\perp)$
	CH_2 antisymmetric stretching, $\nu_a(CH_2)$	0	B_{3g}
		π	$B_{1u}(\perp)$
	CH_2 bending, $\delta(CH_2)$	0	A_g
		π	$B_{2u}(\perp)$
External hydrogen vibrations of CH_2 group	CH_2 wagging, $w(CH_2)$	0	$B_{3u}(\parallel)$
		π	B_{1g}
	CH_2 rocking, $r(CH_2)$	0	B_{3g}
		π	$B_{1u}(\perp)$
	CH_2 twisting, $t(CH_2)$	0	A_u
		π	B_{2g}
Skeletal vibrations (CH_2 group moving as a unit)	x direction	0	$B_{3u}(T_x)$
		π	B_{1g}
	y direction	0	A_g
		π	$B_{2u}(T_y)$
	z direction	0	$B_{3g}(R_x)$
		π	$B_{1u}(T_z)$

[a] Phase difference with respect to $C_2^s(x)$.

in a unit cell, a given vibrational mode of one molecule may be expected to split into two modes according to the phase relations, in-phase and out-of-phase, between the two molecular chains. As is described earlier, the symmetry of an individual molecule in the crystal lattice is lowered from D_{2h} for the isolated molecule to C_{2h}. The set of these symmetry elements forms a subgroup of D_{2h} of the isolated free molecule and is called the site symmetry. The relation between the symmetry species of a given group and those of its subgroup is derived by comparing the characters of the irreducible representations of both groups. In the present case, if the characters for the symmetry operations of the site group are selected from the character table of D_{2h} (enclosed by dashed lines in Table 5.7), it will be clear that the pairs of species A_g and B_{3g} of D_{2h} are associated with the species A_g of C_{2h}, B_{1g} and B_{2g} with B_g, A_u and B_{3u} with A_u, and B_{1u} and B_{2u} with B_u. A similar correlation

Table 5.6 Factor Group Analysis of PE Crystal (From Krimm et al.[142])

D_{2h}^{16}	E	$C_2^s(a)$	$C_2^s(b)$	$C_2^s(c)$	i	$\sigma_g(bc)$	$\sigma_g(ac)$	$\sigma(ab)$	n_i	$n_i(T)$	n'_i	Infrared	Raman
A_g	1	1	1	1	1	1	1	1	6	0	6	Inactive	$\alpha_{aa}, \alpha_{bb}, \alpha_{cc}$
B_{1g}	1	1	−1	−1	1	1	−1	−1	3	0	3	Inactive	α_{bc}
B_{2g}	1	−1	1	−1	1	−1	1	−1	3	0	3	Inactive	α_{ca}
B_{3g}	1	−1	−1	1	1	−1	−1	1	6	0	6	Inactive	α_{ab}
A_u	1	1	1	1	−1	−1	−1	−1	3	0	3	Inactive	Inactive
B_{1u}	1	1	−1	−1	−1	−1	1	1	6	T_a	5	μ_a	Inactive
B_{2u}	1	−1	1	−1	−1	1	−1	1	6	T_b	5	μ_b	Inactive
B_{3u}	1	−1	−1	1	−1	1	1	−1	3	T_c	2	μ_c	Inactive
N_R	12	0	0	0	0	0	0	12					
$(\pm 1 + 2\cos\phi)$	3	−1	−1	−1	−3	1	1	1					
$\chi(R)$	36	0	0	0	0	0	0	12					

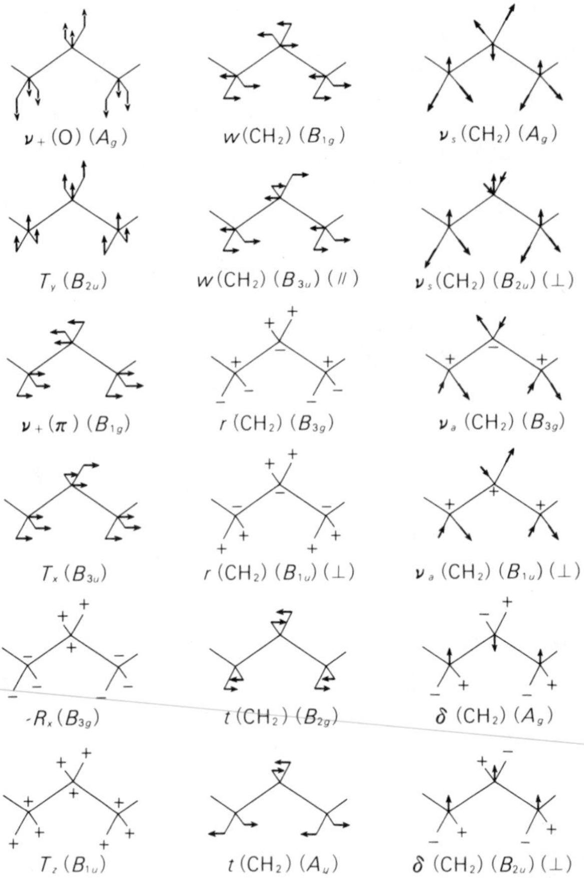

Fig. 5.44 Vibrational modes of a single PE chain (schematic representation in terms of the symmetry coordinates). (From Krimm et al.[142])

Table 5.7 Correlation between the Symmetry Species of the Point Group D_{2h} and One of Its Subgroup C_{2h}

D_{2h}	E	$C_2(y)$	$C_2(z)$	$C_2^s(x)$	i	$\sigma(xy)$	$\sigma(yz)$	$\sigma_g(xz)$	C_{2h}
A_g	1	1	1	1	1	1	1	1	A_g
B_{1g}	1	−1	1	−1	1	1	−1	−1	B_g
B_{2g}	1	1	−1	−1	1	−1	−1	1	B_g
B_{3g}	1	−1	−1	1	1	−1	1	−1	A_g
A_u	1	1	1	1	−1	−1	−1	−1	A_u
B_{1u}	1	−1	1	−1	−1	−1	1	1	B_u
B_{2u}	1	1	−1	−1	−1	1	1	−1	B_u
B_{3u}	1	−1	−1	1	−1	1	−1	1	A_u

holds between the site group and the three-dimensional space group whose factor group is isomorphous to D_{2h}. Consequently the correlations among the molecular, site, and space groups* given in Table 5.8 are obtained. The correlation diagram indicates that an A_g molecular mode, for example, generates two modes A_g and B_{3g} in the crystal. This is manifested in the observed vibrational spectra as a band splitting or an appearance of bands due to modes that are forbidden for the isolated molecule. The vibrational modes in the crystal can be derived in terms of the symmetry coordinates as shown in Fig. 5.45.

A comparison of the 12 infrared active fundamentals predicted for the crystal with those for a single chain shows the following differences: (1) All of the single chain modes that exhibit perpendicular polarization (the B_{1u} and B_{2u} modes of the molecular group) split into two components in the crystal. Of these, the crystal B_{1u} modes are polarized along the a axis of the crystal and the B_{2u} modes are polarized along the b axis. The extent of splitting depends on the intermolecular forces. (2) The parallel polarized $w(CH_2)$ mode (the B_{3u} of the molecular group) splits into B_{3u} and A_u modes, but experimentally no splitting is observed, because A_u is infrared inactive. (3) The $t(CH_2)$ mode inactive in a single chain (A_u) splits into A_u and B_{3u} in the crystal, of which the B_{3u} mode is infrared active and is observed as a parallel band at 1050 cm^{-1}. (4) Pure translations T_x, T_y, and T_z of a single chain split into pure translations and translatory lattice modes. Of these, the B_{1u} and B_{2u} lattice modes are infrared active.

The doublets at about 725 cm^{-1} assigned to $r(CH_2)$ and at about 1460 cm^{-1} assigned to $\delta(CH_2)$ are prominent examples of case 1. From

Table 5.8 Correlation Table for the Molecular, Site, and Space Groups of PE

Molecular group D_{2h}	Site group C_{2h}	Space group D_{2h}^{16}
$A_g(\alpha_{xx}, \alpha_{yy}, \alpha_{zz})$		$A_g(\alpha_{aa}, \alpha_{bb}, \alpha_{cc})$
$B_{1g}(\alpha_{xy})$	$A_g(\alpha_{xx}, \alpha_{yy}, \alpha_{zz}, \alpha_{yz})$	$B_{1g}(\alpha_{bc})$
$B_{2g}(\alpha_{xz})$	$B_g(\alpha_{xy}, \alpha_{xz})$	$B_{2g}(\alpha_{ac})$
$B_{3g}(\alpha_{yz})$		$B_{3g}(\alpha_{ab})$
A_u(forbidden)		A_u(forbidden)
$B_{1u}(\mu_z)$	$A_u(\mu_x)$	$B_{1u}(\mu_a)$
$B_{2u}(\mu_y)$	$B_u(\mu_y, \mu_z)$	$B_{2u}(\mu_b)$
$B_{3u}(\mu_x)$		$B_{3u}(\mu_c)$

* The correlations for other symmetry groups are found in text books on group theory.[4,143,144]

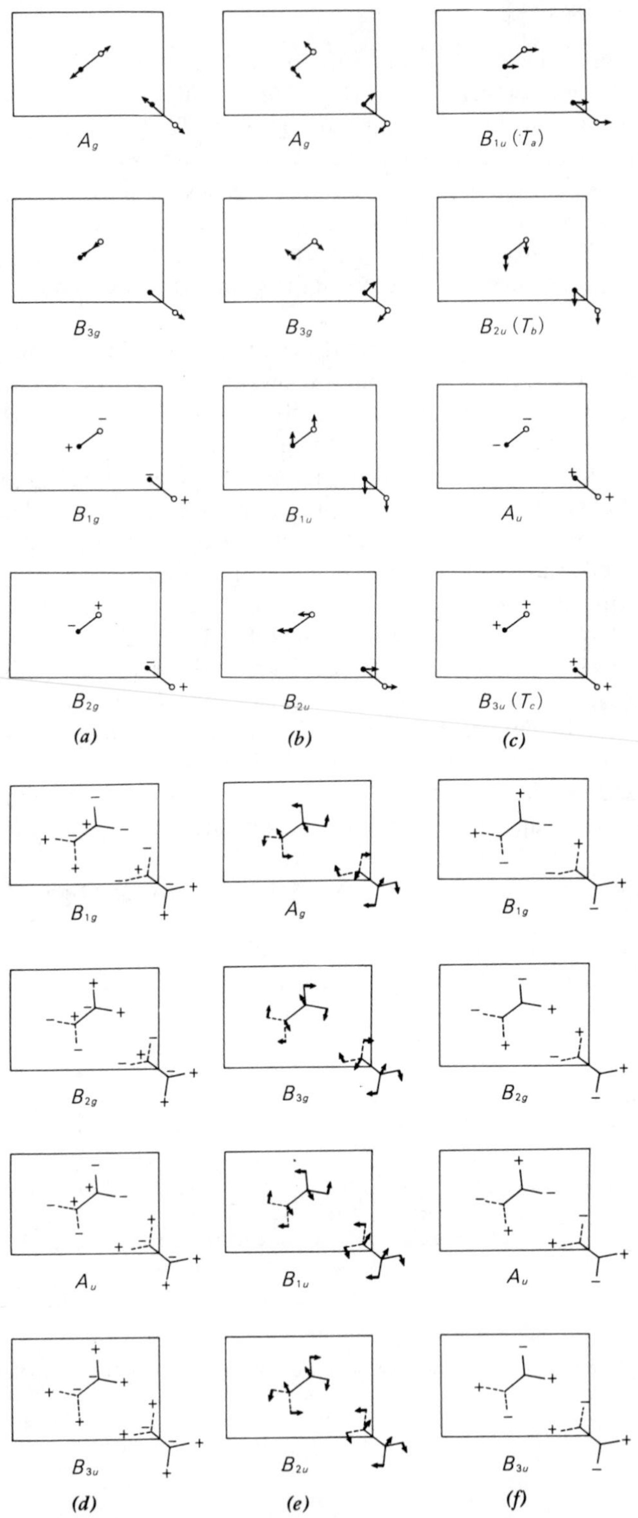

A_g
A_g
B_{1u} (T_a)
B_{3g}
B_{3g}
B_{2u} (T_b)
B_{1g}
B_{1u}
A_u
B_{2g}
B_{2u}
B_{3u} (T_c)
(a)
(b)
(c)
B_{1g}
A_g
B_{1g}
B_{2g}
B_{3g}
B_{2g}
A_u
B_{1u}
A_u
B_{3u}
B_{2u}
B_{3u}
(d)
(e)
(f)

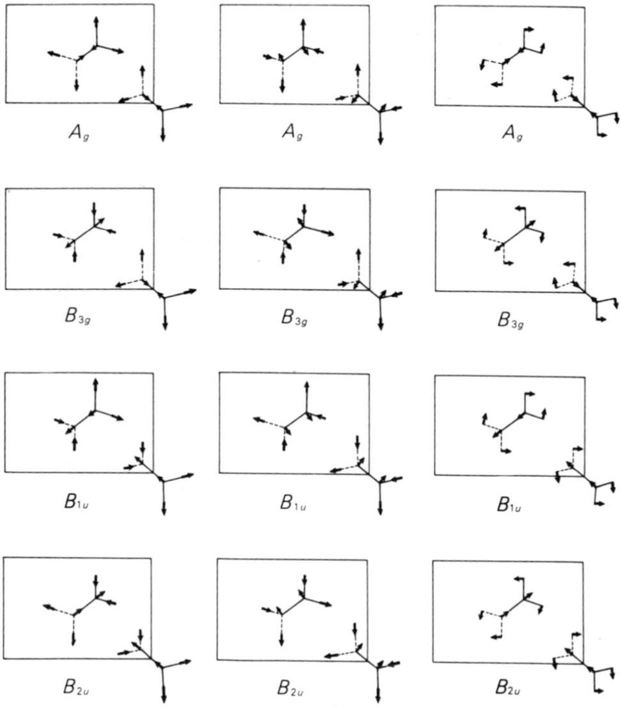

(*g*) (*h*) (*i*)

Fig. 5.45 Vibrational modes of a PE unit cell: (*a*) skeletal modes; (*b* and *c*) lattice modes; (*d*) $w(CH_2)$; (*e*) $r(CH_2)$; (*f*) $t(CH_2)$; (*g*) $\nu_s(CH_2)$; (*h*) $\nu_a(CH_2)$; and (i) $\delta(CH_2)$. (From Krimm et al.[142])

an examination of the polarization of the components in single crystals of n-$C_{36}H_{74}$ and n-$C_{29}H_{60}$ with radiation directed along the c axis, it has been found that the 721 cm^{-1} band polarizes along the b axis and the 731 cm^{-1} band polarizes along the a axis. The polarizations of these bands have also been confirmed on the machine-extruded film of PE as is described in Section 5.3.5.D.

A band assigned to the lattice modes of PE has been observed at 72 cm^{-1} at room temperature[147-149] and shifts by about 6.5 cm^{-1} towards the higher frequency side at 100°K.[147] The corresponding band appears at about 67 cm^{-1} in the case of n-$C_{100}D_{202}$.[149] Bank and Krimm confirmed that this band arises along the a axis (B_{1u}) with the machine-extruded film mentioned above.[150]

Each Raman-active mode splits into two Raman-active modes as shown in Table 5.8. The examples are the doublets of $\delta(CH_2)$ (from A_g of the single chain to A_g and B_{3g} of the crystal), $t(CH_2)$ (from B_{2g} to B_{1g} and B_{2g}), and the C—C stretching modes (from B_{1g} to B_{1g} and B_{2g}) shown in Fig. 5.40.[140]

Table 5.9 Vibrational Spectra of PE (From Tasumi and Shimanouchi[153a])

Species	Observed (cm^{-1}) Infrared	Observed (cm^{-1}) Raman	Calculated (cm^{-1})	Assignment
A_g	Inactive	2883	2899	$\nu_a(CH_2)$
		2848	2845	$\nu_s(CH_2)$
		1418	1437	$\delta(CH_2)$
		1168	1164	$r(CH_2)$
		1131	1127	Skeletal
		137[b]	171	Rotational lattice mode
B_{1g}	Inactive	1370	1408	$w(CH_2)$
		1295	1308	$t(CH_2)$
		1061	1051	Skeletal
B_{2g}	Inactive	1370	1413	$w(CH_2)$
		1295	1303	$t(CH_2)$
		1061	1051	Skeletal
B_{3g}	Inactive	2883	2904	$\nu_a(CH_2)$
		2848	2838	$\nu_s(CH_2)$
		1441	1455	$\delta(CH_2)$
		1168	1164	$r(CH_2)$
		1131	1127	Skeletal
		108[b]	138	Rotational lattice mode
A_u	Inactive	Inactive	1184	$w(CH_2)$
			1053	$t(CH_2)$
			59	Translational lattice mode
B_{1u}	2919	Inactive	2919	$\nu_a(CH_2)$
	2851		2874	$\nu_s(CH_2)$
	1473		1489	$\delta(CH_2)$
	731		749	$r(CH_2)$
	73		76	Translational lattice mode
B_{2u}	2919	Inactive	2917	$\nu_a(CH_2)$
	2851		2877	$\nu_s(CH_2)$
	1463		1479	$\delta(CH_2)$
	720		737	$r(CH_2)$
	109[c]		105	Translational lattice mode
B_{3u}	1175	Inactive	1175	$w(CH_2)$
	1050		1059	$t(CH_2)$

[a] Partly revised (from Carter[133]).
[b] Data at 77°K from Harley et al.[151]
[c] Data at 2°K from Dean and Martin.[154]

The A_g and B_{3g} modes in Fig. 5.45*b* are the Raman-active lattice vibrations and have been observed at 136.5 and 108.0 cm^{-1} in the Raman spectrum of PE taken at 77°K.[151] Detailed information concerning these lines has been obtained for a single crystal of n-$C_{36}H_{74}$ in the range from liquid nitrogen temperature to room temperature.[152]

The band assignments of the vibrational spectra of PE are summarized in Table 5.9, based on factor group analysis, on the normal coordinate treatments of a single chain[155,156] and of crystalline PE,[153,157] and on studies on n-paraffins and related compounds.[156]

5.5.8 Polyoxymethylene

As a typical example of helical molecules, the factor group analysis of the trigonal POM molecule with a (9/5) helix conformation will be considered. As is described in Section 3.3, the ninefold screw axis of the molecule generates nine symmetry operations, which are denoted as C^0 $(=E)$, $C^{\pm 1}$, $C^{\pm 2}$, $C^{\pm 3}$, and $C^{\pm 4}$. There are 18 twofold axes per identity period, and the axes passing through the kth carbon and oxygen atoms are denoted by $C_2(\mathrm{C}_k)$ and $C_2(\mathrm{O}_k)$, respectively. The 18 twofold rotations are not independent of one another. As may be deduced from Fig. 5.46, $C_2(\mathrm{C}_k)$ and $C_2(\mathrm{O}_k)$ produce the following movements of atoms:

$$C_2(\mathrm{C}_k):\mathrm{C}(j) \longrightarrow \mathrm{C}(2k-j), \quad \mathrm{O}(j) \longrightarrow \mathrm{O}(2k-j+1)$$
$$C_2(\mathrm{O}_k):\mathrm{C}(j) \longrightarrow \mathrm{C}(2k-j-1), \quad \mathrm{O}(j) \longrightarrow \mathrm{O}(2k-j)$$

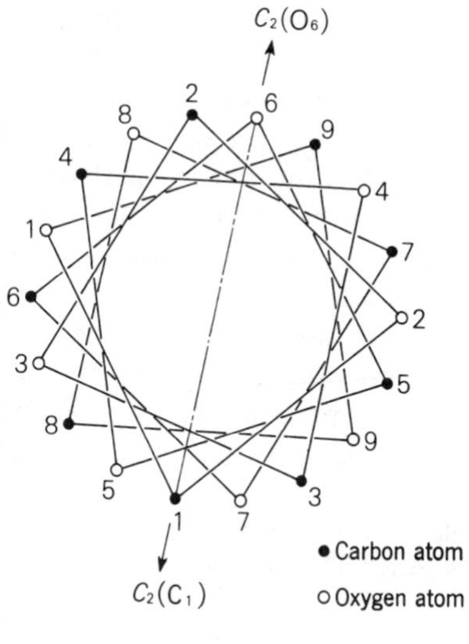

Fig. 5.46 End view of the (9/5) helix of a POM molecule. Only the twofold axis perpendicular to the molecular axis passing through the carbon or oxygen atom is shown.

If $C_2(\mathrm{O}_{k+5})$ is compared with $C_2(\mathrm{C}_k)$, it may be shown that

$$C_2(\mathrm{O}_{k+5}) = T_1 C_2(\mathrm{C}_k) \tag{5.103}$$

Similarly,

$$C_2(\mathrm{C}_{k-5}) = T_{-1} C_2(\mathrm{O}_k) \tag{5.104}$$

Thus an operation $C_2(\mathrm{C}_k)$ can be obtained by the product of a translation and $C_2(\mathrm{O}_{k'})$ and vice versa. Hence there are nine independent twofold rotations per identity period.

These symmetry operations form the one-dimensional space group of the POM molecule. The factor group is isomorphous to the point group D_9 and is represented by the notation $D(10\pi/9)$.*

Each pair of symmetry operations C^{+k} and C^{-k} ($k = 1, 2, 3$, and 4) belongs to a class, since the following relations hold for all values of k.

$$C^{-k'} C^k C^{k'} = C^k \quad \text{and} \quad C_2^{-1} C^k C_2 = C^{-k} \tag{5.105}$$

Nine operations C_2 form another class, since

$$C_2^{-1}(\mathrm{C}_{k'}) C_2(\mathrm{C}_k) C_2(\mathrm{C}_{k'}) = C_2(\mathrm{C}_{2k'-k})$$
$$C^{-k'} C_2(\mathrm{C}_k) C^{k'} = C_2(\mathrm{C}_{k-k'}) \tag{5.106}$$

Thus the 18 elements of the factor group are divided into six classes.

The results of the factor group analysis are shown in Table 5.10. The A_1 modes are only Raman active, and the A_2 modes are only infrared active (parallel bands). The E_1, E_2, E_3, and E_4 species are doubly degenerate, and their irreducible representations are matrices of order 2. The E_1 modes are both infrared (perpendicular) and Raman active. The E_2 modes are only Raman active, and the E_3 and E_4 modes are inactive in both spectra. Five parallel and 11 perpendicular infrared bands should appear, and the parallel bands are expected to be at different frequencies from the Raman bands. This prediction is in a good agreement with the observed spectra as shown in Figs. 5.47 and 5.48. Since the unit cell contains only one molecular chain in trigonal POM, there is no band splitting due to the intermolecular interaction. The results of normal coordinate treatments and band assignments for trigonal POM and POM-d_2 are given in Table 5.11.

The crystal structure of orthorhombic POM has been determined by X-ray analysis as shown in Fig. 5.49. Two molecular chains with a (2/1) helix conformation pass through a unit cell. The vibrations of a single chain are treated by a factor group isomorphous to D_2. The internal rotation angle

* The notation is after Liang and Krimm.[158] The helix including u chemical units and t turns in the identity period is denoted by $C(2t\pi/u)$, and the case with additional twofold axes perpendicular to the helix axis is denoted by $D(2t\pi/u)$.

Table 5.10 Factor Group Analysis of Trigonal POM Molecule[84a]

$D(10\pi/9)$	E	$2C^1$	$2C^2$	$2C^3$	$2C^4$	$9C_2$	n_i	$n_i(T)$	$n_i(R)$	n'_i	Infrared[b]	Raman[b]
A_1	1	1	1	1	1	1	5	0	0	5	Inactive	$\alpha_{xx}+\alpha_{yy}, \alpha_{zz}$
A_2	1	1	1	1	1	-1	7	T_z	R_z	5	μ_z	Inactive
E_1	2	$2\cos\omega$	$2\cos 2\omega$	$2\cos 3\omega$	$2\cos 4\omega$	0	12×2	(T_x, T_y)	0	11×2	(μ_x, μ_y)	$(\alpha_{yz}, \alpha_{zx})$
E_2	2	$2\cos 2\omega$	$2\cos 4\omega$	$2\cos 6\omega$	$2\cos 8\omega$	0	12×2	0	0	12×2	Inactive	$(\alpha_{xx}-\alpha_{yy}, \alpha_{xy})$
E_3	2	$2\cos 3\omega$	$2\cos 6\omega$	$2\cos 9\omega$	$2\cos 12\omega$	0	12×2	0	0	12×2	Inactive	Inactive
E_4	2	$2\cos 4\omega$	$2\cos 8\omega$	$2\cos 12\omega$	$2\cos 16\omega$	0	12×2	0	0	12×2	Inactive	Inactive
N_R	36	0	0	0	0	2						
$(\pm 1+2\cos\phi)$	3	$1+2\cos\omega$	$1+2\cos 2\omega$	$1+2\cos 3\omega$	$1+2\cos 4\omega$	-1		$\omega=\frac{2t\pi}{u}=\frac{10\pi}{9}$				
$\chi(R)$	108	0	0	0	0	-2						

[a] The helix axis is assumed to be the z direction.
[b] Two terms of T, μ, or α in parentheses correspond to the pair of the elements in doubly degenerate species.

Table 5.11 Infrared and Raman Spectra, and Vibrational Assignments of POM and POM-d_2[84] [a]

Species	Observed (cm^{-1}) Infrared	Raman	Calcd. (cm^{-1})	Assignment[b]
POM				
A_1	Inactive	2955 w	2924	$\nu_s(CH_2)$
		1493 m	1508	$\delta(CH_2)$
		1333 s	1330	$t(CH_2)$
		919 s	916	$\nu_s(COC)(79) - \delta(OCO)(13)$
		535 w	587	$\delta(COC)(54) - \delta(OCO)(41)$
A_2	2978(‖) s	Inactive	2977	$\nu_a(CH_2)$
	1381(‖) m		1425	$w(CH_2)$
	1097(‖) vvs		1118	$\nu_a(COC)(77) + r(CH_2)(19)$
	903(‖) vvs		922	$r(CH_2)(75) - \nu_a(COC)(21)$
	235(‖) w		237	$\tau_a(96) + r(CH_2)(3)$
E_1	2980(⊥) s	2997 s	2982	$\nu_a(CH_2)$
	2924(⊥) s	2925 m	2926	$\nu_s(CH_2)$
	1471(⊥) m	—	1506	$\delta(CH_2)$
	1434(⊥) w	—	1407	$w(CH_2)$
	1286(⊥) vw	1298 w	1318	$t(CH_2)$
	1235(⊥) vs	—	1169	$r(CH_2)(46) + \delta(COC)(27) - \nu_s(COC)(23)$
	1091(⊥) vvs	1091 w	1072	$\nu_a(COC)(70) - \delta(OCO)(26)$
	932(⊥) vvs	932 w	930	$\nu_s(COC)(74) + r(CH_2)(23)$
	630(⊥) s	640 w	634	$\delta(OCO)(72) + \nu_a(COC)(20)$
	455(⊥) m	—	483	$\delta(COC)(64) - r(CH_2)(28)$
	—	—	22	$\tau_s(56) - \tau_a(40)$
POM-d_2				
A_1	Inactive	2150 vw	2121	$\nu_s(CD_2)$
		1121 w	1126	$\delta(CD_2)(75) - \nu_s(COC)(17)$
		1015 vw	1013	$t(CD_2)(65) - \nu_s(COC)(21)$
		852 m	828	$\nu_s(COC)(54) + t(CD_2)(25) - \delta(OCO)(11)$
		—	557	$\delta(COC)(51) - \delta(OCO)(39)$
A_2	2263(‖) m	Inactive	2193	$\nu_a(CD_2)$
	1050(‖) vs		1154	$w(CD_2)(49) - \nu_a(COC)(48)$
	970(‖) vvs		1026	$w(CD_2)(50) + \nu_a(COC)(40)$
	835(‖) vs		769	$r(CD_2)(86) - \nu_a(COC)(9)$
	212(—) w		212	$\tau_a(93) + r(CD_2)(5)$

Table 5.11 *(Contd.)*

Species	Observed (cm^{-1}) Infrared	Raman	Calcd. (cm^{-1})	Assignment[b]
E_1	2263(⊥) m	2268 w	2207	$\nu_a(CD_2)$
	2106(⊥) m	2105 vw	2121	$\nu_s(CD_2)$
	1156(⊥) s	—	1178	$\nu_s(COC)(43) - w(CD_2)(43)$
	1128(⊥) s	—	1145	$\delta(CD_2)(45) + \nu_a(COC)(39) - \delta(OCO)(12)$
	1078(⊥) s	—	1047	$\delta(CD_2)(44) - \nu_a(COC)(28) - t(CD_2)(15)$
	1064(⊥) s	—	1044	$r(CD_2)(35) - w(CD_2)(30) + \delta(COC)(27)$
	909(⊥) m	—	916	$t(CD_2)(74) + \delta(OCO)(13)$
	838(⊥) s	—	810	$\nu_s(COC)(55) + r(CD_2)(23) + w(CD_2)(16)$
	618(⊥) m	—	618	$\delta(OCO)(69) + \nu_a(COC)(20)$
	363(—) m	—	372	$\delta(COC)(52) - r(CD_2)(41)$
	—	—	21	$\tau_s(56) - \tau_a(40)$

[a] Raman bands are observed at 70, 53, and 43 cm^{-1} in Fig. 5.48. For assignment of these bands, analysis should be made taking into account the intermolecular interactions (including the E_2 modes).

[b] The potential energy distribution (in percent) among the symmetry coordinates is given in parentheses. The signs denote the phase relations of the coupled coordinates.

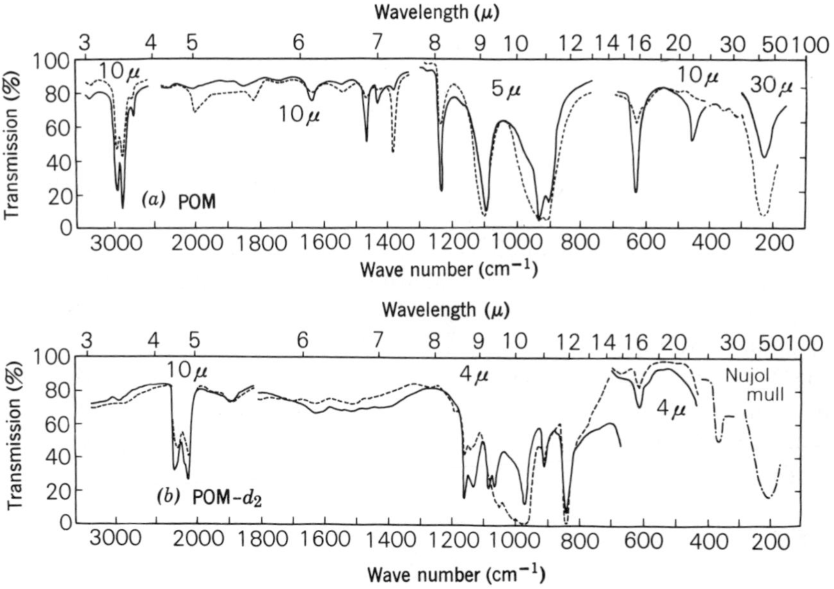

Fig. 5.47 Polarized infrared spectra of (*a*) POM and (*b*) POM-d_2. (———) Perpendicular radiation, (--------) parallel radiation, and (-·-·-·-·-) measurement done with non-polarized light. Film thickness is indicated in the spectra.[84]

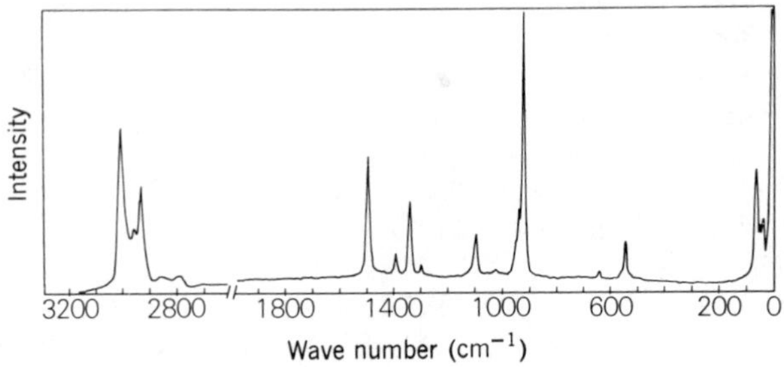

Fig. 5.48 Raman spectrum of POM (paraformaldehyde). The 514.5 nm line of an Ar^+ laser was used.

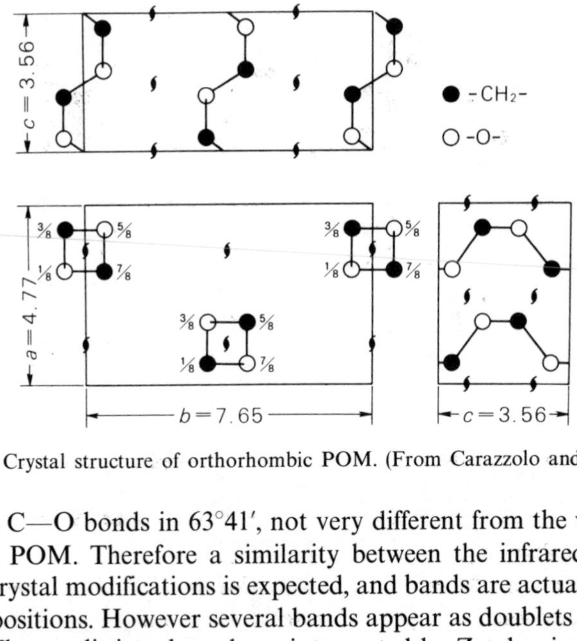

Fig. 5.49 Crystal structure of orthorhombic POM. (From Carazzolo and Mammi.[159])

around the C—O bonds in 63°41′, not very different from the value 78°13′ of trigonal POM. Therefore a similarity between the infrared spectra of these two crystal modifications is expected, and bands are actually observed at similar positions. However several bands appear as doublets as shown in Fig. 5.50. These splittings have been interpreted by Zamboni and Zerbi[160] in terms of the intermolecular interactions between two chains in the unit cell. The correlations among the molecular, site, and space groups are given in Table 5.12.

5.5.9 Translation Group

The vibrational modes where two neighboring unit cells vibrate with a certain phase difference are considered briefly. The group **T** consisting of pure translations T_n in a one-dimensional space group is called a one-

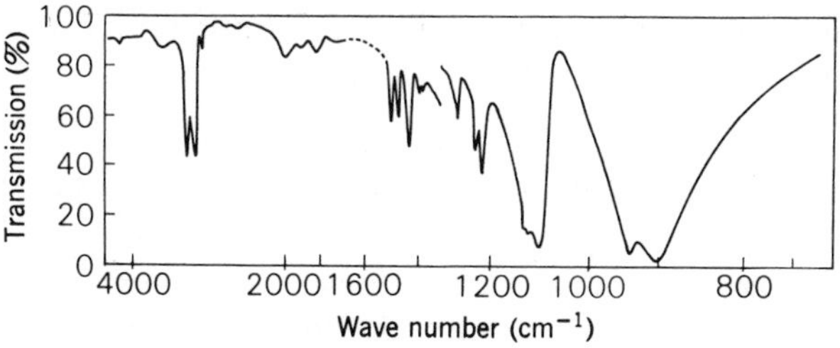

Fig. 5.50 Infrared spectra of orthorhombic POM. (From Carazzolo and Mammi.[159])

Table 5.12 Correlation Table for the Molecular, Site, and Space Groups of Orthorhombic POM (From Zamboni and Zerbi[160])

Molecular group D_2	Site group C_2	Space group $D_2^{\,4}$
A	A	A
$B_1(\mu_z)$	A	$B_1(\mu_c)$
$B_2(\mu_y)$	B	$B_2(\mu_a)$
$B_3(\mu_x)$	B	$B_3(\mu_b)$

dimensional translation group. By application of Born's cyclic boundary condition, this becomes a cyclic group of the order N, and the character of the pth irreducible representation is given for the element $T_n (n = 0, 1, \cdots, N - 1)$ as

$$\chi_n^{(p)} = \exp\left(\frac{2\pi ipn}{N}\right) \qquad p = 0, 1, \cdots, N - 1 \tag{5.107}$$

For the species $\Gamma^{(0)}$, to which the factor group modes belong, the character becomes unity for every element, that is, this species is totally symmetric with respect to all pure translations. All components of the dipole moment and the polarizability tensor have the same symmetry properties and belong to the species $\Gamma^{(0)}$. This is the reason that only the factor group modes are active in infrared and Raman spectra. The vibrational modes belonging to the $\Gamma^{(m)}$ species have a phase difference of $2m\pi/N$ between the two neighboring unit cells. Such vibrational modes are inactive in both spectra, since the changes of the dipole moment or polarizability cancel in the molecular chain.

For a discussion of the selection rules for the combination and overtone bands, see Refs. 4 and 24.

5.6 CALCULATION OF NORMAL MODES OF VIBRATIONS

Before the problems encountered in the study of high polymers are considered it is appropriate to review Wilson's **GF** matrix method[4] for calculating normal modes of vibrations of polyatomic molecules.

5.6.1 Wilson's GF Matrix Method

As is described in Section 5.2.1 the potential energy V and the kinetic energy T under the harmonic approximation may be expressed by Eqs. 5.2 and 5.5. Thus

$$2V = \sum_{i,j=1}^{n} f_{ij} R_i R_j \tag{5.108}$$

$$2T = \sum_{i,j=1}^{n} a_{ij} \dot{R}_i \dot{R}_j \tag{5.109}$$

where $\dot{R}_i$ is the ith internal displacement coordinate. These equations can be transformed into simple forms by using the transformation given by Eq. 5.6.

$$2V = \sum \lambda_k Q_k^{\,2} \tag{5.110}$$

$$2T = \sum \dot{Q}_k^{\,2} \tag{5.111}$$

where Q_k is the kth normal coordinate. If Eqs. 5.108, 5.109, and 5.6 are written in terms of matrix notation as

$$2V = \tilde{\mathbf{R}}\mathbf{F}\mathbf{R} \tag{5.112}$$

$$2T = \tilde{\dot{\mathbf{R}}}\mathbf{A}\dot{\mathbf{R}} \tag{5.113}$$

$$\mathbf{R} = \mathbf{L}\mathbf{Q} \tag{5.114}$$

then Eqs. 5.110 and 5.111 are given by

$$2V = \tilde{\mathbf{Q}}\tilde{\mathbf{L}}\mathbf{F}\mathbf{L}\mathbf{Q} = \tilde{\mathbf{Q}}\mathbf{\Lambda}\mathbf{Q} \tag{5.115}$$

$$2T = \tilde{\dot{\mathbf{Q}}}\tilde{\mathbf{L}}\mathbf{A}\mathbf{L}\dot{\mathbf{Q}} = \tilde{\dot{\mathbf{Q}}}\dot{\mathbf{Q}} \tag{5.116}$$

From these two equations,

$$\tilde{\mathbf{L}}\mathbf{F}\mathbf{L} = \mathbf{\Lambda} \tag{5.117}$$

$$\tilde{\mathbf{L}}\mathbf{A}\mathbf{L} = \mathbf{E} \tag{5.118}$$

where **E** is a unit matrix and **Λ** is the following diagonal matrix.

$$\boldsymbol{\Lambda} = \begin{bmatrix} \lambda_1 & 0 & \cdots & 0 \\ 0 & \lambda_2 & & 0 \\ \cdots & & \ddots & \\ 0 & \cdots & 0 & \lambda_n \end{bmatrix} \tag{5.119}$$

Multiplying both sides of Eq. 5.118 by $\mathbf{L}^{-1}\mathbf{A}^{-1}$ from the right-hand side gives

$$\tilde{\mathbf{L}} = \mathbf{L}^{-1}\mathbf{A}^{-1} \tag{5.120}$$

By substitution of this equation, Eq. 5.117 becomes

$$\mathbf{L}^{-1}\mathbf{A}^{-1}\mathbf{F}\mathbf{L} = \boldsymbol{\Lambda} \tag{5.121}$$

Replacing by

$$\mathbf{G} = \mathbf{A}^{-1} \tag{5.122}$$

gives the important equation

$$\mathbf{L}^{-1}\mathbf{G}\mathbf{F}\mathbf{L} = \boldsymbol{\Lambda} \tag{5.123}$$

When this relation holds, $\lambda_1, \lambda_2, \cdots, \lambda_n$ are the characteristic values of the matrix **GF**, and the column vectors consisting of the matrix **L**

$$\begin{bmatrix} l_{11} \\ l_{21} \\ \vdots \\ l_{n1} \end{bmatrix}, \begin{bmatrix} l_{12} \\ l_{22} \\ \vdots \\ l_{n2} \end{bmatrix}, \ldots, \begin{bmatrix} l_{1n} \\ l_{2n} \\ \vdots \\ l_{nn} \end{bmatrix} \tag{5.124}$$

are the characteristic vectors. The method of obtaining the normal frequencies and normal coordinates is equivalent to the procedure for calculating the **L** matrix that diagonalizes the **GF** matrix by a similarity transformation. **G** is called the kinetic energy matrix, and **F** is the potential energy matrix. Both **G** and **F** matrices are symmetric.

Equation 5.123 is written as

$$\mathbf{H}\mathbf{L} = \mathbf{L}\boldsymbol{\Lambda}, \qquad \mathbf{G}\mathbf{F} = \mathbf{H} \tag{5.125}$$

For this equation, the kth characteristic value and characteristic vector are given by

$$\mathbf{H} \begin{bmatrix} l_{1k} \\ l_{2k} \\ \vdots \\ l_{nk} \end{bmatrix} = \begin{bmatrix} l_{1k} \\ l_{2k} \\ \vdots \\ l_{nk} \end{bmatrix} \lambda_k \tag{5.126}$$

This equation is written as

$$
\begin{aligned}
h_{11}l_{1k} + h_{12}l_{2k} + \cdots + h_{1n}l_{nk} &= \lambda_k l_{1k} \\
h_{21}l_{1k} + h_{22}l_{2k} + \cdots + h_{2n}l_{nk} &= \lambda_k l_{2k} \\
\cdots \\
h_{n1}l_{1k} + h_{n2}l_{2k} + \cdots + h_{nn}l_{nk} &= \lambda_k l_{nk}
\end{aligned} \tag{5.127}
$$

where h_{ij} is the ijth element of **H**. k is omitted and Eq. 5.127 is rearranged to give

$$
\begin{aligned}
(h_{11} - \lambda)l_1 + h_{12}l_2 + \cdots + h_{1n}l_n &= 0 \\
h_{21}l_1 + (h_{22} - \lambda)l_2 + \cdots + h_{2n}l_n &= 0 \\
\cdots \\
h_{n1}l_1 + h_{n2}l_2 + \cdots + (h_{nn} - \lambda)l_n &= 0
\end{aligned} \tag{5.128}
$$

For all l's to be nontrivial, the values of λ satisfying the following secular equation should be found.

$$|\mathbf{H} - \lambda\mathbf{E}| = |\mathbf{GF} - \lambda\mathbf{E}| = 0 \tag{5.129}$$

When a fixed value of λ, say λ_k, is chosen, the coefficients of the unknown $l_i (i = 1, 2, \cdots, n)$ in Eq. 5.128 become fixed, and it is then possible to obtain a solution, l_{ik}, for which the additional subscript is used to indicate the correspondence with the particular value of λ_k. Such a system of equations does not determine the l_{ik} uniquely, but gives only their ratios. Therefore **L** should be normalized so as to satisfy Eq. 5.123. The procedure for solving Eq. 5.123 is equivalent to that employed for Eq. 5.129.

From Eqs. 5.113 and 5.122,

$$2T = \tilde{\dot{\mathbf{R}}}\mathbf{G}^{-1}\dot{\mathbf{R}} \tag{5.130}$$

Hence the momentum P_i conjugate to R_i is given by

$$P_i = \frac{\partial T}{\partial \dot{R}_i} = \sum_{j=1}^{n} (G^{-1})_{ij}\dot{R}_j \tag{5.131}$$

or

$$\mathbf{P} = \mathbf{G}^{-1}\dot{\mathbf{R}} \tag{5.132}$$

which is rewritten as

$$\dot{\mathbf{R}} = \mathbf{GP} \tag{5.133}$$

Substituting this equation to Eq. 5.130 gives

$$2T = \tilde{\mathbf{P}}\mathbf{GP} \tag{5.134}$$

Consequently **G** is the matrix consisting of the coefficients when the kinetic energy is expressed in terms of the quadratic form in P_i instead of $\dot{R}_i$.

5.6.2 Example: Water Molecule

The internal displacement coordinates are

$$\mathbf{R} = \begin{bmatrix} \Delta r_1 \\ \Delta r_2 \\ \Delta \phi \end{bmatrix} \tag{5.135}$$

where r_1, r_2, and ϕ are the bond lengths O—H and the bond angle $\angle$ HOH as shown in Fig. 5.51.

To construct the **F** matrix of a molecule consisting of N atoms the $(3N - 6) \times (3N - 5)/2$ force constants should be determined for the most general case. Although the number of independent constants is reduced if the molecule possesses some kind of symmetry, except for very simple molecules, it is always larger than the number of fundamental frequencies. To reduce the number of independent force constants, various types of force fields are employed. The simplest one is the central force field,[2,4] based on the assumption that the forces act only along the lines joining pairs of atoms. This type of force field yields only diagonal terms in the **F** matrix when the complete set of interatomic distances is adopted as the coordinate system. A force field that is more compatible with the chemical idea of intramolecular forces is the valence force field.[2,4] The force involved here resists the deviation of internal coordinates from their equilibrium values. In the most simple form, this force field also gives a diagonal **F** matrix based on the internal coordinate system. To obtain better agreement with the observed vibrational frequencies, it is necessary to introduce additional interaction terms between different internal coordinates,[156,161] for example, those between two adjacent bonds, a valence angle and one of the bonds enclosing the angle,

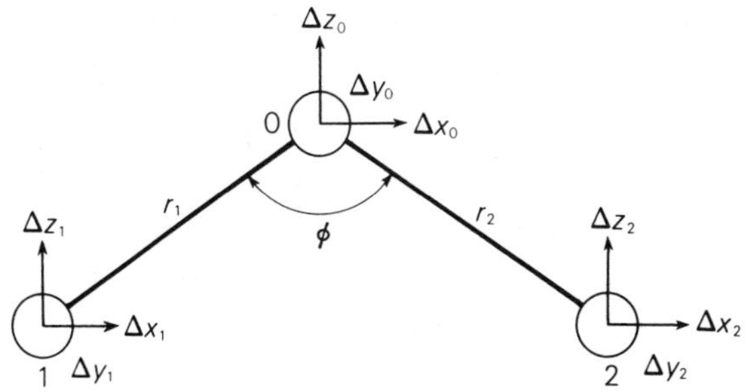

Fig. 5.51 The Cartesian and internal displacement coordinates for a water molecule.

and two valence angles having one or two common atoms. Another important and very useful force field is the Urey-Bradley force field; this force field is discussed in more detail later (Section 5.6.4).

When the potential energy V of the water molecule is expressed using the valence force field, it takes the form

$$V = \tfrac{1}{2}K_r(\Delta r_1)^2 + \tfrac{1}{2}K_r(\Delta r_2)^2 + F_{rr}(\Delta r_1)(\Delta r_2) + \tfrac{1}{2}H_\phi(\Delta\phi)^2 + F_{r\phi}(\Delta r_1)(\Delta\phi) + F_{r\phi}(\Delta r_2)(\Delta\phi) \quad (5.136)$$

Then the **F** matrix in Eq. 5.112 is written as

$$\mathbf{F} = \begin{bmatrix} K_r & \text{symmetric} & \\ F_{rr} & K_r & \\ F_{r\phi} & F_{r\phi} & H_\phi \end{bmatrix} \quad (5.137)$$

The kinetic energy T is conveniently written in terms of the Cartesian displacement coordinates.

$$T = \tfrac{1}{2}m_0[(\Delta\dot{x}_0)^2 + (\Delta\dot{y}_0)^2 + (\Delta\dot{z}_0)^2] + \tfrac{1}{2}m_{\text{H}}[(\Delta\dot{x}_1)^2 + (\Delta\dot{y}_1)^2 + (\Delta\dot{z}_1)^2 + (\Delta\dot{x}_2)^2 + (\Delta\dot{y}_2)^2 + (\Delta\dot{z}_2)^2] \quad (5.138)$$

where Δx_0, Δx_1, and so on are the Cartesian displacement coordinates shown in Fig. 5.51. In the motions described in terms of these nine coordinates, three pure translations (X, Y, Z) and three pure rotations (R_x, R_y, R_z) are involved, in addition to the three internal vibrations. To remove these six coordinates corresponding to the nongenuine vibrations, the Cartesian displacement coordinates are transformed to the internal displacement coordinates **R**, which gives the **G** matrix in terms of **R**. The momenta P_{x_0}, $P_{y_0}, \cdots$ conjugate to $x_0, y_0, \cdots$ are given by

$$P_{x_0} = \frac{\partial T}{\partial \dot{x}_0} = m_0\dot{x}_0 \quad (5.139)$$

This equation is rewritten as

$$P_{x_0} = \frac{\partial T}{\partial \dot{x}_0} = \frac{\partial T}{\partial \Delta\dot{r}_1}\frac{\partial \Delta r_1}{\partial x_0} + \frac{\partial T}{\partial \Delta\dot{r}_2}\frac{\partial \Delta r_2}{\partial x_0} + \frac{\partial T}{\partial \Delta\dot{\phi}}\frac{\partial \Delta\phi}{\partial x_0} + \frac{\partial T}{\partial \dot{X}}\frac{\partial X}{\partial x_0} + \frac{\partial T}{\partial \dot{Y}}\frac{\partial Y}{\partial x_0} + \frac{\partial T}{\partial \dot{Z}}\frac{\partial Z}{\partial x_0} + \frac{\partial T}{\partial \dot{R}_x}\frac{\partial R_x}{\partial x_0} + \frac{\partial T}{\partial \dot{R}_y}\frac{\partial R_y}{\partial x_0} + \frac{\partial T}{\partial \dot{R}_z}\frac{\partial R_z}{\partial x_0}$$

$$= P_{\Delta r_1}\frac{\partial \Delta r_1}{\partial x_0} + P_{\Delta r_2}\frac{\partial \Delta r_2}{\partial x_0} + P_{\Delta\phi}\frac{\partial \Delta\phi}{\partial x_0} + P_X\frac{\partial X}{\partial x_0} + \cdots + P_{R_z}\frac{\partial R_z}{\partial x_0} \quad (5.140)^*$$

* $\dfrac{\partial \Delta\dot{r}_i}{\partial \dot{x}_j} = \dfrac{\partial \Delta r_i}{\partial x_j}$ is obtained by partial differentiation of $\Delta\dot{r}_i = \sum_{j=1}^{n}\left(\dfrac{\partial \Delta r_i}{\partial x_j}\right)\dot{x}_j$ with respect to $\dot{x}_j$.

Here $P_X, \cdots, P_{R_z}$ are to be taken as zero if the molecule is at rest, since these are the momenta of the pure translations and rotations.

Thus the momentum vector written in terms of the internal displacement coordinates is

$$\mathbf{P} = \begin{bmatrix} P_{\Delta r_1} \\ P_{\Delta r_2} \\ P_{\Delta\phi} \end{bmatrix} \tag{5.141}$$

The relation between **R** and the Cartesian displacement coordinates

$$\mathbf{X} = \begin{bmatrix} \Delta x_0 \\ \Delta y_0 \\ \cdot \\ \cdot \\ \cdot \\ \Delta z_2 \end{bmatrix} \tag{5.142}$$

is given by

$$\mathbf{R} = \mathbf{BX} \tag{5.143}$$

The matrix **B** is obtained from the geometry of the molecule shown in Fig. 5.51.*

$$\mathbf{B} = \begin{bmatrix} s & 0 & c & -s & 0 & -c & 0 & 0 & 0 \\ -s & 0 & c & 0 & 0 & 0 & s & 0 & -c \\ 0 & 0 & \dfrac{-2s}{r} & \dfrac{-c}{r} & 0 & \dfrac{s}{r} & \dfrac{c}{r} & 0 & \dfrac{s}{r} \end{bmatrix} \tag{5.144}$$

where r is the equilibrium value of r_1 and r_2, $s = \sin(\phi/2)$, and $c = \cos(\phi/2)$.

Definition

$$\mathbf{P}_X = \begin{bmatrix} P_{x_0} \\ \cdot \\ \cdot \\ \cdot \\ P_{z_2} \end{bmatrix} \tag{5.145}$$

gives

$$\mathbf{P} = \mathbf{BP}_X \qquad \mathbf{P}_X = \tilde{\mathbf{B}}\mathbf{P} \tag{5.146}$$

* Since Δr_1 is the sum of the projections of $\Delta x_0, \Delta z_0, \Delta x_1$, and Δz_1 to the bond r_1, $\Delta r_1 = s\Delta x_0 + c\Delta z_0 - s\Delta x_1 - c\Delta z_1$. For $\Delta\phi$ two components are considered, infinitesimal rotations of the bonds r_1 and r_2 counterclockwise and clockwise, respectively. The former component is obtained by dividing the sum of the projections of $\Delta x_0, \Delta z_0, \Delta x_1$, and Δz_1 to the normal to the bond r_1 by r.

The relations represented by Eq. 5.139 are generally expressed by

$$\mathbf{P}_X = \mathbf{M}\dot{\mathbf{X}}, \qquad \mathbf{M}^{-1}\mathbf{P}_X = \dot{\mathbf{X}} \tag{5.147}$$

where

$$\mathbf{M} = \begin{bmatrix} m_O & & & & & 0 \\ & m_O & & & & \\ & & m_O & & & \\ & & & \cdot & & \\ & & & & \cdot & \\ 0 & & & & & m_H \end{bmatrix}, \mathbf{M}^{-1} = \begin{bmatrix} m_O^{-1} & & & & & 0 \\ & m_O^{-1} & & & & \\ & & m_O^{-1} & & & \\ & & & \cdot & & \\ & & & & \cdot & \\ 0 & & & & & m_H^{-1} \end{bmatrix} \tag{5.148}$$

Substituting Eq. 5.138 expressed by the matrix notation into Eq. 5.147 gives

$$2T = \tilde{\dot{\mathbf{X}}}\mathbf{M}\dot{\mathbf{X}} = \tilde{\mathbf{P}}_X\mathbf{M}^{-1}\mathbf{M}\mathbf{M}^{-1}\mathbf{P}_X = \tilde{\mathbf{P}}_X\mathbf{M}^{-1}\mathbf{P}_X \tag{5.149}$$*

Substituting for $\mathbf{P}_X$ from Eq. 5.146 leads to

$$2T = \tilde{\mathbf{P}}\mathbf{B}\mathbf{M}^{-1}\tilde{\mathbf{B}}\mathbf{P} \tag{5.150}$$

Comparison of this equation with Eq. 5.134 gives

$$\mathbf{G} = \mathbf{B}\mathbf{M}^{-1}\tilde{\mathbf{B}} \tag{5.151}$$

The **G** matrix is obtained from Eqs. 5.144, 5.148, and 5.151 as follows:

$$\mathbf{G} = \begin{bmatrix} \mu_H + \mu_O & & \text{symmetric} \\ \mu_O \cos\phi & \mu_H + \mu_O & \\ -\rho\mu_O \sin\phi & -\rho\mu_O \sin\phi & 2\rho^2[\mu_H + \mu_O(1-\cos\phi)] \end{bmatrix} \tag{5.152}$$

where $\mu_H = 1/m_H, \mu_O = 1/m_O, \rho = 1/r$.

Although the **B** matrix in Eq. 5.144 was derived by the consideration of molecular geometry, the **S** matrix is more conveniently used. In terms of the unit vectors **e** and **p**, Eq. 5.143 is expressed as

$$\begin{bmatrix} \Delta r_1 \\ \Delta r_2 \\ \Delta\phi \end{bmatrix} = \begin{bmatrix} -\mathbf{e}_{01} & \mathbf{e}_{01} & \mathbf{0} \\ -\mathbf{e}_{02} & \mathbf{0} & \mathbf{e}_{02} \\ \dfrac{-\mathbf{p}_{01}+\mathbf{p}_{02}}{r} & \dfrac{\mathbf{p}_{01}}{r} & \dfrac{-\mathbf{p}_{02}}{r} \end{bmatrix} \begin{bmatrix} \boldsymbol{\rho}_0 \\ \boldsymbol{\rho}_1 \\ \boldsymbol{\rho}_2 \end{bmatrix} \tag{5.153}$$

where the vector notation is used, and

$$\boldsymbol{\rho}_\alpha = \begin{bmatrix} \Delta x_\alpha \\ \Delta y_\alpha \\ \Delta z_\alpha \end{bmatrix} \tag{5.154}$$

* $\widetilde{\mathbf{AB}} = \tilde{\mathbf{B}}\tilde{\mathbf{A}}, [\mathbf{AB}]^{-1} = \mathbf{B}^{-1}\mathbf{A}^{-1}$.

Table 5.13 S Matrix Elements (From Mizushima and Shimanouchi[162a])

Bond stretching	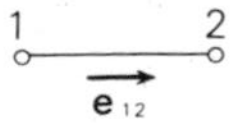	$\Delta r_{12} = -\mathbf{e}_{12}\cdot\boldsymbol{\rho}_1 + \mathbf{e}_{12}\cdot\boldsymbol{\rho}_2$
Valence angle bending	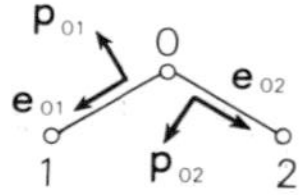	$\Delta\phi_{102} = \left(-\frac{\mathbf{p}_{01}}{r_{01}} + \frac{\mathbf{p}_{02}}{r_{02}}\right)\cdot\boldsymbol{\rho}_0 + \frac{\mathbf{p}_{01}}{r_{01}}\cdot\boldsymbol{\rho}_1 - \frac{\mathbf{p}_{02}}{r_{02}}\cdot\boldsymbol{\rho}_2$
Out-of-plane displacement	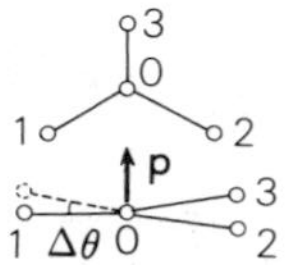	$\Delta\theta_1 \sin\phi_{203} = \Delta\theta_2 \sin\phi_{301} = \Delta\theta_3 \sin\phi_{102} = \left[\left(\frac{\sin\phi_{203}}{r_{01}}\right)(\boldsymbol{\rho}_1 - \boldsymbol{\rho}_0) + \left(\frac{\sin\phi_{301}}{r_{02}}\right)(\boldsymbol{\rho}_2 - \boldsymbol{\rho}_0) + \left(\frac{\sin\phi_{102}}{r_{03}}\right)(\boldsymbol{\rho}_3 - \boldsymbol{\rho}_0)\right]\cdot\mathbf{p}$ $\Delta\theta_1$ denotes the angle between plane (203) and bond (01)
Displacement of internal rotation angle	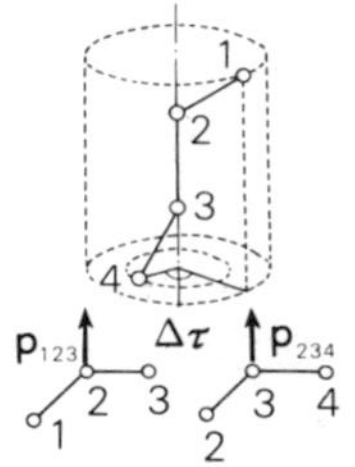	$\Delta\tau = u_{123}\mathbf{p}_{123}\cdot\boldsymbol{\rho}_1 - [(u_{123} - w_{321})\mathbf{p}_{123} - w_{234}\mathbf{p}_{234}]\cdot\boldsymbol{\rho}_2 + [(u_{432} - w_{234})\mathbf{p}_{234} - w_{321}\mathbf{p}_{123}]\cdot\boldsymbol{\rho}_3 - u_{432}\mathbf{p}_{234}\cdot\boldsymbol{\rho}_4$, $u_{ijk} = (r_{ij}\sin\phi_{ijk})^{-1}$, $w_{ijk} = \cos\phi_{ijk}(r_{ij}\sin\phi_{ijk})^{-1}$ $\cos\tau = \mathbf{p}_{123}\cdot\mathbf{p}_{234}$

[a] **e**'s and **p**'s are the unit vectors defined in the second column, and $\boldsymbol{\rho}_i$ is the Cartesian displacement coordinate vector of the ith atom.

Equation 5.153 is written symbolically as

$$\mathbf{R} = \mathbf{S} \cdot \rho \tag{5.155}$$

The **S** matrix expresses the **B** matrix in terms of the vector notation. For example, the first three elements of the first row in Eq. 5.144 sin ($\phi/2$), 0, and cos ($\phi/2$) correspond, respectively, to the x, y, and z components of the vector $-\mathbf{e}_{01}$ in Fig. 5.52. The advantage of the **S** matrix is that its vector components (**S** vectors) can be easily written according to the simple formulas shown in Table 5.13.* Using **S** leads to

$$\mathbf{G} = \mathbf{S}\mathbf{M}^{-1}\tilde{\mathbf{S}}$$

$$= \begin{bmatrix} (\mu_O + \mu_H)\mathbf{e}_{01}^2 & \mu_O \mathbf{e}_{01} \cdot \mathbf{e}_{02} & \dfrac{[\mu_H \mathbf{e}_{01} \cdot \mathbf{p}_{01} + \mu_O \mathbf{e}_{01} \cdot (\mathbf{p}_{01} - \mathbf{p}_{02})]}{r} \\ & (\mu_O + \mu_H)\mathbf{e}_{02}^2 & \dfrac{-[\mu_H \mathbf{e}_{02} \cdot \mathbf{p}_{02} + \mu_O \mathbf{e}_{02} \cdot (\mathbf{p}_{01} - \mathbf{p}_{02})]}{r} \\ \text{symmetric} & & \dfrac{[\mu_H \mathbf{p}_{01}^2 + \mu_H \mathbf{p}_{02}^2 + \mu_O(-\mathbf{p}_{01} + \mathbf{p}_{02})^2]}{r^2} \end{bmatrix} \tag{5.156}$$

Equation 5.152 is obtained from this equation, since

$$\mathbf{e}_{01}^2 = \mathbf{e}_{02}^2 = \mathbf{p}_{01}^2 = \mathbf{p}_{02}^2 = 1, \qquad \mathbf{e}_{01} \cdot \mathbf{p}_{01} = \mathbf{e}_{02} \cdot \mathbf{p}_{02} = 0$$
$$\mathbf{e}_{01} \cdot \mathbf{e}_{02} = \cos\phi, \qquad \mathbf{e}_{01} \cdot \mathbf{p}_{02} = \mathbf{e}_{02} \cdot \mathbf{p}_{01} = \sin\phi,$$
$$(-\mathbf{p}_{01} + \mathbf{p}_{02})^2 = [2\sin(\phi/2)]^2 = 2(1 - \cos\phi)$$

Elements of the **G** matrix can also be readily obtained from the table develop-

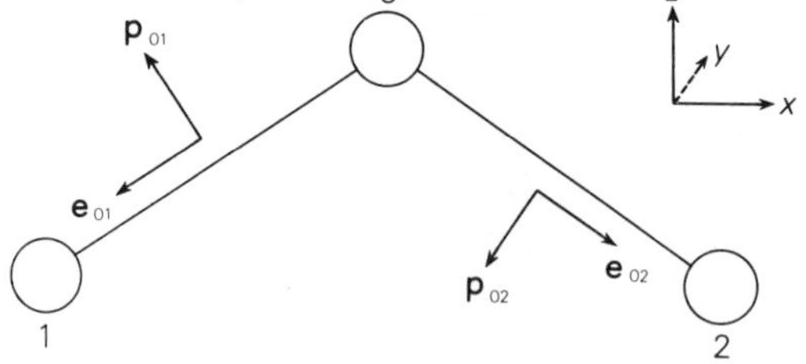

Fig. 5.52 Unit vectors **e** and **p** in the water molecule.

* The components of Eq. 5.153 may be obtained in the following way. $\Delta r_1 = -\mathbf{e}_{01} \cdot \rho_0 + \mathbf{e}_{01} \cdot \rho_1$ results from the assumption $1 \to 0$, $2 \to 1$ in the equation for bond stretching in Table 5.13, Δr_2 is obtained by assuming $1 \to 0$ and $2 \to 2$, $\Delta\phi$ is obtained directly from the equation for valence angle bending.

ed by Decius.[163] The formulas for the present case are given below.

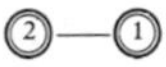
$G_{rr}{}^2 = \mu_1 + \mu_2$

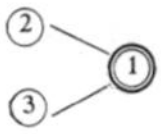
$G_{rr}{}^1 = \mu_1 \cos\phi$

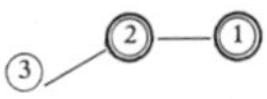
$G_{r\phi}{}^2 = -\rho_{23}\mu_2 \sin\phi$

Atoms common to both coordinates are indicated by double circles. The double subscript indicates the two coordinates involved, and the superscript indicates the number of atoms common to the two coordinates. μ and ρ denote the reciprocals of mass and bond length, respectively. See Refs. 4, 163, and 164.

When the molecule has any symmetry, **G** and **F** matrices can both be reduced to diagonal submatrices of lower orders by a suitable transformation from the internal displacement coordinates to symmetry coordinates. In the case of the water molecule having the C_{2v} symmetry, the following symmetry coordinates are used:*

$$S_1 = \frac{\Delta r_1 + \Delta r_2}{\sqrt{2}} \qquad \text{symmetric stretching } (\nu_s) \tag{5.157}$$

$$S_2 = \Delta\phi \qquad \text{bending } (\delta)$$

$$S_3 = \frac{\Delta r_1 - \Delta r_2}{\sqrt{2}} \qquad \text{antisymmetric stretching } (\nu_a)$$

where S_1 and S_2 belong to the A_1 species, and S_3 belongs to the B_1 species. Using the matrix notation, Eq. 5.157 is written as

$$\mathbf{R}_s = \mathbf{U}\mathbf{R} \tag{5.158}$$

with

$$\mathbf{U} = \begin{bmatrix} \frac{1}{\sqrt{2}} & \frac{1}{\sqrt{2}} & 0 \\ 0 & 0 & 1 \\ \frac{1}{\sqrt{2}} & \frac{-1}{\sqrt{2}} & 0 \end{bmatrix} \tag{5.159}$$

Since **U** is an orthogonal matrix,

$$\mathbf{R} = \tilde{\mathbf{U}}\mathbf{R}_s, \qquad \mathbf{P} = \tilde{\mathbf{U}}\mathbf{P}_s \tag{5.160}$$

* For a description of the construction of the symmetry coordinates, see Ref. 4.

With the **U** matrix, V and T are expressed as

$$2V = \tilde{\mathbf{R}}_s\mathbf{F}_s\mathbf{R}_s \tag{5.161}$$

$$2T = \tilde{\mathbf{P}}_s\mathbf{G}_s\mathbf{P}_s \tag{5.162}$$

where $\mathbf{F}_s$ and $\mathbf{G}_s$ are obtained by the following transformations.

$$\mathbf{F}_s = \mathbf{U}\mathbf{F}\tilde{\mathbf{U}} \tag{5.163}$$

$$\mathbf{G}_s = \mathbf{U}\mathbf{G}\tilde{\mathbf{U}} \tag{5.164}$$

Thus

$$\mathbf{F}_s = \begin{bmatrix} K_r + F_{rr} & \sqrt{2}F_{r\phi} & 0 \\ \sqrt{2}F_{r\phi} & H_\phi & 0 \\ 0 & 0 & K_r - F_{rr} \end{bmatrix} \tag{5.165}$$

$$\mathbf{G}_s = \begin{bmatrix} \mu_O(1+\cos\phi)+\mu_H & -\sqrt{2}\rho\mu_O\sin\phi & 0 \\ -\sqrt{2}\rho\mu_O\sin\phi & 2\rho^2\mu_H+\mu_O(1-\cos\phi) & 0 \\ 0 & 0 & \mu_O(1-\cos\phi)+\mu_H \end{bmatrix} \tag{5.166}$$

From Eqs. 5.151 and 5.164,

$$\mathbf{G}_s = \mathbf{U}\mathbf{B}\mathbf{M}^{-1}\tilde{\mathbf{B}}\tilde{\mathbf{U}} = \mathbf{B}_s\mathbf{M}^{-1}\tilde{\mathbf{B}}_s \tag{5.167}$$

$$\mathbf{B}_s = \mathbf{U}\mathbf{B} \tag{5.168}$$

The $\mathbf{G}_s$ matrix is obtained using $\mathbf{B}_s$. For the water molecule, $m_H = 1.008$, $m_O = 16$, $r = 0.957$ Å, $\phi = 104.5°$, $K_r = 8.453$ mdyn/Å, $H_\phi = 0.694$ mdyn Å, $F_{rr} = -0.105$ mdyn/Å, and $F_{r\phi} = 0.161$ mdyn give

$$\mathbf{G}_s = \begin{bmatrix} 1.0389 & -0.0894 & 0 \\ -0.0894 & 2.3369 & 0 \\ 0 & 0 & 1.0702 \end{bmatrix} \tag{5.169}$$

$$\mathbf{F}_s = \begin{bmatrix} 8.348 & 0.2277 & 0 \\ 0.2277 & 0.694 & 0 \\ 0 & 0 & 8.558 \end{bmatrix} \tag{5.170}$$

From these equations the characteristic values and characteristic vectors are calculated using Eq. 5.123. In the case of the water molecule, the secular equation can be solved easily by hand; the method is explained in Appendix E.

When the order of the determinant is large, electronic computers are used.[4, 165, 166] To diagonalize a matrix by a computer, Jacobi's method is most suitable, but this method is applicable only to a symmetric matrix. Although both the **G** and **F** matrices are symmetric, the **GF** matrix is not necessarily symmetric.* Therefore, in this case, **G** is diagonalized first to obtain $\mathbf{\Gamma}$ and gives an orthogonal matrix **A**.

$$\tilde{\mathbf{A}}\mathbf{G}\mathbf{A} = \mathbf{\Gamma} = \mathbf{\Gamma}^{1/2}\mathbf{\Gamma}^{1/2}, \qquad \mathbf{\Gamma}^{-1/2}\tilde{\mathbf{A}}\mathbf{G}\mathbf{A}\mathbf{\Gamma}^{-1/2} = \mathbf{E} \tag{5.171}$$

where $\mathbf{\Gamma}^{1/2}$ is a diagonal matrix with elements that are the square roots of the corresponding elements of $\mathbf{\Gamma}$. A symmetric matrix $\tilde{\mathbf{C}}\mathbf{F}\mathbf{C}$ is then calculated according to

$$\mathbf{A}\mathbf{\Gamma}^{1/2} = \mathbf{C}, \qquad \mathbf{\Gamma}^{-1/2}\tilde{\mathbf{A}} = \mathbf{C}^{-1}, \qquad \mathbf{C}^{-1}\mathbf{G}\mathbf{F}\mathbf{C} = \mathbf{C}^{-1}\mathbf{G}(\tilde{\mathbf{C}})^{-1}\tilde{\mathbf{C}}\mathbf{F}\mathbf{C} = \tilde{\mathbf{C}}\mathbf{F}\mathbf{C} \tag{5.172}$$

Diagonalizing $\tilde{\mathbf{C}}\mathbf{F}\mathbf{C}$ gives $\mathbf{\Lambda}$ and **L**.

$$\tilde{\mathbf{D}}\tilde{\mathbf{C}}\mathbf{F}\mathbf{C}\mathbf{D} = \mathbf{\Lambda} = \mathbf{L}^{-1}\mathbf{G}\mathbf{F}\mathbf{L}, \qquad \mathbf{C}\mathbf{D} = \mathbf{L} \tag{5.173}$$

In the case of the water molecule, Eqs. 5.169 and 5.170 give

$$\mathbf{\Lambda} = \begin{bmatrix} 8.64708 & 0 & 0 \\ 0 & 1.60676 & 0 \\ 0 & 0 & 9.15877 \end{bmatrix} \tag{5.174}$$

$$\mathbf{L}_s = \begin{bmatrix} 1.01856 & 0.03786 & 0 \\ -0.03097 & -1.52838 & 0 \\ 0 & 0 & 1.03450 \end{bmatrix} \tag{5.175}^{**}$$

Correctness of Eq. 5.175 can be confirmed by the equation

$$\mathbf{L}\tilde{\mathbf{L}} = \mathbf{G} \tag{5.176}^{\dagger}$$

If the units of mass and force constant are atomic weight and mdyn/Å = 10^5 dyn/cm, respectively, $\tilde{\nu}_k(\mathrm{cm}^{-1})$ is given by

$$\tilde{\nu}_k = \frac{1}{2\pi c}\lambda_k^{1/2} = 1302.9\lambda_k^{1/2} \tag{5.177}$$

* Generally speaking, **G** and **F** matrices are Hermitian matrices (described in the case of helical molecules in Section 5.6.5); thus a unitary matrix is to be used for diagonalization. In such a case, Jacobi's method is also applicable.

** From Eqs. 5.114 and 5.158, $\mathbf{R}_s = \mathbf{U}\mathbf{L}\mathbf{Q} = \mathbf{L}_s\mathbf{Q}$.

† From Eqs. 5.117 and 5.123, $\tilde{\mathbf{L}}\mathbf{F}\mathbf{L} = \mathbf{\Lambda} = \mathbf{L}^{-1}\mathbf{G}\mathbf{F}\mathbf{L}$, from which $\tilde{\mathbf{L}} = \mathbf{L}^{-1}\mathbf{G}$ is obtained. Multiplying both sides of this equation by **L** from the left-hand side gives $\mathbf{L}\tilde{\mathbf{L}} = \mathbf{L}\mathbf{L}^{-1}\mathbf{G} = \mathbf{G}$.

From Eq. 5.174

$$\tilde{\nu}_1 = 3831.3\ \text{cm}^{-1}, \qquad \tilde{\nu}_2 = 1651.5\ \text{cm}^{-1}, \qquad \tilde{\nu}_3 = 3943.0\ \text{cm}^{-1} \tag{5.178}$$

These frequencies are in good agreement with the observed frequencies corrected for anharmonicity (3825, 1654, and 3936 cm^{-1}).[167]

The relation between the symmetry coordinates and normal coordinates is given by

$$\begin{aligned} S_1(\nu_s) &= 1.01856Q_1 + 0.03786Q_2 \\ S_2(\delta) &= -0.03097Q_1 - 1.52838Q_2 \\ S_3(\nu_a) &= 1.03450Q_3 \end{aligned} \tag{5.179}$$

From these equations, the form of the normal modes can be obtained. If only the first normal mode is excited and Q_1 takes the value $+0.1$ ($Q_2 = Q_3 = 0$),* then for the symmetry coordinates, $S_1 = (\Delta r_1 + \Delta r_2)/\sqrt{2}$ is about 0.1Å ($\Delta r_1 = \Delta r_2 = +0.07202$Å) and $S_2 = \Delta\phi$ is -0.00310 rad.† This means that the two O—H bond distances increase by the same amount and the bond angle ∠HOH decreases slightly at the same time. The second normal mode consists mainly of the bending motion with a small amount of symmetric stretching. The third normal mode is pure antisymmetric stretching with symmetry species B_1. No mixing of this mode with the other two A_1 modes occurs.

The normal modes can be represented visually by the displacements of atoms using the $\mathbf{L}_X$ matrix.

$$\mathbf{X} = \mathbf{L}_X\mathbf{Q} \tag{5.180}$$

$\mathbf{L}_X$ is derived as follows. From Eq. 5.143,

$$\mathbf{R} = \mathbf{BX} = \mathbf{BL}_X\mathbf{Q} \tag{5.181}$$

From Eq. 5.114,

$$\mathbf{BL}_X = \mathbf{L} \tag{5.182}$$

Equation 5.151 is written as

$$\mathbf{G}^{-1} = (\tilde{\mathbf{B}})^{-1}\mathbf{MB}^{-1} \tag{5.183}$$

from which

$$\tilde{\mathbf{B}}\mathbf{G}^{-1}\mathbf{B} = \tilde{\mathbf{B}}(\tilde{\mathbf{B}})^{-1}\mathbf{MB}^{-1}\mathbf{B} = \mathbf{M} \tag{5.184}$$

* The dimension of Q_k is $[\text{M}^{1/2}\text{L}]$ (see Eq. 5.111) and the unit is (atomic mass unit)$^{1/2}$ Å.

† The dimensions of the $\mathbf{L}$ matrix elements in Eq. 5.179 are $[\text{M}^{-1/2}]$ for l_{11}, l_{12} and $[\text{M}^{-1/2}\text{L}^{-1}]$ for l_{21}, l_{22}.

From Eqs. 5.182 and 5.184,

$$\mathbf{L}_X = \mathbf{M}^{-1}\mathbf{M}\mathbf{L}_X = \mathbf{M}^{-1}\tilde{\mathbf{B}}\mathbf{G}^{-1}\mathbf{B}\mathbf{L}_X = \mathbf{M}^{-1}\tilde{\mathbf{B}}\mathbf{G}^{-1}\mathbf{L} \qquad (5.185)$$

$\mathbf{L}_X$ is also normalized as

$$\mathbf{L}_X\tilde{\mathbf{L}}_X = \mathbf{M}^{-1}\tilde{\mathbf{B}}\mathbf{G}^{-1}\mathbf{L}\tilde{\mathbf{L}}\mathbf{G}^{-1}\mathbf{B}\mathbf{M}^{-1} = \mathbf{M}^{-1}\tilde{\mathbf{B}}\mathbf{G}^{-1}\mathbf{B}\mathbf{M}^{-1} = \mathbf{M}^{-1} \qquad (5.186)$$

The vibrational mode can be represented in terms of a potential energy distribution. From Eq. 5.117, the kth characteristic value λ_k can be written as

$$\lambda_k = \sum_{i,j} l_{ik} l_{jk} f_{ij} \qquad (5.187)$$

If the off-diagonal terms are ignored because of their small values compared to the diagonal terms, an approximate equation is obtained.

$$\lambda_k \cong \sum_{i} l_{ik}^2 f_{ii} \qquad (5.188)$$

It is clear from Eq. 5.110 that λ_k is the contribution of the kth normal mode to the potential energy. Therefore the physical meaning of Eq. 5.188 is that λ_k consists of terms each of which is associated with its respective internal coordinate. The quantity

$$P_{ik} = \frac{l_{ik}^2 f_{ii}}{\lambda_k} \qquad (5.189)$$

represents the contribution of the ith internal coordinate to the potential energy in the kth normal vibration and is called the potential energy distribution.

5.6.3 Factor Group Vibrations of Polymer Chains

For the case of an infinitely long polymer molecule, the displacement coordinates are represented by an infinite column vector. Applying Born's cyclic boundary condition gives the internal and the Cartesian displacement coordinates of the polymer molecule as

$$\mathbf{R} = \begin{bmatrix} \mathbf{R}(1) \\ \vdots \\ \mathbf{R}(j) \\ \vdots \\ \mathbf{R}(N) \end{bmatrix} \quad \text{and} \quad \mathbf{X} = \begin{bmatrix} \mathbf{X}(1) \\ \vdots \\ \mathbf{X}(j) \\ \vdots \\ \mathbf{X}(N) \end{bmatrix} \qquad (5.190)$$

where $\mathbf{R}(j)$ and $\mathbf{X}(j)$ are, respectively, the column vectors composed of the internal and the Cartesian displacement coordinates associated with the jth unit cell of the one-dimensional lattice. Because of the translational symmetry of the molecule, the $\mathbf{B}$ matrix in Eq. 5.143 has the structure

$$
\begin{array}{ccccccccccc}
 & \tilde{\mathbf{X}}(1) & \tilde{\mathbf{X}}(2) & \cdots & \tilde{\mathbf{X}}(j-1) & \tilde{\mathbf{X}}(j) & \tilde{\mathbf{X}}(j+1) & & \cdots & \tilde{\mathbf{X}}(N-1) & \tilde{\mathbf{X}}(N) \\
\mathbf{B} = \ \mathbf{R}(1) & \mathbf{B}_0 & \mathbf{B}_1 & \mathbf{B}_2 & & & & & & \mathbf{B}_{-2} & \mathbf{B}_{-1} \\
\mathbf{R}(2) & \mathbf{B}_{-1} & \mathbf{B}_0 & \mathbf{B}_1 & & & & & & & \mathbf{B}_{-2} \\
\vdots & & \cdots & & & & & & & & \\
\mathbf{R}(j-1) & & & & \mathbf{B}_0 & \mathbf{B}_1 & \mathbf{B}_2 & & & & \\
\mathbf{R}(j) & & & & \mathbf{B}_{-1} & \mathbf{B}_0 & \mathbf{B}_1 & \mathbf{B}_2 & & & \\
\mathbf{R}(j+1) & & & & \mathbf{B}_{-2} & \mathbf{B}_{-1} & \mathbf{B}_0 & \mathbf{B}_1 & & & \\
\vdots & & \cdots & & & & & & & & \\
\mathbf{R}(N-1) & \mathbf{B}_2 & & & & & & & & \mathbf{B}_0 & \mathbf{B}_1 \\
\mathbf{R}(N) & \mathbf{B}_1 & \mathbf{B}_2 & & & & & & & \mathbf{B}_{-1} & \mathbf{B}_0
\end{array}
\tag{5.191*}
$$

From Eq. 5.151 the $\mathbf{G}$ matrix can be expressed as

$$
\mathbf{G} = \mathbf{B}\mathbf{M}^{-1}\tilde{\mathbf{B}} = \begin{bmatrix}
\mathbf{A}_0 & \tilde{\mathbf{A}}_1 & \tilde{\mathbf{A}}_2 & & & & & & \mathbf{A}_2 & \mathbf{A}_1 \\
\mathbf{A}_1 & \mathbf{A}_0 & \tilde{\mathbf{A}}_1 & \tilde{\mathbf{A}}_2 & & & & & & \mathbf{A}_2 \\
\mathbf{A}_2 & \mathbf{A}_1 & \mathbf{A}_0 & \tilde{\mathbf{A}}_1 & \tilde{\mathbf{A}}_2 & & & & & \\
 & & & & \cdots & & & & & \\
 & & & & \mathbf{A}_2 & \mathbf{A}_1 & \mathbf{A}_0 & \tilde{\mathbf{A}}_1 & \tilde{\mathbf{A}}_2 & \\
\tilde{\mathbf{A}}_2 & & & & & \mathbf{A}_2 & \mathbf{A}_1 & \mathbf{A}_0 & \tilde{\mathbf{A}}_1 & \\
\tilde{\mathbf{A}}_1 & \tilde{\mathbf{A}}_2 & & & & & \mathbf{A}_2 & \mathbf{A}_1 & \mathbf{A}_0 &
\end{bmatrix}
\tag{5.192}
$$

* For polymer chains including two main-chain atoms in a unit cell, the $\mathbf{R}$ and $\mathbf{X}$ coordinates can be chosen so that $\mathbf{B}_0$, $\mathbf{B}_1$, and $\mathbf{B}_2$ are nonzero and the other minor matrices are zero. In this case, the stretching and bending coordinates should be taken as shown in Fig. 5.53, and τ_j and τ'_j should be taken at the bonds indicated by r'_j and r_{j+1}, respectively. When the coordinates of internal rotation are not taken into account, $\mathbf{B}_2$ is zero. If three main-chain atoms are included in a unit cell, $\mathbf{B}_0$ and $\mathbf{B}_1$ are nonzero (if the internal rotation is not considered, only $\mathbf{B}_0$ is nonzero). If more than three main-chain atoms are included in a unit cell, only $\mathbf{B}_0$ is nonzero.

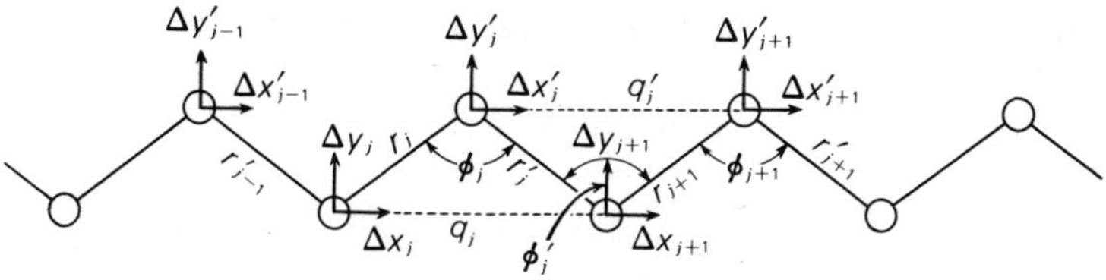

Fig. 5.53 Coordinates for the planar zigzag PE molecule (skeletal model).

with

$$\mathbf{A}_0 = \sum_i \mathbf{B}_i \mathbf{M}^{-1} \widetilde{\mathbf{B}}_i, \qquad \mathbf{A}_1 = \sum_i \mathbf{B}_i \mathbf{M}^{-1} \widetilde{\mathbf{B}}_{i+1}, \qquad \mathbf{A}_2 = \sum_i \mathbf{B}_i \mathbf{M}^{-1} \widetilde{\mathbf{B}}_{i+2} \quad (5.193)$$

$\mathbf{M}^{-1}$ represents the inverse mass matrix associated with one unit cell. The **F** matrix has the same form, where the term $\mathbf{A}_0$ corresponds to the forces acting within one unit cell, $\mathbf{A}_1$ corresponds to the interactions with the nearest neighboring unit cell, $\mathbf{A}_2$ corresponds to those with the next neighboring unit cell, and so on.

As described in Section 5.5.4, the coordinate vectors for the factor group modes are given as

$$\mathbf{R}(k=0) = N^{-1/2} \sum_{j=1}^{N} \mathbf{R}(j) = \mathbf{U}_1 \mathbf{R} \quad (5.194)$$

$$\mathbf{X}(k=0) = N^{-1/2} \sum_{j=1}^{N} \mathbf{X}(j) = \mathbf{U}_2 \mathbf{X} \quad (5.195)$$

where $(k=0)$ indicates that all the unit cells vibrate in phase; k corresponds to the wave vector as is described later (Section 5.6.6.A). The transformation matrices $\mathbf{U}_1$ and $\mathbf{U}_2$ have the same form

$$\mathbf{U}_1 \quad \text{or} \quad \mathbf{U}_2 = N^{-1/2}[\mathbf{E}\,\mathbf{E}\,\mathbf{E} \cdots \mathbf{E}] \quad (5.196)$$

The orders of the unit matrices involved in $\mathbf{U}_1$ and $\mathbf{U}_2$ differ generally, since the numbers of the internal and Cartesian displacement coordinates are not necessarily the same. In terms of these coordinate systems, the **B** matrix for the factor group mode can be defined as

$$\mathbf{R}(k=0) = \mathbf{B}(k=0)\mathbf{X}(k=0) \quad (5.197)$$

and

$$\mathbf{B}(k=0) = \mathbf{U}_1 \mathbf{B} \widetilde{\mathbf{U}}_2 = \mathbf{B}_0 + \sum_j (\mathbf{B}_j + \mathbf{B}_{-j}) \quad (5.198)$$

The corresponding **G** (or **F**) matrix is given by

$$\mathbf{G}(k=0) = \mathbf{U}_1 \mathbf{G} \widetilde{\mathbf{U}}_1 = \mathbf{A}_0 + \mathbf{A}_1 + \widetilde{\mathbf{A}}_1 + \mathbf{A}_2 + \widetilde{\mathbf{A}}_2 + \cdots = \mathbf{B}(k=0)\mathbf{M}^{-1}\widetilde{\mathbf{B}}(k=0) \quad (5.199)$$

5.6.4 In-Plane Skeletal Vibrations of Planar Zigzag Polyethylene Chain

The coordinate systems shown in Fig. 5.53 are used here.

$$\mathbf{R}(j) = \begin{bmatrix} \Delta r_j \\ \Delta\phi_j \\ \Delta r'_j \\ \Delta\phi'_j \end{bmatrix} \quad \text{and} \quad \mathbf{X}(j) = \begin{bmatrix} \Delta x_j \\ \Delta y_j \\ \Delta x'_j \\ \Delta y'_j \end{bmatrix} \tag{5.200}$$

The **B** matrix from Table 5.13 is constructed as

$$\mathbf{B}_0 = \begin{matrix} \\ \Delta r_1 \\ \Delta\phi_1 \\ \Delta r'_1 \\ \Delta\phi'_1 \end{matrix} \begin{matrix} \begin{matrix} \Delta x_1 & \Delta y_1 & \Delta x'_1 & \Delta y'_1 \end{matrix} \\ \begin{bmatrix} -s & -c & s & c \\ -c\rho & s\rho & 0 & -2s\rho \\ 0 & 0 & -s & c \\ 0 & 0 & -c\rho & -s\rho \end{bmatrix} \end{matrix} \qquad \mathbf{B}_1 = \begin{matrix} \begin{matrix} \Delta x_2 & \Delta y_2 & \Delta x'_2 & \Delta y'_2 \end{matrix} \\ \begin{bmatrix} 0 & 0 & 0 & 0 \\ c\rho & s\rho & 0 & 0 \\ s & -c & 0 & 0 \\ 0 & 2s\rho & c\rho & -s\rho \end{bmatrix} \end{matrix}$$

$$\mathbf{B}_2 = \mathbf{B}_3 = \cdots = \mathbf{0} \tag{5.201}$$

where $s = \sin(\phi/2)$, $c = \cos(\phi/2)$, and ρ denotes the reciprocal of the C—C bond length. If $\phi = 109°28'$ is assumed, then

$$\mathbf{B}(k=0) = \mathbf{B}_0 + \mathbf{B}_1$$

$$= \begin{bmatrix} -s & -c & s & c \\ 0 & 2s\rho & 0 & -2s\rho \\ s & -c & -s & c \\ 0 & 2s\rho & 0 & -2s\rho \end{bmatrix} = \begin{bmatrix} -(\frac{2}{3})^{1/2} & -(\frac{1}{3})^{1/2} & (\frac{2}{3})^{1/2} & (\frac{1}{3})^{1/2} \\ 0 & 2\rho(\frac{2}{3})^{1/2} & 0 & -2\rho(\frac{2}{3})^{1/2} \\ (\frac{2}{3})^{1/2} & -(\frac{1}{3})^{1/2} & -(\frac{2}{3})^{1/2} & (\frac{1}{3})^{1/2} \\ 0 & 2\rho(\frac{2}{3})^{1/2} & 0 & -2\rho(\frac{2}{3})^{1/2} \end{bmatrix} \tag{5.202}$$

Equation 5.199 gives

$$\mathbf{G}(k=0) = \mu \begin{bmatrix} 2 & & & \text{symmetric} \\ -(\frac{4}{3})2^{1/2}\rho & (\frac{16}{3})\rho^2 & & \\ -(\frac{2}{3}) & -(\frac{4}{3})2^{1/2}\rho & 2 & \\ -(\frac{4}{3})2^{1/2}\rho & (\frac{16}{3})\rho^2 & -(\frac{4}{3})2^{1/2}\rho & (\frac{16}{3})\rho^2 \end{bmatrix} \tag{5.203}$$

where μ is the reciprocal mass of the so-called "methylene atom."

Here the Urey-Bradley force field is used.[168-171] This force field includes, besides the basic diagonal valence force constants, some central force terms between nonbonded atoms. When the internal coordinate system is used, the latter contribute some diagonal and off-diagonal terms in the **F** matrix. The general formulas describing **F** matrix elements are given in Ref. 171.

For the skeletal model of PE, the potential energy of the Urey-Bradley force field is given by

$$\begin{aligned} 2V(j) = {} & 2K'r_0(\Delta r_j + \Delta r'_j) + K[(\Delta r_j)^2 + (\Delta r'_j)^2] \\ & + 2H'r_0^2(\Delta\phi_j + \Delta\phi'_j) + Hr_0^2[(\Delta\phi_j)^2 + (\Delta\phi'_j)^2] \\ & + 2F'q_0(\Delta q_j + \Delta q'_j) + F[(\Delta q_j)^2 + (\Delta q'_j)^2] \end{aligned} \tag{5.204}$$

where K', K, H', H, F', and F are the force constants. The equilibrium values r_0 and q_0 are introduced so as to express all the force constants in the same units. Δr_j, $\Delta\phi_j$, Δq_j, $\Delta r'_j$, $\Delta\phi'_j$, and $\Delta q'_j$ are not all independent, since Δq_j and $\Delta q'_j$ are expressed in terms of the quadratic terms of the displacement coordinates cited above. Therefore the potential energy equation should include the linear terms of the displacement coordinates as in Eq. 5.204 in order to take the general form of Eq. 5.1. Δq_j is correlated with three other molecular parameters through

$$q_j{}^2 = r_j{}^2 + r'_j{}^2 - 2r_j r'_j \cos\phi_j \tag{5.205}$$

A Taylor power series of Δq_j in terms of the internal displacement coordinates ($R_j = \Delta r_j, \Delta r'_j, \Delta\phi_j$, or $\Delta\phi'_j$) is now considered.

$$\Delta q_j = \frac{1}{1!}\sum_i \left(\frac{\partial q_j}{\partial R_i}\right)_0 R_i + \frac{1}{2!}\sum_{i,k}\left(\frac{\partial^2 q_j}{\partial R_i \partial R_k}\right)_0 R_i R_k + \cdots$$

Neglecting the third order and higher terms leads to

$$\begin{aligned} \Delta q_j = {} & s(\Delta r_j + \Delta r'_j) + t(r_0\Delta\phi_j) + [t^2(\Delta r_j)^2 + t^2(\Delta r'_j)^2 - s^2(r_0\Delta\phi_j)^2 \\ & - 2t^2(\Delta r_j)(\Delta r'_j) + 2ts(\Delta r_j)(r_0\Delta\phi_j) + 2ts(\Delta r'_j)(r_0\Delta\phi_j)]/2q_0 \end{aligned} \tag{5.206}$$

where

$$s = \frac{r_0(1-\cos\phi)}{q_0} \quad \text{and} \quad t = \frac{r_0 \sin\phi}{q_0}$$

In the same way,

$$\begin{aligned} \Delta q'_j = {} & s(\Delta r'_j + \Delta r_{j+1}) + t(r_0\Delta\phi'_j) + [t^2(\Delta r'_j)^2 + t^2(\Delta r_{j+1})^2 - s^2(r_0\Delta\phi'_j)^2 \\ & - 2t^2(\Delta r'_j)(\Delta r_{j+1}) + 2ts(\Delta r'_j)(r_0\Delta\phi'_j) + 2ts(\Delta r_{j+1})(r_0\Delta\phi'_j)]/2q_0 \end{aligned} \tag{5.207}$$

Substitution of these two equations into Eq. 5.204 gives the following expression for the potential energy.

$$\begin{aligned}2V(j) &= 2(K'r_0 + sF'q_0)(\Delta r_j + \Delta r'_j) + (K + 2t^2F' + 2s^2F)[(\Delta r_j)^2 \\ &+ (\Delta r'_j)^2] + 2(H'r_0 + tF'q_0)[(r_0\Delta\phi_j) + (r_0\Delta\phi'_j)] \\ &+ (H - s^2F' + t^2F)[(r_0\Delta\phi_j)^2 + (r_0\Delta\phi'_j)^2] \\ &+ 2(-t^2F' + s^2F)[(\Delta r_j)(\Delta r'_j) + (\Delta r'_j)(\Delta r_{j+1})] \\ &+ 2ts(F' + F)[(\Delta r_j)(r_0\Delta\phi_j) + (\Delta r'_j)(r_0\Delta\phi_j) + (\Delta r'_j)(r_0\Delta\phi'_j) \\ &+ (\Delta r_{j+1})(r_0\Delta\phi'_j)] \qquad (5.208)\end{aligned}$$

Because the only independent variables in this equation are Δr_j, $\Delta\phi_j$, $\Delta r'_j$, and $\Delta\phi'_j$, the linear terms of the displacements can be eliminated in the equilibrium case. Therefore the **F** matrix for the factor group modes is

$$\mathbf{F}(k=0) = \begin{bmatrix} a & \text{symmetric} & & \\ d & b & & \\ 2c & d & a & \\ d & 0 & d & b \end{bmatrix} \qquad (5.209)$$

where

$$a = K + 2t^2F' + 2s^2F, \qquad b = (H - s^2F' + t^2F)r_0^2, \qquad c = -t^2F' + s^2F,$$

$$d = ts(F' + F)r_0$$

When the internal displacement coordinates are transformed into the symmetry coordinates by using the orthogonal matrix

$$\mathbf{U} = \begin{bmatrix} 2^{-1/2} & 0 & 2^{-1/2} & 0 \\ 0 & 2^{-1/2} & 0 & 2^{-1/2} \\ 2^{-1/2} & 0 & -2^{-1/2} & 0 \\ 0 & 2^{-1/2} & 0 & -2^{-1/2} \end{bmatrix} \qquad (5.210)$$

$\mathbf{G}(k=0)$ and $\mathbf{F}(k=0)$ are factored into blocks corresponding to the symmetry species. Thus

$$\mathbf{UG}(k=0)\tilde{\mathbf{U}} = \mu \begin{bmatrix} \frac{4}{3} & -2^{1/2}(\frac{8}{3})\rho & 0 & 0 \\ -2^{1/2}(\frac{8}{3})\rho & \frac{32}{3}\rho^2 & 0 & 0 \\ 0 & 0 & \frac{8}{3} & 0 \\ 0 & 0 & 0 & 0 \end{bmatrix} \begin{matrix} S_1(A_g) \\ S_2(A_g) \\ S_3(B_{1g}) \\ S_4(B_{2u}) \end{matrix} \qquad (5.211)$$

$$\mathbf{UF}(k=0)\tilde{\mathbf{U}} = \left[\begin{array}{cc|c|c} a+2c & 2d & 0 & 0 \\ 2d & b & 0 & 0 \\ \hline 0 & 0 & a-2c & 0 \\ \hline 0 & 0 & 0 & b \end{array}\right] \begin{array}{l} S_1(A_g) \\ S_2(A_g) \\ S_3(B_{1g}) \\ S_4(B_{2u}) \end{array} \tag{5.212}$$

An orthogonal matrix

$$w^{-1/2}\begin{bmatrix} 8^{1/2}\rho & 1 \\ 1 & -8^{1/2}\rho \end{bmatrix}, \qquad w = 1 + 8\rho^2 \tag{5.213}$$

transforms the A_g symmetry coordinates into the following new coordinates:

$$S_R(A_g) = w^{-1/2}(8^{1/2}\rho S_1 + S_2)$$
$$S(A_g) = w^{-1/2}(S_1 - 8^{1/2}\rho S_2) \tag{5.214}$$

The corresponding $\mathbf{G}$ matrix is written as

$$\mathbf{G}(A_g) = \begin{bmatrix} 0 & 0 \\ 0 & \left(\frac{4w}{3}\right)\mu \end{bmatrix} \begin{array}{l} S_R(A_g) \\ S(A_g) \end{array} \tag{5.215}$$

In Eqs. 5.211 and 5.215, the $\mathbf{G}$ matrix elements in the rows and columns corresponding to $S_4(B_{2u})$ and $S_R(A_g)$ are all zero. This means that the vibrations corresponding to these two symmetry coordinates have zero frequency. Such coordinates are called the redundant coordinates. In the vibrational modes for $S_4(B_{2u}) = (\Delta\phi_j - \Delta\phi'_j)/\sqrt{2}$, the angle ϕ_j increases and ϕ'_j decreases with the same magnitude at the same time. This type of vibration cannot satisfy the translational symmetry and therefore does not occur. $S_R(A_g) = w^{-1/2}(8^{1/2}\rho S_1 + S_2) = (w/2)^{1/2}[8^{1/2}\rho(\Delta r_j + \Delta r'_j) + \Delta\phi_j + \Delta\phi'_j]$ corresponds to the mode in which all C—C bonds and all CCC angles increase in phase. Such a vibration cannot occur as the factor group mode. After these redundant coordinates are removed, $S(A_g)$ and $S_3(B_{1g})$ remain, and the corresponding vibrations are the normal vibrations.

As is described in Section 5.2.1, there are $3p-4$ factor group modes. In the present example there are 2, since $p=2$. The calculation described above omits the two internal displacement coordinates of the internal rotation for simplicity. Even if these coordinates are taken into account, four coordinates are removed as redundant coordinates, giving two normal modes. In addition to the above-mentioned redundant coordinates characteristic of polymer molecules, there are many more examples. Another example is that bond angles of the tetrahedral carbon atom vibrate in phase.

By the orthogonal transformation the (2, 2) element of $\mathbf{F}(A_g)$ is obtained as

$$[\mathbf{F}(A_g)]_{2,2} = w^{-1}[(a+2c) - 8\sqrt{2}d\rho + 8b\rho^2] \tag{5.216}$$

If it is assumed that $\phi = 109°28'$, the following characteristic values are obtained.

$$\lambda(A_g) = \tfrac{4}{3}(K + 8H - \tfrac{32}{3}F')\mu \quad \text{and} \quad 0$$
$$\lambda(B_{1g}) = \tfrac{8}{3}(K + \tfrac{4}{3}\ F')$$
$$\lambda(B_{2u}) = 0 \tag{5.217}$$

The observed Raman lines of PE at 1131 and 1061 cm^{-1} are assigned to the A_g and B_{1g} skeletal vibrations, respectively. According to Eq. 5.177 these frequencies lead to the equations

$$K + 8H - \tfrac{32}{3}F' = 7.93 \text{ mdyn/Å}$$
$$K + \tfrac{4}{3}F' = 3.49 \text{ mdyn/Å} \tag{5.218}$$

If it is assumed that $F(CH_2 \cdots CH_2) = 0.33$ mdyn/Å and $F' = -0.1F$, Eq. 5.218 yields values of $K = 3.53$ mdyn/Å and $H = 0.51$ mdyn/Å. More detailed results are obtained by a calculation in which the movements of the hydrogen atoms are taken into consideration (see Section 5.7.1).

Since in this case both A_g and B_{1g} species are one dimensional, the normalized **L** matrices are given simply by the square roots of the corresponding one-dimensional **G**.

$$\mathbf{L}(A_g) = \left(\frac{4w\mu}{3}\right)^{1/2} \quad \text{and} \quad \mathbf{L}(B_{1g}) = \left(\frac{8\mu}{3}\right)^{1/2} \tag{5.219}$$

The corresponding **B** matrices are given according to Eqs. 5.168, 5.210, and 5.213.

$$\mathbf{B}(A_g) = \left(\frac{2w}{3}\right)^{1/2} [\ 0 \quad -1 \quad 0 \quad 1]$$
$$\mathbf{B}(B_{1g}) = \left(\frac{4}{3}\right)^{1/2} [-1 \quad 0 \quad 1 \quad 0] \tag{5.220}$$

The $\mathbf{L}_X$ are derived from Eq. 5.185:

$$\mathbf{L}_X = \left(\frac{\mu}{2}\right)^{1/2} \begin{array}{c} \\ \Delta x \\ \Delta y \\ \Delta x' \\ \Delta y' \end{array} \begin{array}{c} \begin{array}{cccc} Q(A_g) & Q(B_{1g}) & T_x & T_y \end{array} \\ \begin{bmatrix} 0 & -1 & 1 & 0 \\ -1 & 0 & 0 & 1 \\ 0 & 1 & 1 & 0 \\ 1 & 0 & 0 & 1 \end{bmatrix} \end{array} \tag{5.221}$$

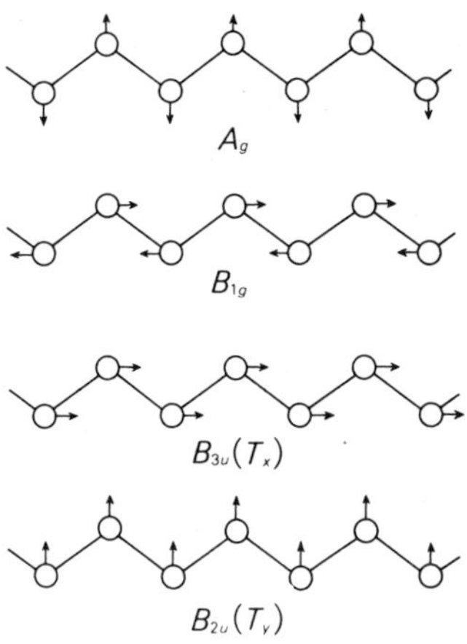

Fig. 5.54 Optically active fundamentals and pure translations of the planar zigzag PE molecule (skeletal mode).

The vibrational modes can be shown schematically by describing the atomic displacement vectors given as the elements of the $\mathbf{L}_X$ matrix. For the present case they are shown in Fig. 5.54.

5.6.5 Vibrations of Helical Polymer Chains

Another typical case is that of vibrations of the helical chain molecules containing u chemical structural units turning t times in the identity period.[172–174] For this case $\mathbf{R}_j (j = 1, 2, \cdots, u)$ are the internal displacement coordinates of the jth structural unit. It should be emphasized that j does not mean the jth identity period (or unit cell) as in Section 5.6.3, but the jth structural unit, one or more being included in an identity period. The $\mathbf{G}$(or $\mathbf{F}$) matrix for the factor group modes of the molecule has the form

$$\begin{matrix}\mathbf{G}(k=0) = \\ \text{or} \\ \mathbf{F}(k=0)\end{matrix} \begin{bmatrix} \alpha & \tilde{\beta} & \tilde{\gamma} & & & & \gamma & \beta \\ \beta & \alpha & \tilde{\beta} & \tilde{\gamma} & & & & \gamma \\ \gamma & \beta & \alpha & \tilde{\beta} & \tilde{\gamma} & & & \\ & & & & & & & \\ \cdots & & & & & & & \\ & & & \gamma & \beta & \alpha & \tilde{\beta} & \tilde{\gamma} \\ \tilde{\gamma} & & & & \gamma & \beta & \alpha & \beta \\ \tilde{\beta} & \tilde{\gamma} & & & & \gamma & \beta & \alpha \end{bmatrix} \qquad (5.222)$$

since the molecule has such a symmetry that any operation of the screw rotation $(C_u)^k$ moves $\mathbf{R}_j$ to $\mathbf{R}_{j+k}$ for any value of j. By transformation by a unitary matrix[‡]

$$\mathbf{U} = \left(\frac{1}{u}\right)^{1/2} \begin{bmatrix} 1 & 1 & 1 & \cdots & 1 \\ 1 & \varepsilon^{+1} & \varepsilon^{+2} & \cdots & \varepsilon^{+(u-1)} \\ 1 & \varepsilon^{-1} & \varepsilon^{-2} & \cdots & \varepsilon^{-(u-1)} \\ \cdots & & & & \\ 1 & \varepsilon^{+k} & \varepsilon^{+2k} & \cdots & \varepsilon^{+k(u-1)} \\ 1 & \varepsilon^{-k} & \varepsilon^{-2k} & \cdots & \varepsilon^{-k(u-1)} \\ \cdots & & & & \end{bmatrix} \times \mathbf{E} \qquad (5.223)$$

with

$$\varepsilon^{\pm k} = \exp\left[\mathrm{i}\left(\pm\frac{2\pi tk}{u}\right)\right]$$

$$k = 1, 2, \cdots, \left(\frac{u-1}{2}\right) \qquad u = \text{odd}$$

$$k = 1, 2, \cdots, \frac{u}{2} \qquad u = \text{even} \qquad (5.224)$$

where **E** is a unit matrix the order of which is the same as that of the submatrices $\boldsymbol{\alpha}$, $\boldsymbol{\beta}$, etc. in Eq. 5.222, and × denotes the direct product.

The **G** or **F** matrix is factored as

$$\mathbf{UGU}^\dagger \text{ or } \mathbf{UFU}^\dagger = \begin{bmatrix} \mathbf{G}(\delta=0) & & & & & \mathbf{0} \\ & \mathbf{G}\left(\delta=\frac{2\pi}{u}\right) & & & & \\ & & \mathbf{G}\left(\delta=-\frac{2\pi}{u}\right) & & & \\ & & & \cdots & & \\ & & & & \mathbf{G}\left(\delta=\frac{2k\pi}{u}\right) & \\ & & & & & \mathbf{G}\left(\delta=-\frac{2k\pi}{u}\right) \\ \mathbf{0} & & & & & & \cdots \end{bmatrix} \qquad (5.225)^{\S}$$

[‡] A matrix consisting of the complex conjugate of the corresponding elements of **A** is denoted by $\mathbf{A}^*$, and its transpose is denoted by $\mathbf{A}^\dagger = \tilde{\mathbf{A}}^*$. The unitary matrix is defined by the relation $\mathbf{A}^\dagger = \mathbf{A}^{-1}$. Real unitary is equivalent to orthogonal ($\tilde{\mathbf{A}} = \mathbf{A}^{-1}$).

[§] δ indicates the phase difference between two adjacent structural units.

Here

$$\mathbf{G}(\delta=0)=\boldsymbol{\alpha}+\boldsymbol{\beta}+\tilde{\boldsymbol{\beta}}+\boldsymbol{\gamma}+\tilde{\boldsymbol{\gamma}}+\ldots$$
$$\mathbf{G}\left(\delta=\pm\frac{2k\pi}{u}\right)=\boldsymbol{\alpha}+\boldsymbol{\beta}\varepsilon^{\pm k}+\tilde{\boldsymbol{\beta}}\varepsilon^{\mp k}+\boldsymbol{\gamma}\varepsilon^{\pm 2k}+\tilde{\boldsymbol{\gamma}}\varepsilon^{\mp 2k}+\cdots \qquad (5.226)$$

The symmetry coordinates belonging to the nondegenerate A symmetry species are the in-phase motions of all the structural units and are given by

$$\mathbf{S}(A)=\left(\frac{1}{u}\right)^{1/2}\sum_{j=1}^{u}\mathbf{R}_j \qquad (5.227)$$

When u is even, there is another nondegenerate B species, the symmetry coordinates and the **G** matrix being given by

$$\mathbf{S}(B)=\left(\frac{1}{u}\right)^{1/2}\sum_{j=1}^{u}(-1)^{j-1}\mathbf{R}_j \qquad (5.228)$$

$$\mathbf{G}(\delta=\pi)=\boldsymbol{\alpha}-(\boldsymbol{\beta}+\tilde{\boldsymbol{\beta}})+(\boldsymbol{\gamma}+\tilde{\boldsymbol{\gamma}})-\cdots \qquad (5.229)$$

The symmetry coordinates

$$\mathbf{S}(E_{\pm k})=\left(\frac{1}{u}\right)^{1/2}\sum_{j=1}^{u}\mathbf{R}_j\varepsilon^{\pm k(j-1)} \qquad (5.230)$$

which belong to the $\mathbf{E}_{\pm k}$ species, represent the vibrational modes with a phase difference of $\delta=\pm(2kt\pi/u)$ between two adjacent structural units. The pair of signs $\pm$ corresponds to the pair of doubly degenerate coordinates.

According to the factor group analysis (Section 5.5.8), the A and E_1 modes are infrared active, and the A, E_1, and E_2 modes are Raman active.

As is seen in Eq. 5.226, the **G** or **F** matrix for the E_k species is Hermitian and so the complex **GF** matrix has real characteristic values λ_s.* The (r, s) element of the **L** matrix is given as a complex number

$$l_{rs}=|l_{rs}|\exp i\delta_{rs} \qquad (5.231)$$

Here

$$|l_{rs}|=\{[\mathrm{Re}(l_{rs})]^2+[\mathrm{Im}(l_{rs})]^2\}^{1/2}$$

and

$$\tan\delta_{rs}=\frac{\mathrm{Im}(l_{rs})}{\mathrm{Re}(l_{rs})}$$

* A matrix defined by $\mathbf{A}^{\dagger}=\mathbf{A}$ is called a Hermitian matrix. Real Hermitian is equivalent to symmetric. The unitary transformation of Hermitian matrix yields a Hermitian matrix.[144] The **GF** matrix is not Hermitian, but the **G** and **F** matrices can be diagonalized separately as shown in Eqs. 5.171–5.173. Therefore the final diagonal matrix is real, that is, the characteristic values are real, since the diagonal elements of a Hermitian matrix are all real.

$\mathrm{Re}(l_{rs})$ and $\mathrm{Im}(l_{rs})$ being the real and imaginary parts of l_{rs}, respectively. l_{rs} represents the amplitude of the rth coordinate in the sth normal vibration and $(\delta_{rs} - \delta_{r's})$ gives the phase difference between the rth and r'th coordinates in the sth normal vibration.

Even in the case of the helical polymer having no symmetry elements other than the u-fold screw axis along the chain, it is convenient to employ local symmetry coordinates associated with each structural unit instead of internal coordinates. The local symmetry coordinates are defined as linear combinations of the internal coordinates and correspond to the vibrational modes of some atomic groups, such as CH_2 symmetric stretching, CH_2 bending, and CH_3 degenerate bending. They are constructed by transforming R_j with a suitable orthogonal matrix. At this stage local redundant coordinates associated with the atomic groups, such as —CH_3, —CH_2—, and >CH, can be removed and therefore the order of the secular determinant to be solved is reduced. Equations 5.222 through 5.231 are also available without any modification if $\mathbf{R}_j$ is replaced by the local symmetry coordinates $\mathbf{s}_j$. An example of local symmetry coordinates is seen in the case of *it*-PP, which has been treated under the factor group isomorphous to C_3.[68]

For the case of a helical molecule with twofold rotation axes perpendicular to the chain axis, that is, having the dihedral symmetry D_u, another set of symmetry coordinates is convenient for factoring the **G** and **F** matrices.[173,174] A typical case is that of trigonal POM (Fig. 5.55).[84] Because of the symmetry with respect to the twofold rotation axes passing through the carbon (referred to as I) and oxygen atoms (II), the local symmetry coordinates associated with each structural unit can be classified into four groups according to whether the coordinates are symmetric or antisymmetric with respect to I

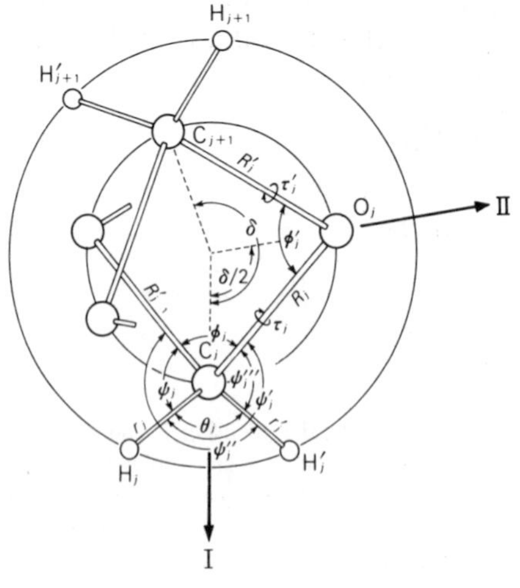

Fig. 5.55 Internal coordinates adopted for the (9/5) helix of trigonal POM.

or II, namely, Is, IIs, Ia, and IIa. Thus the local symmetry coordinates associated with the jth structural unit can be written as

$$\mathbf{s}(j) = \begin{bmatrix} \mathbf{s}(j)_{\mathrm{Is}} \\ \mathbf{s}(j)_{\mathrm{IIs}} \\ \mathbf{s}(j)_{\mathrm{Ia}} \\ \mathbf{s}(j)_{\mathrm{IIa}} \end{bmatrix} \tag{5.232}$$

$$\mathbf{s}(j)_{\mathrm{Is}} = \begin{bmatrix} 2^{-1/2}(\Delta r_j + \Delta r_j') \\ 20^{-1/2}(4\Delta\theta_j - \Delta\psi_j - \Delta\psi_j' - \Delta\psi_j'' - \Delta\psi_j''') \\ 2^{-1}(\Delta\psi_j - \Delta\psi_j' - \Delta\psi_j'' + \Delta\psi_j''') \\ 30^{-1/2}(5\Delta\phi_j - \Delta\theta_j - \Delta\psi_j - \Delta\psi_j' - \Delta\psi_j'' - \Delta\psi_j''') \end{bmatrix} \begin{matrix} \nu_s(\mathrm{CH_2}) \\ \delta(\mathrm{CH_2}) \\ t(\mathrm{CH_2}) \\ \delta(\mathrm{OCO}) \end{matrix}$$

$$\mathbf{s}(j)_{\mathrm{IIs}} = \begin{bmatrix} 2^{-1/2}(\Delta R_j + \Delta R_j') \\ \Delta\phi_j' \\ 2^{-1/2}(\Delta\tau_j + \Delta\tau_j') \end{bmatrix} \begin{matrix} \nu_s(\mathrm{COC}) \\ \delta(\mathrm{COC}) \\ \tau_s \end{matrix}$$

$$\mathbf{s}(j)_{\mathrm{Ia}} = \begin{bmatrix} 2^{-1/2}(\Delta r_j - \Delta r_j') \\ 2^{-1}(\Delta\psi_j - \Delta\psi_j' + \Delta\psi_j'' - \Delta\psi_j''') \\ 2^{-1}(\Delta\psi_j + \Delta\psi_j' - \Delta\psi_j'' - \Delta\psi_j''') \end{bmatrix} \begin{matrix} \nu_a(\mathrm{CH_2}) \\ w(\mathrm{CH_2}) \\ r(\mathrm{CH_2}) \end{matrix}$$

$$\mathbf{s}(j)_{\mathrm{IIa}} = \begin{bmatrix} 2^{-1/2}(\Delta R_j - \Delta R_j') \\ 2^{-1/2}(\Delta\tau_j - \Delta\tau_j') \end{bmatrix} \begin{matrix} \nu_a(\mathrm{COC}) \\ \tau_a \end{matrix} \tag{5.233}$$

After the transformation into the local symmetry coordinates $\mathbf{s}(j)$ from the internal displacement coordinates $\mathbf{R}(j)$ and subsequent unitary transformation, the diagonal blocks of the **G** and **F** matrices (Eq. 5.225) have the general form of Eq. 5.234. Here $\boldsymbol{\alpha}_1, \boldsymbol{\beta}_2, \boldsymbol{\gamma}_3, \cdots$ represent real submatrices. In Eq. 5.234 the eighth and higher terms of Eq. 5.226 are omitted for simplicity.

$$\mathbf{G}\left(\delta = \frac{2k\pi}{u}\right) =$$

$$\begin{matrix} & \tilde{\mathbf{S}}(E_k)_{\mathrm{Is}} & \tilde{\mathbf{S}}(E_k)_{\mathrm{IIs}} \\ \mathbf{S}(E_k)_{\mathrm{Is}} & \boldsymbol{\alpha}_1 & \tilde{\boldsymbol{\beta}}_1(1+\varepsilon^{k}) + \tilde{\boldsymbol{\beta}}_2(\varepsilon^{-k}+\varepsilon^{2k}) \\ \mathbf{S}(E_k)_{\mathrm{IIs}} & \boldsymbol{\beta}_1(1+\varepsilon^{-k}) + \boldsymbol{\beta}_2(\varepsilon^{k}+\varepsilon^{-2k}) & \boldsymbol{\alpha}_2 \\ \mathbf{S}(E_k)_{\mathrm{Ia}} & \boldsymbol{\gamma}_1(\varepsilon^{k}-\varepsilon^{-k}) + \boldsymbol{\gamma}_2(\varepsilon^{2k}-\varepsilon^{-2k}) & \boldsymbol{\beta}_3(1-\varepsilon^{k}) + \boldsymbol{\beta}_4(\varepsilon^{-k}-\varepsilon^{2k}) \\ \mathbf{S}(E_k)_{\mathrm{IIa}} & \boldsymbol{\delta}_1(1-\varepsilon^{-k}) + \boldsymbol{\delta}_2(\varepsilon^{k}-\varepsilon^{-2k}) & \boldsymbol{\gamma}_3(\varepsilon^{k}-\varepsilon^{-k}) + \boldsymbol{\gamma}_4(\varepsilon^{2k}-\varepsilon^{-2k}) \end{matrix}$$

$$\begin{array}{cc} \tilde{\mathbf{S}}(E_k)_{\mathrm{Ia}} & \tilde{\mathbf{S}}(E_k)_{\mathrm{IIa}} \\ \left. \begin{array}{ll} \tilde{\gamma}_1(\varepsilon^{-k}-\varepsilon^{k})+\tilde{\gamma}_2(\varepsilon^{-2k}-\varepsilon^{2k}) & \tilde{\boldsymbol{\delta}}_1(1-\varepsilon^{k})+\tilde{\boldsymbol{\delta}}_2(\varepsilon^{-k}-\varepsilon^{2k}) \\ \tilde{\boldsymbol{\beta}}_3(1-\varepsilon^{-k})+\tilde{\boldsymbol{\beta}}_4(\varepsilon^{k}-\varepsilon^{-2k}) & \tilde{\gamma}_3(\varepsilon^{-k}-\varepsilon^{k})+\tilde{\gamma}_4(\varepsilon^{-2k}-\varepsilon^{2k}) \\ \boldsymbol{\alpha}_3 & \tilde{\boldsymbol{\beta}}_5(1+\varepsilon^{k})+\tilde{\boldsymbol{\beta}}_6(\varepsilon^{-k}+\varepsilon^{2k}) \\ \boldsymbol{\beta}_5(1+\varepsilon^{-k})+\boldsymbol{\beta}_6(\varepsilon^{k}+\varepsilon^{-2k}) & \boldsymbol{\alpha}_4 \end{array} \right] \end{array} \tag{5.234}$$

In the cases of $k = 0\,(\varepsilon^{\pm k} = 1)$ and $k = u/2\,(\varepsilon^{\pm k} = -1)$, which correspond to the A and B species, respectively, Eq. 5.234 can be further factored into the following two real submatrices.

$$\mathbf{G}(A) = \begin{bmatrix} \boldsymbol{\alpha}_1 & 2(\tilde{\boldsymbol{\beta}}_1+\tilde{\boldsymbol{\beta}}_2) & \mathbf{0} & \mathbf{0} \\ 2(\boldsymbol{\beta}_1+\boldsymbol{\beta}_2) & \boldsymbol{\alpha}_2 & \mathbf{0} & \mathbf{0} \\ \mathbf{0} & \mathbf{0} & \boldsymbol{\alpha}_3 & 2(\tilde{\boldsymbol{\beta}}_5+\tilde{\boldsymbol{\beta}}_6) \\ \mathbf{0} & \mathbf{0} & 2(\boldsymbol{\beta}_5+\boldsymbol{\beta}_6) & \boldsymbol{\alpha}_4 \end{bmatrix} = \begin{bmatrix} \mathbf{G}(A_1) & \mathbf{0} \\ \mathbf{0} & \mathbf{G}(A_2) \end{bmatrix} \tag{5.235}$$

$$\mathbf{G}(B) = \begin{bmatrix} \boldsymbol{\alpha}_1 & 2(\tilde{\boldsymbol{\delta}}_1+\tilde{\boldsymbol{\delta}}_2) & \mathbf{0} & \mathbf{0} \\ 2(\boldsymbol{\delta}_1+\boldsymbol{\delta}_2) & \boldsymbol{\alpha}_4 & \mathbf{0} & \mathbf{0} \\ \mathbf{0} & \mathbf{0} & \boldsymbol{\alpha}_2 & 2(\tilde{\boldsymbol{\beta}}_3+\tilde{\boldsymbol{\beta}}_4) \\ \mathbf{0} & \mathbf{0} & 2(\boldsymbol{\beta}_3+\boldsymbol{\beta}_4) & \boldsymbol{\alpha}_3 \end{bmatrix} = \begin{bmatrix} \mathbf{G}(B_1) & \mathbf{0} \\ \mathbf{0} & \mathbf{G}(B_2) \end{bmatrix} \tag{5.236}$$

The symmetry coordinates for the A_1, A_2, B_1, and B_2 symmetry species are written as

$$\begin{aligned} \mathbf{S}(A_1) &= \left(\frac{1}{u}\right)^{1/2} \sum_{j=1}^{u} \begin{bmatrix} \mathbf{s}(j)_{\mathrm{Is}} \\ \mathbf{s}(j)_{\mathrm{IIs}} \end{bmatrix} \\ \mathbf{S}(A_2) &= \left(\frac{1}{u}\right)^{1/2} \sum_{j=1}^{u} \begin{bmatrix} \mathbf{s}(j)_{\mathrm{Ia}} \\ \mathbf{s}(j)_{\mathrm{IIa}} \end{bmatrix} \\ \mathbf{S}(B_1) &= \left(\frac{1}{u}\right)^{1/2} \sum_{j=1}^{u} \begin{bmatrix} (-1)^{j-1}\mathbf{s}(j)_{\mathrm{Is}} \\ (-1)^{j-1}\mathbf{s}(j)_{\mathrm{IIa}} \end{bmatrix} \\ \mathbf{S}(B_2) &= \left(\frac{1}{u}\right)^{1/2} \sum_{j=1}^{u} \begin{bmatrix} (-1)^{j-1}\mathbf{s}(j)_{\mathrm{IIs}} \\ (-1)^{j-1}\mathbf{s}(j)_{\mathrm{Ia}} \end{bmatrix} \end{aligned} \tag{5.237}$$

Because of the twofold rotation symmetry, the phase difference in the E_k modes between $\mathbf{s}(j)_{\mathrm{Is}}$ or $\mathbf{s}(j)_{\mathrm{Ia}}$ and $\mathbf{s}(j)_{\mathrm{IIs}}$ or $\mathbf{s}(j)_{\mathrm{IIa}}$ is $\delta/2 = kt\pi/u$, one half the phase difference δ between adjacent structural units. The final symmetry coordinates are constructed in the following form.

$$\mathbf{S}(E_k)_{\mathrm{Is}} = \left(\frac{2}{u}\right)^{1/2} \sum_{j=1}^{u} \mathbf{s}(j)_{\mathrm{Is}} \begin{matrix}\cos\\ \sin\end{matrix}\, k(j-1)\omega$$

$$\mathbf{S}(E_k)_{\mathrm{IIs}} = \left(\frac{2}{u}\right)^{1/2} \sum_{j=1}^{u} \mathbf{s}(j)_{\mathrm{IIs}} \begin{matrix}\cos\\ \sin\end{matrix}\, k\left(j-\frac{1}{2}\right)\omega$$

$$\mathbf{S}(E_k)_{\mathrm{Ia}} = \left(\frac{2}{u}\right)^{1/2} \sum_{j=1}^{u} \mathbf{s}(j)_{\mathrm{Ia}} \begin{matrix}\sin\\ \cos\end{matrix}\, k(j-1)\omega$$

$$\mathbf{S}(E_k)_{\mathrm{IIa}} = \left(\frac{2}{u}\right)^{1/2} \sum_{j=1}^{u} \mathbf{s}(j)_{\mathrm{IIa}} \begin{matrix}\sin\\ \cos\end{matrix}\, k\left(j-\frac{1}{2}\right)\omega \tag{5.238}$$

where $\omega = 2t\pi/u$. The upper or lower term is to be used for each one of the degenerate pairs. Using these symmetry coordinates, Eq. 4.234 can be transformed to real matrices by the following unitary transformation.

$$\mathbf{u}_k \begin{bmatrix} \mathbf{G}\left(\delta = \dfrac{2k\pi}{u}\right) & \mathbf{0} \\ \mathbf{0} & \mathbf{G}\left(\delta = -\dfrac{2k\pi}{u}\right) \end{bmatrix} \mathbf{u}_k^{\dagger} = \begin{bmatrix} \mathbf{G}(E_k) & \mathbf{0} \\ \mathbf{0} & \mathbf{G}(E_k) \end{bmatrix} \tag{5.239}$$

where

$$\mathbf{u}_k = 2^{-1/2} \begin{bmatrix} 1 & 0 & 0 & 0 & 1 & 0 & 0 & 0 \\ 0 & \varepsilon^{k/2} & 0 & 0 & 0 & \varepsilon^{-k/2} & 0 & 0 \\ 0 & 0 & -\mathrm{i} & 0 & 0 & 0 & \mathrm{i} & 0 \\ 0 & 0 & 0 & -\mathrm{i}\varepsilon^{k/2} & 0 & 0 & 0 & \mathrm{i}\varepsilon^{-k/2} \\ -\mathrm{i} & 0 & 0 & 0 & \mathrm{i} & 0 & 0 & 0 \\ 0 & -\mathrm{i}\varepsilon^{k/2} & 0 & 0 & 0 & \mathrm{i}\varepsilon^{k/2} & 0 & 0 \\ 0 & 0 & 1 & 0 & 0 & 0 & 1 & 0 \\ 0 & 0 & 0 & \varepsilon^{k/2} & 0 & 0 & 0 & \varepsilon^{-k/2} \end{bmatrix} \times \mathbf{E} \tag{5.240}$$

and $\mathbf{G}(E_k)$ is given by Eq. 5.241.

$$\mathbf{G}(E_k)=\begin{bmatrix} \boldsymbol{\alpha}_1 & 2\left(\tilde{\boldsymbol{\beta}}_1\cos\frac{k\omega}{2}+\tilde{\boldsymbol{\beta}}_2\cos\frac{3k\omega}{2}\right) & -2(\tilde{\gamma}_1\sin k\omega+\tilde{\gamma}_2\sin 2k\omega) & 2\left(\tilde{\delta}_1\sin\frac{k\omega}{2}+\tilde{\delta}_2\sin\frac{3k\omega}{2}\right) \\ 2\left(\boldsymbol{\beta}_1\cos\frac{k\omega}{2}+\boldsymbol{\beta}_2\cos\frac{3k\omega}{2}\right) & \boldsymbol{\alpha}_2 & 2\left(\tilde{\boldsymbol{\beta}}_3\sin\frac{k\omega}{2}+\tilde{\boldsymbol{\beta}}_4\sin\frac{3k\omega}{2}\right) & 2(\tilde{\gamma}_3\sin k\omega+\tilde{\gamma}_4\sin 2k\omega) \\ -2(\gamma_1\sin k\omega+\gamma_2\sin 2k\omega) & 2\left(\boldsymbol{\beta}_3\sin\frac{k\omega}{2}+\boldsymbol{\beta}_4\sin\frac{3k\omega}{2}\right) & \boldsymbol{\alpha}_3 & -2\left(\tilde{\boldsymbol{\beta}}_5\cos\frac{k\omega}{2}+\tilde{\boldsymbol{\beta}}_6\cos\frac{3k\omega}{2}\right) \\ 2\left(\delta_1\sin\frac{k\omega}{2}+\delta_2\sin\frac{3k\omega}{2}\right) & 2(\gamma_3\sin k\omega+\gamma_4\sin 2k\omega) & -2\left(\boldsymbol{\beta}_5\cos\frac{k\omega}{2}+\boldsymbol{\beta}_6\cos\frac{3k\omega}{2}\right) & \boldsymbol{\alpha}_4 \end{bmatrix} \tag{5.241}$$

Real **G** and **F** matrices of this type give a real **L** matrix. The phase relation between the local symmetry coordinates in a normal vibration of the E_k species (with $\delta = 2tk\pi/u$) can be derived as follows. A Is local symmetry coordinate is chosen in the jth structural unit $\mathbf{s}_{\mathrm{Is}}(j)$, for example, $\delta(CH_2)$ (Eq. 5.233), as the origin with phase angle zero. The phase angles of the other local symmetry coordinates then can be obtained by the definition of the E_k symmetry coordinates in Eq. 5.238: $\mathbf{s}_{\mathrm{Is}}(j)$, 0 or π; $\mathbf{s}_{\mathrm{IIs}}(j)$, $(\delta/2)$ or $(\delta/2)+\pi$; $\mathbf{s}_{\mathrm{Ia}}(j)$, $-(\pi/2)$ or $(\pi/2)$; $\mathbf{s}_{\mathrm{IIa}}(j)$, $(\delta/2)-(\pi/2)$ or $(\delta/2)+(\pi/2)$. Here the former value of each pair of angles should be taken if the **L** matrix element of the symmetry coordinate in question has the same sign as the **L** element of the Is symmetry coordinate chosen above, that is, $\delta(CH_2)$. The latter value is taken if the above two **L** elements are of opposite signs. This example is given in Ref. 84.

The detailed results of the normal coordinate treatment for POM are given in Table 5.11. The **G** matrix for the skeltal model of the POM molecule is now considered. The methylene group is regarded as a unit atom and is denoted by M. All the bond angles are assumed to be 109°28′. The internal coordinates are taken as shown in Fig. 5.56. Torsional viberations are neglected for simplicity. The minor matrices in Eq. 5.222 are constructed as follows:

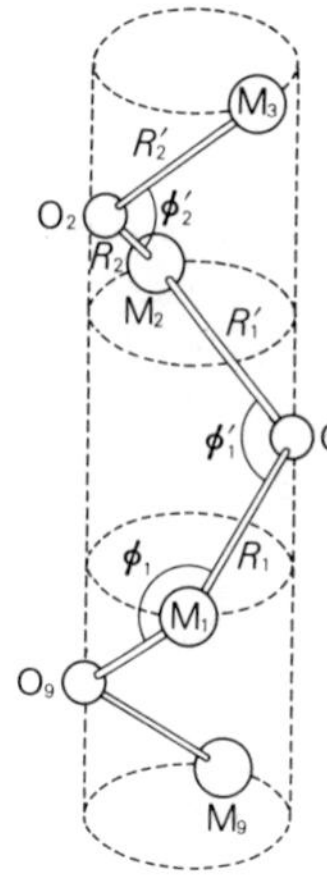

Fig. 5.56 Skeletal model of POM molecule.[173]

$$
\begin{array}{cc|cccc|}
 & & \Delta R_1 & \Delta R'_1 & \Delta\phi_1 & \Delta\phi'_1 \\
\boldsymbol{\alpha} = & \Delta R_1 & a & b & e & f \\
 & \Delta R'_1 & b & a & e_0 & f \\
 & \Delta\phi_1 & e & e_0 & c & g \\
 & \Delta\phi'_1 & f & f & g & d \\
\boldsymbol{\beta} = & \Delta R_2 & 0 & b_0 & 0 & f_0 \\
 & \Delta R'_2 & 0 & 0 & 0 & 0 \\
 & \Delta\phi_2 & e_0 & e & c_0 & g \\
 & \Delta\phi'_2 & 0 & f_0 & 0 & d_0
\end{array}
$$

$$\boldsymbol{\gamma} = \boldsymbol{\delta} = \mathbf{0} \tag{5.242}$$

where for the elements of the **G** matrix $a = \mu_M + \mu_O, b = -\mu_O/3, b_0 = -\mu_M/3$, $c = 2\rho^2\mu_O + \frac{8}{3}\rho^2\mu_M$, $c_0 = [\frac{4}{3}\cos^2\tau - 1]\rho^2\mu_O$, $d = 2\rho^2\mu_M + \frac{8}{3}\rho^2\mu_O$, $d_0 = [\frac{4}{3}\cos^2\tau - 1]\rho^2\mu_M$, $e = -(2\sqrt{2}/3)\rho\mu_M$, $e_0 = (2\sqrt{2}/3)\rho\mu_O\cos\tau$, $f = -(2\sqrt{2}/3)\rho\mu_O$, $f_0 = (2\sqrt{2}/3)\rho\mu_M\cos\tau$, and $g = -\frac{4}{3}\rho^2\cos\tau\ (\mu_M + \mu_O)$, μ_M and μ_O are reciprocal masses of the "methylene atom" and the oxygen atom, respectively, ρ is the reciprocal C—O interatomic distance, and τ is the internal rotation angle. The components of $\mathbf{s}(j)$ are parts of Eq. 5.233 and are given by

$$
\begin{aligned}
\mathbf{s}(j)_{\mathrm{Is}} &= \Delta\phi_j \\
\mathbf{s}(j)_{\mathrm{IIs}} &= \begin{bmatrix} 2^{-1/2}(\Delta R_j + \Delta R'_j) \\ \Delta\phi'_j \end{bmatrix} \\
\mathbf{s}(j)_{\mathrm{IIa}} &= 2^{-1/2}(\Delta R_j - \Delta R'_j)
\end{aligned} \tag{5.243}
$$

G matrices for the A_1, A_2, and E_1 species are given by

$$\mathbf{G}(A_1) = \begin{bmatrix} c + 2c_0 & & \text{symmetric} \\ \sqrt{2}(e + e_0) & a + b + b_0 & \\ 2g & \sqrt{2}(f + f_0) & d + 2d_0 \end{bmatrix} \tag{5.244}$$

$$\mathbf{G}(A_2) = a - b - b_0 \tag{5.245}$$

$$\mathbf{G}(E_1) = \begin{bmatrix} c + 2c_0 \cos\omega & & & \text{symmetric} \\ \sqrt{2}(e + e_0)\cos\frac{\omega}{2} & a + b + b_0 \cos\omega & & \\ 2g\cos\frac{\omega}{2} & \sqrt{2}(f + f_0 \cos\omega) & d + 2d_0 \cos\omega & \\ \sqrt{2}(e - e_0)\sin\frac{\omega}{2} & b_0 \sin\omega & \sqrt{2} f_0 \sin\omega & a - b - b_0 \cos\omega \end{bmatrix} \tag{5.246}$$

The **F** matrix is factored in a corresponding way.

5.6.6 Frequency–Phase Relation and Frequency Distribution

The foregoing sections are concerned with the factor group modes. There are, however, an infinitely large number of modes that have phase differences between two adjacent structural units and do not belong to the factor group modes. These modes must be taken into consideration to interpret some of the physicochemical properties of crystalline polymers, although they are not active in either infrared or Raman spectra. For this purpose it is necessary to investigate how the vibrational frequency depends on the phase difference, in other words, to obtain a frequency dispersion curve.

Studies of this sort were first made on a single chain of PE,[155, 156, 175–179] and then on crystalline PE.[153, 157, 180, 181] Similar treatments have also been carried out on several more complicated polymers such as POM,[182, 183] PEO,[184] *it*-PP,[185] polytetrafluoroethylene,[186, 187] and poly(vinylidene fluoride).[188]

A. *One-Dimensional NaCl-like Lattice.*[189] Considered here is a one-dimensional lattice where two kinds of atoms with masses m_1 and m_2 $(m_1 < m_2)$ are aligned alternately, each interacting only with its nearest

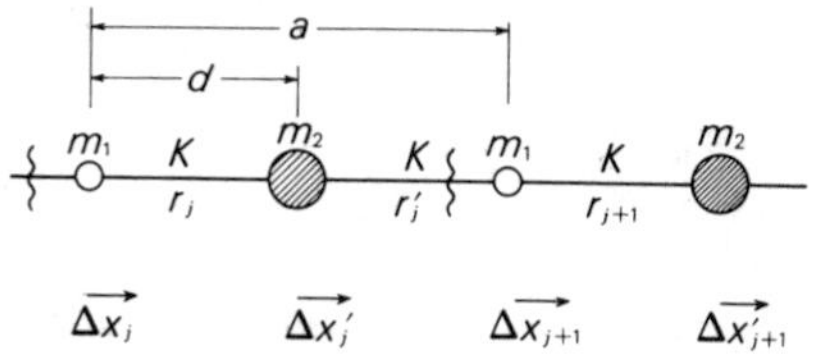

Fig. 5.57 One-dimensional NaCl-like lattice.

neighbors by a harmonic force with a force constant K (Fig. 5.57). The distance between the nearest neighbors is d, which is half of the unit cell length a. If the internal and Cartesian displacement coordinates are taken as

$$\mathbf{R}(j) = \begin{bmatrix} \Delta r_j \\ \Delta r'_j \end{bmatrix} \quad \text{and} \quad \mathbf{X}(j) = \begin{bmatrix} \Delta x_j \\ \Delta x'_j \end{bmatrix} \tag{5.247}$$

only the $\mathbf{B}_0$ and $\mathbf{B}_1$ matrices in Eq. 5.191 are nonzero.

$$\mathbf{B}_0 = \begin{matrix} & \Delta x_j & \Delta x'_j \\ \Delta r_j & -1 & 1 \\ \Delta r'_j & 0 & -1 \end{matrix} \quad \text{and} \quad \mathbf{B}_1 = \begin{matrix} \Delta x_{j+1} & \Delta x'_{j+1} \\ 0 & 0 \\ 1 & 0 \end{matrix} \tag{5.248}$$

According to Eq. 5.191, the elements of Eq. 5.192 are calculated to be

$$\mathbf{A}_0 = \begin{bmatrix} \mu_1 + \mu_2 & -\mu_2 \\ -\mu_2 & \mu_1 + \mu_2 \end{bmatrix} \quad \text{and} \quad \mathbf{A}_1 = \begin{bmatrix} 0 & -\mu_1 \\ 0 & 0 \end{bmatrix} \tag{5.249}$$

and other elements are all zero. μ_1 and μ_2 are the reciprocals of m_1 and m_2, respectively. The vibrational mode with the phase difference between the adjacent unit cells $\delta = 2\pi a/\lambda$ (λ : wavelength) is expressed as*

$$\mathbf{R}(\delta) = \left(\frac{1}{N}\right)^{1/2} \sum_{j=1}^{N} \mathbf{R}(j) \exp\left[\mathrm{i}(j-1)\delta\right] = \mathbf{U}\mathbf{R} \tag{5.250}$$

with

$$\mathbf{U} = \left(\frac{1}{N}\right)^{1/2} [1 \ \ \mathrm{e}^{\mathrm{i}\delta} \ \ \mathrm{e}^{2\mathrm{i}\delta} \ \cdots \ \mathrm{e}^{(N-1)\mathrm{i}\delta}] \tag{5.251}$$

By a unitary transformation, $\mathbf{G}$ in Eq. 5.192 becomes

$$\begin{aligned} \mathbf{G}(\delta) &= \mathbf{U}\mathbf{G}\mathbf{U}^{\dagger} = \mathbf{A}_0 + \mathbf{A}_1 \mathrm{e}^{\mathrm{i}\delta} + \widetilde{\mathbf{A}}_1 \mathrm{e}^{-\mathrm{i}\delta} \\ &= \begin{bmatrix} \mu_1 + \mu_2 & -(\mu_1 \mathrm{e}^{\mathrm{i}\delta} + \mu_2) \\ -(\mu_1 \mathrm{e}^{-\mathrm{i}\delta} + \mu_2) & \mu_1 + \mu_2 \end{bmatrix} \end{aligned} \tag{5.252}$$

*This wavelength λ is the wavelength of the propagating wave of the lattice vibration associated with the phase difference (see Fig. 5.59) and is different from the wavelength of the infrared radiation in Eq. 5.12.

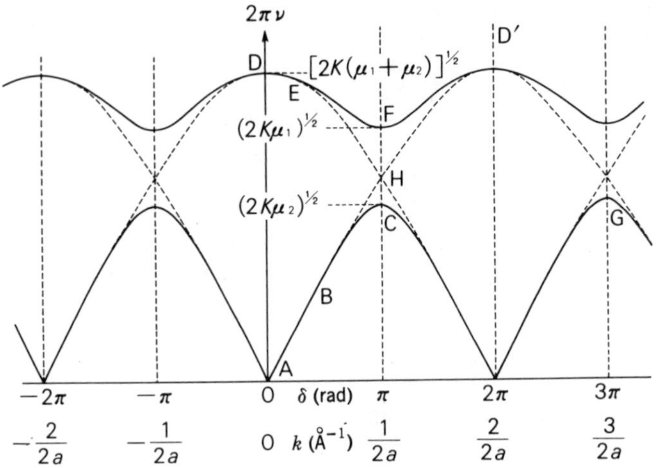

Fig. 5.58 Dispersion curves for a one-dimensional NaCl-like lattice.

The corresponding **F** matrix is given independently of δ as

$$\mathbf{F}(\delta) = \begin{bmatrix} K & 0 \\ 0 & K \end{bmatrix} \tag{5.253}$$

The characteristic value of $\mathbf{G}(\delta)\mathbf{F}(\delta)$ is given by

$$\lambda(\delta) = 4\pi^2\nu^2 = K\{(\mu_1 + \mu_2) \pm [(\mu_1 + \mu_2)^2 - 2\mu_1\mu_2(1 - \cos\delta)]^{1/2}\} \tag{5.254}$$

This equation shows that there are two values of the frequency for each value of δ. ν is plotted against to δ in Fig. 5.58.

The vibrational modes at the points $A, B, \cdots, F$ in Fig. 5.58 are shown in Fig. 5.59, where the displacements in the right (or left) direction are indicated by the up- (or down-) pointing arrows. Mode a is the pure translation of the lattice, and b is similar to the longitudinal wave in an elastic body. Thus the curve passing through the origin is called the acoustical branch. In c only the heavier atoms vibrate with the phase difference π, while the lighter atoms rest. d is the factor group mode, and if the two atoms have opposite charges, infrared absorption occurs. The term "optical branch" for the curve passing through d originates from this fact. In f only the lighter atoms vibrate with the phase difference π.

These dispersion curves are periodic functions of δ, and all the possible chain vibrations may be indicated in terms of δ in the range of $-\pi \leqslant \delta \leqslant \pi$. This range is called the first Brillouin zone. For instance, the mode of $\delta = 3\pi$, G in Fig. 5.58, is indicated by a dotted curve in Fig. 5.59C, which is equivalent to mode C in Fig. 5.58.

To indicate the phase difference, a wave vector or wave number vector

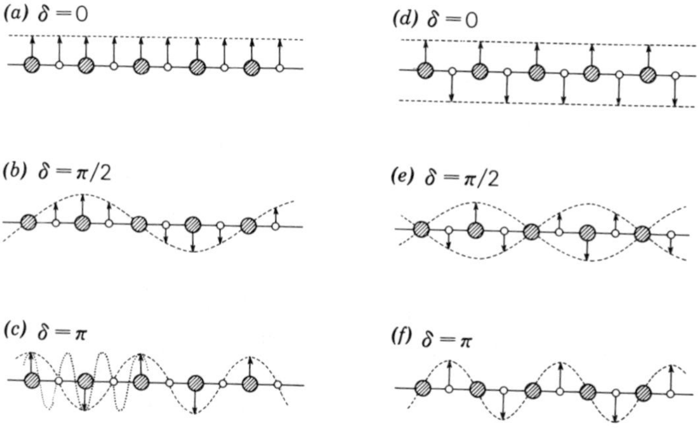

Fig. 5.59 The vibrational modes corresponding to the points on the dispersion curves in Fig. 5.58.

$k = 1/\lambda$ is often used in addition to δ. This vector is defined to be normal to the wave plane in a three-dimensional crystal and corresponds to a point in reciprocal space. If the wave vector is defined as $k = 1/\lambda$, the first Brillouin zone is $-(\frac{1}{2a}) \leqslant k \leqslant (\frac{1}{2a})$.*

In the above treatment, if $m_1 = m_2$, the optical and acoustical branches degenerate at the point $H(\delta = \pi)$ as shown by the dashed line in Fig. 5.58. In this case by taking the unit cell length as $a/2$ (the first Brillouin zone is $-2\pi \leqslant \delta \leqslant 2\pi$), the modes are represented only by an acoustical branch, giving a dispersion curve for a lattice consisting of atoms of one kind ($ABHD'$ in Fig. 5.58).

The frequency distribution $g(\nu)$ is considered next. If the number of k having a frequency between ν and $\nu + \Delta\nu$ is denoted by $g(\nu)\Delta\nu$, $g(\nu)$ is proportional to $\partial k/\partial \nu$. In the case of a one-dimensional lattice, when the number of the unit cell defined by Born's boundary condition is L, the k vector is

$$k = \frac{m}{La} \qquad m = 0, 1, 2, \cdots, L \tag{5.255}$$

and is represented by one of the points dividing the range 0 to $1/a$ into L equal parts. Therefore the density of k in the first Brillouin zone equals La. For the one-dimensional case, $g(\nu)$ is given by[190]

$$g(\nu) = \frac{La}{\partial \nu/\partial k} \tag{5.256}$$

This equation yields the frequency distribution curve shown in Fig. 5.60.

* The definition $k = 2\pi/\lambda$ is also used. In this case the first Brillouin zone is $-(\pi/a) \leqslant k \leqslant (\pi/a)$.

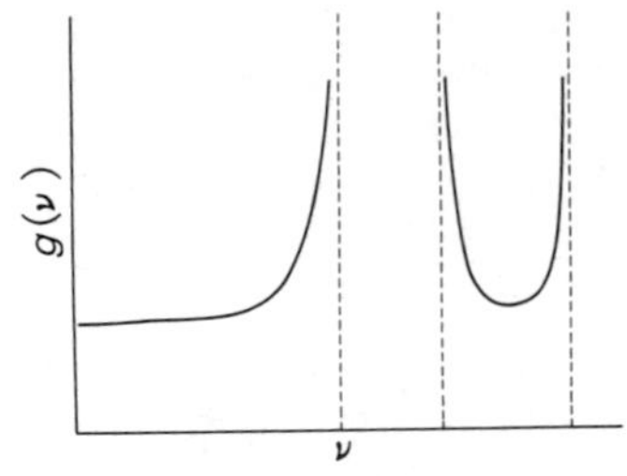

Fig. 5.60 Frequency distribution curve for a one-dimensional NaCl-like lattice.

B. *Polyethylene (Skeletal Vibrations).* The internal coordinates shown in Fig. 5.61, which differ slightly in their numbering from those used in Section 5.6.4, are adopted because it is more convenient to treat this molecule as a twofold helix with twofold rotation symmetry perpendicular to the chain axis. Because of the planar structure the in-plane and out-of-plane modes can be calculated separately.

1. In-Plane Modes. Since Δr_j and $\Delta\phi_j$ belong to the Is and IIs local symmetry coordinates, respectively, the **G** and **F** matrices have the form

$$
\begin{array}{ll}
 & \quad \Delta r_1 \quad \Delta\phi_1 \\
\boldsymbol{\alpha} = \begin{array}{r}\Delta r_1 \\ \Delta\phi_1\end{array} & \begin{bmatrix} a_{11} & a_{21} \\ a_{21} & a_{22} \end{bmatrix} \\
\boldsymbol{\beta} = \begin{array}{r}\Delta r_2 \\ \Delta\phi_2\end{array} & \begin{bmatrix} b_{11} & a_{21} \\ b_{21} & b_{22} \end{bmatrix} \\
\boldsymbol{\gamma} = \begin{array}{r}\Delta r_3 \\ \Delta\phi_3\end{array} & \begin{bmatrix} c_{11} & b_{21} \\ 0 & c_{22} \end{bmatrix}
\end{array}
\tag{5.257}
$$

If 109°28′ is assumed for the valence angles, the elements of **G** are

$$
a_{11} = 2\mu, \quad a_{22} = \left(\frac{14}{3}\right)\mu\rho^2, \quad a_{21} = -\left(\frac{\sqrt{8}}{3}\right)\mu\rho, \quad b_{11} = -\left(\frac{1}{3}\right)\mu,
$$

$$
b_{22} = \left(\frac{8}{3}\right)\mu\rho^2, \quad b_{21} = -\left(\frac{\sqrt{8}}{3}\right)\mu\rho, \quad c_{11} = 0, \quad c_{22} = \left(\frac{1}{3}\right)\mu\rho^2
$$

and the **F** matrix elements of the Urey-Bradley force field are

$$
a_{11} = a, \quad a_{22} = b, \quad a_{21} = d, \quad b_{22} = c, \quad b_{11} = b_{21} = c_{11} = c_{22} = 0
$$

where the notations are the same as in Eq. 5.209. Since Δr_j and $\Delta\phi_j$ belong to the local symmetry coordinates IIs and Is, respectively (Fig. 5.61), the symmetry coordinates with the phase difference δ between two adjacent

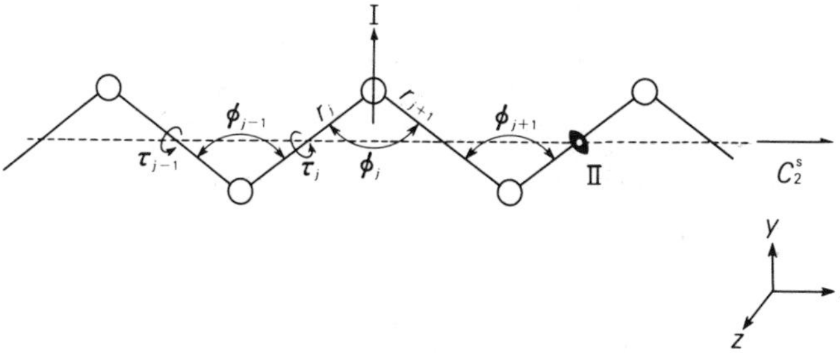

Fig. 5.61 Internal coordinates adopted for the planar zigzag PE molecule (skeletal model).

CH_2 groups may be constructed as

$$S_1(\delta) = \left(\frac{2}{N}\right)^{1/2} \sum \Delta r_j \cos (j-1)\delta$$

$$S_2(\delta) = \left(\frac{2}{N}\right)^{1/2} \sum \Delta \phi_j \cos \left(j - \frac{1}{2}\right)\delta \qquad (5.258)$$

where δ is defined as $\delta = 0$ and $\delta = \pi$ for the modes symmetric and antisymmetric to $C_2{}^s$, respectively. The **G** and **F** matrices are given by

$$\mathbf{G}(\delta) = \begin{bmatrix} 2\mu\left[1 - \left(\frac{1}{3}\right)\cos\delta\right] & \text{symmetric} \\ -\left(\frac{4\sqrt{2}}{3}\right)\mu\rho\left[\cos\left(\frac{\delta}{2}\right) + \cos\left(\frac{3\delta}{2}\right)\right] & \frac{2}{3}\mu\rho^2(7 + 8\cos\delta + \cos 2\delta) \end{bmatrix} \qquad (5.259)$$

$$\mathbf{F}(\delta) = \begin{bmatrix} a + 2c\cos\delta & \text{symmetric} \\ 2d\cos\left(\frac{\delta}{2}\right) & b \end{bmatrix} \qquad (5.260)$$

These matrices give A_g and B_{3u} modes for $\delta = 0$, and B_{1u} and B_{2u} for $\delta = \pi$ (see Table 5.14). Numerical computation using the values of molecular parameters and force constants given in Section 5.6.4 leads to one optical branch (ν_1) and one acoustical (ν_2) branch as shown in Fig. 5.62. $\nu_1(0)$ corresponds to the A_g mode in Fig. 5.54, $\nu_2(0)$ corresponds to $B_{3u}(T_x)$, $\nu_1(\pi)$ corresponds to B_{1g}, and $\nu_2(\pi)$ corresponds to $B_{2u}(T_y)$.

2. Out-of-Plane Modes. One-dimensional $\mathbf{G}(\delta)$ and $\mathbf{F}(\delta)$ for the symmetry coordinate (identical to the normal coordinate in this case)

$$S_3(\delta) = \left(\frac{2}{N}\right)^{1/2} \sum \Delta \tau_j \cos (j-1)\delta \qquad (5.261)$$

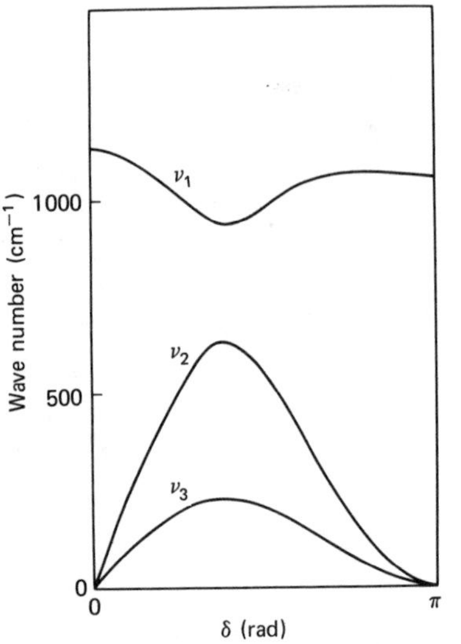

Fig. 5.62 Dispersion curves for the skeletal vibrations of the PE molecule. δ is the phase difference between adjacent CH_2 groups.

are given by

$$\mathbf{G}(\delta) = \left(\tfrac{9}{4}\right)\mu\rho^2(2 + \cos\delta - 2\cos 2\delta - \cos 3\delta)$$

$$\mathbf{F}(\delta) = F_\tau \tag{5.262}$$

where F_τ denotes the force constant for the torsional vibration around the skeletal bonds. For $F_\tau = 0.095$ mdyn Å/rad^2, the dispersion curve ν_3 (acoustical branch) shown in Fig. 5.62 is obtained. $\nu_3(0)$ corresponds to $B_{3g}(R_x)$ and $\nu_3(\pi)$ to $B_{1u}(T_z)$.

If the unit cell is considered to consist of two CH_2 groups, the length of the unit cell doubles, and therefore the length of Brillouin zone becomes half of the original one ($\delta = \pi/2$ in Fig. 5.62). In this case Fig. 5.63 is obtained by folding back the right-hand half of Fig. 5.62. There are four acoustical branches, which correspond to T_x, T_y, T_z, and R_x at $\delta = 0$. In the region of $0 < \delta < \pi$ in Fig. 5.63 ($0 < \delta < \pi/2$ and $\pi/2 < \delta < \pi$ in Fig. 5.62), each branch belongs to one symmetry species of the **k** group.* At the points $\delta = \pi$ ($\delta = \pi/2$ in Fig. 5.62), each branch degenerates into two vibrational modes; examples for ν_2 and ν_3 are shown in Fig. 5.64.

The results of a more detailed analysis of PE are given in Section 5.7.1.

* For the phase-difference vector $\boldsymbol{\delta}$, the **k** group is defined as the ensemble of operations R that satisfies the condition $\mathbf{S}\boldsymbol{\delta} = \boldsymbol{\delta} + 2\pi\mathbf{m}$, where $\mathbf{S}$ is the matrix for symmetry operation defined in Eq. 4.152 and $\mathbf{m}$ is a vector with integral components. **k** groups are subgroups of the factor group. The branches belonging to the same species do not cross each other. For details, see Refs. 191 and 192.

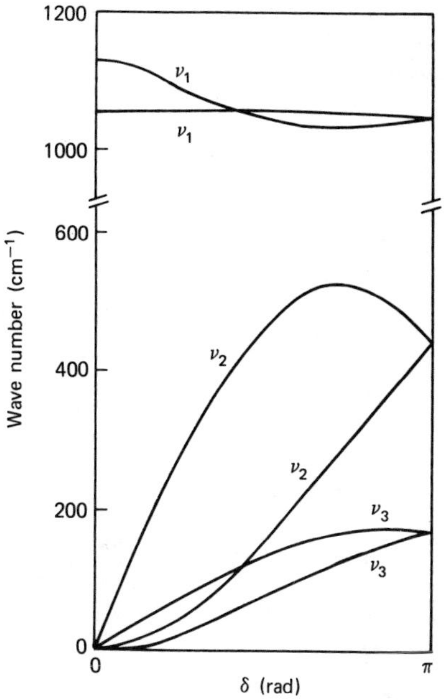

Fig. 5.63 Dispersion curves for the skeletal vibrations of the PE molecule. δ is the phase difference between adjacent unit cells.

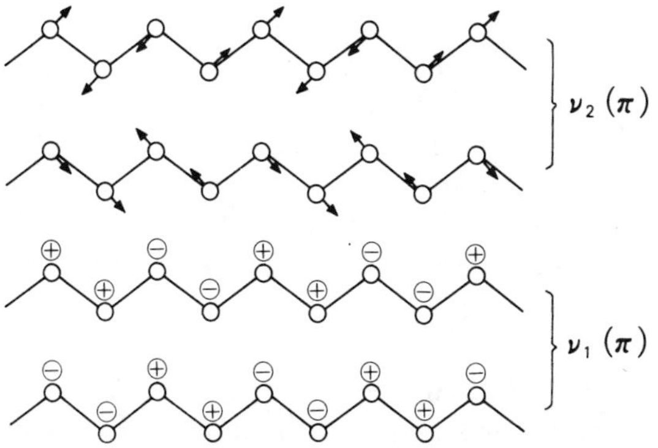

Fig. 5.64 Examples of doubly degenerate vibrations of PE chain (the modes at $\delta = \pi$ in Fig. 5.63).

5.6.7 Vibrations of Three-Dimensional Crystals

The treatment of a one-dimensional lattice can readily be extended to a three-dimensional crystal. If one unit cell contains N molecular chains, and each chain consists of m atoms per unit cell, there are $3mN$ branches. Of these, three are acoustical branches and $3mN - 3$ are optical branches. Dispersion curves for such a case for one direction of the **k** vector are schematically

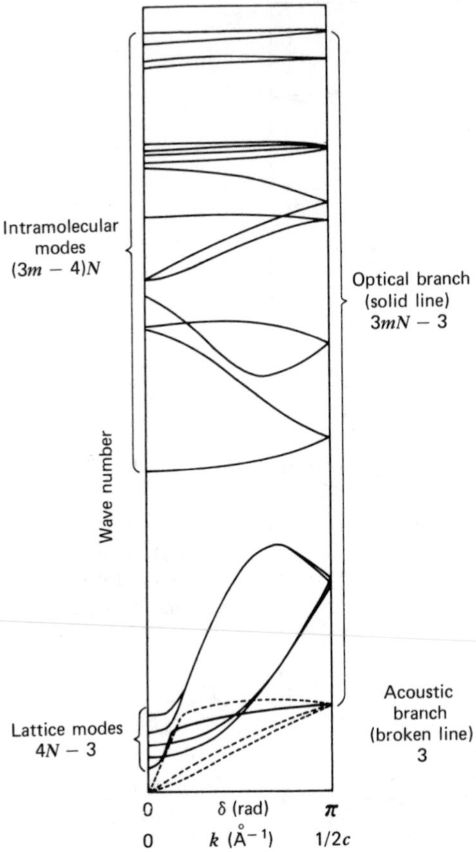

Fig. 5.65 Dispersion curves for a three-dimensional polymer crystal schematically drawn in one direction of **k** vector.

shown in Fig. 5.65. For $\mathbf{k} = \mathbf{0}$, there are $(3m - 4)N$ intramolecular modes, three acoustical modes, and $4N - 3$ lattice modes.* For orthorhombic PE, there are 28 intramolecular modes, five lattice modes, and three acoustical modes. The vibrational modes with $\mathbf{k} = \mathbf{0}$ are schematically shown in Fig. 5.45.

As is described in the preceding chapter (Eq. 4.21), the origin of any unit cell may be given by

$$\mathbf{R}_n = n_1\mathbf{a} + n_2\mathbf{b} + n_3\mathbf{c} \tag{5.263}$$

* In the case of a low-molecular-weight single crystal, in which one unit cell contains Z molecules and a molecule consists of n atoms, there are $(3n - 6)Z$ intramolecular modes, $6Z - 3$ lattice modes, and three acoustical modes.

and the reciprocal lattice vector is expressed by

$$\mathbf{k}_h = h_1\mathbf{a}^* + h_2\mathbf{b}^* + h_3\mathbf{c}^* \tag{5.264}$$

Since n_1, h_1, and so on are integers,

$$\mathbf{R}_n\cdot\mathbf{k}_h = n_1h_1 + n_2h_2 + n_3h_3 \tag{5.265}$$

is also an integer. The phase difference between the vibrations of two unit cells may be expressed as $\exp(2\pi i\mathbf{k}\cdot\mathbf{R}_n)$.* Because of the relation given by Eq. 5.265, the phase difference does not change by adding $\mathbf{k}_h$ to $\mathbf{k}$.

$$\begin{aligned}\exp[2\pi i(\mathbf{k}+\mathbf{k}_h)\cdot\mathbf{R}_n] &= \exp(2\pi i\mathbf{k}\cdot\mathbf{R}_n)\exp(2\pi i\mathbf{k}_h\cdot\mathbf{R}_n)\\ &= \exp(2\pi i\mathbf{k}\cdot\mathbf{R}_n)\end{aligned} \tag{5.266}$$

This condition is equivalent to that described in the case of the one-dimensional lattice, for which it is sufficient to consider only the first Brillouin zone.

In the case of a three-dimensional crystal, N_1, N_2, and N_3 unit cells align in the **a**, **b**, and **c** directions, respectively, for each of them Born's cyclic condition is applied. Under this condition, the following relation holds.

$$\begin{aligned}\exp(2\pi i\mathbf{k}\cdot\mathbf{R}_n) &= \exp[2\pi i\mathbf{k}\cdot(\mathbf{R}_n + N_1\mathbf{a})] = \exp[2\pi i\mathbf{k}\cdot(\mathbf{R}_n + N_2\mathbf{b})]\\ &= \exp[2\pi i\mathbf{k}\cdot(\mathbf{R}_n + N_3\mathbf{c})]\end{aligned} \tag{5.267}$$

Therefore

$$\exp(2\pi i\mathbf{k}\cdot N_1\mathbf{a}) = \exp(2\pi i\mathbf{k}\cdot N_2\mathbf{b}) = \exp(2\pi i\mathbf{k}\cdot N_3\mathbf{c}) = 1 \tag{5.268}$$

To satisfy this relation, m_1, m_2, and m_3 in the following equation must be integers.

$$\mathbf{k} = k_1\mathbf{a}^* + k_2\mathbf{b}^* + k_3\mathbf{c}^* = \frac{m_1}{N_1}\mathbf{a}^* + \frac{m_2}{N_2}\mathbf{b}^* + \frac{m_3}{N_3}\mathbf{c}^* \tag{5.269}$$

In other words, **k** is represented by the intersection of the planes that divide the **a***, **b***, and **c*** axes of the reciprocal lattice, respectively, into N_1, N_2, and N_3 equal spacings. m_1, m_2, and m_3 are integers in the ranges $-\frac{1}{2}N_1 \leqslant m_1 \leqslant \frac{1}{2}N_1$, $-\frac{1}{2}N_2 \leqslant m_2 \leqslant \frac{1}{2}N_2$, and $-\frac{1}{2}N_3 \leqslant m_3 \leqslant \frac{1}{2}N_3$, respectively, since only the first Brillouin zone may be taken into consideration. Figure 5.66 shows this relation. The numbers of m_1, m_2, and m_3 in the first Brillouin zone are N_1, N_2, and N_3, respectively, and therefore the number of **k** is $N_1N_2N_3 = \mathfrak{N}$.

The first Brillouin zone is obtained by drawing a plane passing perpendi-

* Since δ is defined by $\delta = 2\pi a/\lambda = 2\pi ka$ in Eq. 5.250, $\exp(ij\delta) = \exp(2\pi ikja)$. Extension of this to the three-dimensional case gives $\exp(2\pi i\mathbf{k}\cdot\mathbf{R}_n)$.

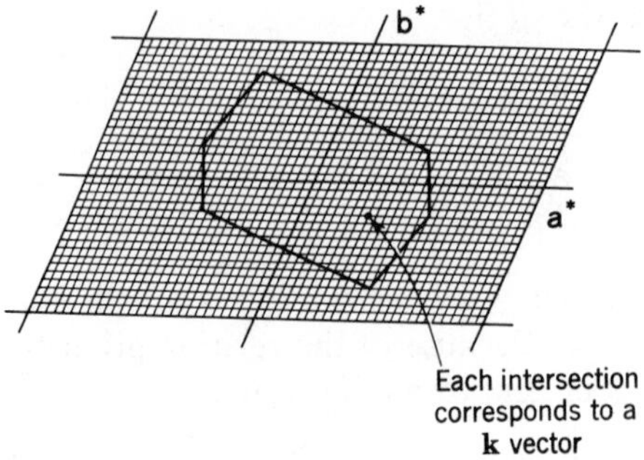

Fig. 5.66 A two-dimensional reciprocal lattice schematically divided with equal spacings. The **k** vector corresponds to each intersection of the lines. The area surrounded by bold lines is the first Brillouin zone.

cularly through the midpoint of the straight line jointing the reciprocal lattice point chosen as the center and each neighboring reciprocal lattice point. The space surrounded by all these planes is the first Brillouin zone.

The derivation of the dynamical equation for a three-dimensional crystal is given in Appendix E.

5.7 EXAMPLES OF STUDIES

5.7.1 Polyethylene

A great deal of experimental and theoretical work has been done on the molecular vibrations of PE and related compounds. The factor group analysis and vibrational assignment are described in Section 5.5.7. Normal coordinate treatments of the PE molecule have been developed by Kirkwood,[193] Whitcomb et al.[194], and Shimanouchi and Mizushima.[169,170] Schachtschneider and Snyder[156] have calculated the fundamental frequencies of a series of fully extended normal paraffins (C_3H_8 through $C_{10}H_{22}$), including crystalline PE, and established a set of 31 independent valence force constants that reproduce 270 observed frequencies with an average error of 0.29%.

The frequency–phase relation of the PE molecule has been investigated in detail by Tasumi and Shimanouchi[153] who used a Urey-Bradley type force field. Figure 5.67 shows the dispersion curves. Table 5.14 gives the vibrational modes on the branches and the symmetry species at $\delta = 0$ and π, which correspond to the optically active fundamentals. The modes in the ranges of $0 < \delta < \pi/2$ and $\pi/2 < \delta < \pi$ are different from those shown in Table 5.14 and belong to the species of the **k** group (see footnote on p. 300). The branches $\nu_1 - \nu_5$ belong to the in-plane modes and $\nu_6 - \nu_9$ to the out-of-plane modes. Two branches, ν_5 and ν_9, are acoustical. The other branches are optical. In the modes ν_7 and ν_8, the coupling between $t(CH_2)$ and $r(CH_2)$

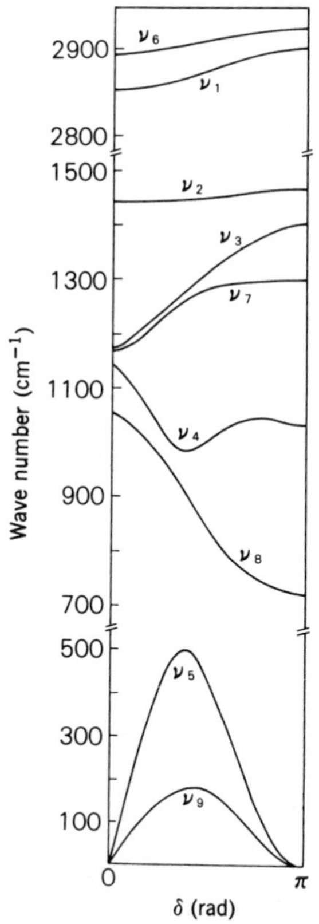

Fig. 5.67 Frequency dispersion curves for PE molecule. (From Tasumi et al.[155])

is remarkable. The potential energy distribution indicates that at $\delta = 0$, ν_7 is pure $r(CH_2)$ and ν_8 is pure $t(CH_2)$. As δ increases a mixing of ν_7 and ν_8 occurs. At $\delta = \pi$, the situation is reversed; ν_7 is $t(CH_2)$ and ν_8 is $r(CH_2)$.

The vibrations of the three-dimensional PE crystal have also been calculated by Tasumi et al.[153, 157] and are expressed in terms of the three-dimensional phase differences δ_a, δ_b, and δ_c. They calculated the frequency dispersion curves for PE and deuterated PE for the case $\delta_a = \delta_b = 0$. Fig. 5.68 shows an example. Here δ_c was taken as the phase difference between adjacent CH_2 groups. The modes at $\delta_c = 0$ and π are optically active. The comparison between the calculated and observed frequencies and band assignments are given in Table 5.9. Each of the nine dispersion curves for a single PE chain splits into two because of intermolecular interactions. ν_a and ν_b in Fig. 5.68 denote the modes symmetric and antisymmetric with respect to $\sigma_g(ac)$, respectively. The ν_5 skeletal bending and the ν_9 skeletal torsional modes mix with the lattice modes. Three of the eight branches are acoustical,

Table 5.14 Vibrational Modes of Dispersion Curves of PE

		Species of factor group modes[a]	
Branches	Vibrational modes	$\delta = 0$	$\delta = \pi$
ν_1	$\nu_s(CH_2)$	A_g	B_{2u}
ν_2	$\delta(CH_2)$	A_g	B_{2u}
ν_3	$w(CH_2)$	B_{3u}	B_{1g}
ν_4	Skeletal	A_g	B_{1g}
ν_5	Skeletal	$B_{3u}(T_x)$	$B_{2u}(T_y)$
ν_6	$\nu_a(CH_2)$	B_{3g}	B_{1u}
ν_7	$t(CH_2)$-$r(CH_2)$	$B_{3g}(r)$	$B_{2g}(t)$
ν_8	$r(CH_2)$-$t(CH_2)$	$A_u(t)$	$B_{1u}(r)$
ν_9	Skeletal torsion	$B_{3g}(R_x)$	$B_{1u}(T_z)$

[a] T_x, T_y, T_z: pure translations along the x (chain direction), y (in-plane), and z (out-of-plane) directions, respectively; R_x: pure rotation about the chain axis; $r:r(CH_2)$; $t:t(CH_2)$.

giving zero frequency (T_a, T_b, T_c) at $\delta_c = 0$ or π. The other five branches give lattice vibrations at $\delta_c = 0$ or π. All the modes, including those of $\delta_a \neq 0$ and $\delta_b \neq 0$, have been calculated by Kitagawa and Miyazawa.[195] The calculated frequency distribution reveals good agreement with the experimental result of neutron inelastic scattering (Fig. 5.69).[196, 197] * Thermodynamic functions, such as specific heat, internal energy, entropy, and Helmholtz energy, can be calculated from the frequency distribution.[180, 181, 191, 197, 198] See also Section 6.6 and Appendix G.

The modes at the points between $\delta_c = 0$ and π are optically inactive for infinite chains, but are active in the case of n-paraffins of finite lengths. In

* Neutron scattering is classified into elastic scattering and inelastic scattering. Elastic scattering does not change the wavelength. Inelastic scattering changes their wavelength as a result of the transfer of vibrational energy within the atoms of the sample. Therefore the frequency distribution in the sample can be obtained from the energy distribution of the scattered neutrons. When the momentum transfer vector is parallel to the vibrational direction of the nucleus, the cross section is maximum. When the vectors are perpendicular, no energy transfer occurs. For Raman scattering, only the factor group modes are active, but no such selection rule exists in neutron scattering, all the vibrations being observed. Inelastic neutron scattering is further classified into incoherent scattering and coherent scattering. In incoherent scattering the cross section of the hydrogen nucleus is very large compared to the scattering cross section of other nuclei. Therefore incoherent inelastic scattering is especially strong for large amplitudes of the hydrogen atoms. In coherent inelastic scattering, the observed data can be plotted on the frequency dispersion curve. In coherent scattering the cross section of deuterium is large, and so deuterated samples are used. For details see Ref. 191.

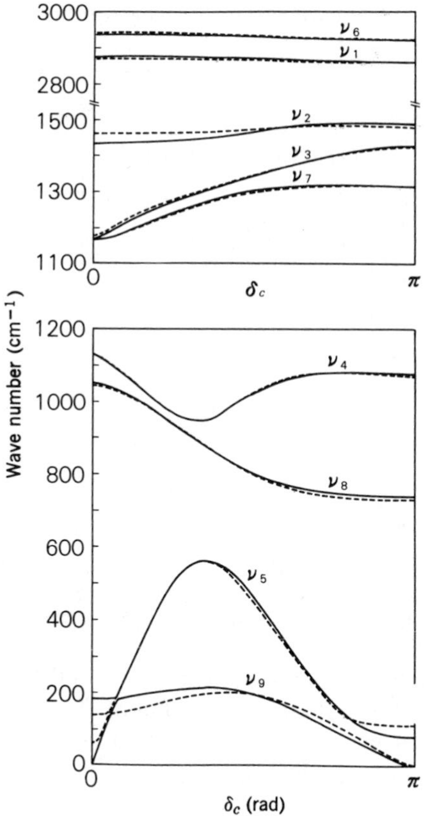

Fig. 5.68 Frequency dispersion curves for a PE crystal ($\delta_a = \delta_b = 0$). The lattice constants at 100°K are used (————) ν_a branch, (- - - - - - - -) ν_b branch. (From Tasumi and Krimm.[157])

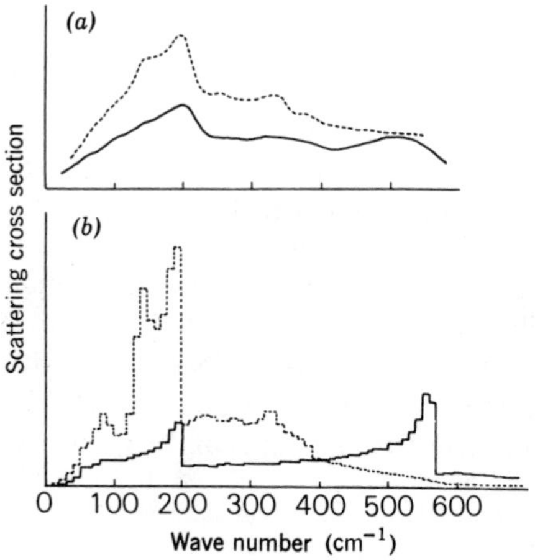

Fig. 5.69 (*a*) Distribution of the cross section of neutron incoherent inelastic scattering by a uniaxially oriented orthorhombic PE (100°K). (————) Momentum transfer vector parallel to the *c* axis and (- - - - - - - -) momentum transfer vector perpendicular to the *c* axis. (Myers et al.[196]) (*b*) Theoretical curve obtained from the frequency distribution by weighting the squared amplitude of the hydrogen atoms. (From Miyazawa and Kitagawa.[197])

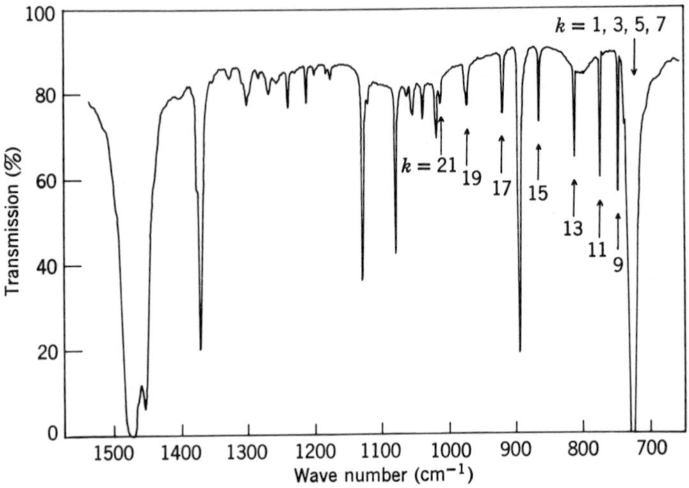

Fig. 5.70 Infrared spectrum of n-$C_{24}H_{50}$ at -196°C. Arrows indicate the $r(CH_2)-t(CH_2)$ progression bands (ν_8). The values of k are shown in the case of n-paraffins with even number carbon atoms (the point symmetry C_{2h}). ν_8 is infrared active only for $k=$ odd (the A_u species).

fact, various groups of progression bands have been observed over some frequency ranges in the infrared spectra of the crystalline n-paraffins as shown in Fig. 5.70 for n-tetracosane. The band progression may be treated using the simple coupled oscillator approximation, in which a vibrational mode of every individual CH_2 group [e.g., $\delta(CH_2)$] is replaced by a harmonic oscillator with the frequency characteristic of the mode under consideration. The frequency ν_k of a linear array of m identical oscillators depends on the phase difference δ_k between adjacent oscillators. The allowed phase difference for the system is given by

$$\delta_k = \frac{k\pi}{m+1} \qquad k = 1, 2, \cdots, m \qquad (5.270)^*$$

The frequencies of the progression bands of n-paraffins correspond approximately (except for some local modes due to terminal methyl groups) to a point on one of the dispersion curves of the infinitely long PE molecule. In Fig. 5.71 the frequencies of the $r(CH_2)$-$t(CH_2)$ mode (ν_8) for C_3H_8 through n-$C_{30}H_{62}$ are plotted against δ. The points for the shorter molecules deviate from the common curve. This deviation is due to interaction with the out-of-plane rocking of the terminal methyl groups. Although only the modes at $\delta = 0$ and π are optically active in PE, the data can be measured

* If the number of carbon atoms is N, $m = N$ for the CH_2 deformation modes, $m = N-2$ for the CCC bending modes, and $m = N-3$ for the skeletal torsional modes.

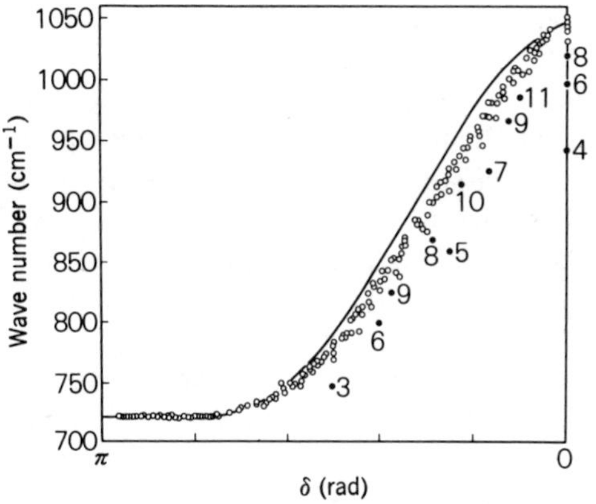

Fig. 5.71 Frequency–phase difference curve for $r(CH_2) - t(CH_2)$ mode for C_3H_8 through n-$C_{30}H_{62}$. Arabic figures indicate the carbon numbers of n-paraffins. The solid line indicates the dispersion curve obtained by taking into account the end effect. (From Matsuda et al.[178])

for the range $0 \leqslant \delta \leqslant \pi$ in the case of polyethers, $[—(CH_2)_m—O—]_n$ ($m = 3$, 4, 6–12), because of the presence of oxygen atoms.[199]

The lowest frequency mode of the ν_5 progression band corresponds to the accordionlike vibration (Fig. 5.72) and varies inversely with the chain length. These relationships were first seen in the n-paraffins (n-$C_{16}H_{34}$ and shorter) by Mizushima and Shimanouchi,[200] and were confirmed by the laser Raman spectra of longer n-paraffins (up to C_{94}) by Schaufele and Shimanouchi.[201]

Peticolas et al.[202,203] observed the Raman lines due to the accordion mode of PE single crystals subjected to heat treatment at various temperatures. They found a parallel relation between the thickness of single crystals calculated from the frequencies and the long spacing estimated by small-angle X-ray scattering.

For an elastic rod model, the frequency ν of the accordion mode (or longitudinal acoustic mode) is related to Young's modulus E by

$$\nu = \left(\frac{m}{2L}\right)\left(\frac{E}{\rho}\right)^{1/2} \tag{5.271}$$

where L is the rod length (lamellar thickness), ρ is the crystalline density, and $m = 1,2,\cdots$ (Raman active only in the case of $m =$ odd).[200] By estimating L

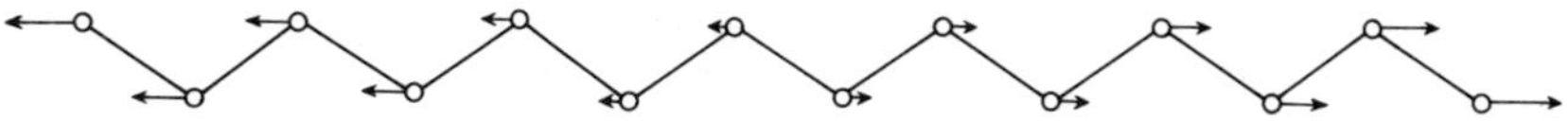

Fig. 5.72 Accordion like vibration.

based on X-ray small angle scattering, Young's moduli have been obtained for PE,[201,204] *it*-PP,[205] PEO,[206] and so on.

Tasumi and Krimm[207] calculated the frequencies for mixed crystal models of PE and deuterated PE with various ratios. The features of the splitting of the CH_2 (or CD_2) scissoring and rocking vibrations are quite different for the two models shown in Fig. 5.73. In model *a*, one $(CH_2)_n$ or $(CD_2)_n$ chain folds along the (110) plane, and the splitting is calculated to be 4.9 cm^{-1} for $\delta(CH_2)$, 4.1 cm^{-1} for $\delta(CD_2)$, 5.2 cm^{-1} for $r(CH_2)$, and 4.6 cm^{-1} for $r(CD_2)$. On the contrary, no such splitting is expected for model *b*, in which the folding occurs along the (200) plane. According to the experimental results,[208] the features of the band splitting are different for solution-cast crystals of the $(CH_2)_n$–$(CD_2)_n$ mixtures at various ratios and melt-prepared samples of corresponding mixtures. In the case of pure $(CH_2)_n$ or $(CD_2)_n$,

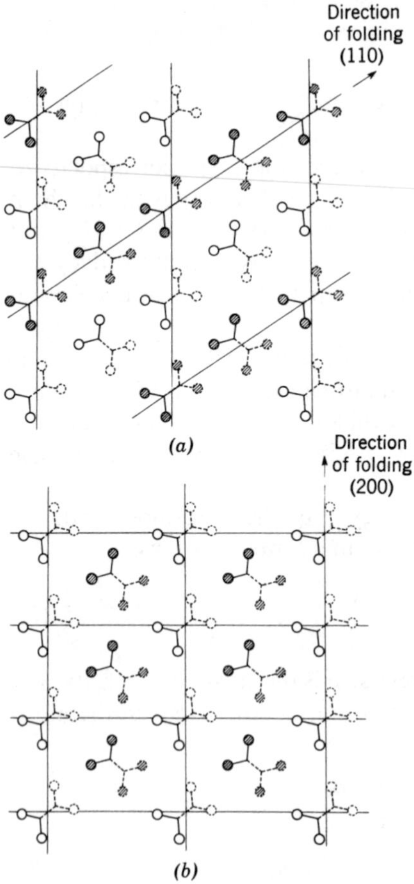

Fig. 5.73 Crystal models of mixtures of PE and deutero-PE. (*a*) Folding along the (110) plane, and (*b*) folding along the (200) plane. (○) H, (●) D. (From Tasumi and Krimm.[207])

no changes are found that depend on the preparation methods. Krimm et al. have considered the preference of the folding of molecular chains along the (110) plane in samples cast from solution, and along the (200) plane in samples from melt. See Refs. 209 and 210 for this series of studies. For spectroscopic evidence of the cis-opening polymerization of ethylene by Ziegler catalyst, see Refs. 64 and 211.

5.7.2 Polyoxacyclobutane

As is described in Section 4.3.3.A, polyoxacyclobutane has four crystal modifications, depending on the preparation condition. Form I is stable only in the presence of water, II is obtained only as oriented samples, and III is thermally most stable. Form IV has been found recently in the highly stretched state. The crystal structures of forms II and III have been determined by X-ray analysis.[212] The molecular models of the first three modifications are reproduced in Fig. 5.74. The molecular conformation of form II is of the $T_3GT_3\bar{G}$ glide type, and that of form III is of the $(T_2G_2)_2$ helix type. The observed fiber period of form I obtained from an X-ray photograph is 4.80 Å, which is in good agreement with the calculated fiber period for the planar zigzag model of 4.85 Å. Thus form I was believed to have a planar zigzag structure: its crystal structure has been determined by a combination of X-ray and infrared methods.[213]

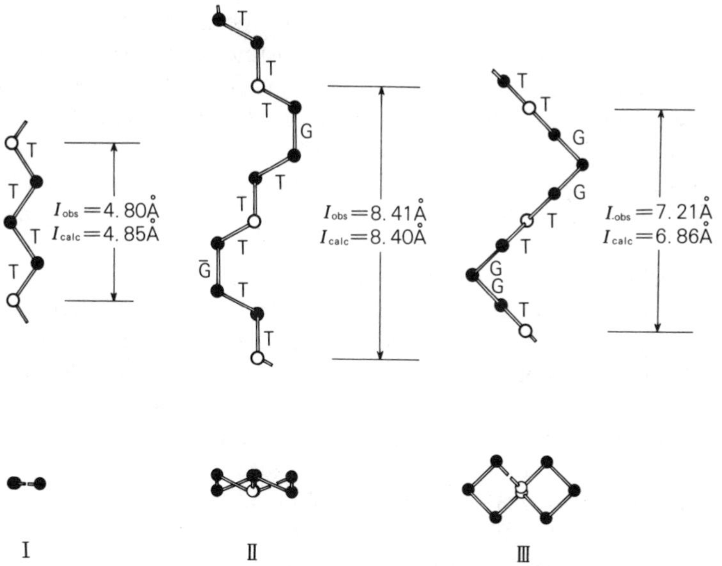

Fig. 5.74 Three types of molecular models of polyoxacyclobutane.[210]

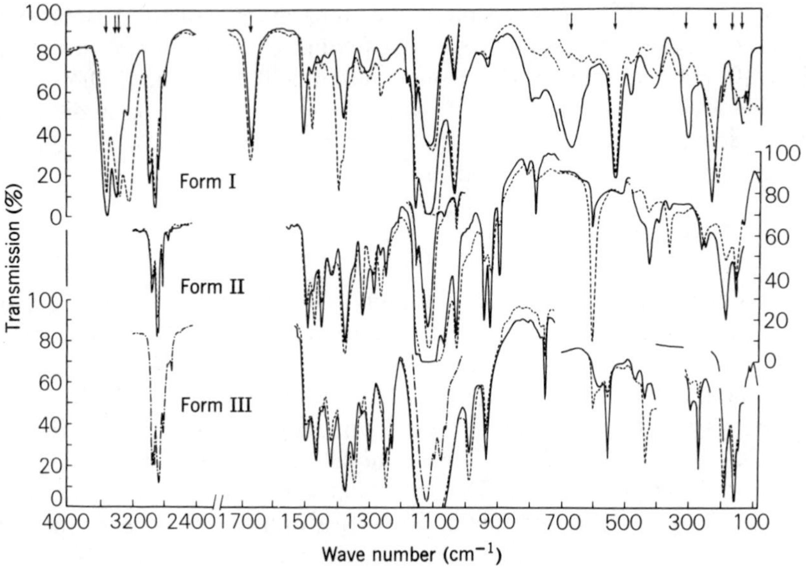

Fig. 5.75 Polarized infrared spectra of the three forms of polyoxacyclobutane. (————) Perpendicular radiation, (---------) parallel radiation, and (-·-·-·-·-) unpolarized spectrum.[213] Arrows indicate the bands considered to be water bands.

Figure 5.75 shows polarized infrared spectra of the three crystal modifications. The bands indicated by arrows in the spectra of form I arise from water molecule vibrations and show very distinct dichroism. These bands can be identified by comparing the normal (upper) and the deuterated (lower) spectra shown in Fig. 5.76. The bands indicated by arrows in Fig. 5.75 shift to the low-frequency side on deuteration; the remaining bands, belonging to the polyoxacyclobutane molecule, do not. Such water bands appear in the OH stretching, the HOH bending, and the H_2O libration and translation regions.

In Fig. 5.77 the spectra in the OH stretching and the HOH bending regions of form I are compared with those of water, ice, and barium chloride dihydrate, $BaCl_2 \cdot 2H_2O$. The OH stretching bands of form I consist of two perpendicular bands on the high frequency side, and two parallel bands on the low frequency side. The HOH bending vibrations produce a pair of bands of opposite dichroism with slightly different frequencies (3 cm^{-1}). The corresponding bands of water and ice are very broad and lack fine structure, but the spectral features of barium chloride dihydrate are similar to those of form I. If a film sample of form I is kept at about $-19°C$ and water is removed under vacuum, it is transformed into form III. As this process continues, the intensity of the OH bands and of the bands due to

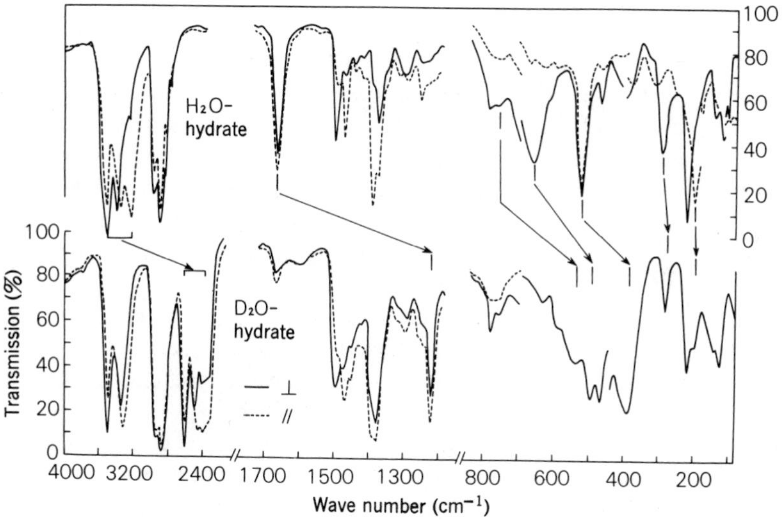

Fig. 5.76 Infrared spectra of form I of polyoxacyclobutane.[213]

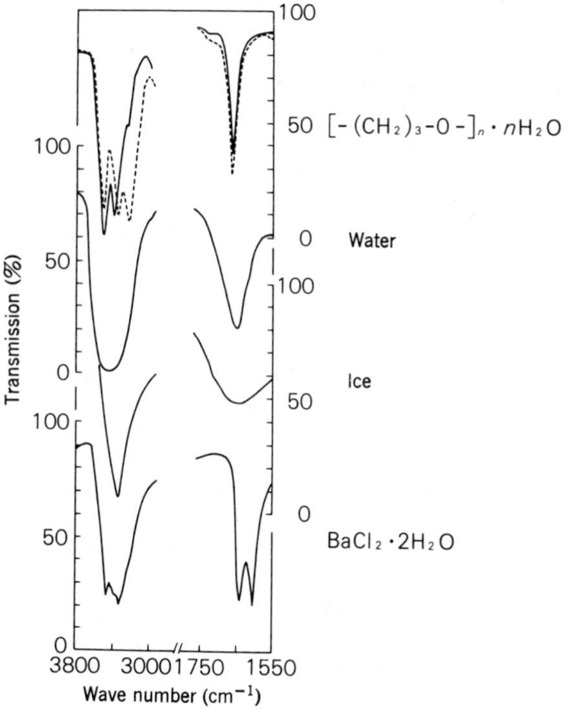

Fig. 5.77 Infrared spectra of the OH stretching and HOH bending of form I of polyoxacyclobutane, water, ice, and barium chloride dihydrate.[213]

form I decreases, and the bands due to form III appear and increase in intensity. As the last OH bands disappear the whole spectrum becomes identical to that of form III. Moreover, polarized spectra of samples in different orientations show that the degrees of dichroism of the water bands vary in the same manner as the polymer bands of form I. From these infrared spectroscopic results, it can be inferred that the water molecules present in the sample of form I are in the crystalline region, the water molecules having definite locations and well defined orientations.

The results of the normal coordinate treatment for the three modifications show good agreement between the observed and calculated frequencies.[214] The molecular symmetries of forms I, II, and III can be treated with the factor groups isomorphous to C_{2v}, C_s, and D_2, respectively.

In the X-ray study a monoclinic cell with the lattice constants given in Table 7.1 was deduced from the data obtained for doubly oriented samples. If it is assumed that the unit cell contains four monomeric units and four water molecules, the calculated density is 1.17 g/cm^3, which is in good agreement with the observed value of 1.11 g/cm^3. The crystal structure is shown in Fig. 5.78. The oxygen atoms of the water molecules are indicated

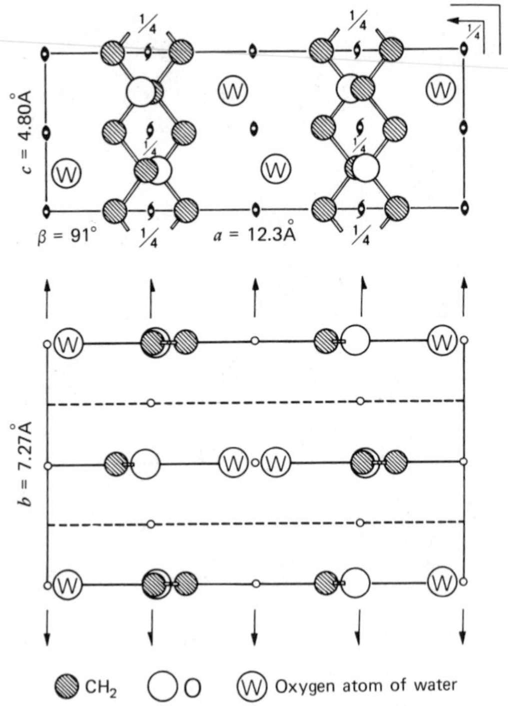

Fig. 5.78 Crystal structure of form I of polyoxacyclobutane.[213]

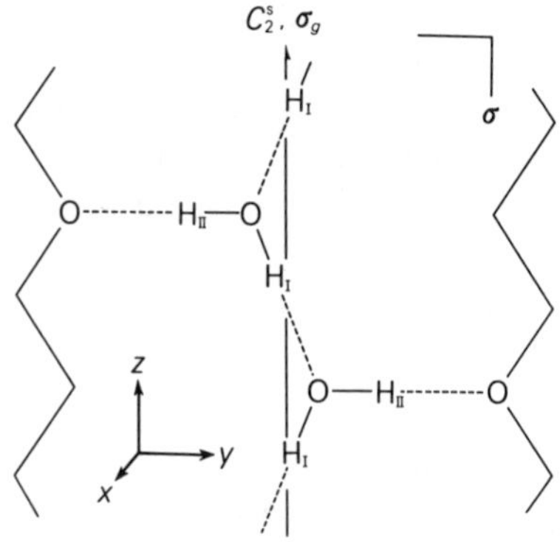

Fig. 5.79 The ribbonlike chain in form I of polyoxacyclobutane with its symmetry elements.[213]

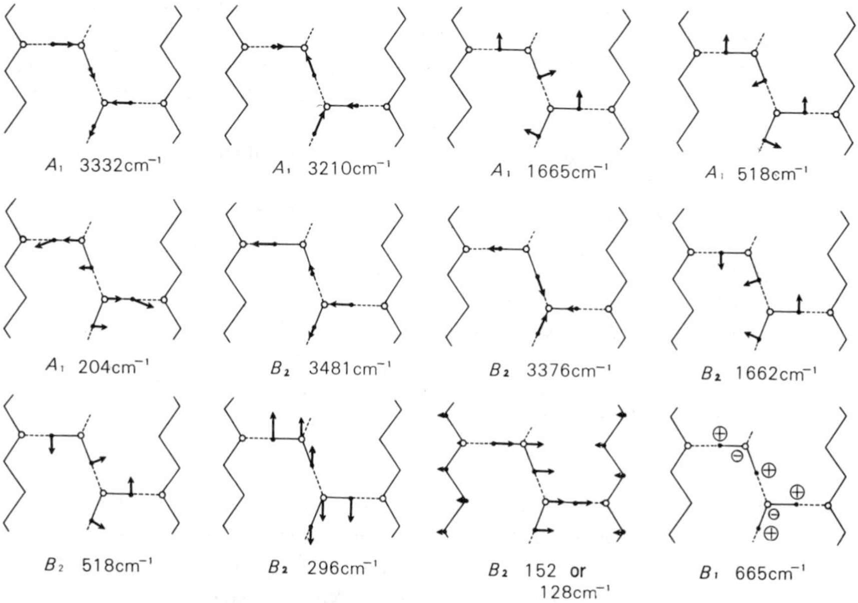

Fig. 5.80 Vibrational modes of form I of polyoxacyclobutane (H_2O hydrate). Arrows, ⊕s or ⊖s represent the displacement of the corresponding atom or group, while their lengths (or sizes) are proportional to the $\mathbf{L}_X$ matrix element.[213]

by the letter W. In this case, the positions of the hydrogen atoms of water molecules cannot be determined by the X-ray method. However the distances between the oxygen atoms determined by X-ray analysis (water–water, about 2.68 Å; water–polymer, about 2.71 Å) are shorter than twice the van der Waals radius of an oxygen atom, 2.80 Å. This fact suggests hydrogen bonding. In the crystal two polymer chains are joined by water molecules through hydrogen bonds. The assumed positions of the hydrogen atoms are shown schematically in Fig. 5.79. This hydrogen-bonded structure has been corroborated by normal coordinate treatments carried out on H_2O and D_2O hydrates. The symmetry may be treated with the factor group isomorphous to the point group C_{2v} (cf. Table 5.1).* The calculated results are shown in Fig. 5.80.

REFERENCES

1. G. Herzberg, *Spectra of Diatomic Molecules*, 2nd ed., Van Nostrand, New York, 1950.
2. G. Herzberg, *Infrared and Raman Spectra of Polyatomic Molecules*, Van Nostrand, New York, 1945.
3. S. Mizushima, *Structure of Molecules and Internal Rotation*, Academic, New York, 1954.
4. E. B. Wilson, Jr., J. C. Decius, and P. C. Cross, *Molecular Vibrations, The Theory of Infrared and Raman Spectra*, McGraw-Hill, New York, 1955.
5. R. N. Jones and C. Sandorfy, The Application of Infrared and Raman Spectroscopy to the Elucidation of Molecular Structure, *Chemical Applications of Spectroscopy*, W. West, Ed., Interscience, New York, 1956, pp. 247–580.
6. W. Brügel, *An Introduction to Infrared Spectroscopy*, translation into English by A. R. Katritzky and A. J. D. Katrizky, Methuen, London, Wiley, New York, 1962.
7. H. A. Szymanski, *IR, Theory and Practice of Infrared Spectroscopy*, Plenum Press, New York, 1964.
8. H. A. Szymanski, Ed., *Raman Spectroscopy, Theory and Practice*, Plenum Press, New York, 1967.
9. H. A. Szymanski, Ed., *Raman Spectroscopy, Theory and Practice*, Vol. 2, Plenum Press, New York, 1970.
10. L. J. Bellamy, *Advances in Infrared Group Frequencies*, Methuen, London, Wiley, New York, 1968.
11. T. R. Gilson and P. J. Hendra, *Laser Raman Spectroscopy*, Wiley-Interscience, New York, 1970.
12. M. C. Tobin, *Laser Raman Spectroscopy*, Wiley-Interscience, New York, 1971.
13. G. Turrell, *Infrared and Raman Spectra of Crystals*, Academic, New York, 1972.
14. J. R. Durig, Ed., *Vibrational Spectra and Structure*, Vol. 1, Dekker, New York, 1972.
15. J. R. Durig, Ed., *Vibrational Spectra and Structure*, Vol. 2, Dekker, New York, 1975.
16. W. G. Fateley, F. R. Dollish, N. T. McDevitt, and F. F. Bentley, *Infrared and Raman Selection Rules for Molecules and Lattice Vibrations: The Correlation Method*, Wiley-Interscience, New York, 1972.
17. H. Poulet and J. P. Mathieu, *Vibration Spectra and Symmetry of Crystals*, Gordon and Breach, New York, 1976.

* Strictly speaking, because of the monoclinic symmetry of the unit cell ($C2/m$-C_{2h}^3), the site group of this molecule is C_s. The vibrations, however, may be reasonably treated under the factor group C_{2v}; since the angle β is 91° (close to 90°) the deviation from precise C_{2v} molecular symmetry is negligibly small.

18. R. G. J. Miller and B. C. Stace, *Laboratory Methods in Infrared Spectroscopy*, 2nd ed., Heyden & Sons, London, 1972.

19. S. Krimm, *Adv. Polym. Sci.*, **2,** 51–172 (1960).

20. M. Tryon and E. Horowitz, Infrared Spectroscopy, *Analytical Chemistry of Polymers. Part II, Analysis of Molecular Structure and Chemical Groups*, G. M. Kline, Ed., Inter-Science, New York, 1962, pp. 291–333.

21. R. Zbinden, *Infrared Spectroscopy of High Polymers*, Academic, New York, 1964.

22. J. Haslam and H. A. Willis, *Identification and Analysis of Plastics*, London Iliffe Book Ltd., London, 1965.

23. D. O. Hummel, *Infrared Spectra of Polymers in the Medium and Long Wave Length Regions*, Wiley-Interscience, New York, 1966.

24. R. G. Zhbankov, *Infrared Spectra of Cellulose and Its Derivatives*, translation into English by A. B. Densham, Consultants Bureau, New York, 1966.

25. C. J. Henniker, *Infrared Spectrometry of Industrial Polymers*, Academic, New York, 1967.

26. D. O. Hummel, *Atlas der Kunststoff-Analyse*, Bd. I. *Hochpolymere und Harze, Spectren und Methoden zur Identifizierung*, Teil 1. *Text* und Teil 2. *Spektren*, Carl Hanser Verlag, München, 1968.

27. A. Elliott, *Infrared Spectra and Structure of Organic Long-Chain Polymers*, Edward Arnold Publishers, London, 1969.

28. D. O. Hummel, Ed., *Polymer Spectroscopy*, Verlag Chemie, Weinheim, 1974.

29. P. J. Hendra, *Adv. Polym. Sci.*, **6,** 151 (1969).

30. R. F. Schaufele, *Macromol. Rev.*, **4,** 67 (1970).

31. D. S. Cain and A. B. Harvey, *Raman Spectroscopy of Polymeric Materials.* Part I, *Selected Commercial Polymers, Developments in Applied Spectroscopy*, Vol. 7B, Plenum Press, New York, 1970, p. 94.

32. J. L. Koenig, *Appl. Spectrosc. Rev.*, **4,** 233 (1971).

33. M. E. A. Cudby, H. A. Willis, P. J. Hendra, and C. J. Peacock, *Chem. Ind. (Lond.)*, **1971,** 531.

34. Butterworths Scientific Publication, *Documentation of Molecular Spectroscopy (DMS)*, 1956–.

35. S. P. Sadtler & Sons, Inc., *Sadtler Standard Spectrograms*, 1957–.

36. *Infrared and Raman Data Committee (IRDC)*, Nankôdô, Tokyo, 1960–.

37. H. M. Hershenson, *Infrared Absorption Spectra Index for 1945–1957*, Academic, New York, 1959, *Infrared Absorption Spectra Index for 1958–1962*, Academic, New York, 1964.

38. F. M. Rugg, J. J. Smith, and L. H. Wartman, *J. Polym. Sci.*, **11,** 1 (1953).

39. U. Baumann, H. Schreiber, and K. Tessmar, *Makromol. Chem.*, **36,** 81 (1960).

40. M. Takeda, K. Iimura, A. Yamada, and Y. Imamura, *Bull. Chem. Soc. Jpn.*, **32,** 1150 (1959); **33,** 1219 (1960); M. Takeda, K. Iimura, and S. Ochiai, *J. Polym. Sci. B*, **4,** 155 (1966).

41. M. Kobayashi, K. Tsumura, and H. Tadokoro, *J. Polym. Sci. A-2*, **6,** 1493 (1968).

42. L. Pauling and E. B. Wilson, Jr., *Introducti n to Quantum Mechanics*, McGraw-Hill, New York, 1935.

43. N. B. Colthup, *J. Opt. Soc. Am.*, **40,** 397 (1950).

44. F. R. Dollish, W. G. Fateley, and F. F. Bentley, *Characteristic Raman Frequencies of Organic Compounds*, Wiley-Interscience, New York, 1974.
45. E. K. Plyler and C. W. Peters, *J. Res. Natl. Bur. Stand. U.S.*, **45,** 462 (1950).
46. H. Tadokoro, S. Seki, and I. Nitta, *Bull. Chem. Soc. Jpn.*, **28,** 559 (1955).
47. A. Weissberger, E. S. Proskauer, J. A. Riddick, and E. E. Toops, Jr., *Organic Solvents*, Interscience, New York, 1955.
48. H. Tadokoro, K. Tatsuka, and S. Murahashi, *J. Polym. Sci.*, **59,** 413 (1962).
49. J. D. Sands and G. S. Turner, *Anal. Chem.*, **24,** 791 (1952).
50. P. Holliday, *Nature*, **163,** 602 (1949).
51. M. Tsuboi, *J. Polym. Sci.*, **25,** 159 (1957).
52. D. Morero, A. Santambrogio, L. Porri, and F. Ciampelli, *Chim. Ind. (Milan)*, **41,** 758 (1959).
53. W. T. M. Johnson, *Off. Dig. Fed. Soc. Paint Technol.*, **32,** 1067 (1960).
54. N. J. Harrick, *Internal Reflection Spectroscopy*, Interscience, New York, 1967.
55. T. Morimoto, Susumu Enomoto, and B. Konishi, *Kôbunshi Kagaku*, **29,** 582 (1972).
56. M. Yokoyama, H. Ochi, H. Tadokoro, and C. C. Price, *Macromolecules*, **5,** 690 (1972).
57. H. J. Marrinan and J. Mann, *J. Appl. Chem.*, **4,** 204 (1954).
58. H. Tadokoro, H. Nagai, S. Seki, and I. Nitta, *Bull. Chem. Soc. Jpn.*, **34,** 1504 (1961).
59. H. Tadokoro, *Bull. Chem. Soc. Jpn.*, **32,** 1252 (1959).
60. L. C. Leitch, P. E. Gagnon, and A. Cambron, *Can. J. Res.*, **B28,** 256 (1950).
61. H. Tadokoro, S. Murahashi, R. Yamadera, and T. Kamei, *J. Polym. Sci. A*, **1,** 3029 (1963).
62. C. Y. Liang, F. G. Pearson, and R. H. Marchessault, *Spectrochim. Acta*, **17,** 568 (1961).
63. R. Yamadera, H. Tadokoro, and S. Murahashi, *J. Chem. Phys.*, **41,** 1233 (1964).
64. M. Tasumi, T. Shimanouchi, H. Tanaka, and S. Ikeda, *J. Polym. Sci.*, **A2,** 1607 (1964).
65. H. Tadokoro, T. Kitazawa, S. Nozakura, and S. Murahashi, *Bull. Chem. Soc. Jpn.*, **34,** 1209 (1961).
66. C. Y. Liang, M. R. Lytton, and C. J. Boone, *J. Polym. Sci.*, **44,** 549 (1960); **47,** 139 (1960); C. Y. Liang and F. G. Pearson, *J. Mol. Spectrosc.*, **5,** 290 (1960).
67. M. Peraldo and M. Farina, *Chim. Ind. (Milan)*, **42,** 1349 (1960).
68. H. Tadokoro, M. Kobayashi, M. Ukita, K. Yasufuku, S. Murahashi, and T. Torii, *J. Chem. Phys.*, **42,** 1432 (1965).
69. G. Natta, *Makromol. Chem.*, **35,** 94 (1960).
70. H. Tadokoro, S. Nozakura, T. Kitazawa, Y. Yasuhara, and S. Murahashi, *Bull. Chem. Soc. Jpn.*, **32,** 313 (1959).
71. H. Tadokoro, Y. Nishiyama, S. Nozakura, and S. Murahashi, *J. Polym. Sci.*, **36,** 553 (1959); *Bull. Chem. Soc. Jpn.*, **34,** 381 (1961).
72. M. Kobayashi, Doctoral thesis, Osaka University, 1968.
73. M. Kobayashi, *Bull. Chem. Soc. Jpn.*, **33,** 1416 (1960); **34,** 560, 1045 (1961).
74. T. Onishi and S. Krimm, *J. Appl. Phys.*, **32,** 2320 (1961).
75. S. Narita, S. Ichinohe, and Satoru Enomoto, *J. Polym. Sci.*, **37,** 281 (1959).
76. S. Krimm, V. L. Folt, J. J. Shipman, and A. R. Berens, *J. Polym. Sci.*, **A1,** 2621 (1963).
77. S. Narita, S. Ichinohe, and Satoru Enomoto, *J. Polym. Sci.*, **37,** 263 (1959).
78. H. Nagai, H. Watanabe, and A. Nishioka, *J. Polym. Sci.*, **62,** S95 (1962); H. Nagai, *J. Appl. Polym. Sci.*, **7,** 1697 (1963).

79. H. Tadokoro, M. Kobayashi, Koh Mori, Y. Takahashi, and S. Taniyama, *J. Polym. Sci. C*, **22,** 1031 (1969).
80. Y. Tanaka, Y. Takeuchi, M. Kobayashi, and H. Tadokoro, *J. Polym. Sci. A-2*, **9,** 43 (1971).
81. M. A. Golub and J. J. Shipman, *Spectrochim. Acta*, **20,** 701 (1964).
82. M. A. Golub and J. J. Shipman, *Spectrochim. Acta*, **16,** 1165 (1960).
83. A. Novak and E. Whalley, *Trans. Faraday Soc.*, **55,** 1484 (1959).
84. H. Tadokoro, M. Kobayashi, Y. Kawaguchi, A. Kobayashi, and S. Murahashi, *J. Chem. Phys.*, **38,** 703 (1963).
85. H. Tadokoro, Y. Chatani, T. Yoshihara, S. Tahara, and S. Murahashi, *Makromol. Chem.*, **73,** 109 (1964).
86. T. Yoshihara, H. Tadokoro, and S. Murahashi, *J. Chem. Phys.*, **41,** 2902 (1964).
87. C. C. Price and R. Spector, *J. Am. Chem. Soc.*, **88,** 4171 (1966).
88. K. Imada, H. Tadokoro, A. Umehara, and S. Murahashi, *J. Chem. Phys.*, **42,** 2807 (1965).
89. M. Yokoyama, H. Ochi, A. M. Ueda, and H. Tadokoro, *J. Macromol. Sci., Phys.*, **7,** 465 (1973).
90. D. Grime and I. M. Ward, *Trans. Faraday Soc.*, **54,** 959 (1958).
91. W. W. Daniels and R. E. Kitson, *J. Polym. Sci.*, **33,** 161 (1958).
92. C. Y. Liang and S. Krimm, *J. Mol. Spectrosc.*, **3,** 554 (1959).
93. A. Miyake, *J. Polym. Sci.*, **38,** 479, 497 (1959); *J. Phys. Chem.*, **64,** 510 (1960).
94. A. H. Kimball, *Bibliography of Research on Heavy Hydrogen Compounds, National Nuclear Energy Series*, Manhattan Project Technical Section Division III, Vol. 4C, McGraw-Hill, New York (1949).
95. L. M. Brown, A. S. Friedman, and C. W. Beckett, *Bibliography of Research on Deuterium and Tritium Compounds 1945 to 1952, Natl. Bur. Stand. Circ.*, **562,** (1956).
96. A. Murray, III and D. L. Williams, *Organic Syntheses with Isotopes, Part II*, Interscience, New York, 1958.
97. IUPAC, Commission on Molecular Structure and Spectroscopy, *Tables of Wavenumbers for the Calibration of Infra-Red Spectrometers*, 2nd ed., Pergamon, Oxford, 1977.
98. K. N. Rao, C. J. Humpherys, and D. H. Rank, *Wavelength Standards in the Infrared*, Academic, New York, 1966.
99. M. Kobayashi, K. Akita, and H. Tadokoro, *Makromol. Chem.*, **118,** 324 (1968).
100. D. Z. Robinson, *Anal. Chem.*, **23,** 273 (1951).
101. E. Charney, *J. Opt. Soc. Am.*, **45,** 980 (1955).
102. G. R. Bird and M. Parrish, Jr., *J. Opt. Soc. Am.*, **50,** 886 (1960).
103. M. Hass and M. O'Hara, *Appl. Opt.*, **4,** 1027 (1965).
104. J. P. Auton and M. C. Hutley, *Infrared Phys.*, **12,** 95 (1972).
105. R. D. B. Fraser, *J. Chem. Phys.*, **21,** 1511 (1953).
106. A. Elliott, E. J. Ambrose, and R. B. Temple, *Nature*, **163,** 567 (1949); E. J. Ambrose, A. Elliott, and R. B. Temple, *Proc. Roy. Soc. (Lond.)*, **A199,** 183 (1949).
107. H. Tadokoro, *Bull. Chem. Soc. Jpn.*, **32,** 1334 (1959).
108. L. Glatt, D. S. Waber, C. Seaman, and J. W. Ellis, *J. Chem. Phys.*, **18,** 413 (1950).
109. R. G. J. Miller and H. A. Willis, *Trans. Faraday Soc.*, **49,** 433 (1953).
110. H. Tadokoro, S. Seki, and I. Nitta, *J. Polym. Sci.*, **22,** 563 (1956).
111. H. Tadokoro, S. Seki, I. Nitta, and R. Yamadera, *J. Polym. Sci.*, **28,** 244 (1958).

112. H. Arimoto, *J. Polym. Sci. A*, **2,** 2283 (1964).
113. K. Imada, T. Miyakawa, Y. Chatani, H. Tadokoro, and S. Murahashi, *Makromol. Chem.*, **83,** 113 (1965).
114. A. Brown, *J. Appl. Phys.*, **20,** 552 (1949).
115. S. Krimm, *J. Chem. Phys.*, **22,** 567 (1954).
116. M. Tasumi and T. Shimanouchi, *Spectrochim. Acta*, **17,** 731 (1961).
117. D. R. Holmes, R. G. Miller, R. P. Palmer, and C. W. Bunn, *Nature*, **171,** 1104 (1953).
118. A. Keller and I. Sandeman, *J. Polym. Sci.*, **15,** 133 (1955).
119. S. L. Aggarwal, G. P. Tilley, and O. J. Sweeting, *J. Appl. Polym. Sci.*, **1,** 91 (1959).
120. R. G. Quynn and R. Steele, *Nature*, **173,** 1240 (1954).
121. R. E. Richards and H. W. Thompson, *Trans. Faraday Soc.*, **41,** 183 (1945).
122. W. H. Cobbs, Jr. and R. L. Burton, *J. Polym. Sci.*, **10,** 275 (1953).
123. M. Shen, W. N. Hansen, and P. C. Romo, *J. Chem. Phys.*, **51,** 425 (1969).
124. R. G. Brown, *J. Chem. Phys.*, **40,** 2900 (1964).
125. C. K. Wu, *J. Polym. Sci., Polym. Phys. Ed.*, **12,** 2493 (1974).
126. A. S. Balchan and H. G. Drickamer, *Rev. Sci. Instrum.*, **31,** 511 (1960).
127. R. P. Wool and W. O. Statton, *J. Polym. Sci., Polym. Phys. Ed.*, **12,** 1575 (1974).
128. R. P. Wool, *J. Polym. Sci., Polym. Phys. Ed.*, **13,** 1795 (1975).
129. V. J. Coates, A. Offner, and E. H. Siegler, Jr., *J. Opt. Soc. Am.*, **47,** 984 (1953).
130. G. Garoti and J. H. Dusenbury, *J. Polym. Sci.*, **22,** 399 (1956).
131. P. D. Vasko and J. L. Koenig, *Macromolecules*, **3,** 597 (1970).
132. T. C. Damen, S. P. S. Porto, and B. Tell, *Phys. Rev.*, **142,** 570 (1966).
133. V. B. Carter, *J. Mol. Spectrosc.*, **34,** 356 (1970).
134. R. T. Bailey, A. J. Hyde, and J. J. Kim, *Spectrochim. Acta*, **30A,** 91 (1974).
135. F. J. Boerio and R. A. Bailey, *J. Polym. Sci. B*, **12,** 433 (1974).
136. J. Purvis, D. I. Bower, and I. M. Ward, *Polymer*, **14,** 298 (1973).
137. J. L. Derouault, P. J. Hendra, M. E. A. Cudby, and H. A. Willis, *Chem. Comm.*, **1972,** 1187.
138. J. Purvis and D. I. Bower, *Polymer*, **15,** 645 (1974).
139. K. Tashiro, M. Kobayashi, and H. Tadokoro, *Polym. Bull.*, **1,** 61 (1978).
140. F. J. Boerio and J. L. Koenig, *J. Chem. Phys.*, **52,** 3425 (1970).
141. M. C. Tobin, *J. Chem. Phys.*, **23,** 891 (1955).
142. S. Krimm, C. Y. Liang, and G. B. B. M. Sutherland, *J. Chem. Phys.*, **25,** 549 (1956).
143. F. A. Cotton, *Chemical Applications of Group Theory*, 2nd ed., Wiley-Interscience, New York, 1971.
144. H. Margenau and G. M. Murphy, *The Mathematics of Physics and Chemistry*, Van Nostrand, New York, 1943.
145. G. M. Barrow, *Introduction to Molecular Spectroscopy*, McGraw-Hill, New York, 1962.
146. S. Bagavantam and T. Venkatarayudu, *Theory of Groups and Its Application to Physical Problems*, 2nd ed., Andhra University, Waltair, India, 1951.
147. J. E. Bertie and E. Whalley, *J. Chem. Phys.*, **41,** 575 (1967).
148. A. O. Frenzel and J. P. Butler, *J. Opt. Soc. Am.*, **54,** 1059 (1964).
149. S. Krimm and M. I. Bank, *J. Chem. Phys.*, **42,** 4059 (1965).

150. M. I. Bank and S. Krimm, *J. Appl. Phys.*, **39,** 4951 (1968).
151. R. T. Harley, W. Hayes, and J. F. Twisleton, *J. Phys. C: Solid State Phys.*, **6,** L167 (1973).
152. M. Kobayashi, T. Uesaka, and H. Tadokoro, *Chem. Phys. Lett.*, **37,** 577 (1976).
153. M. Tasumi and T. Shimanouchi, *J. Chem. Phys.*, **43,** 1245 (1965).
154. G. D. Dean and D. H. Martin, *Chem. Phys. Lett.*, **1,** 415 (1967).
155. M. Tasumi, T. Shimanouchi, and T. Miyazawa, *J. Mol. Spectrosc.*, **9,** 261 (1962); **11,** 422 (1963).
156. J. H. Schachtschneider and R. G. Snyder, *Spectrochim. Acta*, **19,** 117 (1963).
157. M. Tasumi and S. Krimm, *J. Chem. Phys.*, **46,** 755 (1967).
158. C. Y. Liang and S. Krimm, *J. Chem. Phys.*, **25,** 563 (1956).
159. G. Carazzolo and M. Mammi, *J. Polym. Sci. A*, **1,** 965 (1963).
160. V. Zamboni and G. Zerbi, *J. Polym. Sci. C*, **7,** 153 (1964).
161. R. G. Snyder and G. Zerbi, *Spectrochim. Acta*, **23A,** 391 (1967).
162. S. Mizushima and T. Shimanouchi, *Infrared Absorption and Raman Effect* (in Japanese), Kyôritsu Publishing Co., Tokyo, 1958.
163. J. C. Decius, *J. Chem. Phys.*, **16,** 1025 (1948); **17,** 1315 (1949).
164. S. M. Ferigle and A. G. Meister, *J. Chem. Phys.*, **19,** 982 (1951).
165. W. J. Taylor, *J. Chem. Phys.*, **18,** 1301 (1950).
166. T. Miyazawa, *J. Chem. Phys.*, **29,** 246 (1958).
167. D. M. Dennison, *Rev. Mod. Phys.*, **12,** 175 (1940).
168. H. C. Urey and C. A. Bradley, Jr., *Phys. Rev.*, **38,** 1969 (1931).
169. T. Shimanouchi and S. Mizushima, *J. Chem. Phys.*, **17,** 1102 (1949).
170. T. Shimanouchi, *J. Chem. Phys.*, **17,** 245, 734, 848 (1949).
171. J. R. Scherer and J. Overend, *J. Chem. Phys.*, **33,** 1681 (1960).
172. P. W. Higgs, *Proc. Roy. Soc. (Lond.)*, **A220,** 472 (1953).
173. H. Tadokoro, *J. Chem. Phys.*, **33,** 1558 (1960); **35,** 1050 (1961).
174. T. Miyazawa, *J. Chem. Phys.*, **35,** 693 (1961).
175. T. P. Lin and J. L. Koenig, *J. Mol. Spectrosc.*, **9,** 228 (1962).
176. R. G. Snyder and J. H. Schachtschneider, *Spectrochim. Acta*, **19,** 85 (1963).
177. R. G. Snyder, *J. Chem. Phys.*, **47,** 1316 (1967).
178. H. Matsuda, K. Okada, T. Takase, and T. Yamamoto, *J. Chem. Phys.*, **41,** 1527 (1964).
179. K. Okada, *J. Chem. Phys.*, **43,** 2497 (1965).
180. M. Kobayashi and H. Tadokoro, *Macromolecules*, **8,** 897 (1975).
181. M. Kobayashi and H. Tadokoro, *J. Chem. Phys.*, **66,** 1258 (1977).
182. L. Piseri and G. Zerbi, *J. Chem. Phys.*, **48,** 3561 (1968).
183. H. Sugeta and T. Miyazawa, *Rep. Prog. Polym. Phys. Jpn.*, **9,** 177 (1966).
184. H. Matsuura and T. Miyazawa, *Rep. Prog. Polym. Phys. Jpn.*, **10,** 187 (1967).
185. G. Zerbi and L. Piseri, *J. Chem. Phys.*, **49,** 3840 (1968).
186. M. J. Hannon, F. J. Boerio, and J. L. Koenig, *J. Chem. Phys.*, **50,** 2829 (1969).
187. G. Zerbi and M. Sacchi, *Macromolecules*, **6,** 692 (1973).
188. M. Kobayashi, K. Tashiro, and H. Tadokoro, *Macromolecules*, **8,** 158 (1975).
189. L. Brillouin, *Wave Propagation in Periodic Structures*, Dover, New York, 1964.

190. J. M. Ziman, *Principles of the Theory of Solids*, Cambridge University Press, London, 1964.
191. T. Kitagawa and T. Miyazawa, *Adv. Polym. Sci.*, **9,** 335 (1972).
192. J. C. Slater, Symmetry and Energy Bands in Crystals, *Quantum Theory of Molecules and Solids*, Vol. 2, McGraw-Hill, New York, 1965.
193. J. G. Kirkwood, *J. Chem. Phys.*, **7,** 506 (1939).
194. S. E. Whitcomb, H. H. Nielsen, and L. H. Thomas, *J. Chem. Phys.*, **8,** 143 (1940).
195. T. Kitagawa and T. Miyazawa, *Bull. Chem. Soc. Jpn.*, **42,** 3437 (1969); **43,** 372 (1970).
196. W. Myers, G. C. Summerfield, and J. S. King, *J. Chem. Phys.*, **44,** 184 (1966).
197. T. Miyazawa and T. Kitagawa, *J. Polym. Sci. B*, **2,** 395 (1964).
198. T. Kitagawa and T. Miyazawa, *J. Chem. Phys.*, **47,** 337 (1967); *J. Polym. Sci. B*, **6,** 83 (1968).
199. D. Makino, M. Kobayashi, and H. Tadokoro, *Spectrochim. Acta*, **31A,** 1481 (1975).
200. S. Mizushima and T. Shimanouchi, *J. Am. Chem. Soc.*, **71,** 1320 (1949).
201. R. F. Schaufele and T. Shimanouchi, *J. Chem. Phys.*, **47,** 3605 (1967).
202. W. L. Peticolas, G. W. Hibler, J. L. Lippert, A. Peterlin, and H. G. Olf, *Appl. Phys. Lett.*, **18,** 87 (1971).
203. H. G. Olf, A. Peterlin, and W. L. Peticolas, *J. Polym. Sci., Polym. Phys. Ed.*, **12,** 359 (1974).
204. G. R. Strobl and R. Eckel, *J. Polym. Sci., Polym. Phys. Ed.*, **14,** 913 (1976).
205. S. L. Hsu, S. Krimm, S. Krause, and G. S. Y. Yeh, *J. Polym. Sci., Polym. Lett. Ed.*, **14,** 195 (1976).
206. A. J. Hartley, Y. K. Leung, C. Booth, and I. W. Shepherd, *Polymer*, **17,** 354 (1976).
207. M. Tasumi and S. Krimm, *J. Polym. Sci. A-2*, **6,** 995 (1968).
208. M. I. Bank and S. Krimm, *J. Polym. Sci. A-2*, **7,** 1785 (1969).
209. F. C. Stehling, E. Ergos, and L. Mandelkern, *Macromolecules*, **4,** 672 (1971).
210. S. Krimm and J. H. C. Ching, *Macromolecules*, **5,** 209 (1972).
211. M. Tasumi and G. Zerbi, *J. Chem. Phys.*, **48,** 3913 (1968).
212. H. Tadokoro, Y. Takahashi, Y. Chatani, and H. Kakida, *Makromol. Chem.*, **109,** 96 (1967).
213. H. Kakida, D. Makino, Y. Chatani, M. Kobayashi, and H. Tadokoro, *Macromolecules*, **3,** 569 (1970).
214. D. Makino, M. Kobayashi, and H. Tadokoro, *J. Chem. Phys.*, **51,** 3901 (1969).
215. G. Natta and P. Corradini, *Makromol. Chem.*, **16,** 17 (1955).

Chapter 6

Energy Calculations

Energy calculations made for single molecules and crystals serve three purposes: (1) they clarify the factors governing the crystal and molecular structure already determined experimentally, (2) they predict the most stable molecular conformation and its crystal packing starting from the chemical structure (this information may be useful in setting up probable models for structure analyses), and (3) they provide a collection of reliable potential functions and parameters for intra- and intermolecular interactions based on well defined crystal structures. The description in Sections 6.1–6.5 concerns only static potential energy. The contribution of entropy is assumed to be negligibly small in crystals of this description. However, strictly speaking, the stability of states should be discussed in terms of free energy by taking entropy into account. An example of such a treatment is given in Section 6.6. References 1–6 are helpful for energy calculations.

6.1 POTENTIAL ENERGY FUNCTIONS

For potential energy calculations, the internal rotation barrier, the non-bonded atomic interaction, the electrostatic interaction, the energy of hydrogen bonding, and so on are taken into account. Potential energy functions and parameters are necessary and are usually assumed semiempirically.

6.1.1 Potential Barrier Hindering Internal Rotation

The internal rotation potential originates from the exchange repulsion between the electrons belonging to the atoms neighboring the bond. For example, the sinusoidal function

$$V = \frac{V_0(1 \pm \cos n\tau)}{2} \tag{6.1}$$

has been used. Here $n = 3$ for all C—C bonds and the C—O bonds of ethers, and $n = 2$ for the C(O)—O bonds of esters and C(O)—N bonds of amides. The sign is taken as positive or negative, depending on whether V has a maximum or a minimum, respectively, at $\tau = 0°$. The direct determination of the barrier height V_0 is difficult for polymers. V_0(C—C) = 2 kcal/mol has been assumed by Scott and Scheraga[7]; this value was estimated by subtracting the nonbonded atomic interaction energy from the observed internal rotation barriers of ethane, propane, and n-butane.

The relation between V_0 and the internal rotation force constant F_τ (mdyn/Å) obtained from far-infrared spectra is derived below. In the case of the C—C single bond,

$$V = \tfrac{1}{2} V_0(1 + \cos 3\tau) \tag{6.2}$$

If the same kind of bonds are linked to both sides of the central C—C bond, the following relation is obtained.

$$2V = F_\tau(R_0 \sin \phi_0)^2(\Delta\tau)^2 \tag{6.3}$$

from which

$$\left(\frac{\partial^2 V}{\partial \tau^2}\right)_{\tau=0} = F_\tau {R_0}^2 \sin^2 \phi_0 \tag{6.4}$$

where R_0 and ϕ_0 are the equilibrium values of the bond length of the neighboring bonds and bond angle, respectively. From Eq. 6.2,

$$\left(\frac{\partial^2 V}{\partial \tau^2}\right)_{\tau=0} = \tfrac{9}{2} V_0 \tag{6.5}$$

Substitution into Eq. 6.4 yields the relation between V_0 and F_τ

$$V_0 = \tfrac{2}{9} {R_0}^2 \sin^2 \phi_0 F_\tau \tag{6.6}$$

6.1.2 Nonbonded Atomic Interactions

Various functions have been proposed for expressing the potential of nonbonded atomic interactions. Among these, the Lennard-Jones "6–12" type function

$$V_{ij} = \frac{d_{ij}}{{r_{ij}}^{12}} - \frac{e_{ij}}{{r_{ij}}^6} \tag{6.7}$$

and the Buckingham "6–exp" type function

$$V_{ij} = A_{ij} \exp(-B_{ij} r_{ij}) - \frac{C_{ij}}{{r_{ij}}^6} \tag{6.8}$$

have been used widely. The former is especially convenient for calculation,

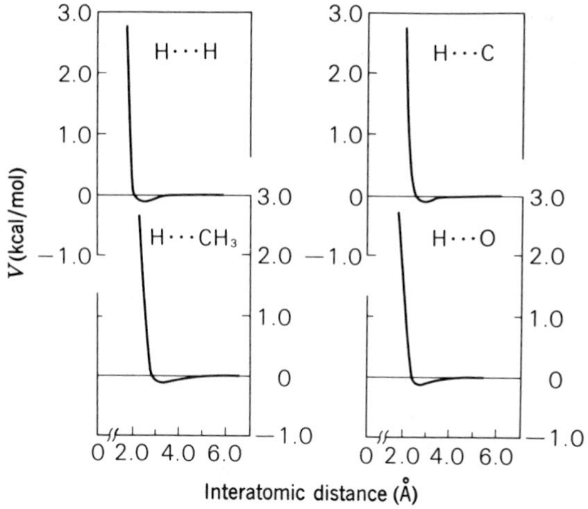

Fig. 6.1 Potential energy curves for nonbonded atomic interactions calculated using parameters from Table 6.1.

since the number of parameters is small. In the above two equations, the first term represents repulsion and the second term represents van der Waals attraction. Here r_{ij} is the interatomic distance. The parameters d_{ij} and e_{ij} are either obtained experimentally from the second virial coefficient of gases or calculated according to the Slater-Kirkwood equations.[7]

$$e_{ij} = \frac{3eh\alpha_i\alpha_j}{[4\pi\sqrt{m}(\sqrt{\alpha_i/N_i} + \sqrt{\alpha_j/N_j})]}$$

$$d_{ij} = \tfrac{1}{2}e_{ij}r_{\min}{}^{6} \tag{6.9}$$

where e is the electronic charge, h is Planck's constant, m is the electronic mass, α_i and N_i are the atomic polarizability and effective number of outer shell electrons of the ith atom, respectively, and $r_{\min}$ is the sum of the van der Waals radii of both atoms. Figure 6.1 shows examples of the potential curves for nonbonded atomic interactions. Table 6.1 gives the parameters used for Eq. 6.9. Most of these values are from Scott and Scheraga,[8] but here the methyl group is considered as a unit and additivity is assumed for α_i and N_i.

6.1.3 Electrostatic Interactions

For electrostatic interaction, the point dipole model and point charge model are generally used. Although true molecules have widely distributed charges in space, in the point dipole model a point dipole is assumed to be at the midpoint of the polar bond. The calculation is made by using bond

Table 6.1 Parameters for Calculating Potentials for Nonbonded Interactions[9] (mainly after Scott and Scheraga[8])

Atom or group	van der Waals radius (Å)	$\alpha_i \times 10^{24}$ (cm^3)	N_i
H	1.20	0.42	0.9
C	1.70	0.93	5.2
C (carbonyl)	1.70	1.30	4.7
CH_3	2.00	2.19	7.9
CH_3 (ester)	2.00	2.19	7.8
O (ether)	1.52	0.64	7.2
O (carbonyl)	1.52	0.84	7.4

moments[10] according to the equation

$$V_{ij} = \frac{(\boldsymbol{\mu}_i \cdot \boldsymbol{\mu}_j) - 3(\boldsymbol{\mu}_i \cdot \mathbf{r}_{ij})(\boldsymbol{\mu}_j \cdot \mathbf{r}_{ij})/r_{ij}^{\,2}}{\varepsilon r_{ij}^{\,3}} \tag{6.10}$$

where $\boldsymbol{\mu}_i$ is the ith point dipole vector, $\mathbf{r}_{ij}$ is the vector jointing the ith and jth point dipoles, and ε is the dielectric constant of the medium. Several methods have been proposed for the estimation of ε for the present purpose.[4,11,12]

In the point charge model, formal charges are assumed in the atom or group so as to fit the observed dipole moments and a Coulomb potential is assumed

$$V_{ij} = \frac{q_i q_j}{\varepsilon r_{ij}} \tag{6.11}$$

where q_i is the formal charge of the ith atom. As seen in Eqs. 6.10 and 6.11, these interactions vary inversely with the third power and first power of the atomic distance, respectively. Therefore the electrostatic interactions are longer range effects than the nonbonded atomic interactions (Eqs. 6.7 and 6.8). Of the two models, the point charge model is a more accurate approximation than the point dipole model, but it requires a longer distance for convergence.

6.1.4 Other Interactions

The energy of a hydrogen bond is several kcal/mol. Rather complicated potential functions have been proposed for expressing the energy.[4,8,12–14]

In some cases the energy of bending of the bond angle has been considered.[14-17]

For quantum mechanical calculations, see Refs. 4 and 6 and references therein.

6.2 CALCULATIONS

6.2.1 Coordinate Transformations

The potential functions stated above, except for the internal rotation potentials, are the functions of interatomic distances (or relative positions between dipoles). Therefore, for calculation of the potential energy, it is convenient to express the atomic positions in terms of the Cartesian coordinate system. Although the conformation is usually given using the internal coordinates, that is, the bond lengths, bond angles, and internal rotation angles, it can also be expressed in terms of the Cartesian coordinate system by the following transformation.

Sets of right-hand Cartesian coordinate systems as defined in Appendix B and also in Section 4.6.2.B are now considered. The coordinates of the ith atom represented by the coordinate system fixed at the jth atom $\mathbf{x}_j(i)$ can be rewritten in terms of the $(j-1)$th coordinates, $\mathbf{x}_{j-1}(i)$, according to Eyring's equation (Appendix **B**):

$$\mathbf{x}_{j-1}(i) = \mathbf{A}_{j-1,j}\mathbf{x}_j(i) + \mathbf{B}_{j-1,j} \tag{6.12}$$

where $\mathbf{A}_{j-1,j}$ and $\mathbf{B}_{j-1,j}$ are the same as in Eqs. 4.59 and 4.60.

The coordinates of the ith atom represented by the jth coordinate system can be expressed in terms of the zeroth coordinate system.

$$\begin{aligned}\mathbf{x}_0(i) = {} & \mathbf{A}_{01}\mathbf{A}_{12}\cdots\mathbf{A}_{j-1,j}\mathbf{x}_j(i) + \mathbf{A}_{01}\mathbf{A}_{12}\cdots\mathbf{A}_{j-2,j-1}\mathbf{B}_{j-1,j} \\ & + \mathbf{A}_{01}\mathbf{A}_{12}\cdots\mathbf{A}_{j-3,j-2}\mathbf{B}_{j-2,j-1} + \cdots + \mathbf{A}_{01}\mathbf{B}_{12} + \mathbf{B}_{01}\end{aligned} \tag{6.13}$$

6.2.2 Summation of Potential Energy Terms

The summation of the potential energy terms for a molecular chain or a crystal is conveniently made according to the following procedure.[18] For a crystal consisting of M polymer molecules, M being infinitely large, the total interaction energy V_{cr} of the whole crystal is given by

$$V_{cr} = MV_{\text{intra}} + \sum_{m=1}^{M-1}\sum_{m'>m}^{M} V_{\text{inter}}(m; m') \qquad (6.14)^*$$

* $m' > m$ means to count the same pair only once.

where V_{intra} is the intramolecular interaction energy of a polymer molecule and $V_{\text{inter}}(m;\, m')$ is the intermolecular interaction energy between the mth and m'th polymer molecules. If a polymer molecule consisting of $2N+1$ translational units $(-N, -N+1, \cdots, 0, \cdots, N)$ and satisfying Born's cyclic boundary condition (Section 3.5.1) is assumed, then

$$V_{\text{inter}}(m;m') = \sum_{u=-N}^{N} \sum_{u'=-N}^{N} V^0_{\text{inter}}(m,u;m',u') \tag{6.15*}$$

where V^0_{inter} is the interaction energy between the uth unit in the mth molecule and the u'th unit in the m'th molecule.

V_{intra} is expressed by

$$V_{\text{intra}} = (2N+1)V^0_{\text{rot}} + \tfrac{1}{2} \sum_{u=-N}^{N} \sum_{u'=-N}^{N} V^0_{\text{intra}}(u;u') \tag{6.16}$$

where V^0_{rot} is the internal rotation potential in a translational unit and $V^0_{\text{intra}}(u;\, u')$ is the interaction energy between the uth and u'th units in a molecule. The van der Waals and electrostatic interactions contribute to the second terms of Eqs. 6.14 and 6.16. $V^0_{\text{intra}}(u;\, u+k)$ is determined only by k, not by u because of the translational symmetry of the crystal. Therefore the intramolecular interaction energy per translational unit is expressed by dividing Eq. 6.16 by $(2N+1)$.

$$V^0_{\text{intra}} = V^0_{\text{rot}} + \tfrac{1}{2} \sum_{k=-N}^{N} V^0_{\text{intra}}(0;k) \tag{6.17}$$

with

$$V^0_{\text{intra}}(0;k) = \sum_{a=1}^{n} \sum_{a'=1}^{n} V^0_{\text{intra}}(0,a;k,a') \tag{6.18}$$

where $V^0_{\text{intra}}(0, a;\, k, a')$ is the van der Waals interaction energy (or the electrostatic interaction energy) between the ath atom (or the point dipole) in the 0th unit and the a'th atom (or the point dipole) in the kth unit. n is the number of the atoms (or the point dipoles) involved in a translational unit. In $V^0_{\text{intra}}(0, a;\, 0, a')$ the terms with $a = a'$ vanish.

The intermolecular interaction energy per translational unit and the potential energy of the crystal per translational unit are given by

$$\begin{aligned} V^0_{\text{inter}} &= \frac{1}{M(2N+1)} \sum_{m=1}^{M-1} \sum_{m'>m}^{M} V_{\text{inter}}(m;m') \\ &= \frac{1}{M(2N+1)} \sum_{m=1}^{M-1} \sum_{m'>m}^{M} \sum_{u=-N}^{N} \sum_{u'=-N}^{N} V^0_{\text{inter}}(m,u;m',u') \end{aligned} \tag{6.19}$$

$$V^0_{\text{cr}} = V^0_{\text{intra}} + V^0_{\text{inter}} \tag{6.20}$$

*For a uniform helix of (u/t), an asymmetric unit is considered instead of the translational unit mentioned above. In this case, $N = N'u$ (N' is an integer) is used.

If a translational unit consists of two or more monomeric units, V_{cr}^0 is divided by the number of monomeric units involved.

Actually, the calculation is made by utilizing the symmetry of the crystal lattice. The energy terms that do not depend on the conformation are omitted in the intramolecular energy calculation.

6.3 INTRAMOLECULAR INTERACTION ENERGY

6.3.1 Cases without Fixing Fiber Identity Period

The intramolecular interaction energy for the chain that takes various internal rotation angles can be calculated without fixing the fiber identity period only under the assumption that (1) the chain forms a helix, (2) the chain has glide-plane symmetry, or (3) the chain has only translational symmetry. This type of calculation is useful in examining the factors governing the stability of special conformations.

A. *Calculations Based on the Assumption of Helix Formation.* The condition that the polymer chain forms a helix is, in other words, that the set of internal rotation angles in a chemical structural unit* repeats along the chain.† Accordingly, it is only necessary to calculate various sets of internal rotation angles involved in a chemical structural unit. For the number of degrees of freedom of internal rotation see Section 4.6.2.C.

Five typical isotactic polymers, namely, *it*-PP, *it*-poly(4-methyl-1-pentene), *it*-poly(3-methyl-1-butene), *it*-polyacetaldehyde, and *it*-poly(methyl methacrylate) (Table 6.2) are considered here.[9] The molecular conformations of the first four polymers have been determined by X-ray analyses to be (3/1),[20] (7/2),[21-23] (4/1),[24] and (4/1) helices,[25] respectively (Fig. 6.2). For *it*-poly(methyl methacrylate),[27] a (5/1) helix was considered reasonable at the time of the energy calculation (in 1970, before the finding of the double helix structure[28]).

The correlation between helical structure and the type of side chains was qualitatively discussed by Bunn and Holmes[29] for the series of isotactic poly-α-olefins having the (3/1), (7/2), and (4/1) helical conformations. The

* Strictly speaking, in the case of polyisobutylene[19] a crystallographic asymmetric unit consists of four chemical structural units.

† A coiled coil is a kind of helix, in which the helix axis of the smaller helix forms a larger helix. An asymmetric unit consists of units contained in some range of the larger helix. The internal rotation angle differs slightly from the corresponding rotation angle of the neighboring chemical structural unit, but it is the same as the corresponding rotation angle of the neighboring asymmetric unit.

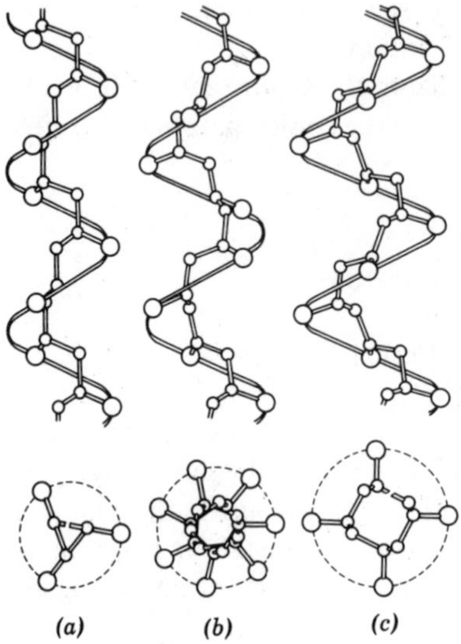

Fig. 6.2 Helical conformations of *it*-polymers. (*a*) (3/1); (*b*) (7/2); and (*c*) (4/1) helices (From Natta and Corradini.[26])

conformational stability of *it*-PP was studied quantitatively by Natta et al.[30] and De Santis et al.[31,32]

The numbering of the atoms and the internal rotation angles is summarized in Table 6.2. For the calculation, the relation between the location of the τ_1 and τ_2 bonds and the configuration of the groups around the carbon atoms should be defined. This relation is indicated in Fig. 6.3 by Fischer projections.

it-PP and *it*-polyacetaldehyde are the simplest cases, having only two internal rotation angles of the main chain, τ_1 and τ_2. Figure 6.4 shows the contour map of the intramolecular potential energy for *it*-PP. The potential minima are indicated by the symbol × and the symbol ● indicates the structure determined by X-ray. The two minima correspond to right- and left-hand helices. On the diagonal line, there is a high energy band,[31,33] where the helix pitch d is 0 Å and the polymer chain forms a closed loop, causing monomer units to overlap (if a long enough chain is used in the calculation).

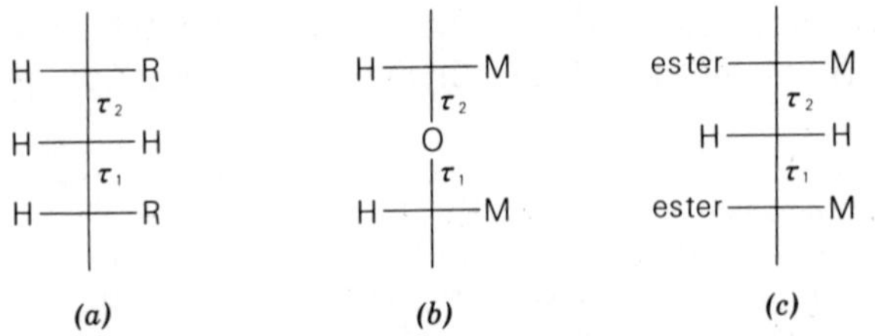

Fig. 6.3 Fischer projections indicating the relation between the rotation angles τ_1 and τ_2 and the configurations of the groups around the carbon atoms.[9]

Table 6.2 Numbering of the Atoms and the Internal Rotation Angles[9]

Polymer	Numbering	
it-Polypropylene	$\text{—}C_{II}MH(-1)\overset{\tau_1}{\text{—}}C_IH_2\overset{\tau_2}{\text{—}}C_{II}MH\text{—}C_IH_2(+1)\text{—}$	$\tau_1[C_I(-1)C_{II}(-1)C_IC_{II}]$, $\tau_2[C_{II}(-1)C_IC_{II}C_I(+1)]$
it-Poly(4-methyl-1-pentene)	$\text{—}C_{II}H_I(-1)\overset{\tau_1}{\text{—}}C_IH_2\overset{\tau_2}{\text{—}}C_{II}H_I\text{—}C_IH_2(+1)\text{—}$ $\tau_3\vert$ $C_{III}H_2$ $\tau_4\vert$ $C_{IV}H_{II}M_2$	$\tau_1[C_I(-1)C_{II}(-1)C_IC_{II}]$, $\tau_2[C_{II}(-1)C_IC_{II}C_I(+1)]$, $\tau_3[H_IC_{II}C_{III}C_{IV}]$, $\tau_4[C_{II}C_{III}C_{IV}H_{II}]$
it-Poly(3-methyl-1-butene)	$\text{—}C_{II}H_I(-1)\overset{\tau_1}{\text{—}}C_IH_2\overset{\tau_2}{\text{—}}C_{II}H_I\text{—}C_IH_2(+1)\text{—}$ $\tau_3\vert$ $C_{III}H_{II}M_2$	$\tau_1[C_I(-1)C_{II}(-1)C_IC_{II}]$, $\tau_2[C_{II}(-1)C_IC_{II}C_I(+1)]$, $\tau_3[H_IC_{II}C_{III}H_{II}]$
it-Polyacetaldehyde	$\text{—}CMH(-1)\overset{\tau_1}{\text{—}}O\overset{\tau_2}{\text{—}}CMH\text{—}O(+1)\text{—}$	$\tau_1[O(-1)C(-1)OC]$, $\tau_2[C(-1)OCO(+1)]$
it-Poly(methyl methacrylate)	$\text{—}C_{II}M_I(-1)\overset{\tau_1}{\text{—}}C_IH_2\overset{\tau_2}{\text{—}}C_{II}M_I\text{—}C_IH_2(+1)\text{—}$ $\tau_3\vert$ $C_{III}=O_{II}$ $\tau_4\vert$ O_IM_{II}	$\tau_1[C_I(-1)C_{II}(-1)C_IC_{II}]$, $\tau_2[C_{II}(-1)C_IC_{II}C_I(+1)]$, $\tau_3[M_IC_{II}C_{III}O_I]$, $\tau_4[C_{II}C_{III}O_IM_{II}]$

Table 6.3 Stable Conformations[9]

	Energy calculation				X-Ray analysis			
Polymer	τ_1(°)	τ_2(°)	N	d(Å)	τ_1(°)	τ_2(°)	N	d(Å)
it-Polypropylene	179	− 56	2.91	2.16	180	− 60	3.0	2.17
it-Poly(4-methyl-1-pentene)	163	− 71	3.52	2.00	162	− 71	3.5	1.98
it-Poly(3-methyl-1-butene)	132	− 83	4.17	1.31	149	− 81	4.0	1.17
it-polyacetaldehyde	131	− 80	3.94	1.17	136	− 83	4.0	1.20

The point with $\tau_1 = 60°$ and $\tau_2 = 300° = -60°$ is the sequence of $G\bar{G}$, giving the so-called "pentane effect." With respect to this problem, the present calculation took into account up to 40 monomeric units for poly(methyl methacrylate) and 20 units for the other polymers.

The other three polymers have additional rotation angles of the side chain, τ_3 and/or τ_4. For *it*-poly(3-methyl-1-butene) the minimum was found in the three-dimensional plot. For *it*-poly(4-methyl-1-pentene) and *it*-poly(methyl methacrylate) the stable conformation of the side chain was first calculated with the main chain fixed in a conformation corresponding to the (7/2) or (5/1) helix with the observed fiber period. The potential energy was calculated against the main-chain rotation angles τ_1 and τ_2 by fixing τ_3 and τ_4 of the side chain at these calculated values. For further confirmation of the result of the simplified calculation, the three-dimensional τ_1, τ_2, and τ_3 map was also examined around the values obtained above by fixing the angle τ_4, and the results showed no essential changes.

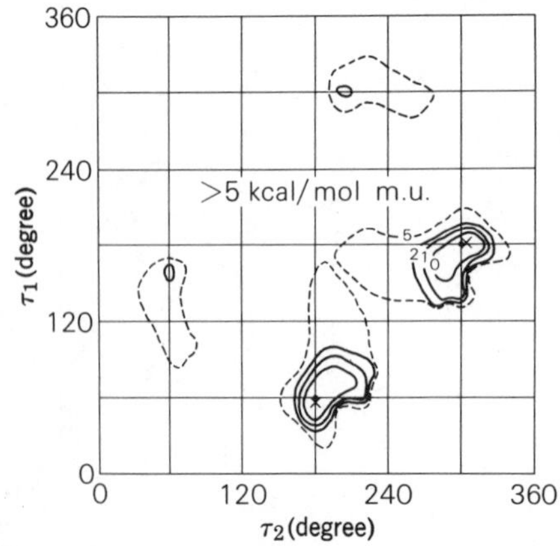

Fig. 6.4 Potential energy map of *it*-PP. The energy values are given in units of kilocalories per mole of monomer unit.[9]

The results for the first four polymers are given in Table 6.3. The internal rotation angles of the main chain τ_1 and τ_2, the number of monomeric units per turn N, and the helix pitch per monomeric unit d for the calculated stable conformations are compared with the values for the structures determined by X-ray analyses. As shown in this table, very good agreement was obtained between the calculated stable conformations and the X-ray results for these four polymers even though only intramolecular interactions were considered. This fact suggests that the helical structures of these four polymers are determined primarily by intramolecular interactions and especially by steric hindrance of the side chains.

The contour map for *it*-poly(methyl methacrylate) is shown in Fig. 6.5. The lowest minimum is found at position A, which corresponds to a (12/1) helix. The minimum corresponding to the (5/1) helix is indicated by B. This map is the result of a calculation in which the bond angle $\angle \mathrm{C{-}CH_2{-}C}$ is assumed to be 124° and the energy difference between the two minima is 3.0 kcal per mole of monomer unit. The present result led to the finding of the double helix structure of this polymer (Section 4.6.6).

Convenient charts for N and d plotted against τ_1 and τ_2 have been reported by De Santis et al.[31]

B. *Calculations for Glide-Plane Symmetry.* As is described in Section 4.6.2. B, the condition required for glide-type conformation is that the signs of the corresponding internal rotation angles of the neighboring asymmetric units be reversed and that the trace of the transformation matrix defined in Eq. 4.62 be 3.[34] The planar zigzag conformation is a special case having both the helix and glide symmetries, that is, a helix with an infinitely large number of monomeric units per turn and with zero pitch, and a glide chain in which all the internal rotation angles are 180°. For the number of degrees of freedom of internal rotation in glide-plane symmetry, see Section 4.6.2.C.

PEO has two crystal modifications, helix[35,36] and planar zigzag,[37]

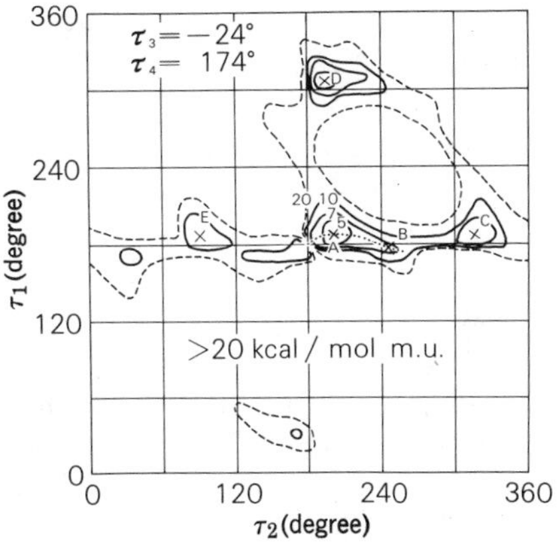

Fig. 6.5 Potential energy map of *it*-poly (methyl methacrylate).[9]

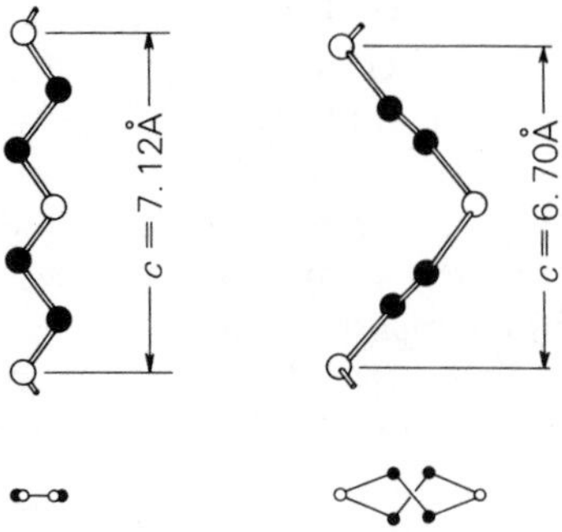

(a) (b)

Fig. 6.6 Molecular structures of (*a*) plannar zigzag PEO and (*b*) poly (ethylene sulfide).[37,38]

while poly(ethylene sulfide)[38] has a glide-plane conformation as shown in Figs. 4.45 and 6.6. The local conformation of poly(ethylene sulfide) is opposite that of the helix of PEO.

$$—CH_2—(—O—CH_2—CH_2—)_n—$$

τ_1	τ_2	τ_3		
T	T	G	$(T_2G)_6$	helix conformation

$$—CH_2—(—S—CH_2—CH_2—)_n—$$

τ_1	τ_2	τ_3		
G	G	T	$\bar{G}_2T$	glide conformation

In PEO, the C—C bond is gauche and the C—O bonds are trans, while in poly(ethylene sulfide), the C—C bond is trans and the C—S bonds are gauche. Figure 6.7 shows results of calculations for these two polymers under glide symmetry, where $\tau_1 = \tau_2$ is assumed. PEO is stable with the T_6 planar zigzag, while poly(ethylene sulfide) is stable with the $G_2T\bar{G}_2T$ glide, corresponding to the structure determined by X-ray analysis. The different conformations may be attributed mainly to the difference in the bond lengths of C—O (1.43Å) and C—S (1.815Å) and the van der Waals radii of the oxygen (1.40 Å) and sulfur (1.85 Å) atoms.

The results of calculations for polyoxacyclobutane with helical and glide symmetries are also in agreement with the three molecular conformations mentioned in Section 5.7.2.[34]

6.3.2 Cases with Fixed Fiber Identity Period

Calculation of the intramolecular interaction energy with a fixed fiber identity period is useful for setting up probable molecular models. This calculation

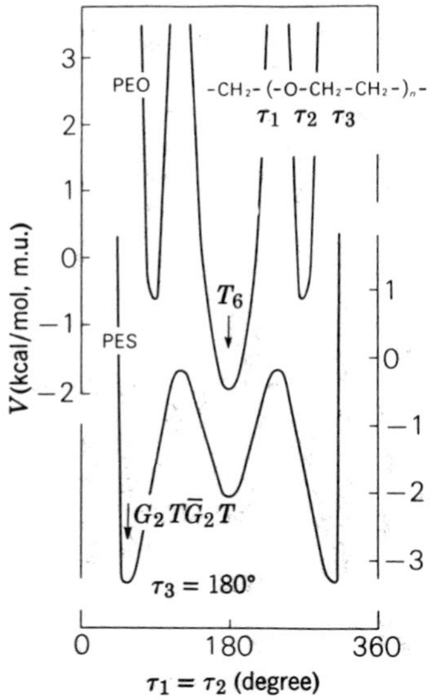

Fig. 6.7 Potential curves for PEO and poly (ethylene sulfide) under glide symmetry.[34]

can be done when the fiber period has been determined by X-ray analysis and some information has been obtained about the molecular symmetry (Section 4.6.3).

A. *Plots against Internal Rotation Angles.* Until recently the structure of polyisobutylene had not been determined. Molecular structures have been proposed by many authors, including Fuller et al.[39] [(8/1) helix], Liquori[40] (8/5), Bunn and Holmes[29] (8/5), and Wasai et al.[41] (8/3). Allegra et al.[16] reported another (8/3) helix from detailed conformational analysis considering the change of the bond angles. They proposed a molecular structure with the bond angles $\angle C{-}CM_2{-}C = 110^\circ$ and $\angle C{-}CH_2{-}C = 124^\circ$ and the internal rotation angles $\tau_1 = -157.7^\circ$ and $\tau_2 = -47.5^\circ$. They compared the diffraction intensities calculated for the molecular model with the observed X-ray data. The crystal structure has been determined in the author's laboratory[19] starting from the (8/3) helix model of Allegra et al. Figure 6.8 shows the potential map reported by Allegra et al. The minima correspond as follows: *A*, model of Fuller et al., *B*, model of Liquori; *C*, model of Bunn and Holmes; *D*, model of Wasai et al.; *E*, the lowest potential minimum; and *F*, model of Allegra et al. In the structure determined by X-ray analysis only a twofold screw symmetry exists and the crystallographic asymmetric unit consists of four monomeric units (see Section 7.5). Hence the molecular symmetry deviates appreciably from the (8/3) uniform helix. Thus it cannot be represented by Fig. 6.8.

The X-ray fiber diagram of the α-form of poly(ethylene oxybenzoate)

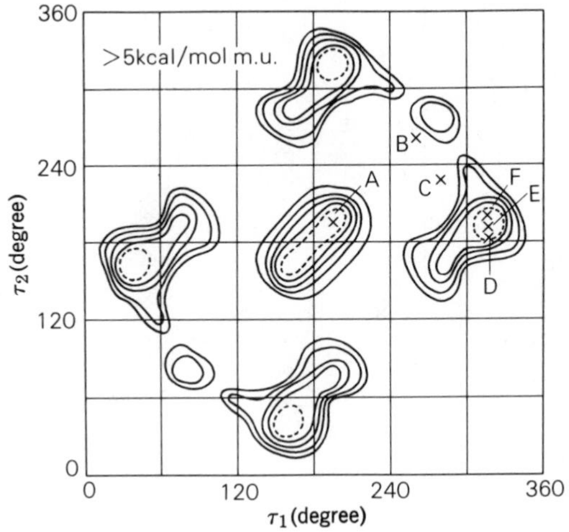

Fig. 6.8 Potential map of polyisobutylene. Reprinted with permission from G. Allegra, E. Benedetti, and C. Pedone, *Macromolecules*, **3**, 727 (1970). Copyright by the American Chemical Society.

suggests that two helical chains, each consisting of two monomeric units, pass through an orthorhombic unit cell of the space group $P2_12_12_1\text{-}D_2^{4}$.[42] The numbering of the internal rotation angles is shown below.

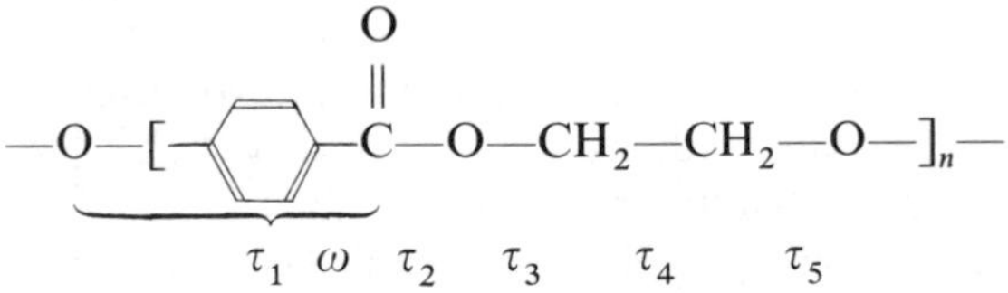

$$-O-[-\underset{\tau_1 \quad \omega}{\underbrace{\bigcirc}}-\overset{\overset{O}{\|}}{\underset{\tau_2}{C}}-O-\underset{\tau_3}{CH_2}-\underset{\tau_4}{CH_2}-\underset{\tau_5}{O-}]_n-$$

Here $O-\bigcirc-C$ is regarded as a virtual bond, and the dihedral angle between the planes of the benzene ring and the ester group is indicated by ω ($-90° < \omega \leqslant 90°$). The internal rotation angles (except for ω) are examined first, and it is assumed that the angle τ_2 of the ester group is essentially $180°$ and that the fiber period is 15.60 Å with the (2/1) helix symmetry. There are two independent variables among τ_1, τ_3, τ_4, and τ_5. The possible conformations are limited on the closed surface of a cube defined by the three-dimensional Cartesian coordinates τ_3, τ_4, and τ_5, each covering from 0 to 360° as shown in Fig. 6.9. If τ_3 and τ_4 are given, τ_5 can assume two values, upper and lower intersecting points, resulting in a pair of values for τ_1. The dihedral angle ω should also be considered. The calculation of the intramolecular potential energy for about 5000 molecular models yielded seven stable molecular models. Starting from these models and examining the azimuthal angle and the z parameter in the unit cell, only one model was found to give

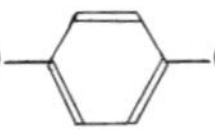

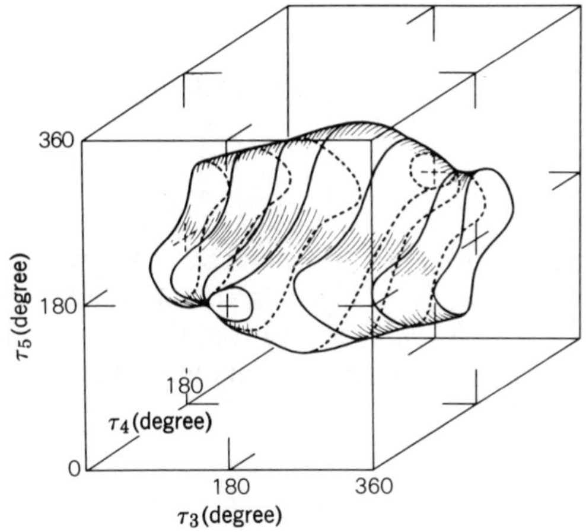

Fig. 6.9 Three-dimensional closed surface for possible conformations of the skeletal chain of poly (ethylene oxybenzoate) with the (2/1) helical symmetry and a fiber period of 15.60 Å.[42]

fairly good agreement between the observed and calculated diffraction intensities ($R = 38\%$). For a further discussion of the process of analysis, see Section 6.5.3.

B. *Plots against Bond Angles.* In the case of poly(vinylidene chloride), if glide-plane symmetry is assumed (the main chain bond angles $\phi_1 = \angle C—CH_2—C$ and $\phi_2 = \angle C—CCl_2—C$ are considered to be variable) and if the fiber period is fixed at 4.68 Å, there are two degrees of freedom (Section 4.6.2.D). In Fig. 6.10 potential energy contours are plotted against ϕ_1 and ϕ_2.[17] The relationship among ϕ_1, ϕ_2, the fiber period I, and the C—C

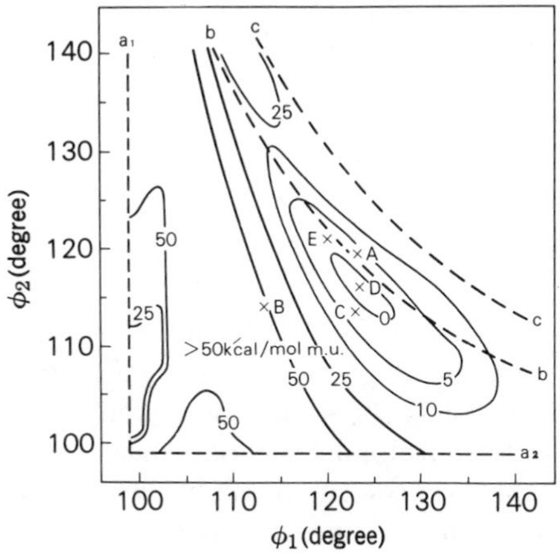

Fig. 6.10 Potential energy map for poly (vinylidene chloride) plotted against the main chain bond angles.[17]

bond distance r is given by

$$(1-\cos\phi_1)(1-\cos\phi_2)-\frac{I^2}{4r^2}\begin{cases}<0 & \text{glide type}\\ =0 & (TC)_2 \text{ type}\\ >0 & (2/1)\text{ helix}\end{cases} \tag{6.21}$$

In the region defined by straight lines a_1 and a_2 and curve b, the value of this equation is negative, and the molecular conformation is the glide type. Between curves b and c, the value is positive [(2/1) helix] and on line b the value is zero. The above-mentioned boundary lines are given by

$$\begin{aligned} a_1 &: \cos\phi_1 = 1-\frac{I^2}{8r^2}\\ a_2 &: \cos\phi_2 = 1-\frac{I^2}{8r^2}\\ c &: \cos\phi_1+\cos\phi_2+\frac{I^2}{8r^2}=0 \end{aligned} \tag{6.22}$$

Outside these boundary lines, no molecular model can be assumed.

The models proposed so far correspond to points on the figure. Fuller's distorted zigzag model[43] lies on line a_2 and Reinhardt's $(TC)_2$ type[44] corresponds to point E. Points A and B correspond to the (2/1) helix (De Santis et al.[31,45,46]) and glide-type model (Miyazawa and Ideguchi[47]), respectively. The lowest potential minimum in this figure is D (glide type). The structure determined by X-ray analysis[17] corresponds to C (see Section 7.5).

Further references concerning intramolecular interaction include: PE and polytetrafluoroethylene, Ref. 48; poly(vinylidene fluoride), Ref. 49; and energy calculations for the folding part of PE, Refs. 50–52.

6.4 INTERMOLECULAR INTERACTION ENERGY

6.4.1 Calculations for Fixed Crystal Structures

Poly(vinylidene fluoride) has three crystal forms (Figs. 6.11). In forms I and III, the molecular conformation is essentially planar zigzag (TT type), while in form II, it is a $TGT\bar{G}$ type.[18,53] Table 6.4 shows the result of potential energy calculations for forms I and II. The intramolecular potential for the TT form is much higher than that for the $TGT\bar{G}$ form, possibly because the F$\cdots$F atomic distance in the TT form, 2.56 Å, is appreciably shorter than

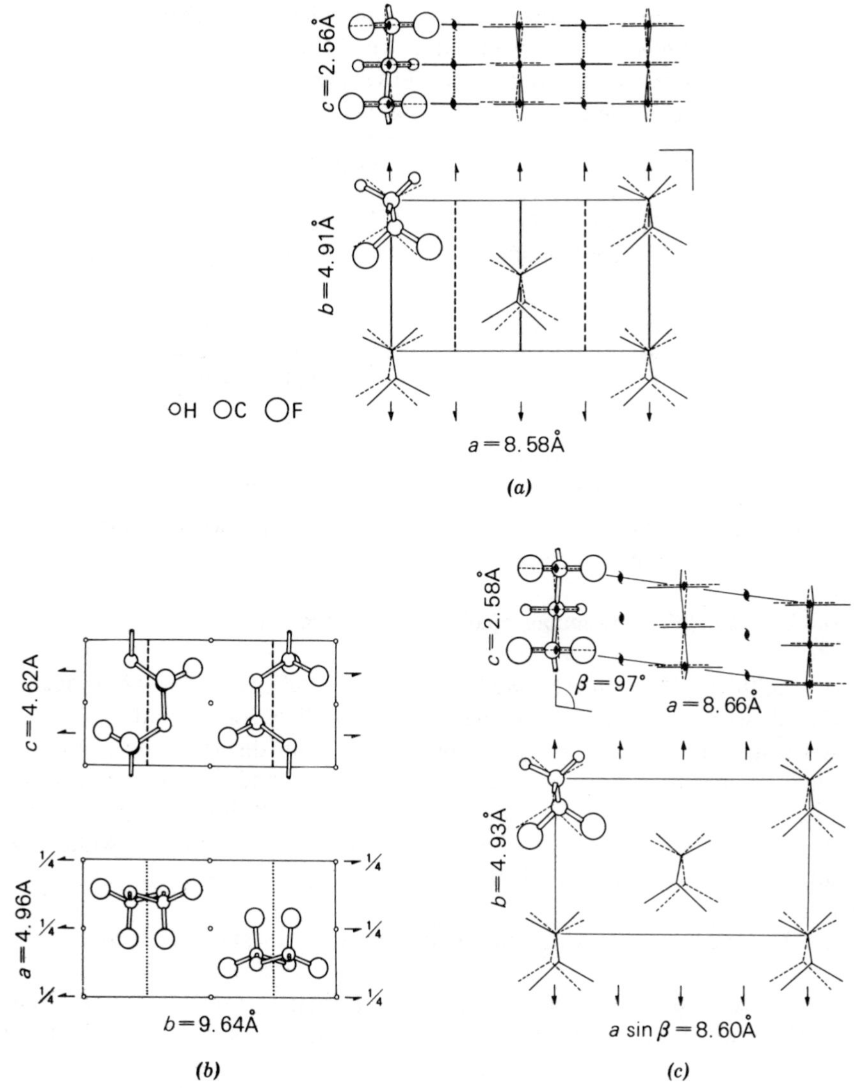

Fig. 6.11 Crystal structures of three modifications of poly (vinylidene fluoride).[18,53]

twice the van der Waals radius of the fluorine atom (2.70 Å). All the dipole moments of the CF_2 groups align in parallel. Taking the intermolecular interaction into account, the total potential energies of the two forms are nearly equal, possibly because of molecular packing. Intermolecular interaction may be more favorable in form I, as the density is 1.97 g/cm^3 for form I and 1.93 g/cm^3 for II. The calculated result could explain the higher potential

Table 6.4 Intra- and Intermolecular Interaction Potential Energy of Forms I and II of Poly(vinylidene Fluoride) (kcalories per mole of monomeric unit)[18]

Form	I (TT)	II ($TGT\bar{G}$)
Intramolecular interaction	− 0.48	− 1.46
(van der Waals)[a]	(− 1.19)	(− 1.57)
(Electrostatic)	(+ 0.71)	(+ 0.11)
Intermolecular interaction	− 5.25	− 4.57
(van der Waals)	(− 5.06)	(− 4.44)
(Electrostatic)	(− 0.19)	(− 0.13)
Total	− 5.73	− 6.03

[a]The values given in parentheses are the components.

of intramolecular interaction of the TT form if some special conditions are satisfied, for instance, high pressure.[18]

6.4.2 Calculation Assuming Space Group Symmetry

Corradini and Avitabile[54] calculated the potential energies of *it*-polyacetaldehyde with the (4/1) helix conformation ($c = 4.79$ Å) for six kinds of tetragonal space groups. The nearest neighboring chain distance, the azimuthal angle of one chain with respect to the other chain, and the relative displacement between two chains along the chain axis were chosen as parameters. The most stable packing was found to be the space group $I4_1/a$, which had already been determined by X-ray analysis.

6.4.3 Calculation Varying Molecular Position and Azimuthal Angle

To clarify the factors governing chain packing in the crystal, the intermolecular interaction energy between two planar PE chains was first calculated

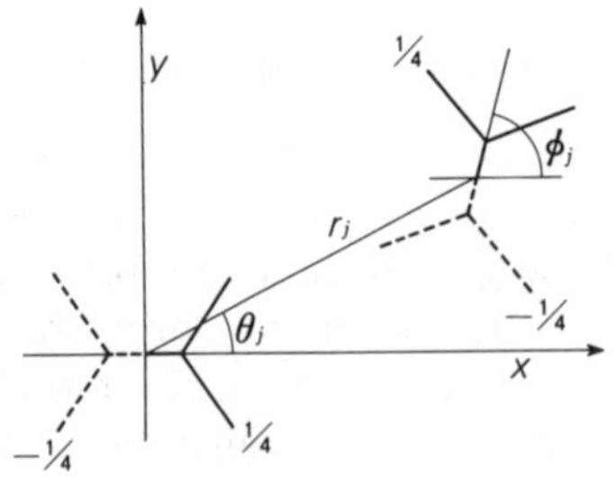

Fig. 6.12 Position parameters for the jth planar zigzag chain of PE.[55]

by varying the position parameters shown in Fig. 6.12.[55]* The first chain is located at the origin. The position parameters of the jth chain are defined by the polar coordinates r_j and θ_j and by the azimuthal angle ϕ_j. For simplicity the relative heights of the chains along the c axis were taken to be the same. The intermolecular interaction energy between the first and second chains was calculated using the three-dimensional position parameters of the second chain. Six potential minima were obtained. The relative positions are shown in Fig. 6.13. The relations A and B correspond to packing in the orthorhombic form; C and D correspond to packing in the monoclinic form[56]; E and F correspond to no actual structure of PE. However the arrangement does appear in a few low-molecular-weight alkane derivatives.[57,58] Then the stable positions were sought by adding the third chain to the two chain models in which the second chain was fixed at the position of an energy minimum A, B, C, or D. The resulting stable relative chain positions correspond to the packings of the two crystal modifications. Further calculations were made by assuming a twofold screw axis along the chain axis of the first chain. The second and third chains are related to each other by the symmetry.

The most stable positions of the first through fifth chains calculated in this way, starting from relation A, are shown in Fig. 6.14 by thick lines. The structure determined by Bunn[59] is shown by thin lines. According to the

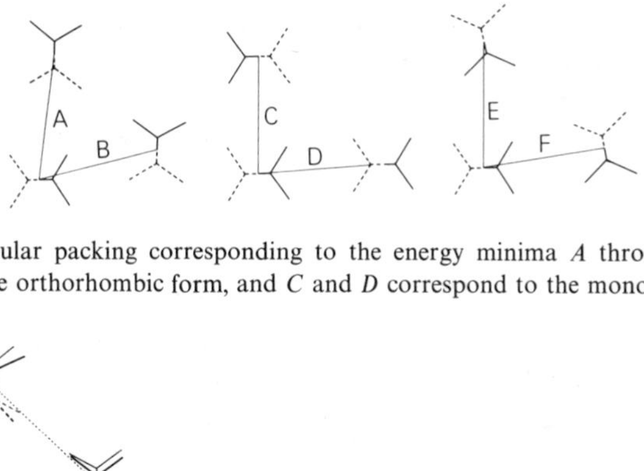

Fig. 6.13 Molecular packing corresponding to the energy minima A through F. A and B correspond to the orthorhombic form, and C and D correspond to the monoclinic form.[55]

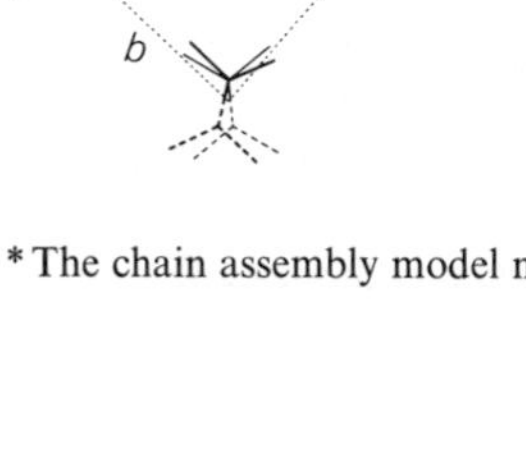

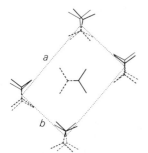

Fig. 6.14 The most stable positions of the first through fifth chains calculated from relation A in Fig. 6.13 (heavy lines)[55] and Bunn's structure (thin lines).[59]

* The chain assembly model method was first proposed by Hopfinger.[4]

calculated result, all the molecular planes of the chains are not exactly parallel, but, generally speaking, the calculated stable packing is quite similar to the X-ray structure. The corresponding stable packing starting from the minima *C* and *D* is a monoclinic structure.

These results indicate that favorable mutual relations between two nearest neighbor chains giving the minimum energy is the dominant chain packing factor. Packing that also satisfies the crystal symmetry then appears.

Further references on intermolecular interaction include: orthorhombic PE, Refs. 51, 60–62; transformation between monoclinic and orthorhombic PE, Ref. 63; poly(amino acid), Refs. 62, 64–66; epitaxial growth of PE, Ref. 67; and molecular motion of PE, Ref. 68.

6.5 PACKING ENERGY MINIMIZATION METHOD

A method yielding the stable molecular conformation and packing in the crystal where the lattice constants, the space group, and rough molecular conformation are known has been developed.[69]

6.5.1 Procedure

For the potential between nonbonded atoms, in principle, the functions described in Section 6.1.2 should be used; however, since the application of these functions to the packing energy minimization requires a very large amount of calculation, the potential energy may be approximated by a parabolic function where only the repulsive term is kept.* According to the procedure developed by Williams,[71]

$$V_{ij} = \begin{cases} \frac{1}{2}k_{ij}(r_{ij}^0 - r_{ij})^2 & r_{ij}^0 \geqslant r_{ij} \\ 0 & r_{ij}^0 < r_{ij} \end{cases} \tag{6.23}$$

where r_{ij}^0 and k_{ij} are the van der Waals distance and the force constant, respectively, between the ith and jth atoms. The constants can be determined so that the parabolic function curves have good fit to the Lennard-Jones 6–12 potential function curves in the range from the minimum to about 3 kcal/mol. The conformational and packing energy E is expressed by

$$E = \frac{1}{2}\sum_{i,j} V_{ij} = \frac{1}{4}\sum_{i,j} k_{ij}(r_{ij}^0 - r_{ij})^2 \tag{6.24}$$

where r_{ij} is any nonbonded interatomic distance in the crystal. If the lattice constants and the space group are known, r_{ij} can be represented as a function

* For a comprehensive summary of multidimensional optimization procedures, see Ref. 70.

of the internal coordinates and the position of a certain asymmetric unit (the fractional coordinates of an atom belonging to the asymmetric unit, and the Eulerian angles indicating the orientation of the asymmetric unit) (see Section 4.9.2). The necessary stereochemistry at the junction of successive asymmetric units during the minimization requires the introduction of the constraining conditions $G_h = 0$ with Lagrange undetermined multipliers λ_h. A stable conformational and packing geometry in the crystal is obtained by least-squares minimization of Φ in terms of the constraints given by

$$\Phi = E + \sum_{h=1}^{H} \lambda_h G_h \tag{6.25}$$

The minimizing method is the same as described in Section 4.9.2.

A measure of the discrepancy between the atomic coordinates obtained by the minimization method and X-ray analysis is the average deviation $\langle d \rangle$ expressed by

$$\langle d \rangle = \left(\frac{\sum_{i=1}^{3N} \Delta_i^2}{3N} \right)^{1/2} \tag{6.26}$$

N is the number of atoms in the asymmetric unit, and Δ_i is the difference between the coordinates obtained by both methods.

6.5.2 Examples of Application to Known Structures

The PEO molecule in the crystal lattice is considerably deformed from the uniform (7/2) helix shown in Section 4.12.2 (Fig. 6.15*c*). Starting from a crystal structure model consisting of the uniform (7/2) helices (Fig. 6.15*a*), the minimization method without the condition of a uniform helix resulted in the model shown in Fig. 6.15*b*. This structure is in fairly good agreement with the structure determined by X-ray analysis (Fig. 6.15*c*). The repulsive energy E is 35.2 and 9.4 kcal per mole of asymmetric unit for the models before and after minimization and $\langle d \rangle$ is 0.34 and 0.23 Å. This suggests that the deformation of the PEO chain from a uniform helix in the crystal is principally due to intermolecular interaction.

Polyisobutylene is another case in which the conformation is considerably deformed from the uniform (8/3) helix (Section 7.5). The structure determined by X-ray analysis can be reproduced very well with the minimization procedure. $E = 23.6$ and 20.9 kcal per mole of asymmetric unit and $\langle d \rangle = 0.43$ and 0.11 Å are obtained for the models before and after minimization. *it*-PP is an example in which the molecular chain is not on the 3_1 screw symmetry element in the unit cell, but X-ray analysis shows almost no deviation from the

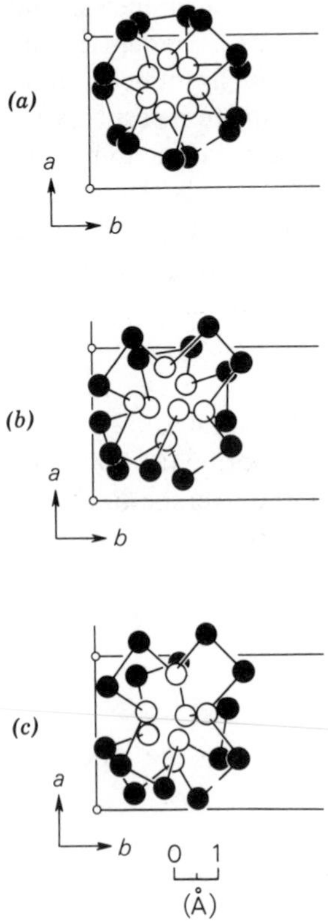

Fig. 6.15 Application of packing energy minimization method to PEO.[69] (*a*) Starting model of uniform helix, (*b*) stable crystal structure model obtained by energy minimization calculations, and (*c*) the structure determined by X-ray analysis.

uniform (3/1) helix.[20] The minimization procedure with nonuniform symmetry gave the same result. Therefore, in the case of *it*-PP, the uniform (3/1) helices predicted by intramolecular interaction are not affected appreciably by intermolecular interaction on the crystal packing. On the contrary, PEO and polyisobutylene show significant deformation due to intermolecular interaction.

6.5.3 Application to X-Ray Structure Analysis

A reasonable model of poly(ethylene oxybenzoate) α-form as obtained from the intramolecular interaction calculation is shown in Table 6.5 (row I). Refinement of the model by the least-squares method did not give a reasonable crystal structure. The result of the application of packing energy minimization is shown in Table 6.5, row II, and the final structure obtained by subsequent refinement is shown in row III. In row IV the differences between the internal rotation angles (rows I and III) are given. The maximum dif-

Table 6.5 Internal Rotation Angles of Poly(ethylene Oxybenzoate) α-Form on the Process of Analysis[42]

Model	τ_1(°)	ω(°)	τ_2(°)	τ_3(°)	τ_4(°)	τ_5(°)	R(%)
I. Molecular model found by intramolecular interaction calculation	−41	10	180	−75	−60	−164	38
II. Model after application of packing energy minimization	−18	−29	−177	−77	−76	−183	32
III. Final model	−13	−15	−172	−102	−59	−186	13
IV. Difference between rows I and III	28	25	8	27	1	22	

ference is as large as 28°. The molecular model (row I) was obtained through the calculation of the intramolecular interaction energy, but the final model could not be obtained without the application of packing energy minimization. This demonstrates the significance and limitations of intramolecular interaction energy calculation for deriving molecular models for structure analysis. The final crystal structure is shown in Fig. 6.16. This method has also been utilized in the analysis of *it*-poly(4-methyl-1-pentene).[23]

It is very difficult using the X-ray least-squares method to refine internal rotation angles as large as 28 (Table 6.5). The merit of the energy minimiza-

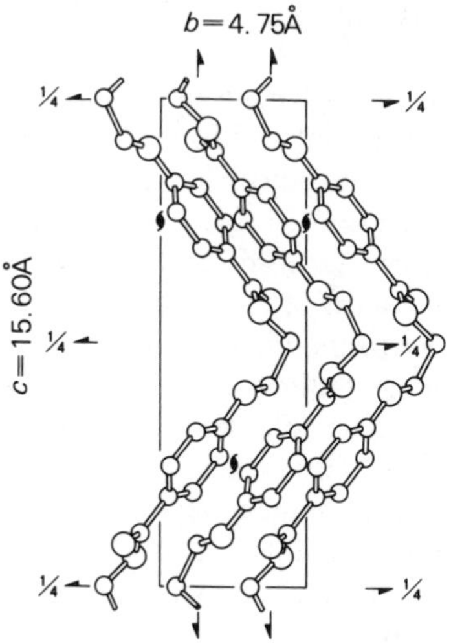

Fig. 6.16 Crystal structure of the α-form of poly (ethylene oxybenzoate).[42]

tion method is its ability to refine such large differences. The packing energy minimization method can be combined with the least-squares method for X-ray analysis according to the equation[69]

$$\Theta = \sum_{m=1}^{M} [|F_m(\mathrm{obs})| - |F_m|]^2 + \omega\Phi \tag{6.27}$$

where the coefficient ω is the weight of the energy constraint to the least-squares refinement of the structure factors. The other symbols are the same as in Eq. 4.108. A similar method was derived independently by Arnott et al.[72]

6.6 CALCULATION OF VIBRATIONAL FREE ENERGY

PE has both orthorhombic[58] and monoclinic forms[56,73,74]; the former is a stable phase and the latter is a metastable phase appearing as a result of stress. Yemni and McCullough[63] tried to clarify the transition mechanism between these two crystal forms by calculating the static potential energy with variation in cell constants and chain orientations. In contrast to the experimental findings, the monoclinic form was found to be more stable by about 0.15 kJ per mole of the CH_2 unit compared to the orthorhombic form. A similar tendency has also been recognized in the calculation described in Section 6.4.3.[55]

So far only the static potential energy has been considered, but the vibrational free energy should play an important role in the stability of the crystalline phase. However detailed calculations of free energy have been made only for inert gas crystals. For *n*-paraffins and PE approximate treatments have been proposed.[75,76]

For calculations for orthorhombic and monoclinic PE made in the author's laboratory,[77] the lattice parameters for orthorhombic PE reported by Smith ($a = 7.47$ Å, $b = 4.97$ Å, and $c = 2.515$ Å)[78] and those for monoclinic PE reported by Seto et al. [$a = 8.09$ Å, b(fiber axis) $= 2.53$ Å, $c = 4.79$ Å, and $\beta = 107.9°$][56] were used. Two types of intermolecular force constants were used: (1) the second derivative of the repulsive term of the Lennard-Jones type H$\cdots$H potential function[7,15] with respect to the interatomic distance (solid line in Fig. 6.17) and (2) the potential for crystalline benzene proposed by Harada and Shimanouchi[79] (dashed line). The following description considers mainly the potential force constant (1). For each crystal form, the normal frequencies were computed for the wave number vector **k** within the first Brillouin zone of the reciprocal space at the interval of 1.5° for δ_c and 90° for δ_a and δ_b. This allows calculation of frequency–phase relations and frequency distributions (see Section 5.6.6.B).

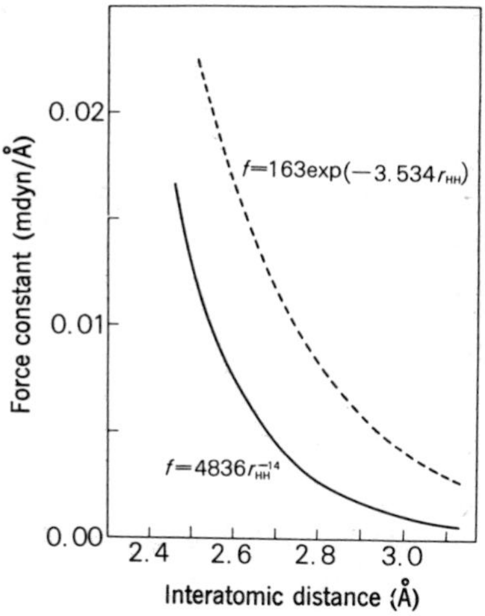

Fig. 6.17 Hydrogen–hydrogen intermolecular potential functions.[77] Solid curve is derived from the repulsive term of a Lennard–Jones type H···H potential. Dashed curve is from the repulsive term of a Buckingham type H···H potential.[79]

Under the harmonic approximation, vibrational internal energy E, entropy S, the Helmholtz energy A, and heat capacity C_V are given by the following equations (see Appendix G for their derivation):[80]

$$E = \left(\frac{1}{\mathfrak{N}}\right)\sum_{\mathbf{k},j} h\nu_j(\mathbf{k})\left\{\left[\exp\left(\frac{h\nu_j(\mathbf{k})}{kT}\right) - 1\right]^{-1} + \frac{1}{2}\right\} \tag{6.28}$$

$$S = \left(\frac{k}{\mathfrak{N}}\right)\sum_{\mathbf{k},j}\left\{\frac{h\nu_j(\mathbf{k})}{2kT}\coth\left(\frac{h\nu_j(\mathbf{k})}{2kT}\right) - \ln\left[2\sinh\left(\frac{h\nu_j(\mathbf{k})}{2kT}\right)\right]\right\} \tag{6.29}$$

$$A = \frac{kT}{\mathfrak{N}}\sum_{\mathbf{k},j}\ln\left[2\sinh\left(\frac{h\nu_j(\mathbf{k})}{2kT}\right)\right] \tag{6.30}$$

$$C_V = \frac{k}{\mathfrak{N}}\sum_{\mathbf{k},j}\frac{(h\nu_j(\mathbf{k})/2kT)^2}{\sinh^2[h\nu_j(\mathbf{k})/2kT]} \tag{6.31}$$

where $\mathfrak{N}$ denotes the number of unit cells (or the number of points in the reciprocal space where frequencies are computed), $\nu_j(\mathbf{k})$ is the normal frequency of the jth branch with the wave number vector $\mathbf{k}$, h is Planck's constant, k is the Boltzmann constant, and T is the absolute temperature.

Figure 6.18 shows the frequency distribution in the range below 600 cm^{-1} for the two forms of PE. The difference between the two forms is remarkable in the region below 200 cm^{-1}. The distribution lies wholly in the orthorhombic form in the lower frequency side. The values E, S, and A calculated for orthorhombic PE using Eqs. 6.28–6.30 are shown in Fig. 6.19 by solid lines. The experimental data by Chang[81] are denoted by the symbol ○. E or A at 0°K means the zero point energy of the crystal. Since the absolute values of E and A cannot be determined from the thermal data, the observed points

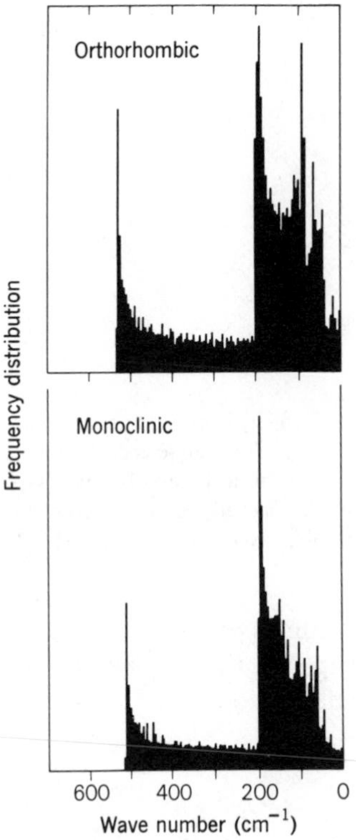

Fig. 6.18 Frequency distribution of orthorhombic PE and monoclinic PE.[77]

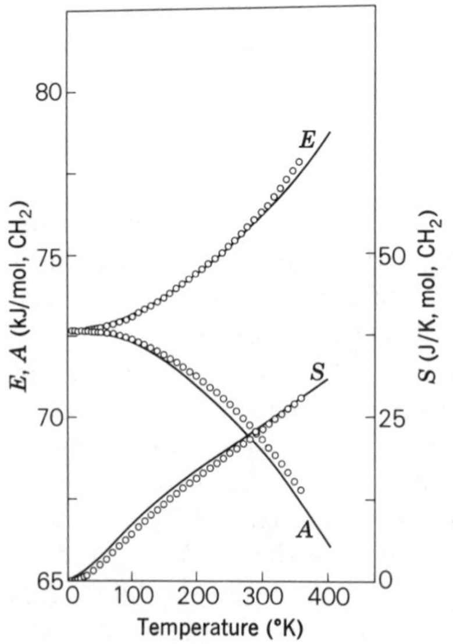

Fig. 6.19 The internal energy E, the entropy S, and the Helmholtz energy A calculated for orthorhombic PE.[77] Open circles denote the experimental data of Chang.[81]

of E and A can be shifted along the ordinate. The zero point energy is chosen so as to fit the calculated curve. The good agreement between the observed and calculated values of the thermodynamic functions suggests that the potential functions assumed in the present calculation are suitable for reproducing the thermodynamic properties of orthorhombic PE.

Figure 6.20 shows the calculated results for orthorhombic (solid line) and monoclinic PE (dashed line). Although the internal energy is not appreciably different for both forms, the entropy is significantly higher for orthorhombic PE. The difference in the zero point energy is less than 0.1 kJ per mole of CH_2 unit. Therefore at 300°K orthorhombic PE has a vibrational free energy of about 0.5 kJ per mole of CH_2 unit less than that of monoclinic PE. This difference is large enough to compensate for the higher static potential energy (estimated as about 0.15 kJ per mole of CH_2 unit) of the orthorhombic form. Thus under normal conditions the vibrational free energy term contributes to the thermodynamic stability of orthorhombic PE.

More accurate treatments require a quasiharmonic approximation and anharmonic calculations. The former treatment has recently been carried out.[82] In this study the frequency distributions were obtained at six temperatures in the range of 0–400°K. Then E, S, and A were calculated and interpolated smoothly in this temperature range. The agreement between the results and the experimental data is superior to that seen in Fig. 6.20.

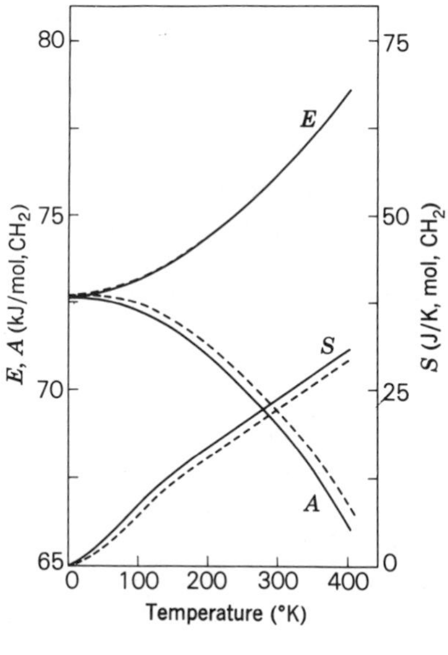

Fig. 6.20 E, S, and A calculated for orthorhombic (———) and monoclinic PE (– – – –).[77]

REFERENCES

1. T. M. Birshtein and O. B. Ptitsyn, *Conformations of Macromolecules*, Interscience, New York, 1966.
2. E. L. Eliel, N. L. Allinger, S. J. Angyal, and G. A. Morrison, *Conformational Analysis*, Interscience, New York, 1965.
3. A. I. Kitaigorodskii, General View on Molecular Packing, *Advances in Structure Research by Diffraction Methods*, R. Brill and R. Mason, Eds., Vol. 3, Pergamon Press, New York, 1970, pp. 173–247.
4. A. J. Hopfinger, *Conformational Properties of Macromolecules*, Academic, New York, 1973.
5. A. J. Hopfinger, *Intermolecular Interactions and Biomolecular Organization*, Wiley-Interscience, New York, 1977.
6. G. A. Segal, Ed., *Semiempirical Methods of Electronic Structure Calculation*, Plenum Press, New York, 1977.
7. R. A. Scott and H. A. Scheraga, *J. Chem. Phys.*, **44,** 3054 (1966).
8. R. A. Scott and H. A. Scheraga, *J. Chem. Phys.*, **45,** 2091 (1966).
9. H. Tadokoro, K. Tai, M. Yokoyama, and M. Kobayashi, *J. Polym. Sci., Polym. Phys. Ed.*, **11,** 825 (1973).
10. C. P. Smyth, *Dielectric Behavior and Structure*, McGraw-Hill, New York, 1955.
11. D. A. Brant and P. J. Flory, *J. Am. Chem. Soc.*, **87,** 2791 (1965).
12. T. Ooi, R. A. Scott, G. Vanderkooi, and H. A. Scheraga, *J. Chem. Phys.*, **46,** 4410 (1967).
13. A. J. Hopfinger and A. G. Walton, *J. Macromol. Sci., Phys.*, **3,** 195 (1969).
14. E. R. Lippincott and R. Schroeder, *J. Chem. Phys.*, **23,** 1099 (1955); R. Schroeder and E. R. Lippincott, *J. Phys. Chem.*, **61,** 921 (1957).
15. R. A. Scott and H. A. Scheraga, *J. Chem. Phys.*, **42,** 2209 (1965).
16. G. Allegra, E. Benedetti, and C. Pedone, *Macromolecules*, **3,** 727 (1970).
17. Y. Chatani, T. Takahagi, T. Kusumoto, and H. Tadokoro, *J. Polym. Sci., Polym. Phys. Ed.*, to be published.
18. R. Hasegawa, M. Kobayashi, and H. Tadokoro, *Polym. J.*, **3,** 591 (1972).
19. T. Tanaka, Y. Chatani, and H. Tadokoro, *J. Polym. Sci., Polym. Phys. Ed.*, **12,** 515 (1974).
20. G. Natta and P. Corradini, *Nuovo Cimento, Suppl.*, **15,** 40 (1960).
21. F. C. Frank, A. Keller, and A. O'Connor, *Phils. Mag.*, **4,** 200 (1959).
22. I. W. Bassi, O. Bonsignori, G. P. Lorenzi, P. Pino, P. Corradini, and P. A. Temussi, *J. Polym. Sci. A-2*, **9,** 193 (1971).
23. H. Kusanagi, Y. Chatani, M. Takase, and H. Tadokoro, *J. Polym. Sci., Polym. Phys. Ed.*, **16,** 131 (1978).
24. P. Corradini, P. Ganis, and V. Petracone, *Eur. Polym. J.*, **6,** 281 (1970).
25. G. Natta, P. Corradini, and I. W. Bassi, *J. Polym. Sci.*, **51,** 505 (1961).
26. G. Natta and P. Corradini, *Nuovo Cimento, Suppl.*, **15,** 9 (1960).
27. H. Tadokoro, Y. Chatani, H. Kusanagi, and M. Yokoyama, *Macromolecules*, **3,** 441 (1970).
28. H. Kusanagi, H. Tadokoro, and Y. Chatani, *Macromolecules*, **9,** 531 (1976).
29. C. W. Bunn and D. R. Holmes, *Discuss. Faraday Soc.*, **25,** 95 (1958).
30. G. Natta, P. Corradini, and P. Ganis, *Makromol. Chem.*, **39,** 238 (1960).
31. P. De Santis, E. Giglio, A. M. Liquori, and A. Ripamonti, *J. Polym. Sci. A*, **1,** 1383 (1963).

32. A. M. Liquori, *J. Polym. Sci. C*, **12,** 209 (1966).
33. C. V. Goebel, W. L. Dimpfl, and D. A. Brant, *Macromolecules*, **3,** 644 (1970).
34. K. Tai and H. Tadokoro, *Macromolecules*, **7,** 507 (1974).
35. H. Tadokoro, Y. Chatani, T. Yoshihara, S. Tahara, and S. Murahashi, *Makromol. Chem.*, **73,** 109 (1964).
36. Y. Takahashi and H. Tadokoro, *Macromolecules*, **6,** 672 (1973).
37. Y. Takahashi, I. Sumita, and H. Tadokoro, *J. Polym. Sci., Polym. Phys. Ed.*, **11,** 2113 (1973).
38. Y. Takahashi, H. Tadokoro, and Y. Chatani, *J. Macromol. Sci., Phys.*, **2,** 361 (1968).
39. C. S. Fuller, C. J. Frosch, and N. R. Pape, *J. Am. Chem. Soc.*, **62,** 1905 (1940).
40. A. M. Liquori, *Acta Crystallogr.*, **8,** 345 (1955).
41. G. Wasai, T. Saegusa, and J. Furukawa, *Makromol. Chem.*, **86,** 1 (1965).
42. H. Kusanagi, H. Tadokoro, Y. Chatani, and K. Suehiro, *Macromolecules*, **10,** 405 (1977).
43. C. S. Fuller, *Chem. Rev.*, **26,** 143 (1940).
44. R. C. Reinhardt, *Ind. Eng. Chem.*, **35,** 422 (1943).
45. V. M. Coiro, P. De Santis, A. M. Liquori, and A. Ripamonti, *Ric. Sci., Rend. Sez. A*, **3,** 1043 (1963).
46. V. M. Coiro, P. De Santis, and A. M. Liquori, *J. Polym. Sci. B.*, **4,** 821 (1966).
47. T. Miyazawa and Y. Ideguchi, *J. Polym. Sci. B*, **3,** 541 (1965).
48. L. D'Ilario and E. Giglio, *Acta Crystallogr. B*, **30,** 372 (1974).
49. G. Cortili and G. Zerbi, *Spectrochim. Acta*, **23A,** 285 (1967).
50. T. Oyama, K. Shiokawa, and I. Ishimaru, *J. Macromol. Sci., Phys.*, **8,** 229 (1973).
51. P. E. McMahon, R. L. McCullough, and A. A. Schlegel, *J. Appl. Phys.*, **38,** 4123 (1967).
52. V. Petraccone, G. Allegra, and P. Corradini, *J. Polym. Sci. C*, **38,** 419 (1972).
53. R. Hasegawa, Y. Takahashi, Y. Chatani, and H. Tadokoro, *Polym. J.*, **3,** 600 (1972).
54. P. Corradini and G. Avitabile, *Eur. Polym. J.*, **4,** 385 (1968).
55. K. Tai, M. Kobayashi, and H. Tadokoro, *J. Polym. Sci., Polym. Phys. Ed.*, **14,** 783 (1976).
56. T. Seto, T. Hara, and K. Tanaka, *Jpn. J. Appl. Phys.*, **7,** 31 (1968).
57. A. I. Kitaigorodskii, *Molecular Crystals and Molecules*, Academic, New York, 1973.
58. E. Segerman, *Acta Crystallogr.*, **19,** 789 (1965).
59. C. W. Bunn, *Trans. Faraday Soc.*, **35,** 482 (1939).
60. A. Odajima and T. Maeda, *J. Polym. Sci. C.*, **15,** 55 (1966).
61. A. Odajima, T. Maeda, M. Seto, T. Kitagawa, and T. Miyazawa, *Rep. Prog. Polym. Phys. Jpn.*, **11,** 209 (1968).
62. D. J. Nelson and J. Hermans, Jr., *Biopolymers*, **12,** 1269 (1973).
63. T. Yemni and L. McCullough, *J. Polym. Sci., Polym. Phys. Ed.*, **11,** 1385 (1973).
64. A. J. Hopfinger and A. G. Walton, *J. Macromol. Sci., Phys.*, **4,** 185 (1970).
65. A. J. Hopfinger, *Biopolymers*, **10,** 1299 (1971).
66. R. F. McGuire, G. Vanderkooi, F. A. Momany, R. T. Ingwall, G. M. Crippen, N. Lotan, R. W. Tuttle, K. L. Kashuba, and H. A. Scheraga, *Macromolecules*, **4,** 112 (1971).
67. K. A. Mauritz, E. Baer, and A. J. Hopfinger, *J. Polym. Sci., Polym. Phys. Ed.*, **11,** 2185 (1973).
68. R. L. McCullough, *J. Macromol. Sci., Phys.*, **9,** 97 (1974).
69. H. Kusanagi, H. Tadokoro, and Y. Chatani, *Polym. J.*, **9,** 181 (1977).

70. S. R. Niketić, *Force Field Calculations on Coordination Compounds*, Ph. D. thesis, The Technical University of Denmark, 1974, Chap. 5.
71. D. E. Williams, *Acta Crystallogr.*, **A25,** 464 (1969).
72. S. Arnott, W. E. Scott, E. A. Rees, and C. G. McNab, *J. Mol. Biol.*, **90,** 253 (1974).
73. H. Kiho, A. Peterlin, and P. H. Geil, *J. Appl. Phys.*, **35,** 1599 (1964).
74. P. W. Teare and D. R. Holmes, *J. Polym. Sci.*, **24,** 496 (1957).
75. A. Warshel and S. Lifson, *J. Chem. Phys.*, **53,** 582 (1970).
76. Y. Wada and R. Hayakawa, *Prog. Polym. Sci. Jpn.*, **3,** 215 (1972).
77. M. Kobayashi and H. Tadokoro, *Macromolecules*, **8,** 897 (1975).
78. A. E. Smith, *J. Chem. Phys.*, **21,** 2229 (1953).
79. I. Harada and T. Shimanouchi, *J. Chem. Phys.*, **44,** 2016 (1966); **46,** 2708 (1967).
80. A. A. Maradudin, E. W. Montroll, and G. H. Weiss, *Theory of Lattice Dynamics in the Harmonic Approximation*, *Solid State Phys.*, *Suppl.*, Vol. 3, Academic, New York, 1963.
81. S. S. Chang, *J. Res. Natl. Bur. Stand.*, **78A,** 387 (1974).
82. M. Kobayashi and H. Tadokoro, *J. Chem. Phys.*, **66,** 1258 (1977).

Chapter 7

Structure of Various Crystalline Polymers

Table 7.1 summarizes the crystallographic data for crystalline polymers considered to be reliable at persent. Further references include Ref. 175 for crystallographic data and Refs. 176–179 for reviews on crystalline polymers.

7.1 POLYETHYLENE

The lattice dimensions of orthorhombic PE are given in Table 7.2. The thermal expansion of orthorhombic PE has been studied in detail by Swan.[182] Expansion occurs primarily along the a axis ($a = 7.414$ Å at 30° C and 7.706 Å at 138° C). The b and c axes do not expand appreciably. The ratio a/b is 1.46 at − 196° C, 1.50 at 30° C, and 1.56 at 138° C. The extrapolated value of $\sqrt{3}$, corresponding to the hexagonal lattice, is too large and would be attained at a temperature higher by 100° C than the melting point.

The angle between the zigzag plane and the bc plane (the setting angle in the orthorhombic form) was first reported to be about 41° by Bunn[1] and has been used for many years. Avitabile et al.[181] have reported this angle to be 49° as determined by neutron diffraction by polydeuteroethylene (prepared by Ziegler-Natta catalyst) at 4 and 90°K. According to their paper, the C—C bond distance is 1.574 Å, appreciably longer than the value 1.54 Å so far reported from X-ray results. Recent X-ray analyses[4] on high density PE have shown that the setting angle is about 45° for nonoriented samples (coinciding with the result of Kavesh and Schultz[3]) and about 46° for oriented samples. In the case of low density PE, the angle changes to 49–51° with the increase of the a axis.

Branching affects the lattice constants.[2,4,183,184] In propylene–ethylene copolymers, the a axis increases distinctly with increasing propylene content, but the b and c axes show almost no change.[2] The methyl side chains can be included up to 13 mole% without destroying the crystal lattice and expand the lattice in the a direction.

Table. 7.1 Crystallographic Data

Polymer	Crystal system, space group, lattice constants, and number of chains per unit cell[a]	Molecular conformation	Crystal density (g/cm³)
Polyethylene			
Polyethylene ($—CH_2—CH_2—)_n$	Stable form,[1–4] orthorhombic, $Pnam\text{-}D_{2h}^{16}$, $a = 7.417$Å, $b = 4.945$Å, $c = 2.547$Å, $N = 2$	Planar zigzag (2/1)	1.00
	Metastable form,[5] monoclinic, $C2/m\text{-}C_{2h}^{3}$, $a = 8.09$Å, b (f.a.) $= 2.53$Å, $c = 4.79$Å, $\beta = 107.9°$, $N = 2$	Planar zigzag (2/1)	0.998
	High pressure form,[6–8] orthohexagonal (assumed), $a = 8.42$Å, $b = 4.56$Å, c (f.a.) has not been determined	—	—
Polytetrafluoroethylene and related polymers			
Polytetrafluoroethylene[9–11] ($—CF_2—CF_2—)_n$	Below 19°C, pseudohexagonal (triclinic), $a' = b' = 5.59$Å, $c = 16.88$Å, $\gamma' = 119.3°$, $N = 1$	Helix (13/6) 163.5°	2.35
	Above 19°C, trigonal, $a = 5.66$Å, $c = 19.50$Å, $N = 1$	Helix (15/7) 165.8°	2.30
	High-pressure form I,[12] orthorhombic, $Pnam\text{-}D_{2h}^{16}$, $a = 8.73$Å, $b = 5.69$Å, $c = 2.62$Å, $N = 2$	Planar zigzag (2/1)	2.55
	High-pressure form II,[193] monoclinic, $B2/m\text{-}C_{2h}^{3}$, $a = 9.50$Å, $b = 5.05$Å, $c = 2.62$Å, $\gamma = 105.5°$, $N = 2$	Planar zigzag (2/1)	2.74

Polychlorotrifluoroethylene[13] (—CFCl—CF_2—)$_n$	Pseudohexagonal, $a = b = 6.438$Å, $c = 41.5$Å, $N = 1$	Helix (16.8/1) (average)	2.10
Ethylene–tetrafluoroethylene alternating copolymer[14] (—$CH_2CH_2CF_2CF_2$—)$_n$	Monoclinic, $a = 9.6$Å, $b = 9.25$Å, $c = 5.0$Å, $\gamma = 96°$, $N = 4$	Planar zigzag	1.93
Poly-α-olefins			
it-Polypropylene (—CH_2—CH—)$_n$ with side group CH_3	α-Form,[15] monoclinic, $C2/c$-$C_{2h}{}^6$ or Cc-$C_s{}^4$, $a = 6.65$Å, $b = 20.96$Å, $c = 6.50$Å, $\beta = 99°20'$, $N = 4$	Helix (3/1) $(TG)_3$	0.936
	β-Form,[16,17] hexagonal, $a = 19.08$Å, $c = 6.49$Å, $N = 9$	Helix (3/1) $(TG)_3$	0.922
	γ-Form,[18] trigonal, $P3_121$-$D_3{}^4$ or $P3_221$-$D_3{}^6$, $a = 6.38$Å, $c = 6.33$Å, $N = 1$	Helix (3/1) $(TG)_3$	0.939
st-Polypropylene	Orthorhombic,[19] $C222_1$-$D_2{}^5$, $a = 14.50$Å, $b = 5.60$Å, $c = 7.40$Å, $N = 2$	Helix (4/1) $(T_2G_2)_2$	0.93
	Second form,[20,21] $c = 5.1$Å	Planar zigzag	—
it-Poly-1-butene (—CH_2—CH—)$_n$ with side group C_2H_5	Form I,[22] trigonal, $R\bar{3}c$-$D_{3d}{}^6$, $a = 17.7$Å, $c = 6.50$Å, $N = 6$	Helix (3/1) $(TG)_3$	0.95
	Form II,[23–25] tetragonal, $P\bar{4}b2$-$D_{2d}{}^7$, $a = 14.5$Å, $c = 21.1$Å, $N = 4$	Helix (11/3)	0.920
it-Poly-1-pentene[26] (—CH_2—CH—)$_n$ with side group C_3H_7	Monoclinic, $a = 11.35$Å, $b = 20.85$Å, $c = 6.49$Å, $\beta = 99.6°$, $N = 4$	Helix (3/1)	0.923

Table 7.1 (*Contd.*)

Polymer	Crystal system, space group, lattice constants, and number of chains per unit cell[a]	Molecular conformation	Crystal density (g/cm^3)
it-Poly-1-hexene[27] $(\text{—CH}_2\text{—CH—})_n$ $\quad\mid$ $\quad C_4H_9$	$c = 13.7$Å	Helix (7/2)	—
it-Poly(3-methyl-1-butene)[28] $(\text{—CH}_2\text{—CH—})_n$ $\quad\mid$ $\quad CH(CH_3)_2$	Monoclinic, $P2_1/b\text{-}C_{2h}{}^5$, $a = 9.55$Å, $b = 17.08$Å, $c = 6.84$Å, $\gamma = 116°30'$, $N = 2$	Helix (4/1)	0.93
it-Poly(4-methyl-1-pentene)[29,30] $(\text{—CH}_2\text{—CH—})_n$ $\quad\mid$ $\quad CH_2CH(CH_3)_2$	Tetragonal, $P\bar{4}b2\text{-}D_{2d}{}^7$ or $P\bar{4}\text{-}S_4{}^1$, $a = 18.63$Å, $c = 13.85$Å, $N = 4$	Helix (7/2)	0.812
it-Poly[(*S*)-4-methyl-1-hexene][29] $(\text{—CH}_2\text{—CH—})_n$ $\quad\mid$ $\quad CH_2CH(CH_3)(C_2H_5)$	Pseudotetragonal, $P1\text{-}C_1{}^1$, $a = b = 19.85$Å, $c = 13.50$Å, $N = 4$	Helix (7/2)	0.86
it-Poly[(*S*), (*R*)-4-methyl-1-hexene][29]	Tetragonal, $P\bar{4}\text{-}S_4{}^1$, $a = 19.85$Å, $c = 13.50$Å, $N = 4$	Helix (7/2)	0.86
it-Poly(5-methyl-1-hexene)[31] $(\text{—CH}_2\text{—CH—})_n$ $\quad\mid$ $\quad (CH_2)_2\text{—}CH(CH_3)_2$	Monoclinic, $P2_1\text{-}C_2{}^2$, $a = 17.62$Å, $b = 10.17$Å, $c = 6.33$Å, $\beta = 90°$, $N = 2$	Helix (3/1)	0.86

it-Poly[(*S*)-5-methyl-1-heptene][31] $(\text{—}CH_2\text{—}CH\text{—})_n$ $|$ $(CH_2)_2\text{—}CH(CH_3)(C_2H_5)$	Monoclinic, $P2_1\text{-}C_2^{\ 2}$, $a = 18.40$Å, $b = 10.62$Å, $c = 6.36$Å, $\beta = 90°$, $N = 2$	Helix (3/1)	0.90
it-Poly[(*S*), (*R*)-5-methyl-1-heptene][31]	Tetragonal, $P\bar{4}\text{-}S_4^{\ 1}$ (probably), $a = b = 20.00$Å, $c = 38.76$Å, $N = 4$	Helix (19/6)	0.91
it-Poly(vinyl cyclohexane)[32] $(\text{—}CH_2\text{—}CH\text{—})_n$ 	Tetragonal, $I4_1/a\text{-}C_{4h}^{\ 6}$, $a = 21.99$Å, $c = 6.43$Å, $N = 4$	Helix (4/1)	0.94
it-Poly(4-phenyl-1-butene)[33] $(\text{—}CH_2\text{—}CH\text{—})_n$ $|$ $(CH_2)_2\text{—}C_6H_5$	Monoclinic, $Pa\text{-}C_s^{\ 2}$, $a = 10.4$Å, $b = 18.0$Å, $c = 6.61$Å, $\beta = 90°$, $N = 2$	Helix (3/1)	1.06
it-Poly(vinyl ethyl silane)[34] $(\text{—}CH_2\text{—}CH\text{—})_n$ $|$ $SiH_2(C_2H_5)$	Trigonal, $R\bar{3}\text{-}C_{3i}^{\ 2}$, $a = 21.60$Å, $c = 6.50$Å, $N = 6$	Helix (3/1)	0.98
it-Poly(allyl cyclopentane)[35] $(\text{—}CH_2\text{—}CH\text{—})_n$ $|$ $CH_2\text{—}C_5H_9$	Tetragonal, $I\bar{4}c2\text{-}D_{2d}^{\ 6}$, $a = 20.30$Å, $c = 47.40$Å, $N = 4$	Helix (24/7)	0.899
it-Polystyrene[36] $(\text{—}CH_2\text{—}CH\text{—})_n$ $|$ C_6H_5	Trigonal, $R3c\text{-}C_{3v}^{\ 6}$ or $R\bar{3}c\text{-}D_{3d}^{\ 6}$, $a = 21.90$Å, $c = 6.65$Å, $N = 6$	Helix (3/1) $(TG)_3$	1.13
it-Poly-*o*-methylstyrene[37]	Tetragonal, $I4_1cd\text{-}C_{4v}^{\ 12}$, $a = 19.01$Å, $c = 8.10$Å, $N = 4$	Helix (4/1)	1.071

Table 7.1 (*Contd.*)

Polymer	Crystal system, space group, lattice constants, and number of chains per unit cell[a]	Molecular conformation	Crystal density (g/cm^3)
it-Poly-*m*-methylstyrene[38–41]	Tetragonal, $P\bar{4}$-$S_4{}^1$, $a = 19.81$Å, $c = 57.1$Å, $N = 4$	Helix (29/8)	1.010
it-Poly-*p*-methylstyrene[41]	$c = 12.9$Å	Helix?	—
it-Poly-*o*-fluorostyrene[42]	Trigonal, $R3c$-$C_{3v}{}^6$, $a = 22.15$Å, $c = 6.63$Å, $N = 6$	Helix (3/1)	1.30
it-Poly-*p*-fluorostyrene[43]	Orthorhombic, $P2_12_12_1$-$D_2{}^4$, $a = 17.6$Å, $b = 12.1$Å, $c = 8.25$Å, $N = 2$	Helix (4/1)	0.918
it-Poly(*o*-methyl-*p*-fluorostyrene)[42]	$c = 8.05$Å	Helix (4/1)	—
it-Poly(α-vinyl naphthalene)[44]	Tetragonal, $I4_1cd$-$C_{4v}{}^{12}$, $a = 21.20$Å, $c = 8.10$Å, $N = 4$	Helix (4/1)	1.125
it-Poly(*p*-trimethyl silylstyrene)[39–41] $(\text{—}CH_2\text{—}CH\text{—})_n$ with side group $C_6H_4\text{—}Si(CH_3)_3$ on CH	$c = 60.4$Å	Helix (29/9)	—
Vinyl polymers			
Poly(vinyl chloride)[45] $(\text{—}CH_2\text{—}CHCl\text{—})_n$	Orthorhombic, $Pcam$-$D_{2h}{}^{11}$, $a = 10.6$Å, $b = 5.4$Å, $c = 5.1$Å, $N = 2$	Planar zigzag	1.42
Poly(vinyl alcohol)[46,47] $(\text{—}CH_2\text{—}CH\text{—})_n$ with OH on CH	Monoclinic, $P2_1/m$-$C_{2h}{}^2$, $a = 7.81$Å, b (f.a.) $= 2.25$Å, $c = 5.51$Å, $\beta = 91.7°$, $N = 2$	Planar zigzag	1.35

Poly (vinyl fluoride)[48] $(\text{—CH}_2\text{—CHF—})_n$	Orthorhombic, $Cm2m\text{-}C_{2v}^{14}$, $a = 8.57$Å, $b = 4.95$Å, $c = 2.52$Å, $N = 2$	Planar zigzag	1.430
it-Poly (vinyl formate)[49] $(\text{—CH}_2-\text{CH—})_n$ with side group OCOH	Trigonal, $R3c\text{-}C_{3v}^{6}$ or $R\bar{3}c\text{-}D_{3d}^{6}$, $a = 15.9$Å, $c = 6.55$Å, $N = 6$	Helix (3/1)	1.50
Vinylidene polymers			
Polyisobutylene[50] $[\text{—CH}_2\text{—C(CH}_3)_2\text{—}]_n$	Orthorhombic, $P2_12_12_1\text{-}D_2^{4}$, $a = 6.88$Å, $b = 11.91$Å, $c = 18.60$Å, $N = 2$	Helix, see text	0.972
Poly (vinylidene chloride)[51] $(\text{—CH}_2\text{—CCl}_2\text{—})_n$	Monoclinic, $P2_1/m\text{-}C_{2h}^{2}$, $a = 6.71$Å, b (f.a.) $= 4.68$Å, $c = 12.51$Å, $\beta = 123°$, $N = 2$	Glide-type $TG\,T\bar{G}$	1.954
Poly (vinylidene fluoride)[52–54] $(\text{—CH}_2\text{—CF}_2\text{—})_n$	Form I or β-form, Orthorhombic, $Cm2m\text{-}C_{2v}^{14}$, $a = 8.58$Å, $b = 4.91$Å, $c = 2.56$Å, $N = 2$	Slightly deflected planar zigzag	1.973
	Form II or α-form, monoclinic, $P2_1/c\text{-}C_{2h}^{5}$, $a = 4.96$Å, $b = 9.64$Å, $c = 4.62$Å, $\beta = 90°$, $N = 2$	Glide-type $TG\,T\bar{G}$	1.925
	Form III, monoclinic, $C2\text{-}C_2^{3}$, $a = 8.66$Å, $b = 4.93$Å, $c = 2.58$Å, $\beta = 97°$, $N = 2$	Slightly deflected planar zigzag	1.944
it-Poly (methyl methacrylate)[55] $[\text{—CH}_2\text{—C(CH}_3)\text{—}]_n$ with side group $COOCH_3$	Orthorhombic, $a = 20.98$Å, $b = 12.06$Å, $c = 10.40$Å, $N = 4$	Double helix (10/1)	1.26
st-Poly (α-methylvinyl methyl ether)[56] $[\text{—CH}_2\text{—C(CH}_3)\text{—}]_n$ with side group OCH_3	Pseudohexagonal ?, $a = b = 9.02$Å, $c = 16.6$Å, $N = 1$	Helix (10/4)	1.02

Table 7.1 (*Contd.*)

Polymer	Crystal system, space group, lattice constants, and number of chains per unit cell[a]	Molecular conformation	Crystal density (g/cm^3)
Polymers containing C=C double bonds			
trans-1, 4-Polybutadiene[57] (—$CH_2CH{=}CHCH_2$—)$_n$	Low-temperature form, monoclinic, $P2_1/a\text{-}C_{2h}{}^5$, $a = 8.63$Å, $b = 9.11$Å, $c = 4.83$Å, $\beta = 114°$, $N = 4$	(1/0) *trans*-$ST\bar{S}$	1.04
cis-1, 4-Polybutadiene[58]	Monoclinic, $C2/c\text{-}C_{2h}{}^6$, $a = 4.60$Å, $b = 9.50$Å, $c = 8.60$Å, $\beta = 109°$, $N = 2$	(2/0) (TT'-*cis*-T') ($T\bar{T}'$-*cis*-$\bar{T}'$)	1.01
it-1, 2-Polybutadiene[59] (—CH_2—CH—)$_n$ with side group $CH{=}CH_2$	Trigonal, $R3c\text{-}C_{3v}{}^6$ or $R\bar{3}c\text{-}D_{3d}{}^6$, $a = 17.3$Å, $c = 6.50$Å, $N = 6$	Helix (3/1)	0.96
st-1, 2-Polybutadiene[45]	Orthorhombic, $Pcam\text{-}D_{2h}{}^{11}$, $a = 10.98$Å, $b = 6.60$Å, $c = 5.14$Å, $N = 2$	Slightly deflected planar zigzag (2/0) $T_2'\bar{T}_2'$	0.964
trans-1, 4-Polyisoprene [—CH_2—$C(CH_3){=}CH$—CH_2—]$_n$	α-Form,[60] monoclinic, $P2_1/c\text{-}C_{2h}{}^5$, $a = 7.98$Å, $b = 6.29$Å, $c = 8.77$Å, $\beta = 102.0°$, $N = 2$	(2/0) *trans*-CTS-*trans*-$CT\bar{S}$	1.05
	β-Form,[61] orthorhombic, $a = 7.78$Å, $b = 11.78$Å, $c = 4.72$Å, $N = 4$	*trans*-$ST\bar{S}$	1.05
cis-1, 4-Polyisoprene[58,61,62]	Monoclinic, $P2_1/a\text{-}C_{2h}{}^5$, $a = 12.46$Å, $b = 8.89$Å, $c = 8.10$Å, $\beta = 92°$, $N = 4$	*cis*-$ST\bar{S}$-*cis*-$\bar{S}TS$, (2/0)	1.02

trans-Polypentenamer[63,b] $[-CH{=}CH-(CH_2)_3-]_n$	Orthorhombic, $Pnam\text{-}D_{2h}^{16}$, $a = 7.28$Å, $b = 4.97$Å, $c = 11.90$Å, $N = 2$	Helix (2/1) $(trans\text{-}ST_4\bar{S})_2$	1.05
trans-Polyoctenamer[c] $[-CH{=}CH-(CH_2)_6-]_n$	Monoclinic,[63] $P2_1/a\text{-}C_{2h}^{5}$, $a = 7.43$Å, $b = 5.00$Å, $c = 9.90$Å, $\beta = 95°10'$, $N = 2$	(1/0) $trans\text{-}ST_5\bar{S}$	1.00
	Triclinic,[64] $P\bar{1}\text{-}C_i^{1}$, $a = 4.34$Å, $b = 5.41$Å, $c = 9.78$Å, $\alpha = 64°25'$, $\beta = 104°50'$, $\gamma = 118°35'$, $N = 1$	(1/0) $trans\text{-}ST_5\bar{S}$	1.01
it-Poly (4-methyl-1, 3-pentadiene)[65] $(-CH_2-CH(-CH{=}C(CH_3)_2)-)_n$	Tetragonal, $I\bar{4}c2\text{-}D_{2d}^{10}$, $a = 17.80$Å, $c = 36.50$Å, $N = 4$	Helix (18/5)	0.85
it-*trans*-1, 4-Poly (1,3-pentadiene)[66] $[-CH_2CH{=}CHCH(CH_3)-]_n$	Orthorhombic, $P2_12_12_1\text{-}D_2^{4}$, $a = 19.80$Å, $b = 4.86$Å, $c = 4.85$Å, $N = 4$	—	0.97
trans-1, 4-Poly-2, 3-dichlorobutadiene[67] $[-CCl{=}CCl-(CH_2)_2-]_n$	Monoclinic, $P2_1/a\text{-}C_{2h}^{5}$, $a = 5.34$Å, $b = 9.95$Å, $c = 4.80$Å, $\beta = 93.5°$, $N = 2$	(1/0) $trans\text{-}ST\bar{S}$	1.60
cis-1, 4-Poly (2-*tert*-butylbutadiene)[68] $[-CH_2C(t\text{-}Bu){=}CHCH_2-]_n$	Triclinic?, $a' = 13.95$Å, $b' = 20.78$Å, $c = 15.3$Å, $N = 2$	Helix (11/3)	0.91
Polyallene[69,70] $[-CH_2-C({=}CH_2)-]_n$	Form I, orthorhombic, $Pnan\text{-}D_{2h}^{6}$, $a = 8.20$Å, $b = 7.81$Å, $c = 3.88$Å, $N = 2$	Helix (2/1) G_4 (65°)	1.07
	Form II, monoclinic, $P2_1\text{-}C_2^{2}$ or $P2_1/m\text{-}C_{2h}^{2}$, $a = 6.37$Å, b (f.a.) $= 3.88$Å, $c = 5.12$Å, $\beta = 96.6°$, $N = 1$	Helix (2/1) G_4(65°)	1.06
	Form III, $c = 3.88$Å, paracrystallinelike	Helix (2/1) G_4 (65°)	—

Table 7.1 (*Contd.*)

Polymer	Crystal system, space group, lattice constants, and number of chains per unit cell[a]	Molecular conformation	Crystal density (g/cm³)
Polyethers			
Polyoxymethylene $(—CH_2O—)_n$	Trigonal,[71–73] $P3_1$-C_3^2 or $P3_2$-C_3^3, $a = 4.47$Å, $c = 17.39$Å, $N = 1$	Helix (9/5) G'_{18} (78°)	1.49
	Orthorhombic,[74] $P2_12_12_1$-D_2^4, $a = 4.77$Å, $b = 7.65$Å, $c = 3.56$Å, $N = 2$	Helix (2/1) G_4	1.54
Poly (ethylene oxide) $(—CH_2CH_2O—)_n$	Form I,[75] monoclinic, $P2_1/a$-C_{2h}^5, $a = 8.05$Å, $b = 13.04$Å, $c = 19.48$Å, $\beta = 125.4°$, $N = 4$	Helix (7/2)	1.228
	Form II,[76] triclinic, $P\bar{1}$-C_i^1, $a = 4.71$Å, $b = 4.44$Å, $c = 7.12$Å, $\alpha = 62.8°$, $\beta = 93.2°$, $\gamma = 111.4°$, $N = 1$	Planar zigzag (2/1)	1.197
Polyoxacyclobutane[77–79] $[—(CH_2)_3—O—]_n$	Form I (hydrate), monoclinic, $C2/m$-C_{2h}^3, $a = 12.3$Å, $b = 7.27$Å, $c = 4.80$Å, $\beta = 91°$, $N = 4$ and 4 H_2O	Planar zigzag (1/0)	1.18
	Form II, trigonal, $R3c$-C_{3v}^6, $a = 14.13$Å, $c = 8.41$Å, $N = 9$	Glide-type (2/0) $T_3GT_3\bar{G}$	1.19
	Form III, orthorhombic, $C222_1$-D_2^5, $a = 9.23$Å, $b = 4.82$Å, $c = 7.21$Å, $N = 2$	Helix (2/1) $(T_2G_2)_2$	1.20
	Form IV, $c = 4.79$Å	Planar zigzag (1/0)	—
Polytetrahydrofuran[80,81] $[—(CH_2)_4—O—]_n$	Monoclinic, $C2/c$-C_{2h}^6, $a = 5.59$Å, $b = 8.90$Å, $c = 12.07$Å, $\beta = 134.2°$, $N = 2$	Planar zigzag (2/1)	1.11

Poly (hexamethylene oxide)[82,d] $[-(CH_2)_6-O-]_n$	Monoclinic, $C2/c\text{-}C_{2h}^6$, $a = 5.64Å$, $b = 8.98Å$, $c = 17.32Å$, $\beta = 134.5°$, $N = 2$	Planar zigzag (2/1)	1.06
Poly (decamethylene oxide)[82,e] $[-(CH_2)_{10}-O-]_n$	Orthorhombic, $Pnam\text{-}D_{2h}^{16}$, $a = 7.40Å$, $b = 4.93Å$, $c = 27.29Å$, $N = 2$	Planar zigzag (2/1)	1.04
Poly-1, 3-dioxolane $[-OCH_2O-(CH_2)_2-]_n$	Form I,[83] triclinic, $P1\text{-}C_1^1$, $a = 12.32Å$, $b = 4.66Å$, $c = 24.74Å$, $\alpha = 90°$, $\beta = 90°$, $\gamma = 100.9°$, $N = 3$	Helix (5/1) $(G_2T\bar{G}S)_5$	1.33
	Form II,[84] orthorhombic, $Pbca\text{-}D_{2h}^{15}$, $a = 9.07Å$, $b = 7.79Å$, $c = 9.85Å$, $N = 4$	Glide-type $(G_2T\bar{G}\bar{S})(\bar{G}_2TGS)$	1.41
	Form III,[83] hexagonal, $a = 8.07Å$, $c = 29.53Å$, $N = 3$	Helix (6/1)	1.33
Poly-1, 3-dioxepane[85] $[-OCH_2O-(CH_2)_4-]_n$	Orthorhombic, $P2_1cn\text{-}C_{2v}^9$, $a = 8.50Å$, $b = 4.79Å$, $c = 13.50Å$, $N = 2$	Glide-type $(G_2T\bar{G}TGT)$ $(\bar{G}_2TGT\bar{G}T)$	1.23
Poly-1, 3-dioxocane[86] $[-OCH_2O-(CH_2)_5-]_n$	Triclinic, $P\bar{1}\text{-}C_i^1$, $a = 8.36Å$, $b = 4.84Å$, $c = 8.15Å$, $\alpha = \beta = \gamma = 90°$, $N = 2$	(1/0) $G_2T\bar{G}T_2\bar{G}T$	1.17
Poly-1, 3-dioxonane[86] $[-OCH_2O-(CH_2)_6-]_n$	Orthorhombic, $P2_1cn\text{-}C_{2v}^9$, $a = 8.4Å$, $b = 4.85Å$, $c = 18.8Å$, $N = 2$	Glide-type $(G_2T\bar{G}T_3GT)$ $(\bar{G}_2TGT_3\bar{G}T)$	1.13
it-Polyacetaldehyde[87,88] $[-CH(CH_3)-O-]_n$	Tetragonal, $I4_1/a\text{-}C_{4h}^6$, $a = 14.63Å$, $c = 4.79Å$, $N = 4$	Helix (4/1)	1.14
it-Polypropionaldehyde[88] $(-CH(CH_2CH_3)-O-)_n$	Tetragonal, $I4_1/a\text{-}C_{4h}^6$, $a = 17.50Å$, $c = 4.8Å$, $N = 4$	Helix (4/1)	1.05
it-Poly-*n*-butylaldehyde[88] $(-CH((CH_2)_2CH_3)-O-)_n$	Tetragonal, $I4_1/a\text{-}C_{4h}^6$, $a = 20.01Å$, $c = 4.78Å$, $N = 4$	Helix (4/1)	0.997

Table 7.1 (*Contd.*)

Polymer	Crystal system, space group, lattice constants, and number of chains per unit cell[a]	Molecular conformation	Crystal density (g/cm³)
it-Polyisobutylaldehyde[88] $(\text{—CH—O—})_n$ with side group $CH(CH_3)_2$	Tetragonal, $c = 5.2$ Å	Helix (4/1)	—
it-Polyisovaleroaldehyde[88] $(\text{—CH—O—})_n$ with side group $CH_2CH(CH_3)_2$	Tetragonal, $a = 20.6$ Å, $c = 5.2$ Å, $N = 4$	Helix (4/1)	1.037
it-Poly (propylene oxide)[89] $[\text{—}CH_2CH(CH_3)\text{—O—}]_n$	Orthorhombic, $P2_12_12_1\text{-}D_2^{\ 4}$, $a = 10.46$ Å, $b = 4.66$ Å, $c = 7.03$ Å, $N = 2$	Slightly distorted planar zigzag (2/1)	1.126
it-Poly (isopropylethylene oxide)[90] $(\text{—}CH_2\text{—CH—O—})_n$ with side group $CH(CH_3)_2$	Orthorhombic, $P2_12_12_1\text{-}D_2^{\ 4}$, $a = 12.85$ Å, $b = 7.52$ Å, $c = 5.55$ Å, $N = 2$	Helix (2/1)	1.07
it-Poly (*tert*-butylethylene oxide) $(\text{—}CH_2\text{—CH—O—})_n$ with side group $C(CH_3)_3$	Form I (racemic),[91] tetragonal, $P\bar{4}n2\text{-}D_{2d}^{\ 8}$, $a = 15.42$ Å, $c = 24.65$ Å, $N = 4$	Helix (9/4)	1.02
	Form II (optically active),[92] $c = 11.05$ Å,	—	—
Poly (isobutylene oxide)[93] $[\text{—}CH_2C(CH_3)_2\text{—O—}]_n$	Orthorhombic, $P2_12_12_1\text{-}D_2^{\ 4}$, $a = 10.76$ Å, $b = 5.76$ Å, $c = 7.00$ Å, $N = 2$	Distorted planar zigzag (2/1) $(TT'\bar{T}\bar{T}')_2$ (180, 153, − 153°)	1.10

Poly (3, 3-dimethyl oxacyclobutane)[79] $[\text{—CH}_2\text{—C(CH}_3)_2\text{—CH}_2\text{—O—}]_n$	Form I, $c = 4.83$ Å	Planar zigzag (1/0)	—
	Form II, monoclinic, $P2_1/c\text{-}C_{2h}{}^5$, $a = 8.93$Å, $b = 7.48$Å, $c = 8.35$Å, $\beta = 97.9°$, $N = 2$	Glide-type (2/0), $T_3GT_3\bar{G}$	1.037
	Form III, orthorhombic, $C222_1\text{-}D_2{}^5$, $a = 15.6$Å, $b = 5.74$Å, $c = 6.51$Å, $N = 2$	Helix (2/1) $(T_2G_2)_2$	0.982
Poly [(3,3-bis(chloromethyl) oxacyclobutane][94] $[\text{—CH}_2\text{—C(CH}_2\text{C1)}_2\text{—CH}_2\text{—O—}]_n$	α-Form, orthorhombic, $Pnam\text{-}D_{2h}{}^{16}$, $a = 17.85$Å, $b = 8.16$Å, $c = 4.67$Å, $N = 4$	Planar zigzag (1/0)	1.514
	β-Form, orthorhombic, $Bb2_1m\text{-}C_{2v}{}^{12}$, $a = 13.01$Å, $b = 11.71$Å, $c = 4.67$Å, $N = 4$	Planar zigzag (1/0)	1.448
Poly(*trans*-2-butene oxide)[95] $[\text{—CH(CH}_3)\text{—CH(CH}_3)\text{—O—}]_n$	Orthorhombic, $P2_12_12_1\text{-}D_2{}^4$, $a = 13.72$Å, $b = 4.60$Å, $c = 6.90$Å, $N = 2$	Slighly distorted planar zigzag (2/1)	1.10
Poly (*cis*-2-butene oxide)[95]	Orthorhombic?, $a = 11.20$Å, $b = 10.44$Å, $c = 7.01$Å, $N = 4$	Planar zigzag?	1.17
Poly (3, 3, 3-trifluoro-1, 2-epoxypropane)[96] $[\text{—CH}_2\text{—CH(CF}_3)\text{—O—}]_n$	Orthorhombic, $P2_12_12_1\text{-}D_2{}^4$, $a = 11.42$Å, $b = 6.26$Å, $c = 6.26$Å, $N = 2$	Distorted planar zigzag (2/1)	1.66
Hexafluoroacetone–ethylene alternating copolymer[97] $[\text{—(CH}_2)_2\text{—O—C(CF}_3)_2\text{—}]_n$	Orthorhombic, $P2_12_12_1\text{-}D_2{}^4$, $a = 10.63$Å, $b = 8.00$Å, $c = 8.01$Å, $N = 2$	Helix (2/1) $(TTTC')_2$	1.89
Poly (*p*-phenylene oxide)[98] 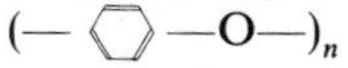	Orthorhombic, $Pbcn\text{-}D_{2h}{}^{14}$, $a = 8.07$Å, $b = 5.54$Å, $c = 9.72$Å, $N = 2$	Helix (2/1)	1.408
Poly (2, 6-diphenyl-*p*-phenylene oxide)[99] 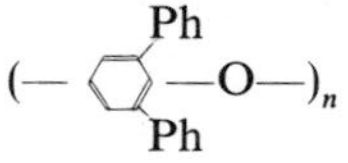	Tetragonal, $P4_12_12\text{-}D_4{}^4$ or $P4_32_12\text{-}D_4{}^8$, $a = 12.51$Å, $c = 17.08$Å, $N = 2$	Helix (4/1)	1.214

Table 7.1 *(Contd.)*

Polymer	Crystal system, space group, lattice constants, and number of chains per unit cell[a]	Molecular conformation	Crystal density (g/cm^3)
it-Poly (vinyl methyl ether)[100] $(\text{—}CH_2\text{—}CH\text{—})_n$ with side group OCH_3	Trigonal, $R3\text{-}C_{3i}^{2}$, $a = 16.25$ Å, $c = 6.50$ Å, $N = 6$	Helix (3/1)	1.168
it-Poly (vinyl isopropyl ether)[101] $(\text{—}CH_2\text{—}CH\text{—})_n$ with side group $OCH(CH_3)_2$	Tetragonal, $a = 17.2$ Å, $c = 35.5$ Å, $N = 4$	Helix (17/5)	0.93
it-Poly (vinyl isobutyl ether)[102,103] $(\text{—}CH_2-CH\text{—})_n$ with side group $OCH_2CH(CH_3)_2$	Orthorhombic, $a = 16.8$ Å, $b = 9.70$ Å, $c = 6.50$ Å, $N = 2$	Helix (3/1)	0.942
it-Poly (vinyl *tert*-butyl ether)[104] $(\text{—}CH_2\text{—}CH\text{—})_n$ with side group $O\text{—}C(CH_3)_3$	Tetragonal, $I4_1/a\text{-}C_{4h}^{6}$, $a = 18.84$ Å, $c = 7.65$ Å, $N = 4$	Helix (4/1)	0.980
it-Poly (vinyl neopentyl ether)[101] $(\text{—}CH_2\text{—}CH\text{—})_n$ with side group $OCH_2\text{—}C(CH_3)_3$	Orthorhombic, $a = 18.2$ Å, $b = 10.5$ Å, $c = 6.50$ Å, $N = 2$	Helix (3/1)	0.916
it-Poly (vinyl benzyl ether)[105] $(\text{—}CH_2\text{—}CH\text{—})_n$ with side group $OCH_2\text{—}C_6H_5$	$c = 6.30$ Å	Helix (3/1)	—

Polythioethers			
Polythiomethylene[106] $(\text{—}CH_2S\text{—})_n$	Trigonal, $P1\text{-}C_1{}^1$, $a = b = 5.07$Å, $c = 36.52$Å, $\gamma = 120°$, $N = 1$	Helix (17/9)	1.60
Poly (ethylene sulfide)[107] $[\text{—}(CH_2)_2\text{—}S\text{—}]_n$	Orthorhombic, $Pbcn\text{-}D_{2h}{}^{14}$, $a = 8.50$Å, $b = 4.95$Å, $c = 6.70$Å, $N = 2$	Glide-type (2/0) $TG_2T\bar{G}_2$	1.416
Poly (trimethylene sulfide)[108] $[\text{—}(CH_2)_3\text{—}S\text{—}]_n$	Monoclinic, $Pc\text{-}C_s{}^2$, $a = 5.16$Å, b (f.a.) $= 4.06$Å, $c = 10.33$Å, $\beta = 120.5°$, $N = 2$	Helix (1/1) G_4	1.32
Poly (pentamethylene sulfide)[109] $[\text{—}(CH_2)_5\text{—}S\text{—}]_n$	Monoclinic, $P2_1/a\text{-}C_{2h}{}^5$, $a = 9.61$Å, $b = 9.78$Å, $c = 7.84$Å, $\beta = 131°$, $N = 4$	Planar zigzag (1/0)	1.223
it-Poly (propylene sulfide)[110] $(\text{—}CH_2\text{—}CH(CH_3)\text{—}S\text{—})_n$	Orthorhombic, $P2_12_12_1\text{-}D_2{}^4$, $a = 9.95$Å, $b = 4.89$Å, $c = 8.20$Å, $N = 2$	Slightly distorted planar zigzag (2/1)	1.234
it-Poly (*tert*-butylethylene sulfide)[111] $(\text{—}CH_2\text{—}CH(C(CH_3)_3)\text{—}S\text{—})_n$	Racemic form, monoclinic, $P2_1/a\text{-}C_{2h}{}^5$, $a = 16.67$Å, $b = 19.27$Å, $c = 6.52$Å, $\beta = 90°$, $N = 4$	Helix (3/1) $(SG\bar{S})_3$	1.10
	Optically active form, trigonal, $P3_1\text{-}C_3{}^2$, $a = 16.91$Å, $c = 6.50$Å, $N = 3$	Helix (3/1) $(SG\bar{S})_3$	1.08
Poly (*p*-phenylene sulfide)[112] (—  —S—)$_n$	Orthorhombic, $Pbcn\text{-}D_{2h}{}^{14}$, $a = 8.67$Å, $b = 5.61$Å, $c = 10.26$Å, $N = 2$	Helix (2/1)	1.44
Polyesters			
Polyglycolide[113] $(\text{—}CH_2COO\text{—})_n$	Orthorhombic, $Pcmn\text{-}D_{2h}{}^{16}$, $a = 5.22$Å, $b = 6.19$Å, $c = 7.02$Å, $N = 2$	Planar zigzag (2/1)	1.70

Table 7.1 (*Contd.*)

Polymer	Crystal system, space group, lattice constants, and number of chains per unit cell[a]	Molecular conformation	Crystal density (g/cm^3)
Poly-β-propiolactone[114] $[-(CH_2)_2-COO-]_n$	Form II, $a' = 7.73$Å, $b' = 4.48$Å, $c = 4.77$Å, $\gamma' = 90°$, $N = 2$, paracrystallinelike	Planar zigzag (1/0)	1.445
Poly-ε-caprolactone[115] $[-(CH_2)_5-COO-]_n$	Orthorhombic, $P2_12_12_1$-D_2^4, $a = 7.47$Å, $b = 4.98$Å, $c = 17.05$Å, $N = 2$	Slightly distorted planar zigzag (2/1)	1.20
it-Poly (β-methyl-β-propiolactone)[116,117] $[-CH(CH_3)CH_2COO-]_n$	Orthorhombic, $P2_12_12_1$-D_2^4, $a = 5.76$Å, $b = 13.20$Å, $c = 5.96$Å, $N = 2$	Helix (2/1) $(\bar{G}_2T_2)_2$	1.26
it-Poly(β-ethyl-β-propiolactone)[118] $[-CH(C_2H_5)-CH_2-COO-]_n$	Orthorhombic, $P2_12_12_1$-D_2^4, $a = 9.32$Å, $b = 10.02$Å, $c = 5.56$Å, $N = 2$	Helix (2/1) $(\bar{G}\bar{C}TS)_2$	1.28
Polypivalolactone[119,120] $[-CH_2-C(CH_3)_2-COO-]_n$	Monoclinic, $P2_1/b$-C_{2h}^5, $a = 9.05$Å, $b = 11.58$Å, $c = 6.03$Å, $\beta = 121.5°$, $N = 2$	Helix (2/1) $(T_2G_2)_2$	1.234
Polydiketene[121] $[-CH(=CH_2)-CH_2-COO-]_n$	Orthorhombic, $P22_12_1$-D_2^3, $a = 5.43$Å, $b = 8.94$Å, $c = 7.75$Å, $N = 2$	Helix (2/1) $(T_2G_2)_2$	1.50
Poly(ethylene oxalate)[122] $[-O-(CH_2)_2-O-COCO-]_n$	Orthorhombic, $Pbcn$-D_{2h}^{14}, $a = 6.44$Å, $b = 6.22$Å, $c = 11.93$Å, $N = 2$	(2/0) $T_5GT_5\bar{G}$	1.613
Poly(ethylene succinate)[122] $[-O-(CH_2)_2-O-CO-(CH_2)_2-CO-]_n$	Orthorhombic, $Pbnb$-D_{2h}^{10}, $a = 7.60$Å, $b = 10.75$Å, $c = 8.33$Å, $N = 4$	(1/0) $T_3GT_3\bar{G}$	1.41
Poly(ethylene adipate)[123] $[-O-(CH_2)_2-O-CO-(CH_2)_4-CO-]_n$	Monoclinic, $P2_1/a$-C_{2h}^5, $a = 5.47$Å, $b = 7.23$Å, $c = 11.72$Å, $\beta = 113°30'$, $N = 2$	Distorted planar zigzag (1/0)	1.345
Poly(ethylene suberate)[123] $[-O-(CH_2)_2-O-CO-(CH_2)_6-CO-]_n$	Monoclinic,$P2_1/a$-C_{2h}^5, $a = 5.51$Å, $b = 7.25$Å, $c = 14.28$Å, $\beta = 114°30'$, $N = 2$	Distorted planar zigzag (1/0)	1.281

Poly(ethylene sebacate)[124, 125] [—O—$(CH_2)_2$—O—CO—$(CH_2)_8$—CO—]$_n$	Monoclinic, $P2_1/a\text{-}C_{2h}{}^5$, $a = 5.52$Å, $b = 7.30$Å, $c = 16.65$Å, $\beta = 115.0°$, $N = 2$	Distorted planar zigzag (1/0)	1.244
Poly(tetramethylene succinate)[126] [—O—$(CH_2)_4$—O—CO—$(CH_2)_2$—CO—]$_n$	Monoclinic, $P2_1/n\text{-}C_{2h}{}^5$, $a = 5.21$Å, $b = 9.14$Å, $c = 10.94$Å, $\beta = 124°$, $N = 2$	(1/0) $T_7GT\bar{G}$	1.32
Poly(ethylene terephthalate)[127] [—O—$(CH_2)_2$—O—CO—C_6H_4—CO—]$_n$	Triclinic, $P\bar{1}\text{-}C_i{}^1$, $a = 4.56$Å, $b = 5.94$Å, $c = 10.75$Å, $\alpha = 98.5°$, $\beta = 118°$, $\gamma = 112°$, $N = 1$	Nearly planar	1.455
Poly(ethylene isophthalate)[128] [—O—$(CH_2)_2$—O—CO—C_6H_4—CO—]$_n$	Form II, triclinic, $P\bar{1}\text{-}C_i{}^1$, $a = 5.41$Å, $b = 6.35$Å, $c = 21.2$Å, $\alpha = 116.5°$, $\beta = 135.5°$, $\gamma = 83.5°$, $N = 1$	Nearly planar	1.470
Poly(trimethylene terephthalate)[129] [—O—$(CH_2)_3$—O—CO—C_6H_4—CO—]$_n$	Triclinic, $P\bar{1}\text{-}C_i{}^1$, $a = 4.58$Å, $b = 6.22$Å, $c = 18.12$Å, $\alpha = 96.9°$, $\beta = 89.4°$, $\gamma = 111.0°$, $N = 1$ (two monomeric units)	—	1.43
Poly(butylene terephthalate) [—O—$(CH_2)_4$—O—CO—C_6H_4—CO—]$_n$	α-Form,[130–132] triclinic, $P\bar{1}\text{-}C_i{}^1$, $a = 4.83$Å, $b = 5.94$Å, $c = 11.59$Å, $\alpha = 99.7°$, $\beta = 115.2°$, $\gamma = 110.8°$, $N = 1$	See text	1.40
	β-Form,[130, 132–134], triclinic, $P\bar{1}\text{-}C_i{}^1$, $c = 12.95$Å, $N = 1$	—	—
Poly(ethylene naphthalate)[135] [—O—$(CH_2)_2$—O—CO—$C_{10}H_6$—CO—]$_n$	Triclinic, $P\bar{1}\text{-}C_i{}^1$, $a = 6.51$Å, $b = 5.75$Å, $c = 13.2$Å, $\alpha = 81.3°$, $\beta = 144°$, $\gamma = 100°$, $N = 1$	—	1.407
Polyamides			
Nylon 3[136,137] [—$(CH_2)_2$—CONH—]$_n$	α-Form, triclinic, $P1\text{-}C_1{}^1$, $a = 9.3$Å, $b = 8.7$Å, $c = 4.8$Å, $\alpha = \beta = 90°$, $\gamma = 60°$, $N = 4$	Planar zigzag (1/0)	1.40
Nylon 4[138] [—$(CH_2)_3$—CONH—]$_n$	α-Form, monoclinic, $P2_1\text{-}C_2{}^2$, $a = 9.29$Å, b(f.a.) $= 12.24$Å, $c = 7.97$Å, $\beta = 114.5°$, $N = 4$	Planar zigzag (2/1)	1.37

Table 7.1 (*Contd.*)

Polymer	Crystal system, space group, lattice constants, and number of chains per unit cell[a]	Molecular conformation	Crystal density (g/cm^3)
Nylon 5[136] $[—(CH_2)_4—CONH—]_n$	α-Form, triclinic, $P1$-$C_1{}^1$, $a = 9.5$ Å, $b = 5.6$ Å, $c = 7.5$ Å, $\alpha = 48°$, $\beta = 90°$, $\gamma = 67°$, $N = 2$	Planar zigzag (1/0)	1.30
Nylon 6 $[—(CH_2)_5—CONH—]_n$	α-Form,[139] monoclinic, $P2_1$-$C_2{}^2$, $a = 9.56$ Å, b (f.a.) = 17.2 Å, $c = 8.01$ Å, $\beta = 67.5°$, $N = 4$	Planar zigzag (2/1)	1.23
	γ-Form,[140] monoclinic, $P2_1/a$-$C_{2h}{}^5$, $a = 9.33$ Å, b(f.a.) = 16.88 Å, $c = 4.78$ Å, $\beta = 121°$, $N = 2$	Helix (2/1) $(T_4 S T \bar{S})_2$	1.17
Nylon 7[136] $[—(CH_2)_6—CONH—]_n$	α-Form, triclinic, $P1$-$C_1{}^1$, $a = 9.8$ Å, $b = 10.0$ Å, $c = 9.8$ Å, $\alpha = 56°$, $\beta = 90°$, $\gamma = 69°$, $N = 4$	Planar zigzag (1/0)	1.19
Nylon 8[141] $[—(CH_2)_7—CONH—]_n$	α-Form, monoclinic, $P2_1$-$C_2{}^2$, $a = 9.8$ Å, b(f.a.) = 22.4 Å, $c = 8.3$ Å, $\beta = 65°$, $N = 4$	Planar zigzag (2/1)	1.14
	γ-Form, $a = b = 4.79$ Å, $c = 21.7$ Å, $\alpha = \beta = 90°$, $\gamma = 60°$, $N = 1$	$(T_6 S T \bar{S})_2$ (2/1)	1.09
Nylon 9[136] $[—(CH_2)_8—CONH—]_n$	α-Form, triclinic, $P1$-$C_1{}^1$, $a = 9.7$ Å, $b = 9.7$ Å, $c = 12.6$ Å, $\alpha = 64°$, $\beta = 90°$, $\gamma = 67°$, $N = 4$	Planar zigzag (1/0)	1.07
Nylon 11[136] $[—(CH_2)_{10}—CONH—]_n$	α-Form, triclinic, $P1$-$C_1{}^1$, $a = 9.5$ Å, $b = 10.0$ Å, $c = 15.0$ Å, $\alpha = 60°$, $\beta = 90°$, $\gamma = 67°$, $N = 4$	Planar zigzag (1/0)	1.09
Nylon 12[142,143] $[—(CH_2)_{11}—CONH—]_n$	γ-Form, monoclinic, $P2_1/c$-$C_{2h}{}^5$, $a = 9.38$ Å, b (f.a.) = 32.2 Å, $c = 4.87$ Å, $\beta = 121.5°$, $N = 2$	Helix (2/1) $(T_{10} S T \bar{S})_2$	1.04

Nylon 66[144] $[-NH-(CH_2)_6-NHCO-(CH_2)_4-CO-]_n$	α-Form, triclinic, $P\bar{1}$-C_i^1, $a = 4.9$Å, $b = 5.4$Å, $c = 17.2$Å, $\alpha = 48.5°$, $\beta = 77°$, $\gamma = 63.5°$, $N = 1$	Planar zigzag (1/0)	1.24
	β-Form, triclinic, $P\bar{1}$-C_i^1, $a = 4.9$Å, $b = 8.0$Å, $c = 17.2$Å, $\alpha = 90°$, $\beta = 77°$, $\gamma = 67°$, $N = 2$	Planar zigzag (1/0)	1.248
Nylon 6 10[144] $[-NH-(CH_2)_6-NHCO-(CH_2)_8-CO-]_n$	α-Form, triclinic, $P\bar{1}$-C_i^1, $a = 4.95$Å, $b = 5.4$Å, $c = 22.4$Å, $\alpha = 49°$, $\beta = 76.5°$, $\gamma = 63.5°$, $N = 1$	Planar zigzag (1/0)	1.157
	β-Form, triclinic, $P\bar{1}$-C_i^1, $a = 4.9$Å, $b = 8.0$Å, $c = 22.4$ Å, $\alpha = 90°$, $\beta = 77°$, $\gamma = 67°$, $N = 2$	Planar zigzag (1/0)	1.196
Nylon 77[145] $[-NH-(CH_2)_7-NHCO-(CH_2)_5-CO-]_n$	γ-Form, pseudohexagonal, Pm-C_s^1, $a = 4.82$Å, b (f.a.) $= 19.0$Å, $c = 4.82$Å, $\beta = 60°$, $N = 1$	(1/0) $T_6ST\bar{S}T_4ST\bar{S}$	1.105
Poly(*m*-xylene adipamide)[146] $[-CH_2-C_6H_4-CH_2-NHCO-(CH_2)_4$ $-CONH-]_n$ 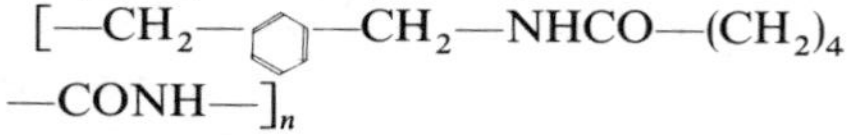	Triclinic, $P\bar{1}$-C_i^1, $a = 12.01$Å, $b = 4.83$Å, $c = 29.8$Å, $\alpha = 75.0°$, $\beta = 26.0°$, $\gamma = 65.0°$, $N = 1$	Slightly distorted planar zigzag (2/0)	1.250
Poly-*p*-benzamide[147] $[-C_6H_4-CONH-]_n$ 	Orthorhombic, $P2_12_12_1$-D_2^4, $a = 7.71$Å, $b = 5.14$Å, $c = 12.8$Å, $N = 2$	$(TC)_2$ (2/1)	1.54
Poly(*p*-phenylene terephthalamide)[147, 148] $[-NH-C_6H_4-NHCO-C_6H_4-CO-]_n$ 	Monoclinic, $P2_1/n$-C_{2h}^5, $a = 7.80$Å, $b = 5.19$Å, $c = 12.9$Å, $\gamma = 90°$, $N = 2$	$TTTT$	1.50
Poly(*m*-phenylene isophthalamide)[149] $[-NH-C_6H_4-NHCO-C_6H_4-CO-]_n$ 	Triclinic, $P1$-C_1^1, $a = 5.27$Å, $b = 5.25$Å, $c = 11.3$Å, $\alpha = 111.5°$, $\beta = 111.4°$, $\gamma = 88.0°$, $N = 1$	$\bar{S}GT\bar{G}ST$ (1/0)	1.45
Other polymers			
Poly(ethylene oxybenzoate) $[-(CH_2)_2-O-C_6H_4-CO-O-]_n$ 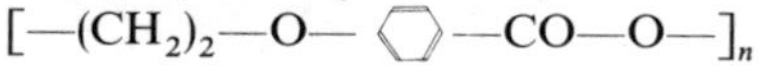	α-Form,[150] orthorhombic, $P2_12_12_1$-D_2^4, $a = 10.49$Å, $b = 4.75$Å, $c = 15.60$Å, $N = 2$	(2/1)	1.41
	β-Form,[151] monoclinic, $P2_1/n$-C_{2h}^5, $a = 8.19$Å, $b = 11.07$Å, $c = 19.05$Å, $\beta = 114.8°$, $N = 4$	Almost extended	1.390

Table 7.1 (*Contd.*)

Polymer	Crystal system, space group, lattice constants, and number of chains per unit cell[a]	Molecular conformation	Crystal density (g/cm^3)
Poly(dimethyl ketene) $[—C(CH_3)_2—CO—]_n$	α-Form,[152] orthorhombic, $P2_1cn$-C_{2v}^{9}, $a = 12.85$ Å, $b = 6.53$ Å, $c = 8.80$ Å, $N = 2$	Glide-type (4/0) $TG_3T\bar{G}_3$	1.31
	β-Form,[153] $c = 4.40$ Å	Helix (2/1) G_4	—
Poly-*p*-xylylene [—⬡—$(CH_2)_2$—]$_n$	α-Form,[154] monoclinic, $C2/m$-C_{2h}^{3}, $a = 5.92$ Å, $b = 10.64$ Å, $c = 6.55$ Å, $\beta = 134.7°$, $N = 2$	(1/0)	1.18
	β-Form,[155] $c = 6.581$ Å	(1/0)	—
Poly(4, 4′-isopropylidene diphenylene carbonate)[156] [—⬡—$C(CH_3)_2$—⬡—O—CO—O—]$_n$	Monoclinic, Pc-C_s^{2}, $a = 12.3$ Å, $b = 10.1$ Å, $c = 20.8$ Å, $\beta = 84°$, $N = 4$	(2/1)	1.315
Poly(4, 4′-thiodiphenylene carbonate)[156] [—⬡—S—⬡—O—CO—O—]$_n$	Orthorhombic, $P2_1cn$-C_{2v}^{9}, $a = 5.6$ Å, $b = 8.7$ Å, $c = 22.2$ Å, $N = 2$	(2/1)	1.50
Poly(4, 4′-methylene diphenylene carbonate)[156] [—⬡—CH_2—⬡—O—CO—O—]$_n$	Orthorhombic, $P2_1cn$-C_{2v}^{9}, $a = 5.0$ Å, $b = 10.5$ Å, $c = 22.0$ Å, $N = 2$	(2/1)	1.303
Polyketone[157, 158] $(—CH_2CH_2—CO—)_n$	Orthorhombic, $Pnam$-D_{2h}^{16}, $a = 7.97$ Å, $b = 4.76$ Å, $c = 7.57$ Å, $N = 2$	Planar zigzag (2/1)	1.296
Poly(phosphonitrile chloride)[159, 160] $(—N{=}PCl_2—)_n$	Orthorhombic, $Pn2_1a$-C_{2v}^{9}, $a = 11.07$ Å, b (f.a.) $= 4.92$ Å, $c = 12.72$ Å, $N = 4$	Helix (2/1) $(T'C')_2$ (156°, 14°)	2.22
Rubber hydrochloride[161] $[—CH_2—C(CH_3)Cl—CH_2—CH_2—]_n$	Monoclinic, $P2_1/c$-C_{2h}^{5}, $a = 5.83$ Å, $b = 10.38$ Å, $c = 8.95$ Å, $\beta = 90°$, $N = 2$	Glide-type (2/0) $T_3GT_3\bar{G}$	1.282

Fibrous sulfur[162] (—S—)$_n$	Monoclinic, $P2\text{-}C_2^{\,1}$, $a = 17.6$Å, $b = 9.25$ Å, $c = 13.8$Å, $\beta = 113°$, $N = 8$	Helix (10/3)	2.06
Poly[1, 2-bis(*p*-tolylsulfonyloxymethylene)-1-butene-3-inylene][163] (—CR=CR—C≡C—)$_n$, R = —CH_2OSO_2—C_6H_4—CH_3	Monoclinic, $P2_1/c\text{-}C_{2h}^{\,5}$, $a = 14.493$Å, b (f.a.) = 4.910Å, $c = 14.936$Å, $\beta = 118.14°$, $N = 2$	Planar	1.483
Ethylene–butene-2 alternating copolymer[164] [—$(CH_2)_2$—$CH(CH_3)$—$CH(CH_3)$—]$_n$	Monoclinic, $P2_1/b\text{-}C_{2h}^{\,5}$, $a = 10.92$Å, $b = 7.75$Å, $c = 9.15$Å, $\gamma = 130°$, $N = 2$	Glide-type (2/0) $T_3GT_3\bar{G}$	0.94
2, 6-Polyurethane[165] [—O—$(CH_2)_2$—O—CONH—$(CH_2)_6$-—NHCO—]$_n$	Form I, triclinic, $P\bar{1}\text{-}C_i^{\,1}$, $a = 4.93$Å, $b = 4.58$Å, $c = 16.8$Å, $\alpha = 113°$, $\beta = 103°$, $\gamma = 109°$, $N = 1$	Planar zigzag (1/0)	1.27
	Form II, triclinic, $P\bar{1}\text{-}C_i^{\,1}$, $a = 4.59$Å, $b = 5.14$Å, $c = 13.9$Å, $\alpha = \beta = 90°$, $\gamma = 119°$, $N = 1$	(1/0)	1.33
3, 6-Polyurethane[165] [—O—$(CH_2)_3$—O—CONH—$(CH_2)_6$-—NHCO—]$_n$	Monoclinic, $A2/m\text{-}C_{2h}^{\,3}$, $a = 4.70$Å, $b = 8.36$Å, $c = 33.9$Å, $\gamma = 115°$, $N = 2$	(2/1)	1.34
Polymer complexes			
Poly(ethylene oxide)-$HgCl_2$ complex	Type I,[166] orthorhombic, $Ccmm\text{-}D_{2h}^{\,17}$, $Ccm2_1\text{-}C_{2v}^{\,12}$, $Cc2m\text{-}C_{2v}^{\,16}$, or $C222_1\text{-}D_2^{\,5}$, $a = 13.55$Å, $b = 8.58$Å, $c = 11.75$Å, 16(CH_2CH_2O) and 4 $HgCl_2$	Glide-type (4/0) $T_5GT_5\bar{G}$	2.18
	Type II,[167] orthorhombic, $Pncm\text{-}D_{2h}^{\,7}$ or $Pnc2\text{-}C_{2v}^{\,6}$, $a = 7.75$Å, $b = 12.09$Å, $c = 5.88$Å, 4 (CH_2CH_2O) and 4 $HgCl_2$	Glide-type (2/0) $TG_2T\bar{G}_2$	3.79
Poly(2, 3-dichlorobutadiene)–thiourea complex[67]	Monoclinic, $P2_1/a\text{-}C_{2h}^{\,5}$, $a = 9.91$Å, $b = 15.85$Å, $c = 12.5$Å, $\beta = 114.1°$, 5.2 monomeric units and 12 thiourea	*trans*-$ST\bar{S}$	1.44

Table 7.1 (*Contd.*)

Polymer	Crystal system, space group, lattice constants, and number of chains per unit cell[a]	Molecular conformation	Crystal density (g/cm^3)
Poly(2,3-dimethylbutadiene)–thiourea complex[168]	Monoclinic, $P2_1/a$-C_{2h}^5, $a = 10.40$Å, $b = 15.47$Å, $c = 12.5$Å, $\beta = 114.4°$, 5.2 monomeric units and 12 thiorea	*trans*-$ST\bar{S}$	1.22
trans-1, 4-Polybutadiene–urea complex[169]	Hexagonal, $P6_122$-D_6^2, $a = 8.22$Å, $c = 11.01$ Å, 2.28 monomeric units and 6 urea	*trans*-$ST\bar{S}$	—
trans-1, 4-Polybutadiene-PHTP complex[170]	Hexagonal, $P6_3/m$-C_{6h}^2, $a = 14.26$Å, $c = 4.78$Å, 1 monomeric unit and 2 PHTP	*trans*-$ST\bar{S}$	1.08
Polyethylene–PHTP complex[171–174]	Hexagonal, $P6_3/m$-C_{6h}^2, $a = 14.34$Å, $c = 4.78$Å, 2 PHTP	TT	—

[a] f.a. indicates fiber axis.
[b] The structure of *trans*-polyheptenamer is similar to that of *trans*-polypentenamer with $c = 17.10$ Å.
[c] The structures of *trans*-polydecenamer and polydodecenamer are similar to those of *trans*-polyoctenamer with different c values.
[d] $[-(CH_2)_8-O-]_n$ has a structure similar to $[-(CH_2)_6-O-]_n$ with $c = 22.44$ Å.
[e] $[-(CH_2)_{12}-O-]_n$ has a structure similar to $[-(CH_2)_{10}-O-]_n$ with $c = 32.50$ Å.
[f] PHTP is *trans*-anti-*trans*-anti-*trans*-perhydrotriphenylene

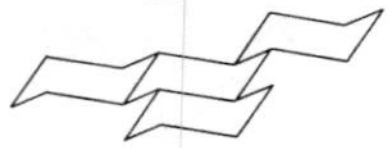

Table 7.2 Lattice Constants of Orthorhombic PE at Various Temperatures

Temperature (°K)[a]	Lattice constants (Å) *a*	*b*	*c*	Specific volume (cm^3/g)
4[b]	7.121	4.851	2.548	0.945
10[c]	7.16	4.86	2.534	0.947
77[c]	7.18	4.88	2.534	0.953
77[d]	7.155	4.899	2.5473	0.959
90[b]	7.161	4.866	2.546	0.953
195[c]	7.27	4.91	2.534	0.971
293[e]	7.399 (7.432)	4.946 (4.945)	2.543 (2.543)	0.999 (1.003)
297[c]	7.42	4.96	2.534	1.001
297[f]	7.40	4.93	2.534	0.993
297[g]	7.36	4.92	2.534	0.985
303[d]	7.414	4.942	2.5473	1.002

[a] Data obtained by X-ray diffraction studies unless otherwise indicated.
[b] Data obtained by neutron diffraction by Avitable et al.[181]
[c] Data from Shen et al.[180]
[d] Data from Swan.[182]
[e] The data of nonoriented samples are given without parentheses and those of oriented samples are given with parentheses, both for high-density PE as determined by Chatani et al.[4]
[f] Data from Bunn.[1]
[g] Data from Walter and Reding.[183]

The monoclinic form is obtained by rolling or biaxial stretching, but this form is metastable and transforms just below the melting point into the orthorhombic form on heating.[5,185,186] Recently a high pressure form was found (5000 kg/cm^2, 240°C).[6–8]

Further references include: infrared bands due to folded parts, Ref. 187; infrared studies of lamellar linking by cilia in PE single crystal mats, Ref. 188; specific heat and root-mean-squared displacements, Ref. 189; influence of defects on the infrared spectra, Ref. 190; and PE sample with high elastic modulus (70×10^{10} dyn/cm^2), Refs. 191 and 192.

7.2 POLYTETRAFLUOROETHYLENE AND RELATED POLYMERS

The structure (Section 2.3) and transition (Section 4.10.5.C) of polytetrafluoroethylene are described earlier. In addition to the usual crystal forms, Brown[192a] found a planar zigzag form from infrared studies under a high pressure, and Flack[193] determined the crystal structure of the monoclinic form. They used anvil-type high-pressure cells. Thereafter Takemura et al.[7,12] determined an orthorhombic structure stable under a hydrostatic pressure. Zerbi et al.[194,195] have reported the existence of planar zigzag parts in the structure under atmospheric pressure as determined by infrared spectroscopic studies.

7.3 POLY-α-OLEFINS

7.3.1 Isotactic Polymers

The crystal structure of *it*-PP is discussed in Section 4.10.5.B. There are restrictions on the formation of helices of *it*-polymers due to steric hindrance. A hypothetically extended planar zigzag chain may be placed in two ways if the zigzag plane is coincident to the page and the side chains are towards the left-hand side. The possible structures are shown in Fig. 7.1. In Fig. 7.1*a* the side chains are below the page, and in *b* they are above the page. When chain *a* is rotated by 180° around an axis on the page and perpendicular to the chain axis, it turns into *b*. *it*-Polymers, for example, *it*-PP, cannot take the planar zigzag conformation owing to the steric hindrance between side chains. Instead it forms a (3/1) helix as shown in Fig. 4.86*a*, which is identical to Fig. 7.1*c*. In this case the bonds $R_2—C_4—C_5—C_6$ take the trans form and cannot take the gauche form. The conformation shown in Fig. 7.1*c*′ cannot occur, as the distance (1.7 Å) between R_2 and one hydrogen atom attached to the atom C_1 is too short.[196] Therefore a left-hand helix with the CH_3 group pointing upward as shown in Fig. 7.1*c* or a right-hand downward helix *d* results from the planar zigzag *a*, and a left-hand downward helix *e* or a right-hand upward helix *f* results from *b*. In case *a*, for example, the conformation *c* or *d* occurs according to whether the part $C_1—C_2—C_3—C_4$ takes the gauche or trans form, respectively. Thus the right- or left-hand helix does not correspond to case *a* or *b*, but depends on the manner of helix formation discussed above.

The structure of form II of *it*-poly-1-butene has been studied by many authors.[23–25,197] Corradini et al.[25] prepared form II using a copolymer with 3-methyl-1-butene (4%) and reported a rather large value of $a = 15.42$ Å. The values cited in Table 7.1 are those measured on a sample of pure form II

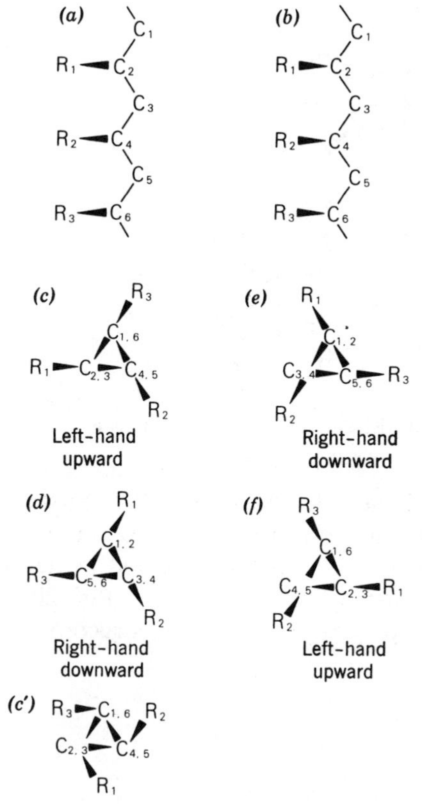

Fig. 7.1. Relationship of the planar zigzag to the helical conformation in an *it*-polymer. Helix *c* or *d* results from case *a*, and helix *e* or *f* results from *b*. Conformation *c′* cannot occur.

prepared by the method described in Section 4.3.1.C from a homopolymer.[23]

The crystal structure of *it*-poly(3-methyl-1-butene) was determined by Corradini et al.[28] as shown in Fig. 7.2. (4/1) helix molecules are packed in a monoclinic lattice instead of a tetragonal lattice.

The analysis of *it*-poly(4-methyl-1-pentene) was made independently by Corradini et al.[29] and Kusanagi et al.[30] Both molecular conformations are essentially the same. Kusanagi et al. introduced the disorder of anticlined chains and a small deviation from the uniform helix. For the special density phenomenon of this polymer, see Section 4.3.3.B.

Natta et al.[65] developed the close-packing consideration applied to helical molecules. When D_1 and D_2 are the helix radii of the main chain and the side group, respectively, and the ratio D_1/D_2 lies between 0.3 and 0.8, the most efficient way of packing is with each right-hand helix surrounded by four left-hand helices and vice versa. This rule was applied to the analyses of *it*-poly[(*R*), (*S*)-5-methyl-1-heptene][31] and *it*-1, 2-poly(4-methyl-1, 3-pentadiene).[65]

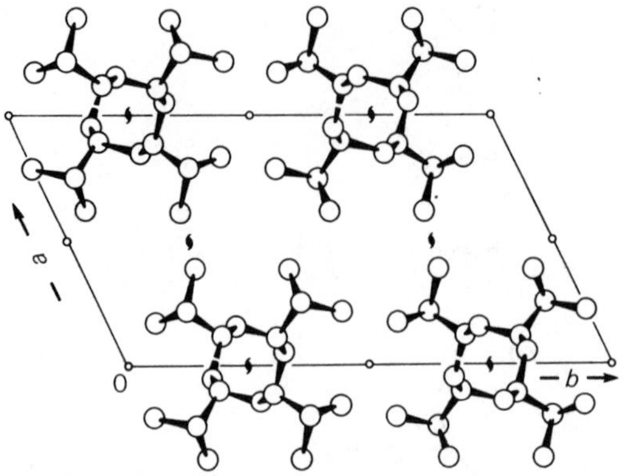

Fig. 7.2 Crystal structure of *it*-poly (3-methyl-1-butene). (From Corradini et al.[28])

Further references include: relation between melting point and crystal structure of *it*-poly-α-olefins with *n*-alkane side chains, Ref. 198; crystallizable and noncrystallizable *it*-polystyrene derivatives, Ref. 199; infrared bands characteristic of the helical structure of *it*-polystyrene and its derivatives, Ref. 200; infrared spectra of *it*-polystyrene solutions, Ref. 201; helix length and infrared spectra of *it*-polystyrene, Ref. 202; structure of

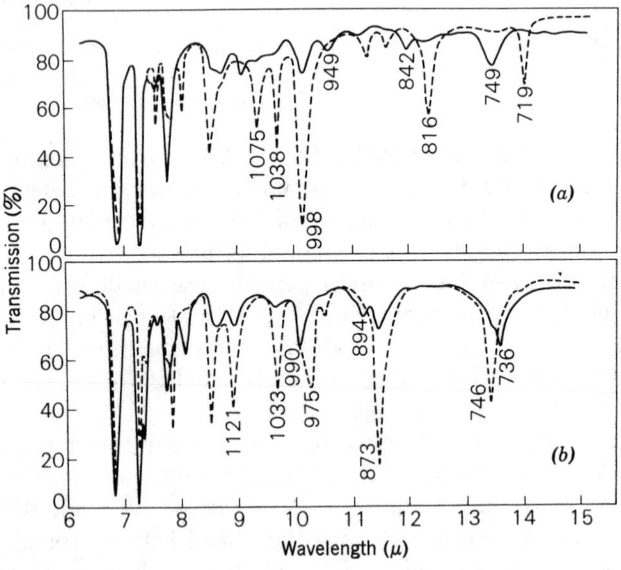

Fig. 7.3 Polarized infrared spectra of (*a*) poly-*trans*-1-deuteropropylene and (*b*) poly-*cis*-1-deuteropropylene. (———) Perpendicular radiation, (– – – –) parallel radiation. (From Peraldo and Farina.[204])

it-polymers with side chains involving asymmetric carbon atoms, Refs. 29 and 31; and conformational energy of *it*-PP, Ref. 203.

7.3.2 Diisotactic Polymers

Natta[204a] and Peraldo and Farina[204b] have reported that the diisotactic

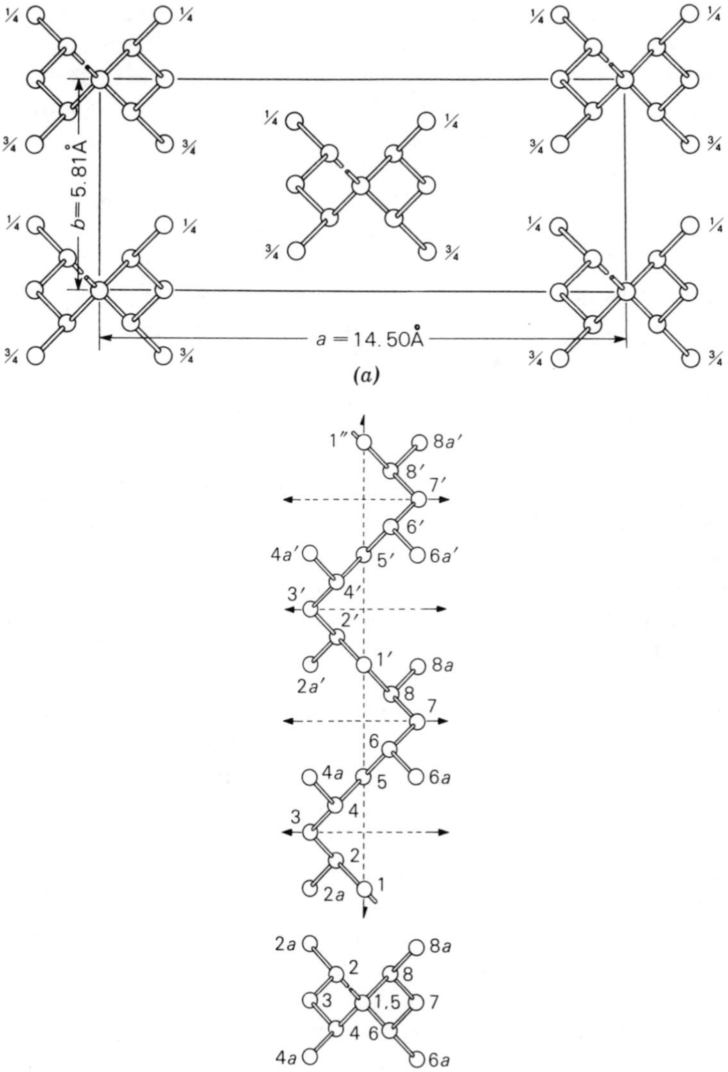

(*b*)

Fig. 7.4 (*a*) Crystal structure and (*b*) molecular structure of *st*-PP. (From Corradini et al.[19])

polymers obtained from *trans*- and *cis*-1-deuteropropylene by a Ziegler-type

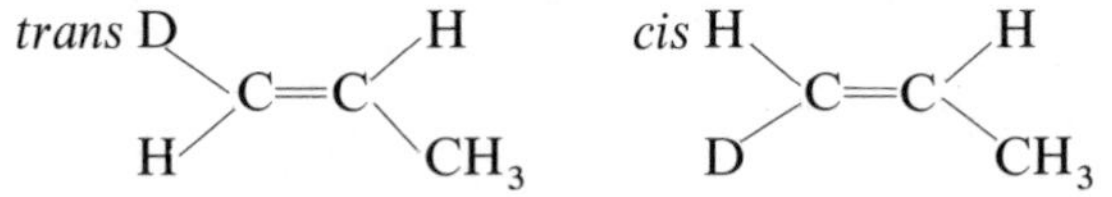

catalyst show similar X-ray diagrams, but give infrared spectra remarkably different from each other (Fig. 7.3). This fact suggested two types of structures: threo and erythro (Section 2.1). However which type of polymer was obtained from which monomer was not determined at that time. The correspondence was clarified by normal coordinate treatments for models of threo- and erythro-*dit*-poly-1-deuteropropylenes.[205,206] The study showed that the *trans*- and *cis*-polymers have the structures of threo- and erythro-*dit*-poly-1-deuteropropylenes, respectively. This conclusion suggests the cis-opening mechanism of the coordinated anionic polymerization.[207]

Natta et al.[207] prepared several *dit*-poly(β-substituted vinyl ethers), [—CHR—CH(OR′)—]$_n$, by cationic polymerization and determined their structures by X-ray analysis. The results also indicate the cis-opening mechanism.

7.3.3 Syndiotactic Polymers

Polymers of both *it*- and *st*-configurations are obtained in the 1,2-addition of butadiene. Although 1,2-*it*-polybutadiene is a (3/1) helix,[59] its *st*-polymer is a slightly distorted planar zigzag.[45]

For *st*-PP the structure of a $(T_2G_2)_2$ type helix was first found (Fig. 7.4). According to the conformational energy calculation by Natta et al.[208] the planar zigzag conformation was expected to be equally stable. The planar zigzag form was prepared by quenching the melt in ice water and subsequent stretching.[20,21] This form is metastable and transforms into the helix form by heat treatment or holding at room temperature. For infrared spectra and the normal coordinate treatment of *st*-PP, see Refs. 21 and 209–211.

7.4 VINYL POLYMERS

Poly(vinyl chloride) does not give a sharp X-ray diagram. Natta and Corradini[45] proposed a crystal structure (Table 7.1) based on the analogy to *st*-poly-1,2-butadiene. Krimm et al.[212] also obtained a similar structure from the analysis of a single crystal mat. The infrared spectrum of poly(vinyl chloride) is a case for which the following experimental rule is applicable. If the C—H bond is trans to the C—Cl bond, the C—Cl stretching bands should appear at 605–650 cm^{-1}. If the C—C bond is at the trans position, the C—Cl bands appear at 670–700 cm^{-1}.[213] In poly(vinyl chloride) the 638 and 603 cm^{-1} bands are assigned to the C—Cl stretching vibration of the *st* parts, and the 690 and 680 cm^{-1} bands are assigned to the *it* and *at* parts, that is, the latter bands originate from the C—Cl stretching in the amorphous regions.[214,215] Poly(vinyl chloride) samples prepared by radiation polymerization of a urea complex show only strong 638 and 603 cm^{-1} bands and are considered to be highly syndiotactic.[216] For band assignments of this polymer, see Refs. 217–219.

PVA prepared from poly(vinyl acetate) is considered to be atactic, and its structure is discussed in Section 4.10.5.A. *it*- PVA is prepared through *it*-poly(vinyl ethers), $[—CH_2—CH(OR)—]_n$ $[R = -CH_2—C_6H_5, —C(CH_3)_3, -Si(CH_3)_3]$, by low-temperature polymerization.[49,105,220,221] Polarized infrared spectra of *it*- and *at*-PVAs are, as a whole, similar to each other as shown in Fig. 7.5. However the spectra exhibit several differences in the detailed features. The 3340 and 1460 cm^{-1} bands show parallel dichroism in the *it*-polymer, while the 1141 and 909 cm^{-1} bands are observed only in the *at*-polymer.[105] The formate of *it*-PVA has a crystal structure similar to that of *it*-polystyrene.[49] The X-ray diagram of *it*-PVA shows a lower degree of crystallinity than *at*-PVA. From the fiber period (2.51 Å) and infrared spectroscopic evidence, a planar zigzag molecular model with intramolecular hydrogen bonds was proposed (Fig. 7.6).[105]

The X-ray diagram of polyacrylonitrile is not well defined, giving reflections

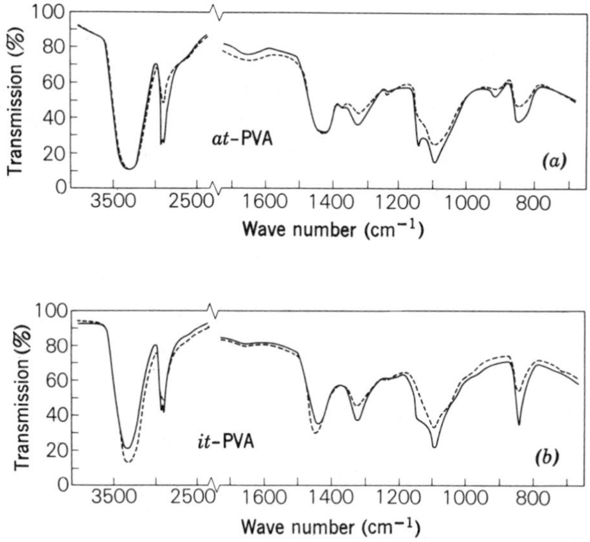

Fig. 7.5 Polarized infrared spectra of (*a*) *at*-PVA and (*b*) *it*-PVA. (————) Perpendicular radiation, (– – – – –) parallel radiation.[105]

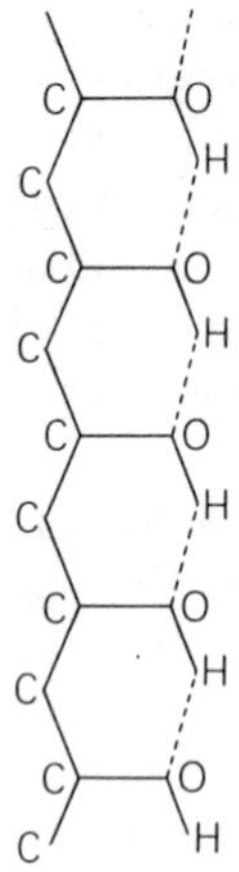

Fig. 7.6 Molecular model of *it*-PVA.[105]

only on the equator and few in number. Hence the crystal structure has not yet been determined.[222] References for this polymer include: laterally ordered structure, Ref. 223; studies by electron microscope and electron diffraction, Ref. 224; treatment as a paracrystal, Ref. 225; and infrared spectra and normal coordinate treatment, Ref. 226.

7.5 VINYLIDENE POLYMERS

The crystal structure of polyisobutylene has recently been determined as shown in Fig. 7.7 (Section 6.3.2.A).[50] The helical molecule has a fiber period consisting of two asymmetric units each containing four monomeric units. The helix axis coincides with the 2_1 screw axis in the lattice. The averaged bond angle C—C$(CH_3)_2$—C is about 110°, but the angle C—CH_2—C is much larger, about 128°.

For poly(vinylidene chloride) a statistically disordered crystal structure has been suggested, in which upward and downward molecular chains of $TGT\bar{G}$ type are included in the cell with $\frac{1}{2}$ probability (Fig. 7.8).[51] A copolymer including 20% vinyl chloride gives an X-ray diagram essentially the same as that of the homopolymer. Such a copolymer can be doubly oriented by rolling, where the rolled plane is parallel to the (100) plane.

For the structure of poly(vinylidene fluoride), see Section 6.5.1 and Ref. 227. Oriented poly(vinylidene fluoride) films show distinct piezoelectricity.[228–230] This phenomenon may be associated with the polar crystal structures of forms I and III, or partially oriented molecules adjacent to the crystallites.

For poly(methyl methacrylate), see Section 4.6.6.

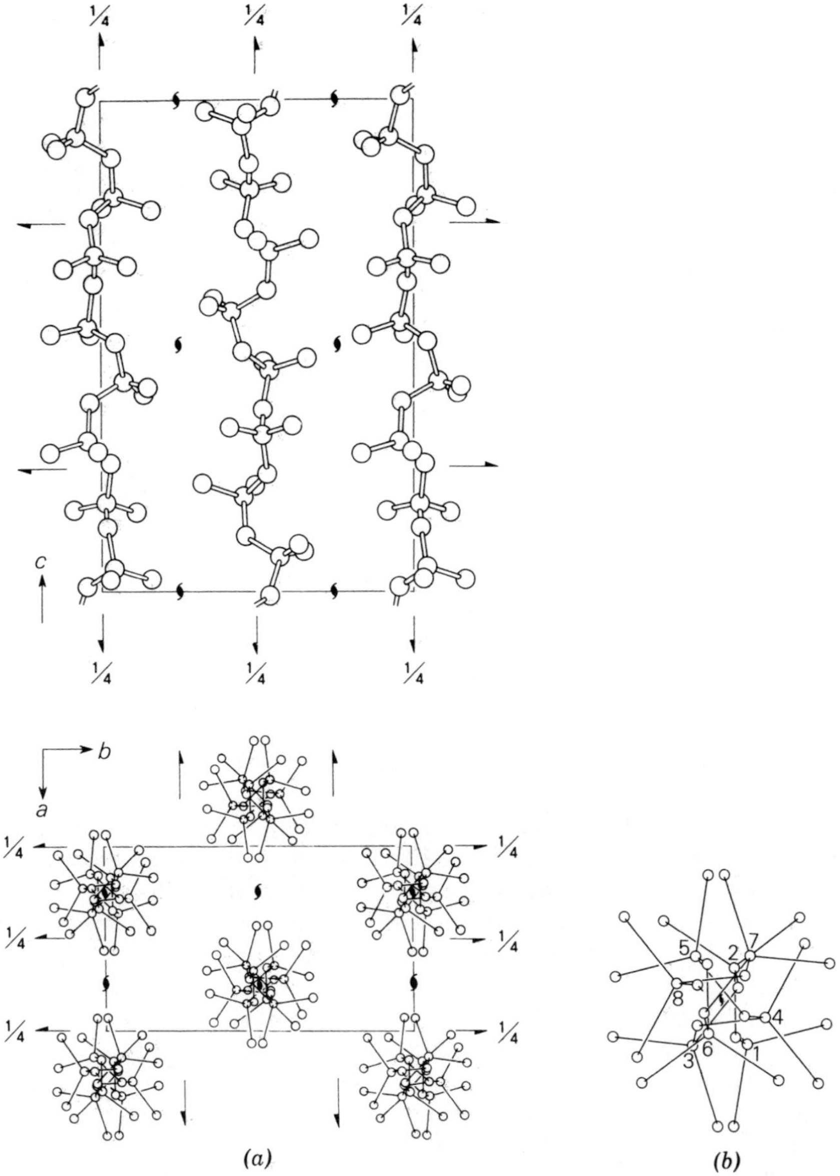

Fig. 7.7 (*a*) Crystal structure and (*b*) molecular structure of polyisobutylene.[50]

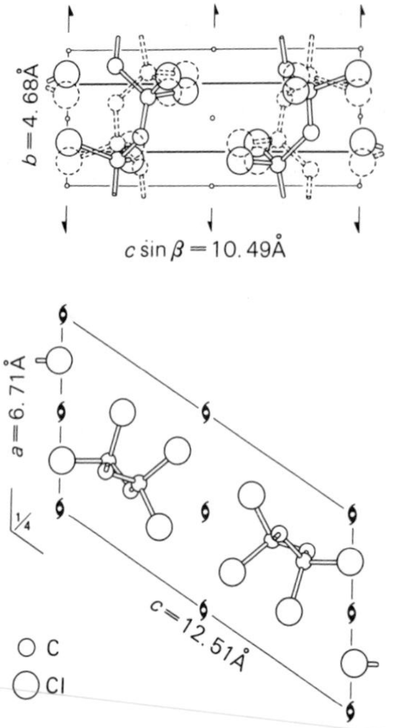

Fig. 7.8 Crystal structure of poly (vinylidene chloride).[51]

7.6 POLYMERS CONTAINING C=C DOUBLE BONDS

trans-1,4-Polybutadiene has two crystal modifications, one above and one below the transition temperature of 75°C. The molecular conformation of the low temperature form is a *trans*-$ST\bar{S}$ type.[57] The structure of the high temperature form has not yet been established, although several disordered structures have been proposed.[57,231,232]

As early as 1942, Bunn determined that the crystal structure of the β-form of *trans*-1,4-polyisoprene (gutta percha) has a *trans*-$ST\bar{S}$ conformation; the α-form was analyzed only recently[60] and was formed to have a glide-type conformation as shown in Fig. 3.9.

Natural rubber is *cis*-1,4-polyisoprene. By the development of stereoregular polymerization, polyisoprene consisting of almost only *cis*-1,4-bonding, "synthetic natural rubber," has been prepared. Its infrared spectrum shows characteristic features clearly different from the polybutadiene prepared so far.[233,234] Nyburg[62] proposed a statistically disordered crystal structure in which the molecular chain having the conformation denoted by the solid line in Fig. 7.9 and its mirror image, denoted by the broken line, exist with equal probability. Natta and Corradini[58] reported the crystal structure shown in Fig. 7.10, where, for example, A and A′ indicate the mole-

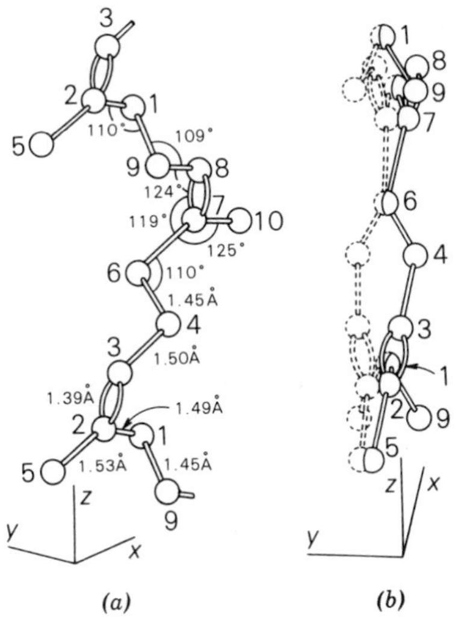

Fig. 7.9 Molecular structure of natural rubber. (*a*) A single molecule and (*b*) a pair of chains that are mirror images. (From Nyburg.[62])

cular chains that are mirror images. They concluded that the array of the A—B—A′—B′ ⋯ type does not occur, but that either the C—D—C—D ⋯ or the C′—D′—C′—D′ ⋯ type alignment may pack adjacent to the A—B—A—B ⋯ type alignment; statistical disorder occurs only along the *a* axis.

Polyalkenamers are prepared from cycloolefins. Odd-numbered members have the orthorhombic space group *Pnam*. The molecular packing is similar

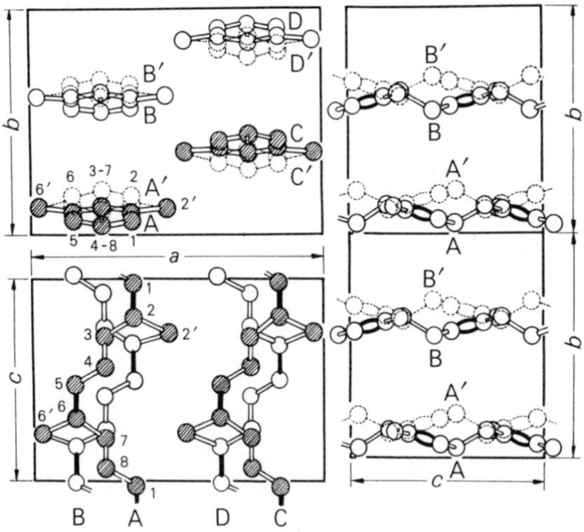

Fig. 7.10 Crystal structure of *cis*-1, 4-polyisoprene (From Natta and Corradini.[58])

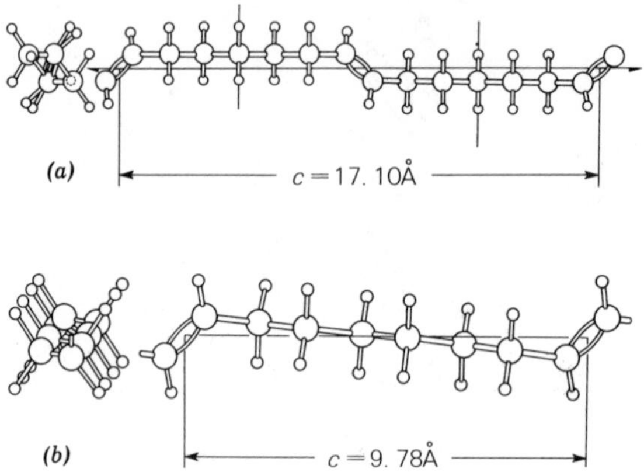

Fig. 7.11 Molecular models of (*a*) *trans*-polyheptenamer and (*b*) *trans*-polyoctenamer. (From Natta et al.[63])

to that of orthorhombic PE.[63] The torsional angles around the single bonds adjacent to the double bonds are about $\pm 120°$ (Fig. 7.11). Even-numbered members crystallize in monoclinic or triclinic form depending on the conditions of preparation.[63]

The molecular structure of 1, 2-*it*-poly(4-methyl-1, 3-pentadiene) is a (18/5) helix as shown in Fig. 7.12.[65] The calculated cyrstalline density of this

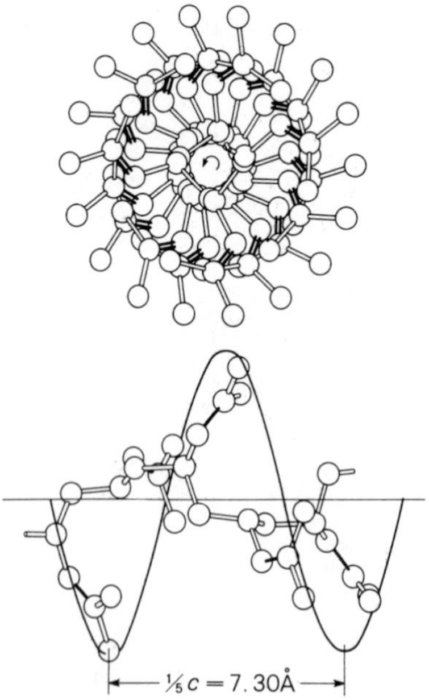

Fig. 7.12 Molecular structure of *it*-1, 2-poly (4-methyl-1, 3-pentadiene). (From Natta et al.[65])

polymer (0.85 g/cm^3) was found to be lower than the observed density, just as in the case of *it*-poly(4-methyl-1-pentene).

7.7 POLYETHERS

7.7.1 $[-(CH_2)_m-O-]_n$ Type Polyethers

This series of polyethers and also polythioethers show remarkable variations of physical properties as m increases. POM ($m = 1$) has a high melting point (Fig. 7.13) and a high density. This polymer is very hard and is insoluble in ordinary solvents at room temperature (except for hexafluoroacetone sesquihydrate). In contrast, PEO ($m = 2$) has a lower melting point, has a lower density, is not as hard, and is soluble in water at room temperature. The members with $m = 3$, 4 have much lower melting points and are very soft polymers. These remarkable differences in properties cannot be explained solely in terms of different chemical structures, that is, the one-to-one increase of methylene groups between oxygen (or sulfur) atoms. The molecular conformation and crystal structure must play an important role. The structures of POM (Sections 4.6.4.A, 5.5.8, and 5.6.5) and PEO (Sections 4.6.5, 4.12.2, and 6.5.2) are described earlier. Although polyoxacyclobutane has four crystal modifications (Sections 4.3.3.A and 5.7.2), polytetrahydrofuran and higher members are all planar zigzag.[80,82] Depending on the conditions of preparation the members with $m = 6-12$ crystallize into one of two modifications, PE type or polytetrahydrofuran type (Fig. 7.14).

Flory and Mark[236,237] have calculated the characteristic ratio $C_\infty = (\langle r^2\rangle_0/nl^2)_\infty$ and the mean-square dipole moment $(\langle\mu^2\rangle/nm^2)_\infty$ for

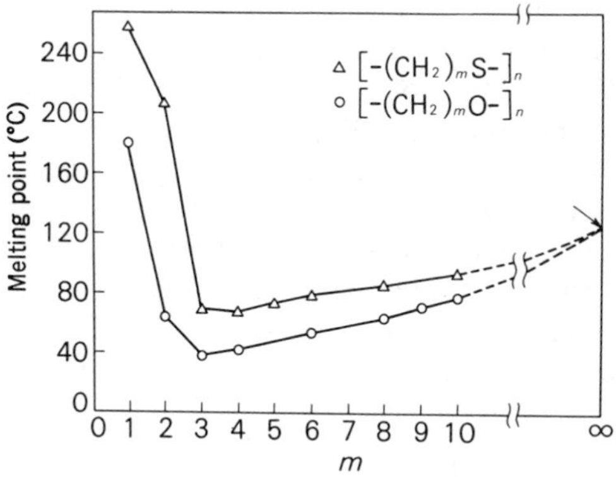

Fig. 7.13 Melting temperature of polyethers and polythioethers. (Modified from the original by Lal and Trick.[235])

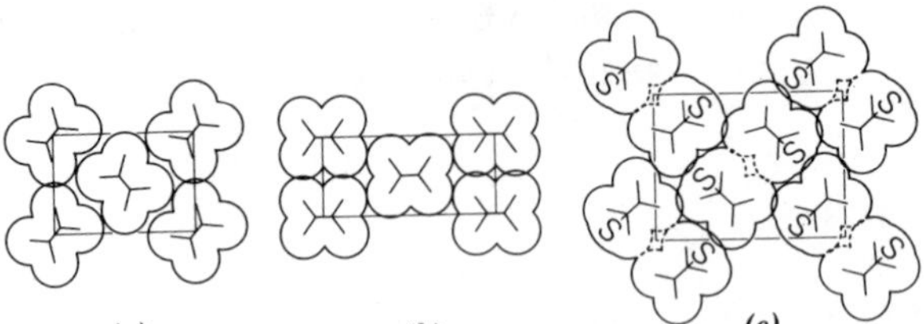

(a) (b) (c)

Fig. 7.14 Molecular packings of (*a*) PE,[1] (*b*) polytetrahydrofuran,[80,81] and (*c*) poly (pentamethylene sulfide).[109]

various polymers. They have shown by comparing the calculated and observed values for PEO that this molecule tends to take the *TTG* conformation locally in solution as in the case of the crystalline state. Here n and l are, respectively, the number and the bond length of the backbone bonds in the molecule, and $\langle r^2 \rangle_0$ is the mean-square end-to-end distance of the unperturbed chain. The subscript ∞ means the limit $n \to \infty$. $\langle \mu^2 \rangle$ is the mean-square dipole moment and m is the bond dipole moment of the backbone bonds.

Further references include: solid-state radiation polymerization of ring oligomers, trioxane and tetraoxane, Ref. 238; pentoxane, Ref. 239, and hexoxane, Ref. 240; infrared spectroscopic evidence for inversion opening polymerization of ethylene oxide, Ref. 241; infrared studies of chain folding of PEO, Ref. 242; skeletal vibrations of planar zigzag polyethers, Ref. 243; and normal vibrations of orthorhombic POM, Ref. 244.

7.7.2 Polyformals $[-OCH_2-O-(CH_2)_m-]_n$

Poly-1,3-dioxolane ($m = 2$) has three crystal modifications, among which form II has been analyzed in detail.[84] The molecular conformation is of the glide type (Fig. 7.15*a*). The molecule of poly-1,3-dioxepane ($m = 4$) is also glide type, but rather complicated (Fig. 7.15*b*).[85]

In cationic ring-opening polymerization of cyclic formals, the question arises as to whether the bond scission occurs exclusively at the same type of bond, for example, bond I, or randomly at both bonds I and II. If scission occurs at random, the resultant polymer should not have a well defined crystal structure. Yamashita et al.[245] found, using a chemical method, that the ring opening occurs only at bond I.

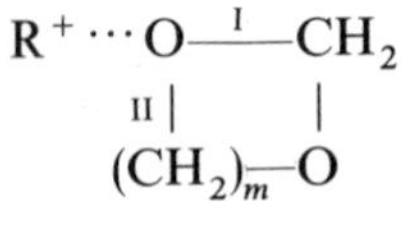

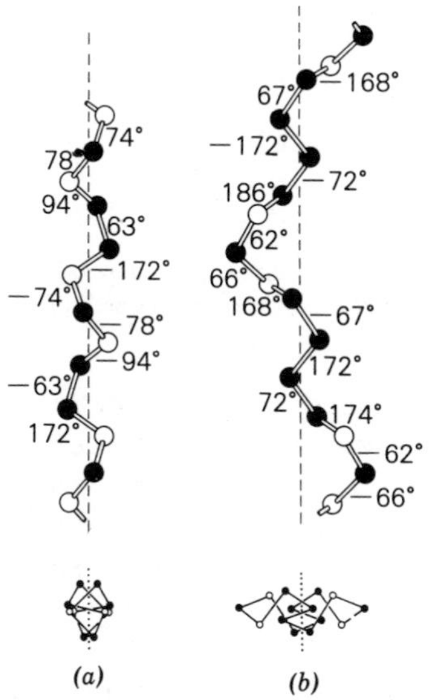

Fig. 7.15 Molecular conformations of (*a*) poly-1, 3-dioxolane[84] and (*b*) poly-1, 3-dioxepane.[85]

The above results of X-ray analyses show regular head-to-tail sequences, confirming that the ring opening occurs exclusively at the same type of bond, although it is not clear whether the scission occurs at bond I or II.

For vibrational analysis of crystalline poly-1, 3-dioxolane, see Ref. 246.

7.7.3 Substituted Polyethers

In addition to *it*-polyacetaldehyde, *it*-polymers of various aldehydes have been prepared. All these polyaldehydes have tetragonal lattices with chains of (4/1) helix conformations. The *a* axis increases with increasing length of the side chains.

Poly(propylene oxide) has true asymmetric carbon atoms (Section 2.1). Cesari et al.[89] analyzed the racemic sample and found that the molecular chains are isotactic and have a slightly distorted planar zigzag conformation.

When *R* and *S* *it*-polymers mix in a 1 : 1 ratio and give a racemic crystalline sample, the following three kinds of optical compensation can be considered (Fig. 7.16).[110] (*a*) Optical compensation in a unit cell. The optical isomers are included in the unit cell pairwise. (*b*) Compensation in a crystallite. Equal amounts of *S* and *R* polymers are included in a crystallite randomly, that is, they form a statistically disordered structure. (*c*) Intercrystallite compensation. Each crystallite consists only of *R* polymers or only of *S* polymers. Because of the presence of equal amounts of enantiomeric crystallites, the bulk polymer is optically inactive. *it*-Poly(*tert*-butylethylene oxide) form I (Section 4.12.3)[91] and *it*-poly(*tert*-butylethylene sulfide)[111] are examples of

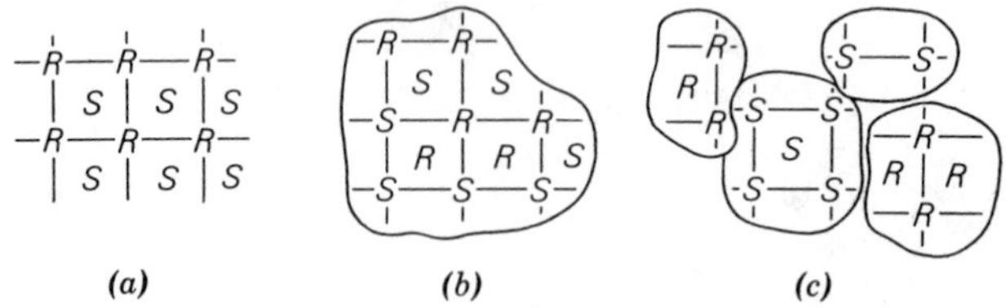

Fig. 7.16 Optical compensation of racemic polymers: (*a*) compensation in a unit cell; (*b*) compensation in a crystallite; (*c*) intercrystallite compensation. *R* and *S* denote rectus and sinister polymer chains, respectively.[110]

case *a*, that is, two pairs of optical isomers are included regularly in a unit cell. *it*-Poly(isopropylethylene oxide) and *it*-poly(propylene sulfide) correspond to case *c*, that is, intercrystallite compensation. The size of the crystallites considered here is based on X-ray diffraction and is associated with the number of regularly repeating planes that produce well defined X-ray spots. Consequently the crystallite size may range from several tens to 100 Å.

it-Poly(propylene oxide)[89] has a crystal structure quite similar to that of *it*-poly(propylene sulfide), and the space group is the same. Thus case *c* is reasonable. However a clear conclusion, as in the case of poly(propylene sulfide), is difficult, since the scattering powers of the oxygen atom and the CH_2 group are similar. The possibility of case *b* cannot be excluded on the basis of the X-ray diffraction data alone. This means that the structure factors calculated for a model in which the position of CH_2 is occupied by an oxygen atom with equal probability cannot be distinguished experimentally from the structure factor for the regular structure.

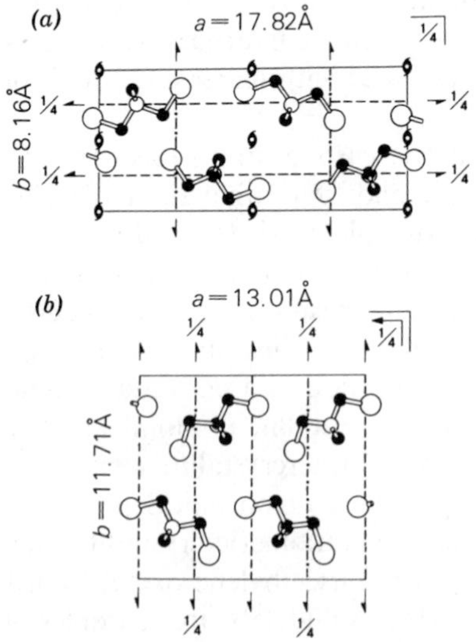

Fig. 7.17 Crystal structures of (*a*) α-form and (*b*) β-form of poly [3,3-bis (chloromethyl) oxacyclobutane].[94]

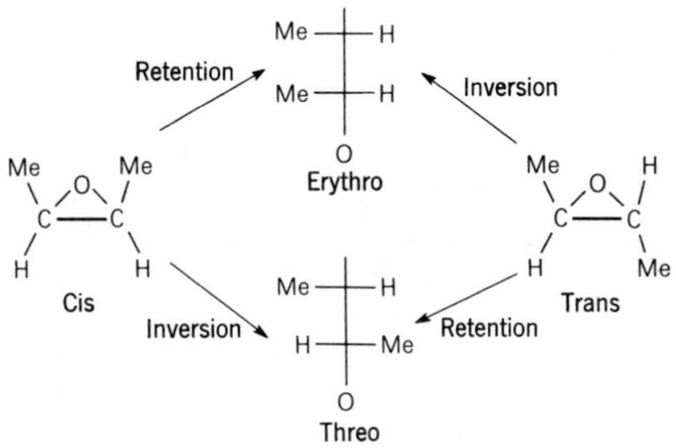

Fig. 7.18 Relationship among the configurations of 2-butane oxide monomers and polymers.

According to the X-ray analyses of poly[3, 3-bis(chloromethyl) oxacyclobutane], the difference between the two crystal modifications is due to the difference between the modes of packing of the molecules having the same conformation. For this polymer the main chain is a planar zigzag with *TT* side chains (Fig. 7.17).[94] The polar structure of the β-form suggests the possibility of piezoelectricity. See also the electron diffraction study of the α-form.[247]

Barlow[95] analyzed the structure of poly(2-butene oxide) prepared from the *trans* monomer and found that it corresponds to an erythro-*dit*-polymer. This fact confirms the inversion-opening mechanism proposed by Vandenberg (Fig. 7.18).[248]

7.8 POLYTHIOETHERS

In the polymerization of propylene sulfide, optically active and racemic polymers were obtained from optically active and racemic monomers, respectively, under suitable condition. Both polymers give essentially the same X-ray diagrams and infrared spectra. According to the X-ray analysis, the unit cell is chiral (the space group: $P2_12_12_1$) and the optical compensation is type *c*. These facts indicate a stereoselective mechanism of polymerization.

Poly(pentamethylene sulfide) has the planar zigzag conformation, but the molecular packing is different from that of both the PE type and the polytetrahydrofuran type as shown in Fig. 7.14*c*.[109] A pair of chains are close together with the CSC dipoles antiparallel. Such pairs of molecules, as a whole, pack in the lattice just as the molecular chains of orthorhombic PE.

7.9 POLYESTERS

7.9.1 Aliphatic Polyesters

The members [—$(CH_2)_m$—COO—$]_n$ with $m = 1$, 2, and 5 have a planar zigzag conformation (Sections 4.7.4.B, 4.8.1.A, and 4.10.4). An exceptional case is poly-β-propiolactone, which has two crystal modifications. Form II is planar zigzag, while the structure of form I has not yet been clarified.

The far-infrared spectra of this series of polyesters have been analyzed in terms of normal coordinate treatments.[249] Figure 7.19 shows polarized infrared spectra of poly-ε-caprolactone and poly-ε-caproamide (nylon 6). These spectra have quite different features in the region above 700 cm^{-1}, but are very similar in the region below 700 cm^{-1}. This can be understood by comparing the ester and amide groups. The spectra are different in the region where the hydrogen atoms of the N—H groups move, but they are similar in the region where the N—H group moves as a unit.

Substituted poly-β-propiolactones assume the (2/1) helix conformation with internal rotation angles as listed below.

	—C—	—C—	—CO—	—O—
Polymer	τ_1(°)	τ_2(°)	τ_3(°)	τ_4(°)
[—C*H(CH_3)CH_2COO—$]_n$[116]	−52	−42	−175	162
[—C*H(C_2H_5)CH_2COO—$]_n$[118]	−60	−21	179	136
[—CH_2C$(CH_3)_2$COO—$]_n$[119]	−41	−61	−164	178
[—C(=CH_2)CH_2COO—$]_n$[121]	−74	−147	−177	−84

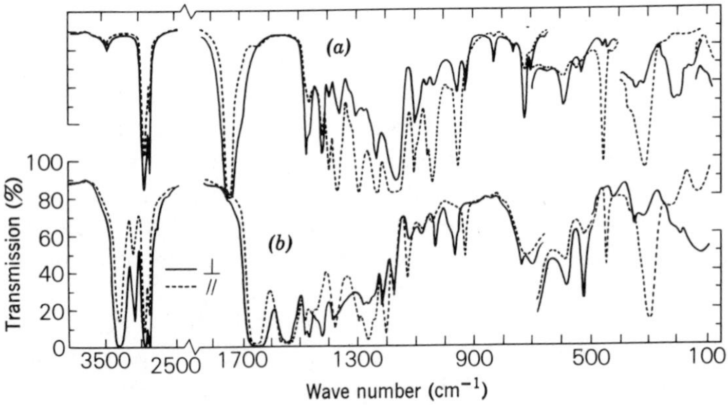

Fig. 7.19 Polarized infrared spectra of (*a*) poly-ε-caprolactone and (*b*) poly-ε-caproamide.[249] (————) Perpendicular radiation, (- - - - - - -) parallel radiation.

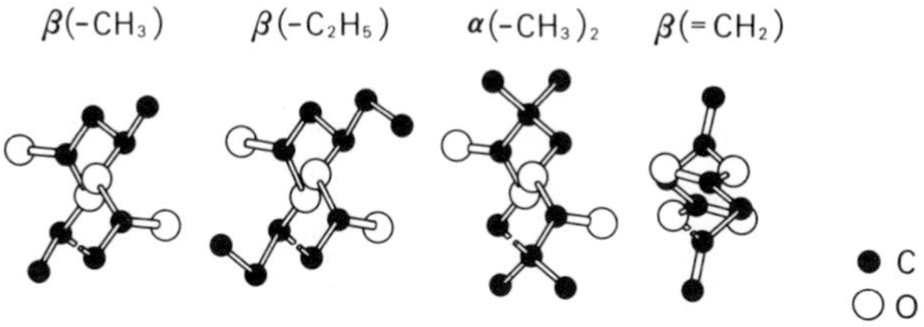

Fig. 7.20 Projections of substituted poly-β-propiolactone molecules viewed along the chain axis.[121]

Figure 7.20 shows a projection viewed along the molecular axis. The β-methyl and β-ethyl derivatives have asymmetric carbon atoms (indicated by*). Naturally occurring optically active poly(D-β-hydroxybutyrate) (this is identical to poly-β-methyl-β-propiolactone) is a rectus *it*-polymer as shown in the Fischer projection.

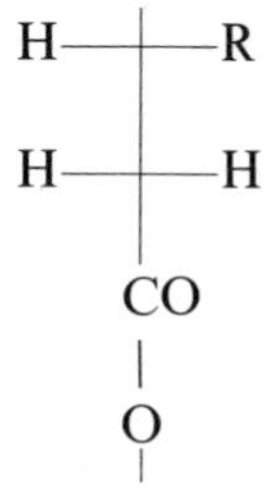

This polymer forms a left-hand helix. Therefore the above-cited angles of internal rotation for these four polymers are all for left-hand helices. The conformations for all of these four polymers are nearly $(T_2G_2)_2$ type helices. Except for polydiketene, the skeletal oxygen atoms are between the nearly trans bonds. In polydiketene the carbon atoms of the carbonyl groups are between the nearly trans bonds. If the oxygen atoms were between the trans bonds, the repulsion between the oxygen atoms of the carbonyl groups and the methylene groups of the vinylidene portion would be very strong. The optical compensation for β-methyl and β-ethyl derivative polymers is case *c* as shown in Fig. 7.16.

See Refs. 250 and 251 for α-β transition of polypivalolactone.

7.9.2 Polyesters Containing Aromatic Rings

For PET see Section 4.12.1. Poly(butylene terephthalate) has two crystal modifications.[130,252,253] The α-form is stable (Fig. 7.21), while the β-form is reversibly formed by stretching. The macroscopic elongation (12%) is roughly the same as the elongation of the *c* axis. Below this elongation ratio both forms coexist and no intermediate form occurs. The elongation is

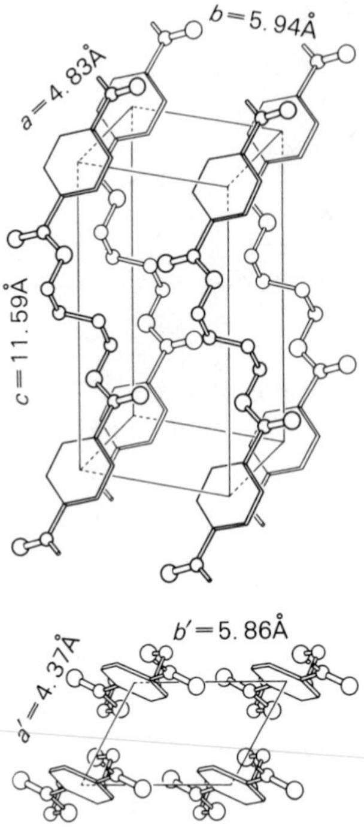

Fig. 7.21 Crystal structure of poly (butylene terephthalate) α-form.[130]

mainly due to the conformational change of the methylene part; both infrared and Raman spectra show distinct variations.

7.10 POLYAMIDES

7.10.1 Aliphatic Polyamides

Figure 7.22 shows the crystal structure of nylon 66 (α-form).[144] Planar zigzag chains form a sheet by hydrogen bonding, and such sheets stack parallel to the *ac* plane. In Fig. 7.23 fiber periods of polyamides determined by X-ray diffraction are plotted against the number of CH_2 groups per monomeric unit. Two lines are obtained.[145] The solid line corresponds to a fully extended planar zigzag, for example, nylon 66, 610, and 6 (α-form). The dashed line corresponds to slightly contracted chains, for example, nylon 77 has an observed fiber period of 18.95 Å. This is a little shorter than the calculated value for a planar zigzag, 19.7 Å. The members on the dashed line are called the γ-form. In the γ-form the bonds next to the amide group assume

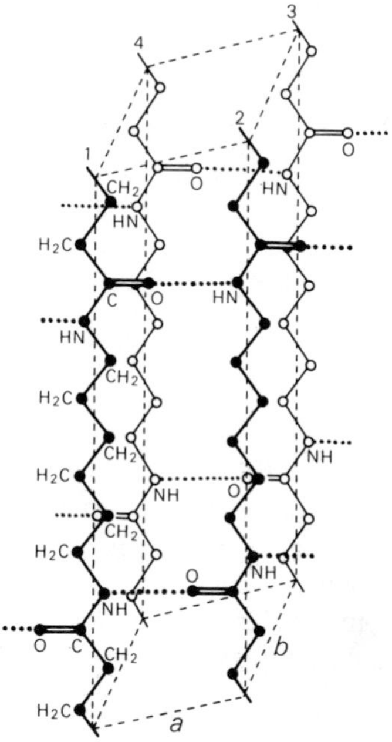

Fig. 7.22 Crystal structure of nylon 66. (From Bunn and Garner.[144])

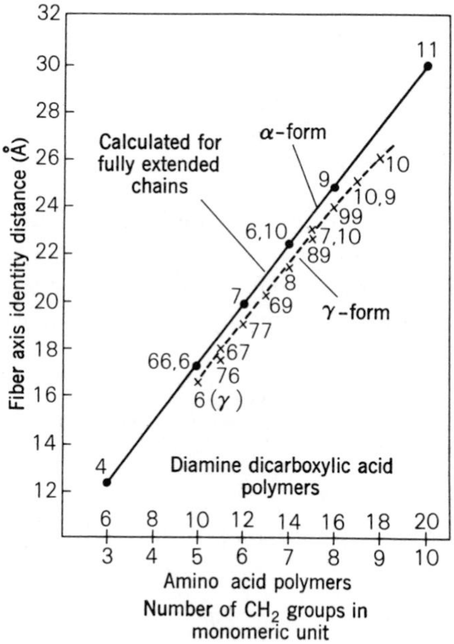

Fig. 7.23 Fiber period of polyamides. For the members $[-NH-(CH_2)_x-NHCO-(CH_2)_{y-2}-CO-]_n$ with x = even and y = odd and with x = odd and y = even, half of the fiber period is plotted on the ordinate, and for the members $[-NH-(CH_2)_{z-1}-CO-]_n$ with z = even, twice the value of the fiber period is plotted. (From Kinoshita.[145])

a conformation shown roughly below.

$$-CH_2-CO-NH-CH_2-$$

$$\quad\quad\; S \quad\;\; T \quad\; \bar{S}$$

or

$$\bar{S} \quad\;\; T \quad\;\; S$$

Which modification is more stable, α or γ, is known for the various nylons, but the reasons have not yet been clarified.

7.10.2 Aromatic Polyamides

Poly(p-phenylene terephthalamide) (Kevlar or Fiber B) and poly-p-benzamide (PRD-49 Type I) have high tenacities and high elastic moduli, and are stable at high temperatures as shown in Table 7.3. Poly(m-phenylene isophthalamide) (Nomex) has a tenacity and modulus closer to usual fibers, but is very stable at high temperatures. The molecular conformations of Kevlar and PRD-49 in the crystal lattice can be represented as $TTTT$ and $TCTC$, respectively. Here N—⟨ ⟩—N, and so on are considered as virtual bonds. The internal rotation angles of the benzene ring from the amide planes are about $30°$.

$$-[-NH-⟨ ⟩-NH-CO-⟨ ⟩-CO-]-NH-$$

$$\quad\quad T \quad\quad\;\; T \quad\quad T \quad\quad\; T$$

$$-[-NH-⟨ ⟩-CO-NH-⟨ ⟩-CO-]-NH-$$

$$\quad\quad C \quad\quad\; T \quad\quad C \quad\quad T$$

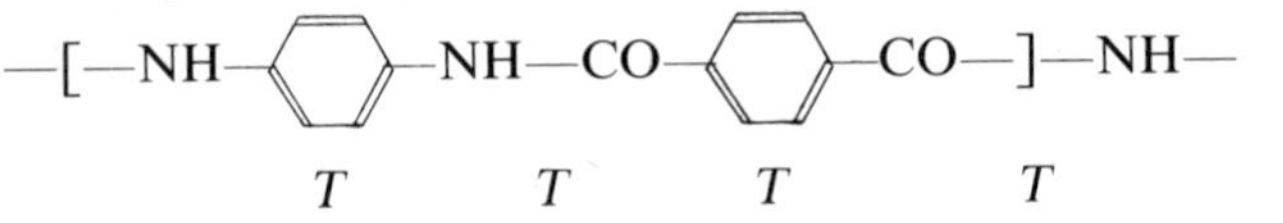

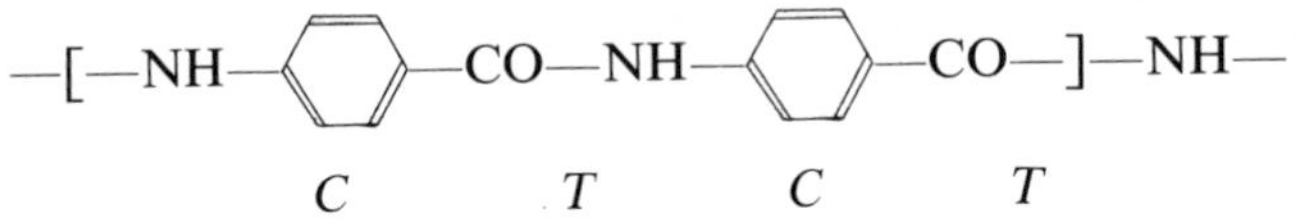

In the crystal lattices of these two polymers (Fig. 7.24) molecular chains form hydrogen-bonded sheets, as in the case of α-nylons. These sheets stack along the a axis. The Nomex molecules are contracted from the fully extended conformation by about 1 Å per fiber period. The internal rotation angles are shown below.

$$-150° \quad 30° \quad T \quad -30° \quad 150° \quad T$$

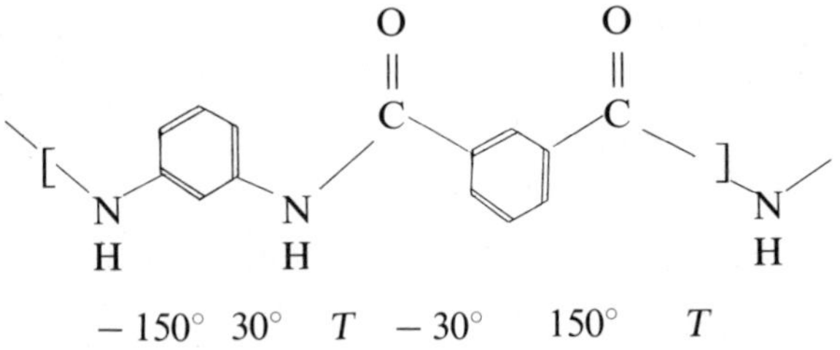

Table 7.3 Properties of Various Fibers[254]

	Tenacity		Elongation	Macroscopic modulus		Crystallite modulus[b]		Mp
Polymer[a]	(g/denier)	(kg/mm^2)	(%)	(g/denier)	(dyn/cm^2)	Obsd.[c] (dyn/cm^2)	Calcd. (dyn/cm^2)	(decomposition point) (°C)
Kevlar	25	330	5	850	111×10^{10}	153×10^{10}	182×10^{10}	(500)
PRD-49	15	200	3	1050	134	—	163	(500)
Nomex	5.5	68	35	82	10	88	90	(415)
PET	9.0	110	7	160	19.5	108	124[d]	260
it-PP	9.0	74	15	120	9.6	34	28[e]	170
PE	9.0	78	8	100	8.5	235	296[e]	135
Nylon 6 (α)	9.5	97	16	50	5.0	165	244[f]	225
PEOB (α)	5.3	64	25	75	8.9	5.9	2.4	225

[a] PEOB: poly(ethylene oxybenzoate).
[b] Crystallite moduli quoted from Ref. 262 unless otherwise noted.
[c] Data from Sakurada et al.[255–258]
[d] Data from Treloar.[259]
[e] Data from Miyazawa.[260]
[f] Data from Manley and Martin.[261]

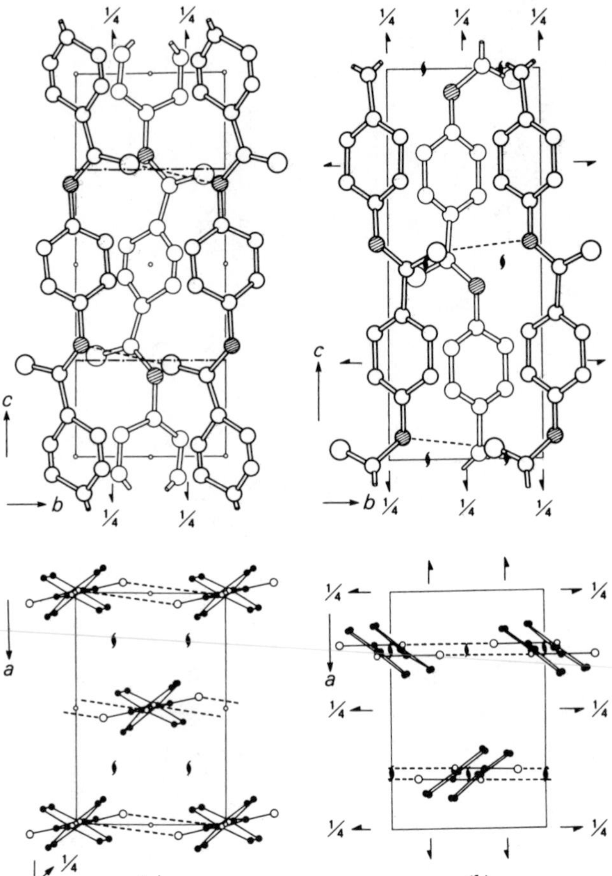

Fig. 7.24 Crystal structures of (*a*) poly (*p*-phenylene terephthalamide) and (*b*) poly-*p*-benzamide.[146]

As the *c* projection shows (Fig. 7.25), unique hydrogen bonds are formed in a check-patterned shape. Rolling gives a doubly oriented sample in which the (101) plane (approximately coincident with the benzene planes) is parallel to the rolled plane. As can be understood from Fig. 7.25, this crystal structure has a pronounced polarity perpendicular to the (101) plane, suggesting future applications as piezoelectric and pyroelectric elements.

The factor governing the internal rotation angles around the bonds —C and 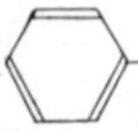—N may be classified into two components as shown in Fig. 7.26.[262] One component is the tendency to maintain the planarity of the benzene ring and the amide groups owing to the partial double bond charac-

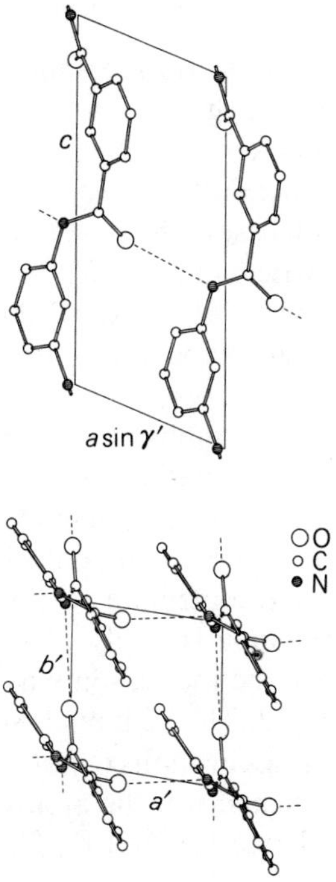

Fig. 7.25 Crystal structure of poly (*m*-phenylene isophthalamide).[149]

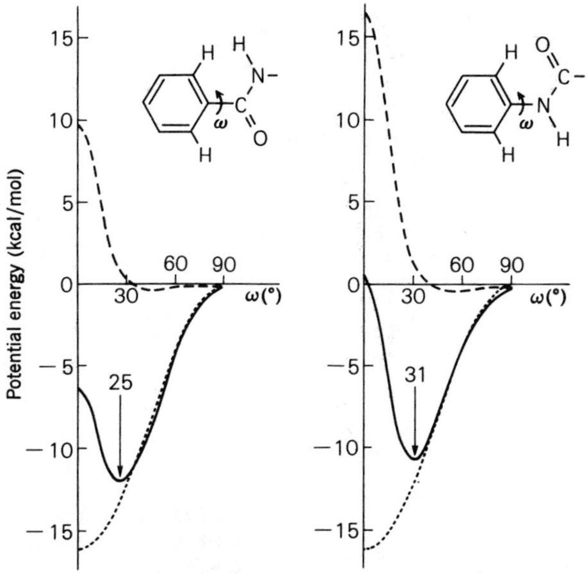

Fig. 7.26 Potential energies versus dihedral angles (ω) of —C and —N.[262]

ter (indicated by a dotted line). The second component is the steric interaction between the hydrogen atom of the benzene ring and the oxygen atom of the amide group, as well as between the hydrogen atoms belonging to benzene ring and amide group (dashed line). The solid line is the sum of these two components. The minimum is found at $\omega \cong 30°$ for both cases. The potential barriers are much higher (6–12 kcal/mol) than the barriers of PET (3 kcal/mol).[263] Also, the potential barrier of the amide C—N bond is around 20 kcal/mol.[264] These results suggest that these three aromatic polyamide molecules should have very low flexibility.

Therefore extended chain structures for the amorphous regions of Kevlar and PRD-49 may be assumed. Also, the liquid crystal formation in concentrated solutions and the difficulty of chain folding for these polymers can now be understood. Because Kevlar has a *TTTT* type molecular conformation, the changes of internal rotation angles cannot contribute to the elongation of the molecular chain. The elongation can arise only from changes of the bond angles and bond lengths. This may be the reason why Kevlar has the high crystallite modulus given in Table 7.3. The high macroscopic modulus may be due to extended molecules in the amorphous regions. Figure 7.27 shows schematically the calculated displacements of the atoms based on the assumption of a hypothetically large elongation of 10% for

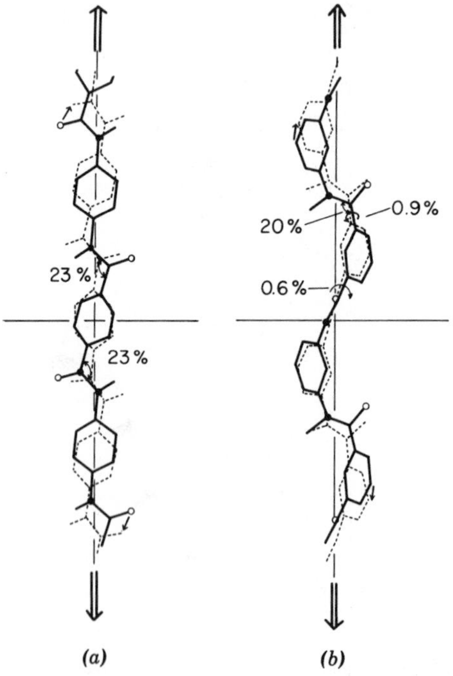

Fig. 7.27 Atomic displacements and potential energy distributions for Kevlar and Nomex molecules when the chains are stretched 10%.[262]

Kevlar and Nomex molecules. The solid lines denote the positions before elongation and the dashed lines indicate the location after elongation. The distribution of the potential energy of strain to the internal coordinates can be calculated. In the figure some of those with large values are shown in percent. The contribution of amorphous regions is, in general, very important for macroscopic modulus of various fibers. Clarification of this mechanism is a future problem.

The melting point of a crystal is given by

$$T_m = \frac{\Delta H}{\Delta S} \tag{7.1}$$

where ΔH is the enthalpy of melting and ΔS is the entropy of melting. In the case of these three aromatic polyamides, ΔH should be large owing to the large intermolecular interaction due to the good packing of the benzene rings and hydrogen bonding, especially the check-patterned hydrogen bonding in Nomex. The low flexibility leads to a small value for ΔS. This suggests how the high melting points of these polymers are related to the crystal structures.

7.11 OTHER POLYMERS

Poly(ethylene oxybenzoate) α-form has a very low crystallite modulus (Table 7.3), possibly because of the deformation of the large-scale zigzag (one monomeric unit corresponds to one zigzag unit) caused by internal rotation.

Radiation-induced copolymerization of ethylene and carbon monoxide under high pressure produces polyketones with the formula $[\text{—}(CH_2CH_2)_m\text{-—}CO\text{—}]_n$ ($m \geqslant 1$). These are all crystalline irrespective of the m values[157] (random copolymers except for the case $m = 1$). The molecular packing in the case of $m = 1$ is similar to that in PE. This phenomenon may be explained by the fact that van der Waals radii of the CH_2 and CO groups are not very different.

The unit cell of fibrous sulfur includes four pairs of right- and left-hand (10/3) helical chains[162] (also see Ref. 265).

Further references include: X-ray studies, Ref. 266; Raman spectra of carbon fibers, Ref. 267; and the four-center type polymer of 2, 5-distyrylpyrazine obtained by solid-state polymerization and its related compounds, Ref. 268.

7.12 POLYMER COMPLEXES

When an oriented sample of PEO (Fig. 4.1*a*) is immersed in a methanolic solution of urea at room temperature, an oriented sample of PEO–urea

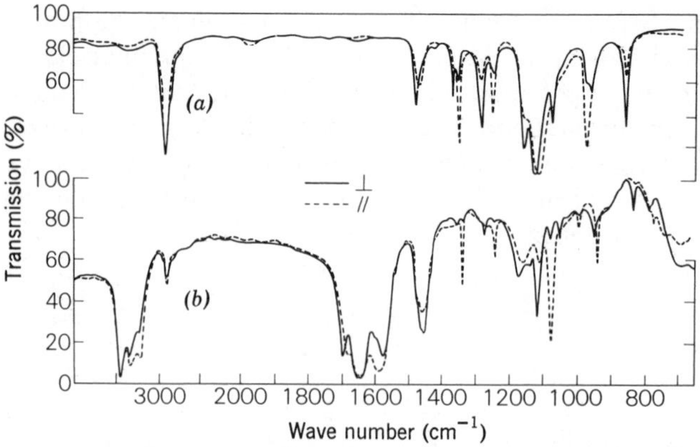

Fig. 7.28 Polarized infrared spectra of (*a*) PEO and (*b*) PEO–urea complex. (————) Perpendicular radiation, (--------) parallel radiation.[269]

complex (Fig. 4.1*b*) is obtained.[269] Moreover, when the latter sample is immersed in methanol, an oriented sample quite similar to the original one is again obtained through the loss of the urea molecules. Figure 7.28 shows the polarized infrared spectra of PEO and the complex. The spectra of the complex may be interpreted as a superposition of the spectra of the two components, that is, urea and PEO, since the majority of the bands of the components appear at different positions. This feature of the spectrum is quite different from that of both urea adducts[270] and tetragonal urea.[271] The spectral bands due to PEO in the complex are similar to those of the ordinary PEO. This similarity suggests that the PEO molecule in the complex does not have a new conformation, such as planar zigzag, but has a structure

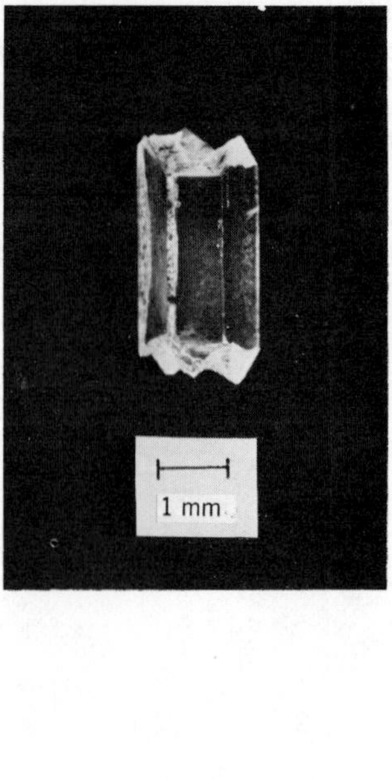

Fig. 7.29 A single crystal of PEO–urea complex.[269]

similar to the (7/2) helix. The complex can be obtained in the form of a large single crystal from the ethanolic solution of urea and PEO with a molecular weight of about 1000 (Fig. 7.29). The crystal gives an X-ray photograph as shown in Fig. 4.1*c*. PEO forms two types of crystalline complexes with thiourea, one is stable at room temperature and the other is stable above 45°C.

The formation of two types of PEO–mercuric chloride complex (Section 4.3.5) and the analysis of the complex type I (Sections 4.7.1.C, 4.7.4.A, and 4.8.1.B) are described earlier. The (7/2) helix of ordinary PEO and the T_5-$GT_5\bar{G}$ conformation in complex type I appear to be quite different, but are transformed as shown below by changing the conformations only of a limited number of bonds in the chain.

PEO

$$\begin{array}{l} \quad T \qquad T \quad G \quad T \quad T \quad G \quad T \quad T \quad G \quad T \quad T \quad G \\ \text{—C—(—O—C—C—O—C—C—O—C—C—O—C—C—)}_n\text{—} \\ \quad T \qquad T \quad T \quad T \quad T \quad G \quad T \quad T \quad T \quad T \quad T \quad \bar{G} \end{array}$$

PEO–mercuric chloride complex type I

When two G's in the upper conformation are transformed to T and one G to $\bar{G}$, the lower conformation is obtained.

In the case of type II, the contribution of the X-ray scattering of the mercury (at. no. 80) and chlorine atoms (at. no. 17) is much larger than for type I.

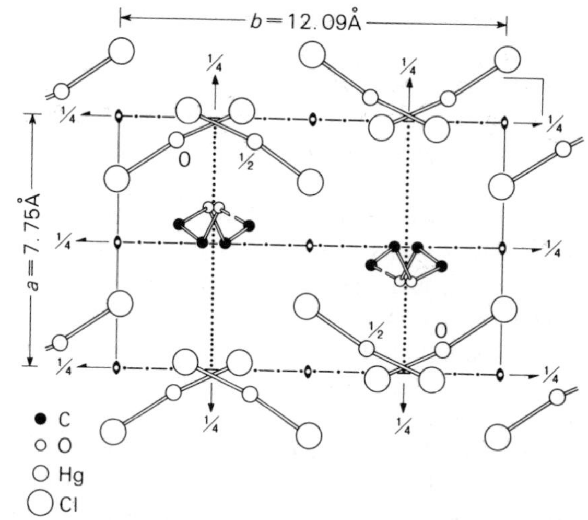

Fig. 7.30 Crystal structure of PEO–mercuric chloride complex type II.[167]

For this reason the conformation of PEO was determined through the analysis of far-infrared spectra (normal coordinate treatment), although the positions of mercuric chloride molecules were determined by the Patterson method.[167] The molecular conformation of PEO thus obtained is

$$\begin{array}{cccccc} —O—(—CH_2 & —CH_2 & —O & —CH_2 & —CH_2 & —O—)_n— \\ T' & G' & G' \quad \bar{T}' & & \bar{G}' & \bar{G}' \\ 167^\circ & 77^\circ & 81^\circ & & & \end{array}$$

Figure 7.30 shows the crystal structure. The agreement between the observed and calculated reflection intensities with this structure was much improved relative to the case of packing only mercuric chloride even though the scattering power of PEO is small: $[f_{Hg}{}^2 + 2f_{Cl}{}^2(=6978) \gg 2f_C{}^2 + f_O{}^2 (=136)]$; from $R = 15\%$ to $R = 13\%$.

A further reference for the PE–urea complex is Ref. 272.

REFERENCES

1. C. W. Bunn, *Trans. Faraday Soc.*, **35,** 482 (1939).
2. P. R. Swan, *J. Polym. Sci.*, **56,** 409 (1962).
3. S. Kavesh and J. M. Schultz, *J. Polym. Sci. A-2*, **8,** 243 (1970).
4. Y. Chatani, Y. Ueda, and H. Tadokoro, Annual Meeting of the Society of Polymer Science, Japan, Tokyo, Preprint p. 1326, 1977.
5. T. Seto, T. Hara, and K. Tanaka, *Jpn. J. Appl. Phys.*, **7,** 31 (1968).
6. D. C. Bassett, S. Block, and G. J. Piermarini, *J. Appl. Phys.*, **45,** 4146 (1974).
7. M. Yasuniwa, Y. Enoshita, and T. Takemura, *Jpn. J. Appl. Phys.*, **15,** 1421 (1976).
8. T. Yamamoto, H. Miyaji, and K. Asai, *Rep. Prog. Polym. Phys. Jpn.*, **19,** 191 (1976).
9. C. A. Sperati and H. W. Starkweather, Jr., *Adv. Polym. Sci.*, **2,** 465 (1961).
10. C. W. Bunn and E. R. Howells, *Nature*, **174,** 549 (1954).
11. E. S. Clark and L. T. Muus, *Z. Krist.*, **117,** 119 (1962).
12. C. Nakafuku and T. Takemura, *Jpn. J. Appl. Phys.*, **14,** 599 (1975).
13. Z. Mencik, *J. Polym. Sci., Polym. Phys. Ed.*, **11,** 1585 (1973).
14. F. C. Wilson and H. W. Starkweather, Jr., *J. Polym. Sci., Polym. Phys. Ed.*, **11,** 919 (1973).
15. G. Natta and P. Corradini, *Nuovo Cimento, Suppl.*, **15,** 40 (1960).
16. A. Turner-Jones and A. J. Cobbold, *J. Polym. Sci. B*, **6,** 539 (1968).
17. R. J. Samuels and R. Y. Yee, *J. Polym. Sci. A-2*, **10,** 358 (1972).
18. E. J. Addink and J. Beintema, *Polymer*, **2,** 185 (1961).
19. P. Corradini, G. Natta, P. Ganis, and P. A. Temussi, *J. Polym. Sci. C*, **16,** 2477 (1967).
20. G. Natta, M. Peraldo, and G. Allegra, *Makromol. Chem.*, **75,** 215 (1964).
21. H. Tadokoro, M. Kobayashi, Shōdō Kobayashi, K. Yasufuku, and Koh Mori, *Rep. Prog. Polym. Phys. Jpn.*, **9,** 181 (1966).

22. G. Natta, P. Corradini, and I. W. Bassi, *Nuovo Cimento, Suppl.*, **15,** 52 (1960).
23. T. Miyashita, M. Yokouchi, Y. Chatani, and H. Tadokoro, Annual Meeting of the Society of Polymer Science, Japan, Tokyo, Preprint p. 453, 1974.
24. A. Turner-Jones, *J. Polym. Sci. B*, **1,** 455 (1963).
25. V. Petraccone, B. Pirozzi, A. Frasci, and P. Corradini, *Eur. Polym. J.*, **12,** 323 (1976).
26. A. Turner-Jones and J. M. Aizlewood, *J. Polym. Sci. B*, **1,** 471 (1963).
27. A. Turner-Jones, *Makromol. Chem.*, **71,** 1 (1964).
28. P. Corradini, P. Ganis, and V. Petraccone, *Eur. Polym. J.*, **6,** 281 (1970).
29. I. W. Bassi, O. Bonsignori, G. P. Lorenzi, P. Pino, P. Corradini, and P. A. Temussi, *J. Polym. Sci. A-2*, **9,** 193 (1971).
30. H. Kusanagi, Y. Chatani, M. Takase, and H. Tadokoro, *J. Polym. Sci., Polym. Phys. Ed.*, **16,** 131 (1978).
31. P. Corradini, E. Martuscelli, A. Montagnoli, and V. Petraccone, *Eur. Polym. J.*, **6,** 1201 (1970).
32. G. Natta, P. Corradini, and I. W. Bassi, *Makromol. Chem.*, **33,** 247 (1959).
33. F. J. Golemba, J. E. Guillet, and S. C. Nyburg, *J. Polym. Sci. A-1*, **6,** 1341 (1968).
34. I. W. Bassi and G. Chioccola, *Eur. Polym. J.*, **5,** 163 (1969).
35. V. Petraccone, B. Pirozzi, and P. Corradini, *Eur. Polym. J.*, **8,** 107 (1972).
36. G. Natta and P. Corradini, *Makromol. Chem.*, **16,** 77 (1955); G. Natta, P. Corradini, and I. W. Bassi, *Nuovo Cimento, Suppl.*, **15,** 68 (1960).
37. P. Corradini and P. Ganis, *Nuovo Cimento, Suppl.*, **15,** 96 (1960).
38. G. Natta and P. Corradini, *J. Polym. Sci.*, **39,** 29 (1959).
39. I. Nitta, Y. Chatani, and Y. Sakata, *Bull. Chem. Soc. Jpn.*, **33,** 125 (1960).
40. Y. Chatani, *J. Polym. Sci.*, **47,** 491 (1960).
41. S. Murahashi, S. Nozakura, and H. Tadokoro, *Bull. Chem. Soc. Jpn.*, **32,** 534 (1959).
42. G. Natta, P. Corradini, and I. W. Bassi, *Nuovo Cimento, Suppl.*, **15,** 83 (1960).
43. B. L. Farmer and J. B. Lando, *J. Macromol. Sci., Phys.*, **10,** 403 (1974).
44. P. Corradini and P. Ganis, *Nuovo Cimento, Suppl.*, **15,** 104 (1960).
45. G. Natta and P. Corradini, *J. Polym. Sci.*, **20,** 251 (1956).
46. C. W. Bunn, *Nature*, **161,** 929 (1948).
47. I. Nitta, I. Taguchi, and Y. Chatani, *Annu. Rep. Inst. Fiber Res. Osaka Univ.*, **10,** 1 (1957).
48. G. Natta, I. W. Bassi, and G. Allegra, *Atti Accad. Naz. Lincei Rend., Cl. Sci. Fis. Mat. Nat.*, **31,** 350 (1961).
49. K. Fujii, T. Mochizuki, S. Imoto, J. Ukita, and M. Matsumoto, *Makromol. Chem.*, **51,** 225 (1962).
50. T. Tanaka, Y. Chatani, and H. Tadokoro, *J. Polym. Sci., Polym. Phys. Ed.*, **12,** 515 (1974).
51. Y. Chatani, T. Takahagi, T. Kusumoto, and H. Tadokoro, *J. Polym. Sci., Polym. Phys. Ed.*, to be published.
52. R. Hasegawa, M. Kobayashi, and H. Tadokoro, *Polym. J.*, **3,** 591 (1972).
53. R. Hasegawa, Y. Takahashi, Y. Chatani, and H. Tadokoro, *Polym. J.*, **3,** 600 (1972).
54. J. B. Lando, H. G. Olf, and A. Peterlin, *J. Polym. Sci. A-1*, **4,** 941 (1966)
55. H. Kusanagi, H. Tadokoro, and Y. Chatani, *Macromolecules*, **9,** 531 (1976).
56. V. Y. Chen, G. Allegra, P. Corradini, and M. Goodman, *Macromolecules*, **3,** 274 (1970).

57. S. Iwayanagi, I. Sakurai, T. Sakurai, and T. Seto, *J. Macromol. Sci., Phys.*, **2,** 163 (1968).
58. G. Natta and P. Corradini, *Angew. Chem.*, **68,** 615 (1956).
59. G. Natta, *Angew. Chem.*, **68,** 393 (1956).
60. Y. Takahashi, T. Sato, H. Tadokoro, and Y. Tanaka, *J. Polym. Sci., Polym. Phys. Ed.*, **11,** 233 (1973).
61. C. W. Bunn, *Proc. Roy. Soc. (Lond.)*, **A180,** 40 (1942).
62. S. C. Nyburg, *Acta Crystallogr.*, **7,** 385 (1954).
63. G. Natta, I. W. Bassi, and G. Fagherazzi, *Eur. Polym. J.*, **5,** 239 (1969).
64. G. Fagherazzi and I. W. Bassi, *Eur. Polym. J.*, **4,** 151 (1968).
65. G. Natta, P. Corradini, I. W. Bassi, and G. Fagherazzi, *Eur. Polym. J.*, **4,** 297 (1968).
66. I. W. Bassi, G. Allegra, and R. Scrodamaglia, *Macromolecules*, **4,** 575 (1971).
67. Y. Chatani, S. Nakatani, and H. Tadokoro, *Macromolecules*, **3,** 481 (1970).
68. M. Cesari, *J. Polym. Sci. B*, **2,** 453 (1964).
69. H. Tadokoro, Y. Takahashi, S. Otsuka, Kan Mori, and F. Imaizumi, *J. Polym. Sci. B*, **3,** 697 (1965).
70. H. Tadokoro, M. Kobayashi, Koh Mori, Y. Takahashi, and S. Taniyama, *J. Polym. Sci. C*, **22,** 1031 (1969).
71. H. Tadokoro, T. Yasumoto, S. Murahashi, and I. Nitta, *J. Polym. Sci.*, **44,** 266 (1960).
72. T. Uchida and H. Tadokoro, *J. Polym. Sci. A-2*, **5,** 63 (1967).
73. G. Carazzolo, *J. Polym. Sci. A*, **1,** 1573 (1963).
74. G. Carazzolo and M. Mammi, *J. Polym. Sci. A*, **1,** 965 (1963).
75. Y. Takahashi and H. Tadokoro, *Macromolecules*, **6,** 672 (1973).
76. Y. Takahashi, I. Sumita, and H. Tadokoro, *J. Polym. Sci., Polym. Phys. Ed.*, **11,** 2113 (1973).
77. H. Tadokoro, Y. Takahashi, Y. Chatni, and H. Kakida, *Makromol. Chem.*, **109,** 96 (1967).
78. H. Kakida, D. Makino, Y. Chatani, M. Kobayashi, and H. Tadokoro, *Macromolecules*, **3,** 569 (1970).
79. Y. Takahashi, Y. Osaki, and H. Tadokoro, *J. Polymer Sci., Polym. Phys. Ed.*, to be published.
80. K. Imada, T. Miyakawa, Y. Chatani, H. Tadokoro, and S. Murahashi, *Makromol. Chem.*, **83,** 113 (1965).
81. M. Cesari, G. Perego, and A. Mazzei, *Makromol. Chem.* **83,** 196 (1965).
82. Shōdō Kobayashi, H. Tadokoro, and Y. Chatani, *Makromol. Chem.*, **112,** 225 (1968).
83. S. Sasaki, Y. Takahashi, and H. Tadokoro, Discussion Meeting of the Society of Polymer Science, Japan, Osaka, Preprint p. 491, 1972.
84. S. Sasaki, Y. Takahashi, and H. Tadokoro, *J. Polym. Sci., Polym. Phys. Ed.*, **10,** 2363 (1972).
85. S. Sasaki, Y. Takahashi, and H. Tadokoro, *Polym. J.*, **4,** 172 (1973).
86. S. Sasaki and H. Tadokoro, *J. Polym. Sci., Polym. Phys. Ed.*, **11,** 1985 (1973).
87. G. Natta, G. Mazzanti, P. Corradini, and I. W. Bassi, *Makromol. Chem.*, **37,** 156 (1960).
88. G. Natta, P. Corradini, and I. W. Bassi, *J. Polym. Sci.*, **51,** 505 (1961).
89. M. Cesari, G. Perego, and W. Marconi, *Makromol. Chem.*, **94,** 194 (1966).
90. Y. Takahashi, H. Tadokoro, T. Hirano, A. Sato, and T. Tsuruta, *J. Polym. Sci., Polym. Phys. Ed.*, **13,** 285 (1975).

91. H. Sakakihara, Y. Takahashi, H. Tadokoro, N. Oguni, and H. Tani, *Macromolecules*, **6,** 205 (1973).

92. H. Kusanagi, M. Takase, H. Tadokoro, H. Tani, and N. Oguni, to be published.

93. K. Kaji and I. Sakurada, *Makromol. Chem.*, **148,** 261 (1971).

94. Y. Takahashi, H. Tsugaya, and H. Tadokoro, Discussion Meeting of the Society of Polymer Science, Japan, Kyoto, Preprint p. 743, 1970.

95. M. Barlow, *J. Polym. Sci. A-2*, **4,** 121 (1966).

96. I. V. Kumpanenko, K. S. Kazanskii, N. V. Ptitsyna, and M. Ya. Kushnerev, *Polym. Sci. USSR*, **12,** 930 (1970).

97. H. Matsubayashi, Y. Chatani, H. Tadokoro, Y. Tabata, and W. Ito, *Polym. J.*, **9,** 145 (1977).

98. J. Boon and E. P. Magré, *Makromol. Chem.*, **126,** 130 (1969).

99. J. Boon and E. P. Magré, *Makromol. Chem.*, **136,** 267 (1970).

100. P. Corradini and I. W. Bassi, *J. Polym. Sci. C*, **16,** 3233 (1968).

101. G. Dall'Asta and N. Oddo, *Chim. Ind. (Milan)*, **42,** 1234 (1960).

102. G. Natta, I. W. Bassi, and P. Corradini, *Makromol. Chem.*, **18/19,** 455 (1956).

103. I. Nitta, I. Taguchi, and Y. Chatani, *Ann. Rep. Inst. Fiber Res. Osaka Univ.*, **12,** 92 (1959).

104. I. W. Bassi, G. Dall'Asta, U. Campigli, and E. Strepparola, *Makromol. Chem.*, **60,** 202 (1963).

105. S. Murahashi, H. Yuki, T. Sano, U. Yonemura, H. Tadokoro, and Y. Chatani, *J. Polym. Sci.*, **62,** S77 (1962).

106. G. Carazzolo and G. Valle, *Makromol. Chem.*, **90,** 66 (1966).

107. Y. Takahashi, H. Tadokoro, and Y. Chatani, *J. Macromol. Sci., Phys.*, **2,** 361 (1968).

108. H. Sakakihara, Y. Takahashi, and H. Tadokoro, Discussion Meeting of the Society of Polymer Science, Japan, Tokyo, Preprint p. 407, 1969.

109. Y. Gotoh, H. Sakakihara, and H. Tadokoro, *Polym. J.*, **4,** 68 (1973).

110. H. Sakakihara, Y. Takahashi, H. Tadokoro, P. Sigwalt, and N. Spassky, *Macromolecules*, **2,** 515 (1969).

111. H. Matsubayashi, Y. Chatani, H. Tadokoro, P. Dumas, N. Spassky, and P. Sigwalt, *Macromolecules*, **10,** 996 (1977).

112. B. J. Tabor, E. P. Magré, and J. Boon, *Eur. Polym. J.*, **7,** 1127 (1971).

113. Y. Chatani, K. Suehiro, Y. Okita, H. Tadokoro, and K. Chujo, *Makromol. Chem.*, **113,** 215 (1968).

114. K. Suehiro, Y. Chatani, and H. Tadokoro, *Polym. J.*, **7,** 352 (1975).

115. Y. Chatani, Y. Okita, H. Tadokoro, and Y. Yamashita, *Polym. J.*, **1,** 555 (1970).

116. M. Yokouchi, Y. Chatani, H. Tadokoro, K. Teranishi, and H. Tani, *Polymer*, **14,** 267 (1973).

117. J. Cornibert and R. H. Marchessault, *J. Mol. Biol.*, **71,** 735 (1972).

118. M. Yokouchi, Y. Chatani, H. Tadokoro, and H. Tani, *Polym. J.*, **6,** 248 (1974).

119. G. Carazzolo, *Chim. Ind. (Milan)*, **46,** 525 (1964).

120. G. Perego, A. Melis, and M. Cesari, *Makromol. Chem.*, **157,** 269 (1972).

121. M. Yokouchi, Y. Chatani, and H. Tadokoro, *J. Polym. Sci., Polym. Phys. Ed.*, **14,** 81 (1976).

122. A. S. Ueda, Y. Chatani, and H. Tadokoro, *Polym. J.*, **2,** 387 (1971).

123. A. Turner-Jones and C. W. Bunn, *Acta Crystallogr.*, **15,** 105 (1962).
124. T. Kanamoto, K. Tanaka, and H. Nagai, *J. Polym. Sci. A-2*, **9,** 2043 (1971).
125. S. Y. Hobbs and F. W. Billmeyer, Jr., *J. Polym. Sci. A-2*, **7,** 1119 (1969).
126. Y. Chatani, R. K. Hasegawa, and H. Tadokoro, Annual Meeting of the Society of Polymer Science, Japan, Tokyo, Preprint p. 420, 1971.
127. R. de P. Daubeny, C. W. Bunn, and C. J. Brown, *Proc. Roy. Soc. (Lond.)*, **A226,** 531 (1954).
128. Shigeo Kobayashi and M. Hachiboshi, *Rep. Prog. Polym. Phys. Jpn.*, **13,** 157 (1970).
129. Y. Chatani, Y. Higashihata, M. Takase, H. Tadokoro, and T. Hirahara, Annual Meeting of the Society of Polymer Science, Japan, Kyoto, Preprint p. 427, 1977.
130. M. Yokouchi, Y. Sakakibara, Y. Chatani, H. Tadokoro, T. Tanaka, and K. Yoda, *Macromolecules*, **9,** 266 (1976).
131. Z. Menzik, *J. Polym. Sci., Polym. Phys. Ed.*, **13,** 2173 (1975).
132. I. H. Hall and M. G. Pass, *Polymer*, **17,** 807 (1976).
133. I. J. Desborough and I. H. Hall, *Polymer*, **18,** 825 (1977).
134. Y. Chatani, Y. Higashihata, and H. Tadokoro, to be published.
135. Z. Menčik, *Chem. Průmysl.*, **17,** 78 (1967).
136. R. K. Hasegawa, K. Kimoto, Y. Chatani, H. Tadokoro, and H. Sekiguchi, Discussion Meeting of the Society of Polymer Science, Japan, Tokyo, Preprint p. 713, 1974.
137. J. Masamoto, K. Sasaguri, C. Ohizumi, and H. Kobayashi, *J. Polym. Sci. A-2*, **8,** 1703 (1970).
138. R. J. Fredericks, T. H. Doyne, R. S. Sparague, *J. Polym. Sci. A-2*, **4,** 899 (1966).
139. D. R. Holmes, C. W. Bunn, and D. J. Smith, *J. Polym. Sci.*, **17,** 159 (1955).
140. H. Arimoto, *J. Polym. Sci. A*, **2,** 2283 (1964).
141. D. C. Vogelsong, *J. Polym. Sci.*, **A1,** 1055 (1963).
142. K. Inoue and S. Hoshino, *J. Polym. Sci., Polym. Phys. Ed.*, **11,** 1077 (1973).
143. G. Cojazzi, A. Fichera, C. Garbuglio, V. Malta, and R. Zannetti, *Makromol. Chem.*, **168,** 289 (1973).
144. C. W. Bunn and E. V. Garner, *Proc. Roy. Soc. (Lond.)*, **A189,** 39 (1947).
145. Y. Kinoshita, *Makromol. Chem.*, **33,** 1, 21 (1959).
146. T. Ota, M. Yamashita, O. Yoshizaki, and E. Nagai, *J. Polym. Sci. A-2*, **4,** 959 (1966).
147. R. K. Hasegawa, Y. Chatani, and H. Tadokoro, Annual Meeting of the Society of Crystallography of Japan, Osaka, Preprint p. 21, 1973.
148. M. G. Northolt and J. J. van Aartsen, *J. Polym. Sci. B*, **11,** 333 (1973); M. G. Northolt, *Eur. Polym. J.*, **10,** 799 (1974).
149. H. Kakida, Y. Chatani, and H. Tadokoro, *J. Polym. Sci., Polym. Phys. Ed.*, **14,** 427 (1976).
150. H. Kusanagi, H. Tadokoro, Y. Chatani, and K. Suehiro, *Macromolecules*, **10,** 405 (1977).
151. Y. Takahashi, T. Kurumizawa, H. Kusanagi, and H. Tadokoro, *J. Polym. Sci., Polym. Phys. Ed.*, **16,** 1989 (1978).
152. I. W. Bassi, P. Ganis, and P. A. Temussi, *J. Polym. Sci. C*, **16,** 2867 (1967).
153. P. Ganis and P. A. Temussi, *Eur. Polym. J.*, **2,** 401 (1966).
154. R. Iwamoto and B. Wunderlich, *J. Polym. Sci., Polym. Phys. Ed.*, **11,** 2403 (1973).
155. W. D. Niegisch, *J. Appl. Phys.*, **38,** 4110 (1967).
156. R. Bonart, *Makromol. Chem.*, **92,** 149 (1966).

157. Y. Chatani, T. Takizawa, S. Murahashi, Y. Sakata, and Y. Nishimura, *J. Polym. Sci.*, **55,** 811 (1961).

158. Y. Chatani, T. Takizawa, and S. Murahashi, *J. Polym. Sci.*, **62,** S27 (1962).

159. E. Giglio, F. Pompa, and A. Ripamonti, *J. Polym. Sci.*, **59,** 293 (1962).

160. K. H. Meyer, W. Lotmar, and G. W. Pankov, *Helv. Chim. Acta*, **19,** 930 (1936).

161. C. W. Bunn and E. V. Garner, *J. Chem. Soc.*, **1942,** 654.

162. M. D. Lind and S. Geller, *J. Chem. Phys.*, **51,** 348 (1969).

163. D. Kobelt and E. F. Paulus, *Acta Crystallogr.*, **B30,** 232 (1974).

164. P. Corradini and P. Ganis, *Makromol. Chem.*, **62,** 97 (1963).

165. Y. Saito, S. Nansai, and S. Kinoshita, *Polym. J.*, **3,** 113 (1972).

166. R. Iwamoto, Y. Saito, H. Ishihara, and H. Tadokoro, *J. Polym. Sci. A-2*, **6,** 1509 (1968).

167. M. Yokoyama, H. Ishihara, R. Iwamoto, and H. Tadokoro, *Macromolecules*, **2,** 184 (1969).

168. Y. Chatani and S. Nakatani, *Macromolecules*, **5,** 597 (1972).

169. Y. Chatani and S. Kuwata, *Macromolecules*, **8,** 12 (1975).

170. A. Colombo and G. Allegra, *Macromolecules*, **4,** 579 (1971).

171. M. Farina, G. Allegra, and G. Natta, *J. Am. Chem. Soc.*, **86,** 516, (1964).

172. G. Allegra, M. Farina, A. Immirzi, A. Colombo, U. Rossi, R. Broggi, and G. Natta, *J. Chem. Soc. (B)*, **1967,** 1020.

173. G. Allegra, M. Farina, A. Colombo, G. Casagrande-Tettamanti, U. Rossi, and G. Natta, *J. Chem. Soc. (B)*, **1967,** 1028.

174. M. Farina, G. Natta, G. Allegra, and M. Löffelholz, *J. Polym. Sci. C*, **16,** 2517 (1967).

175. R. L. Miller, Crystallographic Data for Various Polymers, J. Brandrup and E. H. Immergut, Eds., *Polymer Handbook* 2nd ed., Wiley, New York, 1975, pp. III-3 to III-137.

176. P. Corradini, New Results of Structural Researches on High Polymers, *Advances in Structure Research by Diffraction Methods*, R. Brill and R. Mason, Eds., Vol. 2, pp. 141–163, Friedrich Vieweg & Sohn, Brawnschweig, Interscience, New York, 1966.

177. H. Tadokoro, Structure of Crystalline Polymers, *Macromol. Rev.*, **1,** pp. 119–172, Wiley-Interscience, New York, 1967.

178. H. Tadokoro, Structure of Crystalline Polymers, *Molecular Structure and Properties*, Vol. 2, *MTP International Review of Science*, *Physical Chemistry Series I*, G. Allen, Ed., Butterworths, London, University Park Press, Baltimore, 1972, pp. 45–89.

179. G. Natta and G. Allegra, *Tetrahedron*, **30,** 1987 (1974).

180. M. Shen, W. N. Hansen, and P. C. Romo, *J. Chem. Phys.*, **51,** 425 (1969).

181. G. Avitabile, R. Napolitano, B. Pirozzi, K. D. Rouse, M. W. Thomas, and B. T. M. Willis, *J. Polym. Sci. B*, **13,** 351 (1975).

182. P. R. Swan, *J. Polym. Sci.*, **56,** 403 (1962).

183. E. R. Walter and F. P. Reding, *J. Polym. Sci.*, **21,** 561 (1956).

184. E. A. Cole and D. R. Holmes, *J. Polym. Sci.*, **46,** 245 (1960).

185. P. W. Teare and D. R. Holmes, *J. Polym. Sci.*, **24,** 496 (1957).

186. A. Turner-Jones, *J. Polym. Sci.*, **62,** S53 (1962).

187. J. L. Koenig and D. E. Witenhafer, *Makromol. Chem.*, **99,** 193 (1966).

188. M. I. Bank and S. Krimm, *J. Appl. Phys.*, **40,** 4248 (1969).

189. T. Kitagawa and T. Miyazawa, *Rep. Prog. Polym. Phys. Jpn.*, **11,** 219 (1968).

190. C. G. Opasker and S. Krimm, *J. Polym. Sci. A-2*, **7,** 57 (1969).
191. N. E. Weeks and R. S. Porter, *J. Polym. Sci., Polym. Phys. Ed.*, **12,** 635 (1974).
192. C. Capacio and I. M. Ward, *Polymer*, **15,** 233 (1974).
192a. R. G. Brown, *J. Chem. Phys.*, **40,** 2900 (1964).
193. H. D. Flack, *J. Polym. Sci., Polym. Phys. Ed.*, **10,** 1799 (1972).
194. G. Zerbi and M. Sacchi, *Macromolecules*, **6,** 692 (1973).
195. G. Masetti, F. Cabassi, G. Morelli, and G. Zerbi, *Macromolecules*, **6,** 700 (1973).
196. C. W. Bunn and E. R. Howells, *J. Polym. Sci.*, **18,** 307 (1955).
197. V. F. Holland and R. L. Miller, *J. Appl. Phys.*, **35,** 3241 (1964).
198. K. J. Clark, A. Turner-Jones, and D. J. Sandiford, *Chem. Ind.*, **1962,** 2010; A. Turner-Jones, *Makromol. Chem.*, **71,** 1 (1964).
199. F. Danusso, *Polymer*, **3,** 423 (1962).
200. H. Tadokoro, S. Nozakura, T. Kitazawa, Y. Yasuhara, and S. Murahashi, *Bull. Chem. Soc. Jpn.*, **32,** 313 (1959).
201. M. Kobayashi, K. Tsumura, and H. Tadokoro, *J. Polym. Sci. A-2*, **6,** 1493 (1968).
202. M. Kobayashi, K. Akita, and H. Tadokoro, *Makromol. Chem.*, **118,** 324 (1968).
203. R. H. Boyd and S. M. Breitling, *Macromolecules*, **5,** 279 (1972).
204. (a) G. Natta, *Makromol. Chem.*, **35,** 94 (1960); (b) M. Peraldo and M. Farina, *Chim. Ind. (Milan)*, **42,** 1349 (1960).
205. H. Tadokoro, M. Ukita, M. Kobayashi, and S. Murahashi, *J. Polym. Sci. B*, **1,** 405 (1963); H. Tadokoro, M. Kobayashi, M. Ukita, K. Yasufuku, S. Murahashi, and T. Torii, *J. Chem. Phys.*, **42,** 1432 (1965).
206. T. Miyazawa and Y. Ideguchi, *J. Polym. Sci. B*, **1,** 389 (1963); *Makromol. Chem.*, **79,** 89 (1964).
207. G. Natta, M. Peraldo, M. Farina, and G. Bressan, *Makromol. Chem.*, **55,** 139 (1962).
208. G. Natta, P. Corradini, and P. Ganis, *Makromol. Chem.*, **39,** 238 (1960).
209. M. Peraldo and M. Cambini, *Spectrochim. Acta*, **21,** 1509 (1965).
210. J. H. Schachtschneider and R. G. Snyder, *Spectrochim. Acta*, **21,** 1527 (1965).
211. T. Miyazawa, *J. Chem. Phys.*, **43,** 4030 (1965).
212. C. E. Wilkes, V. L. Folt, and S. Krimm, *Macromolecules*, **6,** 235 (1973).
213. T. Shimanouchi, S. Tsuchiya, and S. Mizushima, *J. Chem. Phys.*, **30,** 1365 (1959).
214. M. Tasumi and T. Shimanouchi, *Spectrochim. Acta*, **17,** 731 (1961).
215. M. Tasumi and T. Shimanouchi, *Polym. J.*, **2,** 62 (1971).
216. I. Sakurada and K. Nambu, *Nihon Kagak u Zasshi* (*J. Chem. Soc. Jpn.*), **80,** 307 (1959).
217. J. J. Shipman, V. L. Folt, and S. Krimm, *Spectrochim. Acta*, **18,** 1603 (1962).
218. S. Krimm, V. L. Folt, J. J. Shipman, and A. R. Berens, *J. Polym. Sci. B*, **2,** 1009 (1964).
219. A. Rubčić and G. Zerbi, *Macromolecules*, **7,** 754, 759 (1974).
220. S. Okamura, T. Kodama, T. Higashimura, *Makromol. Chem.*, **53,** 180 (1962).
221. S. Murahashi, S. Nozakura, M. Sumi, H. Yuki, and K. Hatada, *J. Polym. Sci. B*, **4,** 65 (1966).
222. R. Stefani, M. Chevreton, M. Garnier, and C. Eyraud, *C. R.*, **251,** 2174 (1960).
223. C. R. Bohn, J. R. Schaefgen, and W. O. Statton, *J. Polym. Sci.*, **55,** 531 (1961).
224. V. F. Holland, S. B. Mitchell, W. L. Hunter, and P. H. Lindenmeyer, *J. Polym. Sci.*, **62,** 145 (1962).

225. P. H. Lindenmeyer and R. Hosemann, *J. Appl. Phys.*, **34,** 42 (1963).

226. H. Tadokoro, S. Murahashi, R. Yamadera, and T. Kamei, *J. Polym. Sci. A*, **1,** 3029 (1963); R. Yamadera, H. Tadokoro, and S. Murahashi, *J. Chem. Phys.*, **41,** 1233 (1964).

227. M. Kobayashi, K. Tashiro, and H. Tadokoro, *Macromolecules*, **8,** 158 (1975).

228. H. Kawai, *Jpn. J. Appl. Phys.*, **8,** 975 (1969).

229. K. Nakamura and Y. Wada, *J. Polym. Sci. A-2*, **9,** 161 (1971).

230. M. Oshiki and E. Fukada, *J. Mater. Sci.*, **10,** 1 (1975).

231. G. Natta and P. Corradini, *Nuovo Cimento, Suppl.*, **15,** 9 (1960).

232. K. Suehiro and M. Takayanagi, *J. Macromol. Sci., Phys.*, **4,** 39 (1970).

233. F. W. Stavely, et al., *Ind. Eng. Chem.*, **48,** 778 (1956).

234. S. H. Horne, Jr., et al., *Ind. Eng. Chem.*, **48,** 784 (1956).

235. J. Lal and G. S. Trick, *J. Polym. Sci.*, **50,** 13 (1960).

236. P. J. Flory, *Statistical Mechanics of Chain Molecules*, Wiley-Interscience, New York, 1969.

237. J. E. Mark and P. J. Flory, *J. Am. Chem. Soc.*, **87,** 1415 (1965); **88,** 3702 (1966).

238. Y. Chatani, T. Uchida, H. Tadokoro, K. Hayashi, M. Nishii, and S. Okamura, *J. Macromol. Sci., Phys.*, **2,** 567 (1968).

239. Y. Chatani, K. Kitahama, H. Tadokoro, T. Yamauchi, and Y. Miyake, *J. Macromol. Sci., Phys.*, **4,** 61 (1970).

240. Y. Chatani, T. Ohno, T. Yamauchi, and Y. Miyake, *J. Polym. Sci., Polym. Phys. Ed.*, **11,** 369 (1973).

241. M. Yokoyama, H. Ochi, H. Tadokoro, and C. C. Price, *Macromolecules*, **5,** 690 (1972).

242. A. Angood and J. L. Koenig, *J. Appl. Phys.*, **39,** 4985 (1968).

243. D. Makino, M. Kobayashi, and H. Tadokoro, *Spectrochim. Acta*, **31A,** 1481 (1975).

244. F. J. Boerio and D. D. Cornell, *J. Chem. Phys.*, **56,** 1516 (1972).

245. M. Okada, Y. Yamashita, and Y. Ishii, *Makromol. Chem.*, **80,** 196 (1964).

246. M. Kobayashi, I. Okamoto, and H. Tadokoro, *Spectrochim. Acta*, **31A,** 1799 (1975).

247. W. Claffey, K. Gardner, J. Blackwell, J. Lando, and P. H. Geil, *Philos. Mag.*, **30,** 1223 (1974).

248. E. J. Vandenberg, *J. Am. Chem. Soc.*, **83,** 3538 (1961); *J. Polym. Sci. B*, **2,** 1085 (1964).

249. H. Tadokoro, M. Kobayashi, H. Yoshidome, K. Tai, and D. Makino, *J. Chem. Phys.*, **49,** 3359 (1968).

250. R. E. Prud'homme and R. H. Marchessault, *Macromolecules*, **7,** 541 (1974).

251. R. E. Prud'homme, *J. Polym. Sci., Polym. Phys. Ed.*, **12,** 2455 (1974).

252. R. Jakeways, I. M. Ward, M. A. Wilding, I. H. Hall, I. J. Desborough, and M. G. Pass, *J. Polym. Sci., Polym. Phys. Ed.*, **13,** 799 (1975).

253. C. A. Boye, Jr. and J. R. Overton, *Bull. Am. Phys. Soc., Ser.* 2, **19,** 352 (1974).

254. Japanese Chemical Fibers Association, *Jpn. Chem. Fibers Monthly*, **11** (1974).

255. I. Sakurada and K. Kaji, Meeting of the Society of Polymer Science, Japan, Kobe, Preprint p. 56, 1975.

256. I. Sakurada and K. Kaji, *Kobunshi Kagaku (Chem. High Polym.)*, **26,** 817 (1969).

257. I. Sakurada and K. Kaji, *J. Polym. Sci. C*, **31,** 57 (1970).

258. I. Sakurada, K. Nakamae, K. Kaji, and S. Wadano, *Kobunshi Kagaku (Chem. High Polym.)*, **26,** 561 (1969).

259. L. R. G. Treloar, *Polymer*, **1,** 279 (1960).

260. T. Miyazawa, *Rep. Prog. Polym. Phys. Jpn.*, **8,** 47 (1965).

261. T. R. Manley and C. G. Martin, *Polymer*, **14,** 632 (1973).

262. K. Tashiro, M. Kobayashi, and H. Tadokoro, *Macromolecules*, **10,** 413 (1977).

263. A. E. Tonelli, *J. Polym. Sci. B*, **11,** 441 (1973).

264. W. E. Stewart and T. H. Sidall, III, *Chem. Rev.*, **70,** 517 (1970).

265. F. Tuinstra, *Acta Crystallogr.*, **20,** 341 (1966).

266. W. Ruland, *J. Appl. Phys.*, **38,** 3585 (1967); *J. Polym. Sci. C*, **28,** 143 (1969); W. Ruland and H. Tompa, *Acta Crystallogr.*, **A24,** 93 (1968).

267. F. Tuinstra and J. L. Koenig, *J. Chem. Phys.*, **53,** 1126 (1970).

268. M. Hasegawa, Y. Suzuki, H. Nakanishi, and F. Nakanishi, *Prog. Polym. Sci. Jpn.*, **5,** 143, Kodansha-Wiley, Tokyo, 1973.

269. H. Tadokoro, T. Yoshihara, Y. Chatani, and S. Murahashi, *J. Polym. Sci. B*, **2,** 363 (1964).

270. W. Kutzelnigg and R. Mecke, *Z. Elektrochem.*, **65,** 109 (1961).

271. A. Yamaguchi, *Nihon Kagaku Zasshi (J. Chem. Soc. Jpn.)*, **78,** 1467 (1957).

272. K. Monobe and F. Yokoyama, *J. Macromol. Sci., Phys.*, **8,** 277 (1973).

Chapter 8

Concluding Remarks

The methods available for structural studies of high polymers have made characteristic developments associated with the methods for low-molecular-weight compounds, especially in the last 10 years. The crystal and molecular structures of many polymers, including several polymers whose structures have been unresolved for a long time, have now been determined. Analyses of vibrational spectra have been made and they have been related to thermodynamic functions and elastic moduli. Energy analyses have also become more useful with the clarification of its limitations.

8.1 FACTORS GOVERNING THE STERIC STRUCTURE

1. Molecular conformations are primarily determined by internal rotation potentials and nonbonded interactions. So far it has mainly been considered from the results for free molecules that $T(180°)$, $G(60°)$, and $\bar{G}(-60°)$ are the predominant conformations and that they are due to the internal rotation potentials. However nonbonded interactions are equally important in crystalline polymers. Conformations remarkably deviating from T, G, and $\bar{G}$ have been found for many polymers. Some examples are given below.

it-Poly(4-methyl-1-pentene), $—(—CH_2—CHR—)_n—$: 162°, −71°

it-Poly(3-methyl-1-butene), $—(—CH_2—CHR—)_n—$: 149°, −81°

Polyisobutylene, $—(—CM_2—CH_2—)_n—$: 54°, 164° averaged values

Polyoxymethylene, $—(—CH_2—O—)_n—$: 78°

Poly-1, 3-dioxolane, $—(—O——CH_2——O——CH_2——CH_2—)_n—$

$-94°$ $79°$ $74°$ $173°$ $-63°$

it-Polyacetaldehyde, $—(—O——CHM—)_n—$

$136°$ $-83°$

it-Poly(*tert*-butylethylene oxide), $—(—O——CH_2——CHR—)_n—$

$-97°$ $180°$ $73°$

Polytetrafluoroethylene, $—(—CF_2——CF_2—)_n—$

above 19°C 166°

below 19°C 164°

it-Poly(β-methyl-β-propiolactone), $—(—CHM—CH_2—CO—O—)_n—$

$162°$ $-52°$ $-42°$ $-175°$

For polymers the contribution of electrostatic interactions seems relatively unimportant compared with the two factors mentioned above. In addition to these factors, hydrogen bonding plays an important role.

2. The molecular conformation can be predicted to some extent by considering only intramolecular interactions, but it is modified in crystals by intermolecular interactions. Although the potential functions and parameters for energy calculations are not yet well established, a good example does exist in poly(ethylene oxybenzoate). According to calculations (Table 6.5), the stable internal rotation angles predicted by considering only intramolecular interactions are changed by at most 28° when intermolecular interactions are taken into account.

3. For chain packing in the crystalline regions (Section 6.4.3) the favorable mutual relation between nearest neighbors giving energy minima appears to be the most dominant factor. Packings satisfying three-dimensional crystal symmetries may actually appear.

4. Strictly speaking, a comparison of stability of crystal modifications should not be made merely in terms of static potential energy, but should consider the free energy. Theoretical calculations of the vibrational free energies predict that orthorhombic PE is more stable than monoclinic PE (Section 6.6).

8.2 RELATION OF STRUCTURE TO PROPERTIES AND SYNTHESES

1. The Relation of Structure to Electric Properties such as Piezoelectricity

and Pyroelectricity. According to X-ray analyses, poly(vinylidene fluoride) forms I and III, poly[3, 3-bis(chloromethyl) oxacyclobutane] β-form, poly(trimethylene sulfide), and poly(*m*-phenylene isophthalamide) have polar crystal structures. In the future their practical usage may be expected as piezoelectric and pyroelectric elements.

2. The Relation of Structure to High Modulus, High Tenacity, and High Thermal Stability. These distinguished properties of poly(*p*-phenylene terephthalamide) may be interpreted in terms of the all-trans extended-type molecular conformation and the very low molecular flexibility.

3. The Relation of Structure to the Field of Syntheses. The following polymerization mechanisms have been clarified from structural studies: the cis-opening mechanism of coordinated anionic polymerization (ethylene and propylene) and cationic polymerization (β-substituted vinyl ethers), the inversion-opening polymerization mechanism of ethylene oxide and 1-butene oxide, and the stereoselective polymerization mechanism of monomers containing asymmetric carbon atoms, such as *tert*-butylethylene oxide, isopropylethylene oxide, propylene sulfide, and β-substituted-β-propiolactones.

4. Contribution to the Field of Biopolymers. Biopolymers may be regarded generally as copolymers having side chains with complicated sequences. Therefore the accumulation of reliable data and development of methods of analyses for the structure of synthetic polymers will supply useful information for the study of biopolymers, since the synthetic polymers consist of relatively simple monomeric units compared to biopolymers. Cooperative efforts in these areas are already fruitful.

8.3 FUTURE ASPECTS OF STRUCTURAL STUDIES

The chemical structure, molecular conformation, and crystal structure of naturally occurring polymers have been established by Nature. In the case of synthetic polymers chemists can control the chemical structures, but the molecular conformations and crystal structures depend on the work of Nature. The conformations and crystal structures of many polymers have been determined experimentally along with their chemical structures. Because of these studies, prediction of both stable conformations and crystal structures has become possible empirically and theoretically. Basic studies can supply information regarding what types of chemical structures are needed to obtain polymers having certain conformations and crystal structures.

In the near future the structure–property relationship will probably be clarified, especially for those properties closely related to crystal structure, such as melting point, phase transition, flexibility of molecules, piezoelectricity, pyroelectricity, crystallite modulus, and crystallization behavior. Elucidation of the relation of molecular conformation and crystal structure to the fine texture, especially the structure in a wider sense including the structure of the amorphous regions, should provide a deeper understanding of the relationship between polymer structure and properties involving mechanical behavior.

Appendix A

Fourier Transform, δ-Function, and Convolution

A.1 FOURIER TRANSFORM

The equation expressing the scattering amplitude of X-rays in terms of electron distribution, such as Eq. 4.8, is called a Fourier transform in mathematics.

$$F(\mathbf{S}) = \int_{-\infty}^{\infty} \rho(\mathbf{r}) \exp(2\pi i \mathbf{S}\cdot\mathbf{r})\, dv \qquad \text{(A.1)}$$

A one-dimensional case is considered for simplicity. If the length of the axis is a and the fractional coordinate is x, r is denoted by

$$r = xa$$

In the reciprocal lattice, the unit of the axis is $1/a$, and S is given by

$$S = \frac{h}{a} \qquad \text{(A.2)}$$

Then Eq. A.1 has the form

$$F(h) = a\int_{-\infty}^{\infty} \rho(x) \exp(2\pi i h x)\, dx \qquad \text{(A.3)}$$

For extension to three-dimensions, Eqs. 4.31 and 4.32 may be used.

The existence of an inverse transformation is an important property of the Fourier transform. That is, if Eq. A.1 or A.3 holds, the following relation must hold.

$$\rho(\mathbf{r}) = \int_{-\infty}^{\infty} F(\mathbf{S}) \exp(-2\pi i \mathbf{S}\cdot\mathbf{r})\, dv_s \qquad \text{(A.4)}$$

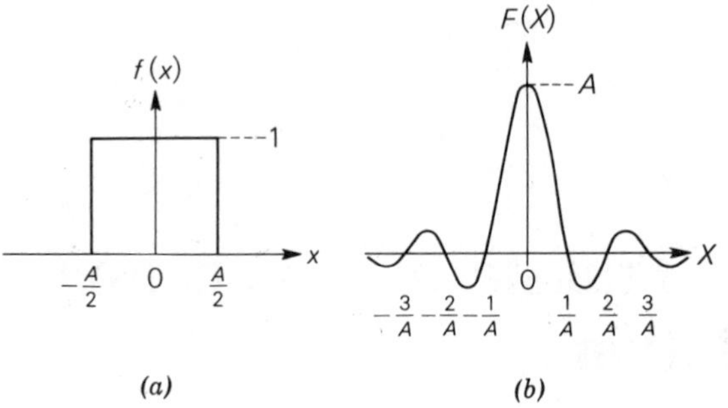

Fig. A.1 Simple example of a Fourier transform.

or

$$\rho(x) = \frac{1}{a} \int_{-\infty}^{\infty} F(h) \exp(-2\pi i h x) \, dh \tag{A.5}$$

where dv_s is the volume element in reciprocal space. Figure A.1 shows a simple example of a one-dimensional Fourier transform in which $f(x) = 1$ for $-\frac{1}{2}A < x < \frac{1}{2}A$ and $f(x) = 0$ for $x \leqslant -\frac{1}{2}A$ or $x \geqslant \frac{1}{2}A$.

Applying Eq. 4.40 to Eq. A.1 gives

$$F(X) = \int_{-(1/2)A}^{(1/2)A} \exp(2\pi i X x) \, dx = \frac{\sin(\pi A X)}{\pi X} \tag{A.6}$$

For Eq. A.6 the height at $X = 0$ is A, since $\lim_{x \to 0} \sin(\pi A X)/\pi X = A$. The width of the central peak is $2/A$, because the first $F(X) = 0$ appears at sin $(\pi A X_1) = 0$, $\pi A X_1 = \pi$, or $X_1 = 1/A$. The sharpness of $F(X)$ increases with increasing A. The above discussion can be readily extended to three dimensions. As the extreme case, the Fourier transform of an infinitely wide plane coinciding with the xy plane is an infinitely long straight line along the z axis.

A.2 δ-FUNCTION

The δ-function is defined by

$$\delta(x) = 0 \qquad x \neq 0$$

$$\int_{-\infty}^{\infty} \delta(x) \, dx = 1 \tag{A.7}$$

and has the following properties.

1. For any function $f(x)$

$$\int_{\alpha}^{\beta} f(x)\,\delta(x-x_0)\,\mathrm{d}x = \begin{cases} f(x_0) & \alpha < x_0 < \beta \\ 0 & x_0 \leqslant \alpha \quad \text{or} \quad \beta \leqslant x_0 \end{cases} \tag{A.8}$$

2. $\delta(cx) = (1/c)\delta(x)$ (A.9)
3. $\delta(x)$ is an even function.

The following two δ-functions are important in X-ray crystallography. If A is assumed as infinite in the function in Eq. A.6, then

$$\lim_{A\to\infty} \frac{\sin \pi A X}{\pi X} = \int_{-\infty}^{\infty} \exp(2\pi \mathrm{i} X x)\,\mathrm{d}x = \delta(X) \tag{A.10}$$

If M is taken as infinite in the Laue function, $\sin \pi M\xi/\sin \pi\xi$ (Eq. 4.18), the function has large values only in the case when ξ is an integer (denoted by h) and becomes zero otherwise.

$$\sum_{h=-\infty}^{\infty} \exp(2\pi \mathrm{i} h x) = \sum_{n=-\infty}^{\infty} \delta(x-n) \tag{A.11}$$

The Fourier transform of an infinite number of δ-functions aligned with spacing a is also an alignment of δ-functions with the spacing $1/a$.

$$\mathfrak{F} \sum_{n} \delta(x-na) = \frac{1}{a} \sum_{h} \delta\left(X - \frac{h}{a}\right) \tag{A.12}$$

where $\mathfrak{F}$ denotes the Fourier transform.

A.3 CONVOLUTION

The convolution or fold (*Faltung* in German) of two functions $f_1(x)$ and $f_2(x)$ is defined by

$$\widehat{f_1(x) f_2(x)} = C(x) = \int_{-\infty}^{\infty} f_1(y) f_2(x-y)\,\mathrm{d}y \tag{A.13*}$$

Also,

$$\widehat{f_1(x) f_2(x)} = \widehat{f_2(x) f_1(x)} \tag{A.14}$$

A simple example is the case in which $f_1(y)$ in Eq. A.13 is a δ-function (Fig. A.2*a*).

$$f_1(y) = A\,\delta(y-a) \tag{A.15}$$

*The symbol $f_1(x) * f_2(x)$ is also used.

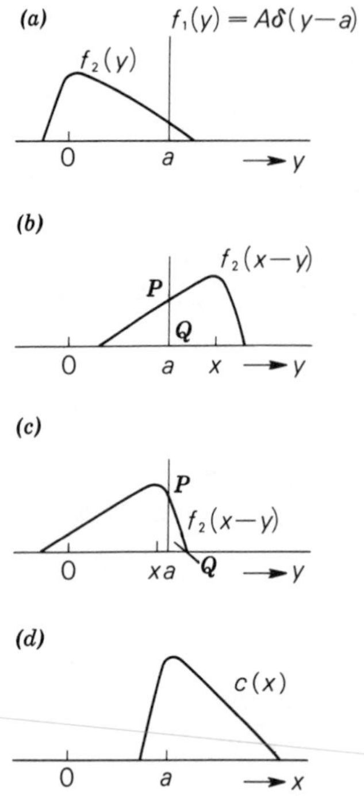

Fig. A.2 (*a*) $f_1(y)$ and $f_2(y)$. (*b* and *c*) $f_2(x-y)$ for fixed x. PQ is the intercept of $f_2(x-y)$ and δ-function. (*d*) $C(x)$.

Using Eq. A.8 gives

$$C(x) = \int_{-\infty}^{\infty} A\,\delta(y-a) f_2(x-y)\,dy$$
$$= A f_2(x-a) \qquad \text{(A.16)}$$

This function is shown in Fig. A.2*d*; $C(x)$ is obtained by shifting $f_2(x)$ by a and then multiplying by A.

These equations are explained by the following consideration. If $f_2(x-y)$ is plotted against y by fixing x to any particular value, the curve shown in Fig. A.2*b* or A.2*c* is obtained. The length of the intercept PQ is the value of $\int_{-\infty}^{\infty} \delta(y-a) f_2(x-y)\,dy$ in the case when x is taken as shown in this figure. If the length PQ is plotted for all x values after it is multiplied by A, $C(x)$ is obtained as shown in Fig. A.2*d*.

A more complicated example is considered here using another method where $f_1(y)$ and $f_2(y)$ are given in Fig. A.3*a*. The integral of Eq. A.13 can be

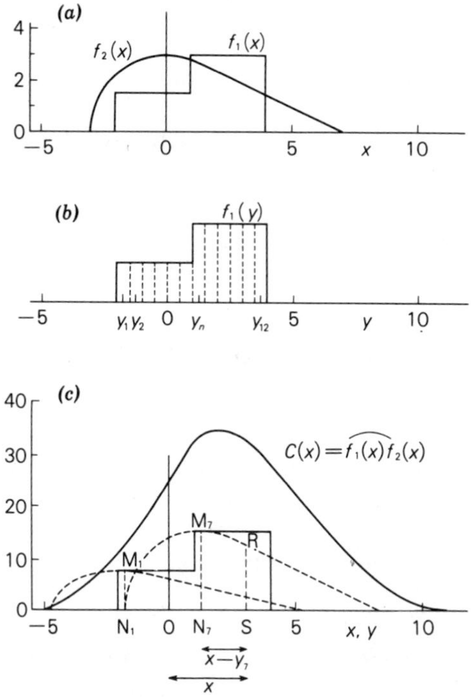

Fig. A.3 (*a*) $f_1(x)$ and $f_2(x)$. $f_2(x)$ consists of a part of a circle with radius 3 and a straight line tangential to it. (*b*) The division of $f_1(y)$ into 12 strips of equal width $\delta y = 0.5$, $\delta A_n = f_1(y_n)\,\delta y$ is 0.75 for $n = 1$–6, and 1.50 for $n = 7$–12. (*c*) $f_1(y_n)f_2\ (x - y_n)\,\delta y$ is shown for $n = 1$ and 7 by the dashed line. M_nN_n is equal to $f_2(0)\,\delta A_n$; $M_1N_1 = 2.25$ and $M_7N_7 = 4.50$. Summation of RS for all n gives $C(x)$. The scale of the ordinate is for $C(x)$.

expressed as a summation

$$C(x) = \int_{-\infty}^{\infty} f_1(y) f_2(x - y)\,\mathrm{d}y$$

$$= \lim_{\delta y \to 0} \sum_{n=1}^{N} f_1(y_n) f_2(x - y_n)\,\delta y \tag{A.17}$$

Here the summation is for the n values of y_n at the centers of the strips of width δy which cover the nonzero range of $f_1(y)$. The following procedure is used: (1) Divide $f_1(y)$ into N strips of equal width δy, each center being y_n. (2) Estimate the area of the strips $\delta A_n = f_1(y_n)\delta y$. (3) Plot $f_2(y)\delta A_n$ with the origin at y_n for all n (Fig. A.3*c*), the height M_nN_n at y_n is equal to $f_2(0)\,\delta A_n$.

The height RS of such curves at any x equals $f_1(y_n)\,f_2(x-y_n)\,\delta y$. (4) Add the resultant curves to obtain $C(x)$.*

Considered next is a Fourier transform of a convolution.

$$\int_{-\infty}^{\infty}\left[\int_{-\infty}^{\infty} f_1(y)f_2(x-y)\,\mathrm{d}y\right]\exp(2\pi\mathrm{i}Xx)\,\mathrm{d}x \tag{A.18}$$

The substitution $x-y=z$ gives

$$\int_{-\infty}^{\infty}\int_{-\infty}^{\infty} f_1(y)f_2(z)\exp[2\pi\mathrm{i}X(y+z)]\,\mathrm{d}y\,\mathrm{d}z$$
$$=\int_{-\infty}^{\infty} f_1(y)\exp(2\pi\mathrm{i}Xy)\,\mathrm{d}y\int_{-\infty}^{\infty} f_2(z)\exp(2\pi\mathrm{i}Xz)\,\mathrm{d}z \tag{A.19}$$

This is simply the product of the Fourier transforms of the two functions. Hence the following theorem is obtained.

The Fourier transform of the convolution of two functions is the product of the Fourier transforms of each function.

$$\mathfrak{F}[\widehat{f_1(x)f_2(x)}]=\mathfrak{F}[f_1(x)]\,\mathfrak{F}[f_2(x)]=F_1(X)F_2(X) \tag{A.20}$$

In other words, the Fourier transform of the product of two functions is the convolution of the transforms of each function.

$$\mathfrak{F}^{-1}[F_1(X)F_2(X)]=\widehat{\mathfrak{F}^{-1}[F_1(X)]\,\mathfrak{F}^{-1}[F_2(X)]}=\widehat{f_1(x)f_2(x)} \tag{A.21}$$

where $\mathfrak{F}^{-1}$ denotes the inverse Fourier transform. This theorem is further written by another expression: the procedure of convolution in direct space corresponds to the procedure of multiplication in reciprocal space (of the Fourier transforms of each function), and vice versa.

In addition to the convolution $C(x)$, P-convolution is defined by

$$P(x)=\widehat{f_1(x)f_2}(-x)=\int_{-\infty}^{\infty} f_1(y)f_2(y-x)\,\mathrm{d}y \tag{A.22}$$

When $f_2(x)$ is an even function, that is, $f_2(x)$ is centrosymmetric with respect to the origin, $P(x)$ is equal to $C(x)$. P-convolution of a function with itself is called self-convolution or convolution square and is denoted by the symbol $\underset{\sim}{2}$. For a three-dimensional distribution of electron density,

$$P(\mathbf{u})=\overset{2}{\tilde{\rho}}(\mathbf{u})=\int\rho(\mathbf{r})\rho(\mathbf{r}-\mathbf{u})\,\mathrm{d}v_{\mathbf{r}} \tag{A.23}$$

* The curve $f_2(x-y)$ is reversed from the original $f_2(y)$ (Fig. A.2*a*) in Figs. A.2*b* and A.2*c*, but it is not reversed in Fig. A.3*c*. This difference may be explained by the fact that $f_2(x-y)$ is plotted against y, fixing x in Fig. A.2, while it is plotted against x, fixing y_n in Fig. A.3.

where $dv_{\mathbf{r}}$ is the volume element in direct space. Since the value of $P(\mathbf{u})$ is unaffected, rewriting the two functions as $\rho(\mathbf{r})\rho(\mathbf{r}+\mathbf{u})$ gives

$$P(\mathbf{u}) = \int \rho(\mathbf{r})\rho(\mathbf{r}+\mathbf{u})\, dv_{\mathbf{r}} \tag{A.24}$$

This function is the Patterson function discussed in Section 4.7.4.

Appendix B

Derivation of Eyring's Transformation Matrix[1]

Sets of right-hand Cartesian coordinate systems as shown in Fig. B.1 are now considered. The origin of the coordinate system $\mathbf{x}_j$ coincides with the position of the jth atom with its x_j axis on the bond $r_{j,j+1}$. The y_j axis lies on the plane determined by two bonds $r_{j-1,j}$ and $r_{j,j+1}$ in such a way that the angle between the $r_{j-1,j}$ and the positive direction of y_j is acute. The coordinate system $\mathbf{x}_j$ is rotated about the z_j axis by $\pi - \phi_j$ (ϕ_j is the bond angle) so as to make the x_j axis parallel to $r_{j-1,j}$, and the new coordinate system is called $\mathbf{x}_j'$. The y_j axis moves to the y_j' axis, which is perpendicular to $r_{j-1,j}$. If the coordinates of the ith atom are expressed by $\mathbf{x}_j(i)$ and $\mathbf{x}_j'(i)$ using these two coordinate systems, then

$$\begin{bmatrix} x_j'(i) \\ y_j'(i) \\ z_j'(i) \end{bmatrix} = \begin{bmatrix} -\cos\phi_j & -\sin\phi_j & 0 \\ \sin\phi_j & -\cos\phi_j & 0 \\ 0 & 0 & 1 \end{bmatrix} \begin{bmatrix} x_j(i) \\ y_j(i) \\ z_j(i) \end{bmatrix} \tag{B.1}$$

Next the $\mathbf{x}_j'$ system is rotated about the x_j' axis by the internal rotation angle $\tau_{j-1,j}$ so as to make the y_j' axis parallel to the y_{j-1} axis. The relation of $x_j'(i)$ to $x_{j-1}(i)$,

$$x_{j-1}(i) = x_j'(i) + r_{j-1,j} \tag{B.2}$$

gives

$$\begin{bmatrix} x_{j-1}(i) \\ y_{j-1}(i) \\ z_{j-1}(i) \end{bmatrix} = \begin{bmatrix} 1 & 0 & 0 \\ 0 & \cos\tau_{j-1,j} & -\sin\tau_{j-1,j} \\ 0 & \sin\tau_{j-1,j} & \cos\tau_{j-1,j} \end{bmatrix} \begin{bmatrix} x_j'(i) \\ y_j'(i) \\ z_j'(i) \end{bmatrix} + \begin{bmatrix} r_{j-1,j} \\ 0 \\ 0 \end{bmatrix}$$

$$= \begin{bmatrix} 1 & 0 & 0 \\ 0 & \cos\tau_{j-1,j} & -\sin\tau_{j-1,j} \\ 0 & \sin\tau_{j-1,j} & \cos\tau_{j-1,j} \end{bmatrix} \begin{bmatrix} -\cos\phi_j & -\sin\phi_j & 0 \\ \sin\phi_j & -\cos\phi_j & 0 \\ 0 & 0 & 1 \end{bmatrix} \begin{bmatrix} x_j(i) \\ y_j(i) \\ z_j(i) \end{bmatrix} + \begin{bmatrix} r_{j-1,j} \\ 0 \\ 0 \end{bmatrix}$$

$$= \begin{bmatrix} -\cos\phi_j & -\sin\phi_j & 0 \\ \cos\tau_{j-1,j}\sin\phi_j & -\cos\tau_{j-1,j}\cos\phi_j & -\sin\tau_{j-1,j} \\ \sin\tau_{j-1,j}\sin\phi_j & -\sin\tau_{j-1,j}\cos\phi_j & \cos\tau_{j-1,j} \end{bmatrix} \begin{bmatrix} x_j(i) \\ y_j(i) \\ z_j(i) \end{bmatrix} + \begin{bmatrix} r_{j-1,j} \\ 0 \\ 0 \end{bmatrix} \tag{B.3}$$

Using matrix notation this equation is written as

$$\mathbf{x}_{j-1}(i) = \mathbf{A}_{j-1,j}\mathbf{x}_j(i) + \mathbf{B}_{j-1,j} = \mathbf{A}^{\tau}{}_{j-1,j}\mathbf{A}^{\phi}{}_j\mathbf{x}_j(i) + \mathbf{B}_{j-1,j} \tag{B.4}$$

where

$$\mathbf{A}^{\tau}{}_{j-1,j} = \begin{bmatrix} 1 & 0 & 0 \\ 0 & \cos\tau_{j-1,j} & -\sin\tau_{j-1,j} \\ 0 & \sin\tau_{j-1,j} & \cos\tau_{j-1,j} \end{bmatrix} \tag{B.5}$$

$$\mathbf{A}^{\phi}_j = \begin{bmatrix} -\cos\phi_j & -\sin\phi_j & 0 \\ \sin\phi_j & -\cos\phi_j & 0 \\ 0 & 0 & 1 \end{bmatrix} \tag{B.6}$$

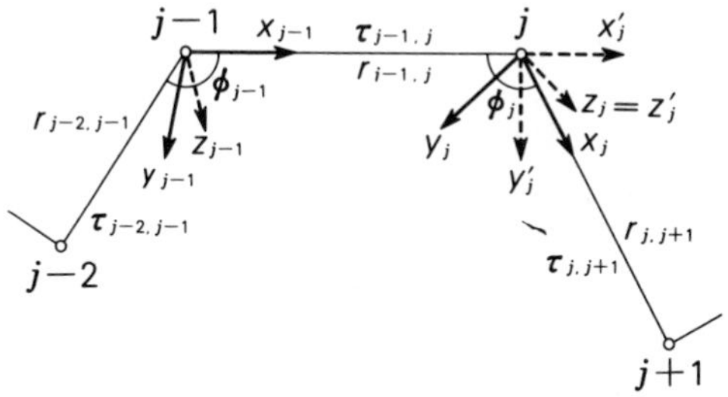

Fig. B.1 Numbering of atoms, molecular parameters, and Cartesian coordinate systems fixed to atoms.

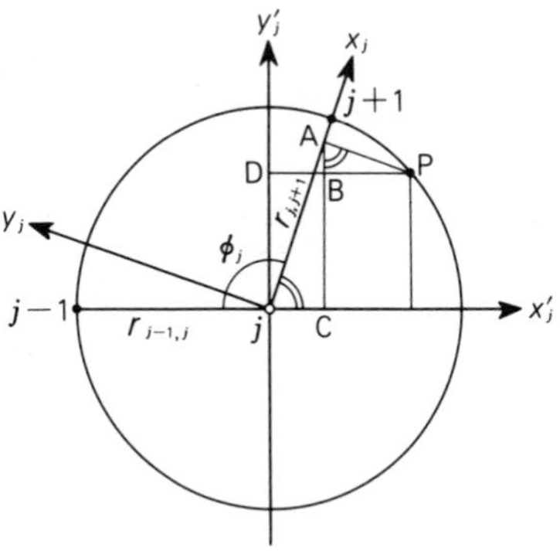

Fig. B.2 Transformation of coordinates.

Equation B.1 is derived as follows. The direction of the projection of the molecule along the z_j axis (equals the z_j' axis) is upward from the page (Fig. B.2). If point P on this projection is expressed by (x_j, y_j) or (x_j', y_j') using the two coordinate systems, then

$$\begin{aligned}
x_j' &= DB + BP = x_j \cos(\pi - \phi_j) - y_j \sin(\pi - \phi_j) \\
&= -x_j \cos\phi_j - y_j \sin\phi_j \\
y_j' &= AC - AB = x_j \sin(\pi - \phi_j) + y_j \cos(\pi - \phi_j) \\
&= x_j \sin\phi_j - y_j \cos\phi_j \\
z_j' &= z_j
\end{aligned} \tag{B.7}$$

These equations are written as Eq. B.1 using matrix notation.

REFERENCE

1. H. Eyring, *Phys. Rev.*, **39,** 746 (1932).

Appendix C

Derivation of the Fourier Transform of Helical Molecules

The method of derivation of Naya et al.,[2] which is different from that of Cochran et al.,[1] is described here. The molecular structure factor of a helical polymer is given by

$$F(\mathbf{X}) = \sum_j f_j \xi_j(\mathbf{X}) \tag{C.1}$$

where f_j is the atomic scattering factor of the jth atom and $\mathbf{X}$ is the Cartesian coordinates of a point in reciprocal space.

$$\mathbf{X} = \begin{bmatrix} R\cos\psi \\ R\sin\psi \\ \dfrac{l}{c} \end{bmatrix} \tag{C.2}$$

where R, ψ, and l/c are the cylindrical coordinates. $\xi(\mathbf{X})$ is the geometrical structure factor and is given by

$$\xi(\mathbf{X}) = \sum_{k=-\infty}^{\infty} \exp(2\pi i \tilde{\mathbf{X}} \mathbf{x}^{(k)}) = \sum_k \exp(2\pi i \tilde{\mathbf{X}} \mathbf{S}_k \mathbf{x}) \tag{C.3}$$

where

$$\mathbf{x} = \begin{bmatrix} r\cos\varphi \\ r\sin\varphi \\ z \end{bmatrix} \tag{C.4}$$

represents a general position of an atom in the direct space in which r, φ, and z are its cylindrical coordinates. $\mathbf{S}_k$ is a screw symmetry operator that shifts the general position $\mathbf{x}$ to the kth equivalent position $\mathbf{x}^{(k)}$ moved along the helix and is expressed by

$$\mathbf{x}^{(k)} = \mathbf{S}_k \mathbf{x} = \mathbf{R}_k \mathbf{x} + \mathbf{t}_k \qquad (\mathbf{x}^{(0)} \equiv \mathbf{x}) \tag{C.5}$$

For a simple right-hand helix with pitch (the translation along the helix axis per turn) P, $\mathbf{R}_k$ and $\mathbf{t}_k$ are given by

$$\mathbf{R}_k = \begin{bmatrix} \cos\frac{2\pi kp}{P} & -\sin\frac{2\pi kp}{P} & 0 \\ \sin\frac{2\pi kp}{P} & \cos\frac{2\pi kp}{P} & 0 \\ 0 & 0 & 1 \end{bmatrix} \quad \text{and} \quad \mathbf{t}_k = \begin{bmatrix} 0 \\ 0 \\ kp \end{bmatrix} \tag{C.6}$$

where $p = c/u$ is the spacing in the z direction between two adjacent equivalent atoms along the helix, c is the fiber period, and u is the number of equivalent units involved in it. Substitution of Eqs. C.2 and C.4–C.6 into Eq. C.3 gives

$$\begin{aligned}\xi(\mathbf{X}) &= \sum_{k=-\infty}^{\infty} \exp(2\pi i\tilde{\mathbf{X}}\mathbf{R}_k\mathbf{x}) \exp(2\pi i\tilde{\mathbf{X}}\mathbf{t}_k) \\ &= \sum_k \exp\left\{2\pi i\left[Rr\cos\left(\psi - \frac{2\pi kp}{P} - \varphi\right)\right]\right\} \exp\left[2\pi i\left(\frac{lz}{c} + \frac{lkp}{c}\right)\right]\end{aligned} \tag{C.7}$$

For summation with respect to k in Eq. C.7, the following expansion formula for the Bessel function is used.

$$\exp(ix\cos\phi) = \sum_{n=-\infty}^{\infty} J_n(x)\exp\left[in\left(\phi + \frac{\pi}{2}\right)\right] \tag{C.8}$$

By setting $x = 2\pi Rr$ and $\phi = \psi - \varphi - 2\pi kp/P$, Eq. C.7 becomes

$$\begin{aligned}\xi(\mathbf{X}) &= \sum_{n=-\infty}^{\infty}\sum_{k=-\infty}^{\infty} J_n(2\pi Rr)\exp\left[in\left(\psi - \varphi - \frac{2\pi kp}{P} + \frac{\pi}{2}\right)\right] \\ &\quad \times \exp\left[2\pi i\left(\frac{lz}{c} + \frac{lkp}{c}\right)\right] \\ &= \sum_n J_n(2\pi Rr)\exp\left[in\left(\psi - \varphi + \frac{\pi}{2}\right)\right]\exp\left(\frac{2\pi ilz}{c}\right) \\ &\quad \times \sum_k \exp\left[2\pi ik\left(-\frac{np}{P} + \frac{lp}{c}\right)\right]\end{aligned} \tag{C.9}$$

The summation with respect to k in Eq. C. 9 is a Laue function (Eq. A. 11). If $-np/P + lp/c = \zeta$, this function becomes a δ-function that is nonzero only when $\zeta = m$, where m is any integer.

$$\sum_{k=-\infty}^{\infty}\exp\left[2\pi ik\left(-\frac{np}{P} + \frac{lp}{c}\right)\right] = \sum_k \exp(2\pi ik\zeta) = \sum_{m=-\infty}^{\infty}\delta(\zeta - m) \tag{C.10}$$

Using Eq. A.9 this is rewritten as

$$\sum_{m=-\infty}^{\infty} \delta\left(-\frac{np}{P}+\frac{lp}{c}-m\right)=\sum_{m}\frac{1}{p}\delta\left[\frac{1}{c}\left(-\frac{nc}{P}+l-\frac{mc}{p}\right)\right]$$
$$=\sum_{m}\frac{c}{p}\delta\left(-\frac{nc}{P}+l-\frac{mc}{p}\right) \tag{C.11}$$

Hence Eq. C.10 has nonzero values only when

$$l=\frac{nc}{P}+\frac{mc}{p}=nt+mu \tag{C.12}$$

holds. Here $t=c/P$ is the number of turns per fiber period. Equation C.12 is the selection rule for the lth layer line with respect to n and is equal to Eq. 4.67. Consequently Eq. C.9 becomes

$$\xi(\mathbf{X})=u\sum_{n=-\infty}^{\infty}{}^{*}\, J_n(2\pi Rr)\exp\left[\mathrm{i}n\left(\psi-\varphi+\frac{\pi}{2}\right)\right]\exp\left(\frac{2\pi\mathrm{i}lz}{c}\right) \tag{C.13}$$

where $\sum_n^*$ means the summation with respect to n, satisfying the selection rule Eq. C.12. From Eqs. C.13 and C.1,

$$F(R,\psi,l/c)=u\sum_j\sum_n{}^{*} f_j J_n(2\pi Rr_j)\exp\left[\mathrm{i}n\left(\psi-\varphi_j+\frac{\pi}{2}\right)+\frac{2\pi\mathrm{i}lz_j}{c}\right] \tag{C.14}$$

which is the same as Eq. 4.65.

Equation C.14 is the equation for a right-hand helix. To obtain the equation for a left-hand helix, φ and $2\pi kp/P$ in Eqs. C.4 and C.6 are replaced by $-\varphi$ and $-2\pi kp/P$. This corresponds to the reflection with respect to the xz plane in Fig. 4.43a and corresponds to taking minus pitch. Then Eq. C.9 becomes

$$\sum_n J_n(2\pi Rr)\exp\left[\mathrm{i}n\left(\psi+\varphi+\frac{\pi}{2}\right)\right]\exp\left(\frac{2\pi\mathrm{i}lz}{c}\right)\sum_k\exp\left[2\pi\mathrm{i}k\left(\frac{np}{P}+\frac{lp}{c}\right)\right] \tag{C.15}$$

Hence Eq. C.12 should be replaced by $l=-nt+mu$. However, since the usual selection rule (Eq. C.12) is more convenient, n is replaced by $-n$ in Eq. C.14. Using the relations $J_{-n}(z)=(-1)^n J_n(z)$ and $(-1)^n=\mathrm{i}^{2n}=\mathrm{e}^{\mathrm{i}n\pi}$ gives the following equation for the left-hand helix.*

$$F\left(R,\psi,\frac{l}{c}\right)=u\sum_j\sum_n{}^{*} f_j J_n(2\pi Rr_j)\exp\left[\mathrm{i}n\left(-\psi-\varphi_j+\frac{\pi}{2}\right)+\frac{2\pi\mathrm{i}lz_j}{c}\right] \tag{C.16}$$

Equations C.14 and C.16 may be considered to be the equations for upward

* $\mathrm{e}^{\mathrm{i}n\pi/2}=(\cos\pi/2+\mathrm{i}\sin\pi/2)^n=\mathrm{i}^n$.

helices. If the helix is rotated around the x axis by 180° in Fig. 4.43a, a downward helix is obtained; the position (r_j, φ_j, z_j) is transformed into $(r_j, -\varphi_j, -z_j)$. Therefore the equations for downward helices can be derived by replacing φ_j and z_j by $-\varphi_j$ and $-z_j$ in Eqs. C.14 and C.16.

REFERENCES

1. W. Cochran, F. H. C. Crick, and V. Vand, *Acta Crystallogr.*, **5,** 581 (1952).
2. S. Naya and I. Nitta, *Kwansei Gakuin Univ. Annu. Stud.*, **15,** 1 (1966); S. Tanaka and S. Naya, *J. Phys. Soc. Jpn.*, **26,** 982 (1969).

Appendix D

Derivation of Molecular Structure Factor of a Nonhelical Molecule in Terms of Cylindrical Coordinates

If cylindrical coordinates in direct and reciprocal spaces are defined as shown in Fig. D.1, then

$$x_j = r_j \cos\varphi_j, \qquad y_j = r_j \sin\varphi_j, \qquad z_j = z_j$$

$$X = R\cos\psi, \qquad Y = R\sin\psi, \qquad Z = \frac{l}{c} \tag{D.1*}$$

With the respective coordinates denoted by $\mathbf{r}_j$ and $\mathbf{S}$, the diffraction intensity is given by

$$\begin{aligned}|F(\mathbf{S})|^2 &= \sum_j \sum_{j'} f_j f_{j'} \exp[2\pi i \mathbf{S}\cdot(\mathbf{r}_j - \mathbf{r}_{j'})] \\ &= \sum_j \sum_{j'} f_j f_{j'} \exp(2\pi i \mathbf{S}\cdot\mathbf{r}_j)\exp(-2\pi i \mathbf{S}\cdot\mathbf{r}_j)\end{aligned} \tag{D.2}$$

where

$$\begin{aligned}\exp(2\pi i\mathbf{S}\cdot\mathbf{r}_j) &= \exp[2\pi i(Xx_j + Yy_j + Zz_j)] \\ &= \exp\left[2\pi i\left(R\cos\psi\, r_j\cos\varphi_j + R\sin\psi\, r_j\sin\varphi_j + \frac{lz_j}{c}\right)\right] \\ &= \exp\left\{2\pi i\left[Rr_j\cos(\varphi_j - \psi) + \frac{lz_j}{c}\right]\right\}\end{aligned} \tag{D.3}$$

Therefore

$$\begin{aligned}|F(\mathbf{S})|^2 = \sum_j \sum_{j'} f_j f_{j'} &\exp\left\{2\pi i\left[Rr_{j'}\cos(\varphi_{j'} - \psi) + \frac{lz_{j'}}{c}\right]\right\} \\ \times &\exp\left\{-2\pi i\left[Rr_{j'}\cos(\varphi_{j'} - \psi) + \frac{lz_{j'}}{c}\right]\right\}\end{aligned}$$

* Here x_j, y_j, and z_j are not fractional coordinates.

$$= \sum_j \sum_{j'} f_j f_{j'} \exp\left[2\pi i l \frac{z_j - z_{j'}}{c}\right] \exp\{2\pi i R[(r_j \cos \varphi_j - r_{j'} \cos \varphi_{j'}) \cos \psi$$

$$+ (r_j \sin \varphi_j - r_{j'} \sin \varphi_{j'}) \sin \psi]\}$$

$$= \sum_j \sum_{j'} f_j f_{j'} \exp\left[2\pi i l \frac{z_j - z_{j'}}{c}\right] \exp[2\pi i R(r_{jj'} \cos \varphi_0 \cos \psi - r_{jj'} \sin \varphi_0$$

$$\sin \psi)]$$

$$= \sum_j \sum_{j'} f_j f_{j'} \exp\left[2\pi i l \frac{z_j - z_{j'}}{c}\right] \exp\{2\pi i[R r_{jj'} \cos(\varphi_0 + \psi)]\} \qquad \text{(D.4)*}$$

For uniaxially oriented samples, Eq. D.4 should be averaged with respect to ψ. Then

$$\left\langle \left| F\left(R, \psi, \frac{l}{c}\right) \right|^2 \right\rangle_\psi = \frac{1}{2\pi} \sum_j \sum_{j'} f_j f_{j'} \exp\left[2\pi i l \frac{z_j - z_{j'}}{c}\right] \int_0^{2\pi} \exp\{2\pi i[R r_{jj'}$$

$$\times \cos(\varphi_0 + \psi)]\} \, d\psi \qquad \text{(D.5)}$$

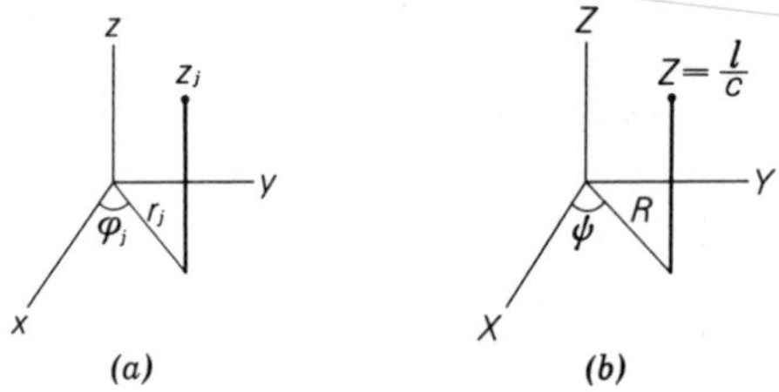

Fig. D.1 Cylindrical coordinates in (*a*) direct and (*b*) reciprocal spaces.

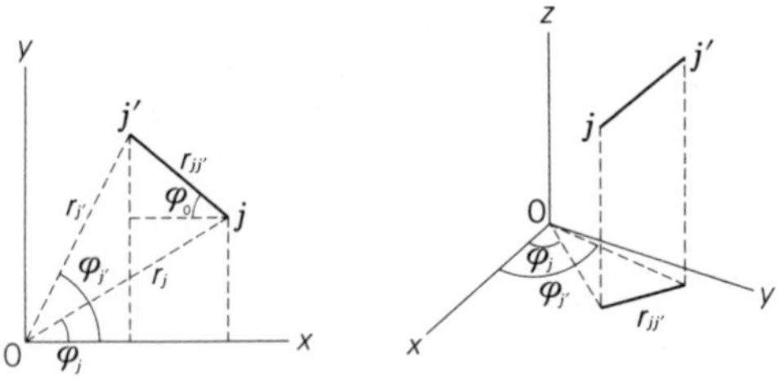

Fig. D.2 Definition of $r_{jj'}$, and φ_0.

* $r_{jj'}$ and φ_0 are defined as shown in Fig. D.2. $r_{jj'} = [(x_j - x_{j'})^2 + (y_j - y_{j'})^2]^{1/2}$.

Using the formula for the Bessel function

$$\int_0^{2\pi} \exp(\mathrm{i}\mu\cos\phi)\exp(\mathrm{i}n\phi)\,\mathrm{d}\phi = 2\pi\mathrm{i}^n J_n(\mu) = 2\pi J_n(\mu)\exp\left(\frac{\mathrm{i}n\pi}{2}\right) \quad \text{(D.6)}$$

gives

$$\int_0^{2\pi} \exp\{2\pi\mathrm{i}[Rr_{jj'}\cos(\varphi_0+\psi)]\}\,\mathrm{d}\psi = 2\pi J_0(2\pi Rr_{jj'}) \quad \text{(D.7)}$$

Hence

$$\left\langle \left| F\left(R,\psi,\frac{l}{c}\right)\right|^2 \right\rangle_\psi = \sum_j \sum_{j'} f_j f_{j'} J_0(2\pi Rr_{jj'})\exp\left[2\pi\mathrm{i}l\frac{z_j - z_{j'}}{c}\right] \quad \text{(D.8)}$$

Appendix E

A Simple Example of Solving the Secular Equation (Water Molecule)

From Eqs. 5.169 and 5.170, $\mathbf{G}_s\mathbf{F}_s = \mathbf{H}_s$ is given by

$$\mathbf{H}_s = \left[\begin{array}{cc|c} 8.6523809 & 0.1745139 & 0 \\ -0.2141991 & 1.6014523 & 0 \\ \hline 0 & 0 & 9.1587716 \end{array}\right] \quad \text{(E.1)}$$

For the block of order 2 (the A_1 species) in Eq. E.1, $|\mathbf{H}_s - \lambda\mathbf{E}| = 0$ is expanded to give

$$\lambda^2 - (h_{11} + h_{22})\lambda + h_{11}h_{22} - h_{12}h_{21} = \lambda^2 - 10.253833\lambda + 13.893755 = 0 \quad \text{(E.2)}$$

from which $\lambda_1 = 8.6470750$ and $\lambda_2 = 1.6067580$ are obtained. From the block of order 1 (the B_1 species), $\lambda_3 = 9.1587716$.

For calculating the ratio of the **L** matrix elements, that is, the $\mathbf{L}^0$ matrix elements for λ_1, the equation

$$\begin{bmatrix} h_{11} & h_{12} \\ h_{21} & h_{22} \end{bmatrix} \begin{bmatrix} l_{11}{}^0 \\ l_{21}{}^0 \end{bmatrix} = \lambda_1 \begin{bmatrix} l_{11}{}^0 \\ l_{21}{}^0 \end{bmatrix} \quad \text{(E.3)}$$

is employed. Then

$$l_{11}{}^0 = h_{12} \quad \text{and} \quad l_{21}{}^0 = \lambda_1 - h_{11}$$

or

$$l_{11}{}^0 = \lambda_1 - h_{22} \quad \text{and} \quad l_{21}{}^0 = h_{21} \quad \text{(E.4)}$$

In the same way, for λ_2

$$l_{12}{}^0 = h_{12} \quad \text{and} \quad l_{22}{}^0 = \lambda_2 - h_{11}$$

or

$$l_{12}{}^0 = \lambda_2 - h_{22} \quad \text{and} \quad l_{22}{}^0 = h_{21} \tag{E.5}$$

From Eqs. E.4 and E.5,

$$\mathbf{L}^0 = \begin{bmatrix} l_{11}{}^0 & l_{12}{}^0 \\ l_{21}{}^0 & l_{22}{}^0 \end{bmatrix} = \begin{bmatrix} 7.0456227 & 0.1745139 \\ -0.2141991 & -7.0456229 \end{bmatrix} \tag{E.6}$$

The $\mathbf{L}^0$ matrix is normalized according to the equation

$$\begin{bmatrix} n_1 & 0 \\ 0 & n_2 \end{bmatrix} \begin{bmatrix} l_{11}{}^0 & l_{21}{}^0 \\ l_{12}{}^0 & l_{22}{}^0 \end{bmatrix} \begin{bmatrix} f_{11} & f_{12} \\ f_{12} & f_{22} \end{bmatrix} \begin{bmatrix} l_{11}{}^0 & l_{12}{}^0 \\ l_{21}{}^0 & l_{22}{}^0 \end{bmatrix} \begin{bmatrix} n_1 & 0 \\ 0 & n_2 \end{bmatrix} = \begin{bmatrix} \lambda_1 & 0 \\ 0 & \lambda_2 \end{bmatrix} \tag{E.7}$$

Using Eq. E.7 gives

$$\mathbf{L} = \begin{bmatrix} 1.0185589 & 0.0378565 \\ -0.0309659 & -1.5283766 \end{bmatrix} \tag{E.8}$$

These values can be confirmed using $\widetilde{\mathbf{L}}\mathbf{L} = \mathbf{G}$. The one-dimensional $\mathbf{L}$ matrix element for the B_1 species is the square root of $\mathbf{G}$.

Appendix F

Dynamical Equation for a Three-Dimensional Crystal

The derivation of the dynamical equation for a three-dimensional crystal is explained here in terms of the **GF** matrix method.[1] The Cartesian displacement coordinates and the internal displacement coordinates for the **n**th unit cell are denoted by $\mathbf{X}_\mathbf{n}$ and $\mathbf{R}_\mathbf{n}$, respectively; **n** is a vector consisting of n_1, n_2, and n_3. Both $\mathbf{X}_\mathbf{n}$ and $\mathbf{R}_\mathbf{n}$ are $3mN$ dimensional vectors; the definitions of m and N are the same as in Section 5.6.7. The total kinetic energy T and potential energy V of the crystal under the harmonic approximation are given by

$$2T = \sum_\mathbf{n} \tilde{\dot{\mathbf{X}}}_\mathbf{n} \mathbf{M} \dot{\mathbf{X}}_\mathbf{n} \tag{F.1}$$

$$2V = \sum_\mathbf{n} \sum_{\mathbf{n}'} \tilde{\mathbf{R}}_\mathbf{n} \mathbf{F}_{\mathbf{nn}'} \mathbf{R}_{\mathbf{n}'} \tag{F.2}$$

$\mathbf{F}_{\mathbf{nn}'}$ corresponds to the interactions within one unit cell if $\mathbf{n} = \mathbf{n}'$, and to the interactions between the atoms belonging to the **n** and **n**′th unit cells if $\mathbf{n} \neq \mathbf{n}'$. Since $\mathbf{F}_{\mathbf{nn}'}$ depends only on the three parameters $|n_1 - n_1'| = t_1$, $|n_2 - n_2'| = t_2$, and $|n_3 - n_3'| = t_3$ owing to the translational symmetry of the crystal, then $\mathbf{F}_{\mathbf{nn}'} = \mathbf{F}_\mathbf{t}$ and $\mathbf{F}_{\mathbf{n}'\mathbf{n}} = \tilde{\mathbf{F}}_\mathbf{t}$; **t** is a vector consisting of t_1, t_2, and t_3.

Born's cyclic boundary condition is assumed to be satisfied in the present system in the same way as in Section 5.6.7; N_1, N_2, and N_3 are the numbers of unit cells along three axes and $\mathfrak{N}$ is the total number of unit cells in the system. Therefore a set of $\mathbf{X}_\mathbf{n}$, $\{\mathbf{X}_0, \mathbf{X}_1, \cdots, \mathbf{X}_{\mathfrak{N}-1}\}$, also satisfies Born's condition; the subscript indicates the numbering of unit cells. The vibrational form with the phase difference corresponding to the wave vector **k** between neighboring unit cells may be expressed by

$$\mathbf{X}(\mathbf{k}) = \mathfrak{N}^{-1/2} \sum_\mathbf{n} \mathbf{X}_\mathbf{n} \exp(-2\pi i \mathbf{k} \cdot \mathbf{r}_\mathbf{n}) \tag{F.3}$$

or in a similar way by

$$\mathbf{S}(\mathbf{k}) = \mathfrak{N}^{-1/2} \sum_\mathbf{n} \mathbf{R}_\mathbf{n} \exp(-2\pi i \mathbf{k} \cdot \mathbf{r}_\mathbf{n}) \tag{F.4}$$

where $\mathbf{r}_n$ indicates the position of the $\mathbf{n}$th unit cell. Using $\mathbf{B}$ matrix of Eq. 5.191, $\mathbf{R}_n$ is expressed by

$$\mathbf{R}_n = \mathbf{B}_0\mathbf{X}_n + \sum_{l}(\mathbf{B}_{-l}\mathbf{X}_{n-l} + \mathbf{B}_l\mathbf{X}_{n+l}) = \sum_{l=-m}^{m} \mathbf{B}_l\mathbf{X}_{n+l} \tag{F.5}$$

Substituting Eq. F.5 into Eq. F.4 gives

$$\begin{aligned}\mathbf{S}(\mathbf{k}) &= \mathfrak{N}^{-1/2}\sum_{n}\sum_{l}\mathbf{B}_l\mathbf{X}_{n+l}\exp(-2\pi i\mathbf{k}\cdot\mathbf{r}_n)\\ &= \sum_{l}\mathbf{B}_l\Big[\mathfrak{N}^{-1/2}\sum_{n}\mathbf{X}_{n+l}\exp(-2\pi i\mathbf{k}\cdot\mathbf{r}_n)\Big]\\ &= \sum_{l}\mathbf{B}_l\exp(2\pi i\mathbf{k}\cdot\mathbf{r}_l)\Big[\mathfrak{N}^{-1/2}\sum_{n}\mathbf{X}_{n+l}\exp(-2\pi i\mathbf{k}\cdot\mathrm{r}_{n+l})\Big]\\ &= \sum_{l}\mathbf{B}_l\exp(2\pi i\mathbf{k}\cdot\mathbf{r}_l)\mathbf{X}(\mathbf{k})\end{aligned} \tag{F.6}$$

Consequently $\mathbf{X}(\mathbf{k})$ and $\mathbf{S}(\mathbf{k})$ are related to each other by

$$\begin{aligned}\mathbf{S}(\mathbf{k}) &= \mathbf{B}(\mathbf{k})\mathbf{X}(\mathbf{k})\\ \mathbf{B}(\mathbf{k}) &= \sum_{l}\mathbf{B}_l\exp(2\pi i\mathbf{k}\cdot\mathbf{r}_l)\end{aligned} \tag{F.7}$$

Equation F.2 is rewritten as

$$\begin{aligned}2V &= \sum_{n}\sum_{n'}\widetilde{\mathbf{R}}_n\mathbf{F}_{nn'}\mathbf{R}_{n'}\\ &= \sum_{n}\sum_{t}\widetilde{\mathbf{R}}_n\mathbf{F}_t\mathbf{R}_{n+t}\end{aligned} \tag{F.8}$$

The inverse Fourier transform of Eq. F.4 is considered next:

$$\mathbf{R}_n = \mathfrak{N}^{-1/2}\sum_{k}\mathbf{S}(\mathbf{k})\exp(2\pi i\mathbf{k}\cdot\mathbf{r}_n) \tag{F.9}$$

$$\widetilde{\mathbf{R}}_n = \mathfrak{N}^{-1/2}\sum_{k}\mathbf{S}^{\dagger}(\mathbf{k})\exp(-2\pi i\mathbf{k}\cdot\mathbf{r}_n) \tag{F.10}$$

where the summation is made over the first Brillouin zone. Substituting Eqs. F.9 and F.10 into Eq. F.8 gives

$$\begin{aligned}2V &= \sum_{n}\sum_{t}\Big[\mathfrak{N}^{-1/2}\sum_{k}\mathbf{S}^{\dagger}(\mathbf{k})\exp(-2\pi i\mathbf{k}\cdot\mathbf{r}_n)\Big]\mathbf{F}_t\Big[\mathfrak{N}^{-1/2}\sum_{k'}\mathbf{S}(\mathbf{k}')\\ &\qquad\times\exp(2\pi i\mathbf{k}'\cdot\mathbf{r}_{n+t})\Big]\\ &= \mathfrak{N}^{-1}\sum_{k}\sum_{k'}\mathbf{S}^{\dagger}(\mathbf{k})\Big[\sum_{t}\mathbf{F}_t\exp(2\pi i\mathbf{k}'\cdot\mathbf{r}_t)\Big]\mathbf{S}(\mathbf{k}')\sum_{n}\exp[2\pi i(\mathbf{k}'-\mathbf{k})\cdot\mathbf{r}_n]\end{aligned} \tag{F.11}$$

where $\mathbf{r}_{n+t} = \mathbf{r}_n + \mathbf{r}_t$ is taken into account. Using the formula

$$\mathfrak{N}^{-1} \sum_{\mathbf{n}} \exp\left[2\pi i(\mathbf{k}' - \mathbf{k})\cdot\mathbf{r}_{\mathbf{n}}\right] = \delta(\mathbf{k} - \mathbf{k}') = \begin{cases} 1 & \mathbf{k} = \mathbf{k}' \\ 0 & \mathbf{k} \neq \mathbf{k}' \end{cases} \tag{F.12}$$

gives

$$\begin{aligned} 2V &= \sum_{\mathbf{k}} \mathbf{S}^{\dagger}(\mathbf{k}) \left[\sum_{t} \mathbf{F}_t \exp(2\pi i \mathbf{k}\cdot\mathbf{r}_t) \right] \mathbf{S}(\mathbf{k}) \\ &= \sum_{\mathbf{k}} \mathbf{S}^{\dagger}(\mathbf{k}) \mathbf{F}_s(\mathbf{k}) \mathbf{S}(\mathbf{k}) \end{aligned} \tag{F.13}$$

where

$$\mathbf{F}_s(\mathbf{k}) = \sum_{t} \mathbf{F}_t \exp(2\pi i \mathbf{k}\cdot\mathbf{r}_t) \tag{F.14}$$

is the **F** matrix based on the internal symmetry coordinates **S**(**k**).

$\mathbf{F}_s(\mathbf{k})$ is transformed into $\mathbf{F}_X(\mathbf{k})$ based on the Cartesian symmetry coordinates **X**(**k**) by

$$\mathbf{F}_X(\mathbf{k}) = \widetilde{\mathbf{B}}(\mathbf{k}) \mathbf{F}_s(\mathbf{k}) \mathbf{B}(\mathbf{k}) \tag{F.15}$$

Therefore frequencies are calculated from $\mathbf{G}_X \mathbf{F}_X(\mathbf{k})$. Since $\mathbf{G}_X = \mathbf{M}^{-1}$ is a diagonal matrix, the following Hermitian matrix is obtained.

$$\mathbf{GF}(\mathbf{k}) = \mathbf{M}^{-1/2} \mathbf{F}_X(\mathbf{k}) \mathbf{M}^{-1/2} = \mathbf{D}(\mathbf{k}) \tag{F.16}$$

The secular equation has the form

$$|\mathbf{D}(\mathbf{k}) - \mathbf{E}\lambda(\mathbf{k})| = 0 \tag{F.17}$$

REFERENCE

1. L. Piseri and G. Zerbi, *J. Mol. Spectrosc.*, **26**, 254 (1968)

Appendix G

Thermodynamic Functions Derived under the Harmonic Approximation

A point with wave number vector **k** on the jth branch in the frequency–phase relations for a crystal is now considered. The frequency of the vibrational mode corresponding to the point is designated by $\nu_j(\mathbf{k})$ (Section 5.6.6). The energy levels for the vibrational mode are denoted by the quantum numbers $n_j(\mathbf{k}) = 0, 1, 2, \cdots$. The whole set of quantum numbers may be represented by $\{n_j(\mathbf{k})\}$. Under the harmonic approximation, the total energy for the energy levels given by the set of quantum numbers is expressed by

$$E\{n_j(\mathbf{k})\} = \sum_{\mathbf{k}}\sum_{j}[n_j(\mathbf{k}) + \tfrac{1}{2}]h\nu_j(\mathbf{k}) \qquad \text{(G.1)}$$

If $n_j(\mathbf{k})$ and $\nu_j(\mathbf{k})$ are written, respectively, as v_i and ν_i for simplicity, Eq. G.1 is expressed by

$$E\{n_j(\mathbf{k})\} = \sum_{i}(v_i + \tfrac{1}{2})h\nu_i \qquad \text{(G.2)}$$

If $1/kT = \beta$ (k is Boltzman's constant and T the absolute temperature), the partition function for the whole system is given by

$$\begin{aligned} Z &= \sum_{v_i}\exp\left[-\beta\sum_{i}(v_i + \tfrac{1}{2})h\nu_i\right] \\ &= \prod_{i}\sum_{v_i}\exp\left[-\beta(v_i + \tfrac{1}{2})h\nu_i\right] \\ &= \prod_{i}\exp(-\tfrac{1}{2}\beta h\nu_i)\sum_{v_i}\exp(-\beta v_i h\nu_i) \\ &= \prod_{i}\exp(-\tfrac{1}{2}\beta h\nu_i)/[1 - \exp(-\beta h\nu_i)] \qquad \text{(G.3)} \end{aligned}$$

The Helmholtz energy A is expressed by

$$A = -kT\ln Z = -kT\sum_i \ln\left[\frac{\exp(-\frac{1}{2}\beta h\nu_i)}{1-\exp(-\beta h\nu_i)}\right]$$

$$= kT\sum_i \ln\left[2\sinh\left(\frac{h\nu_i}{2kT}\right)\right] \tag{G.4}$$

where the formula

$$\sinh x = \frac{e^x - e^{-x}}{2} \tag{G.5}$$

was employed. According to the formulas of thermodynamics, the entropy S, the internal energy E, and the specific heat under constant volume C_V are given by

$$S = -\left(\frac{\partial A}{\partial T}\right)_V = k\sum_i\left\{\frac{h\nu_i}{2kT}\coth\left(\frac{h\nu_i}{2kT}\right) - \ln\left[2\sinh\left(\frac{h\nu_i}{2kT}\right)\right]\right\} \tag{G.6}$$

$$E = A + TS = kT\sum_i \frac{h\nu_i}{2kT}\coth\left(\frac{h\nu_i}{2kT}\right)$$

$$= kT\sum_i \frac{h\nu_i}{2kT}\frac{\exp(h\nu_i/kT)+1}{\exp(h\nu_i/kT)-1}$$

$$= \sum_i\left(\frac{h\nu_i}{2} + \frac{h\nu_i}{\exp(h\nu_i/kT)-1}\right) \tag{G.7}$$

$$C_V = \left(\frac{\partial E}{\partial T}\right)_V = k\sum_i\left(\frac{h\nu_i}{kT}\right)^2\frac{\exp(h\nu_i/kT)}{[\exp(h\nu_i/kT)-1]^2}$$

$$= k\sum_i\left[\left(\frac{h\nu_i}{2kT}\right)^2\Big/\sinh^2\left(\frac{h\nu_i}{2kT}\right)\right] \tag{G.8}$$

In Eq. G.7, the formula

$$\coth x = \frac{e^{2x}+1}{e^{2x}-1} \tag{G.9}$$

is used. If ν_i is replaced by $\nu_j(\mathbf{k})$ in Eqs. G.4 and G.6–G.8, Eqs. 6.28–6.31 are obtained.

Appendix H

Conversion Table for SI and cgs Unit Systems

A summary is given here for units used in this book. For details of SI (Le Systéme International d'Unites), see Refs. 1 and 2.

H.1 SI Prefixes

T (tera, 10^{12}), G (giga, 10^{9}), M (mega, 10^{6}), k (kilo, 10^{3}), h (hecto, 10^{2}), da (deca, 10), d (deci, 10^{-1}), c (centi, 10^{-2}), m (milli, 10^{-3}), μ (micro, 10^{-6}), n (nano, 10^{-9}), p (pico, 10^{-12}).

H.2 Conversion of Units

Physical quantity	cgs units and equivalents	SI
Length	Å	10^{-1} nm
	mμ	10^{-9} m = 1nm
	μ	10^{-6} m = 1 μm
Plane angle	°	$\pi/180$ rad = 1.74533×10^{-2} rad
Force[a]	dyn	10^{-5} N, N (newton) = kg m s^{-2}
Pressure[b]	dyn cm^{-2}	10^{-1} Pa, Pa (pascal) = N m^{-2}
	torr $\simeq$ mmHg	1333.322 Pa
	kg cm^{-2} $= 980 \times 10^{3}$ dyn cm^{-2}	
	$\simeq 10^{6}$ dyn cm^{-2}	$\simeq 10^{5}$ Pa
	atm	1.01325×10^{5} Pa
	bar	10^{5} Pa
	Psi	6.89476×10^{3} Pa
Frequency		Hz = s^{-1}
Temperature	°C, °K	K, t/°C = T/K − 273.15 T, thermodynamic temperature, t, Celsius temperature

H.2 Conversion of Units (*Contd.*)

Physical quantity	cgs units and equivalents	SI
Energy	erg	10^{-7} J, J (joule) = Nm = kg m^2s^{-2}
	cal_{th}(thermochemical calorie)	4.184 J
	eV/molecule	96.48669 kJ mol^{-1}
	cm^{-1}	11.9629 J mol^{-1}
Entropy, heat capacity	e.u. = cal deg^{-1} mol^{-1}	4.184 J K^{-1} mol^{-1}
Concentration	M (mol l^{-1})[c]	1 mol dm^{-3}
Electric current		A (ampere)
Electric charge	C (coulomb) = 3×10^9 cgs esu	A s
Electric potential difference	V (volt)	J A^{-1} s^{-1}
Dipole moment	D (debye) = 10^{-18} esu cm	$10^{-18} \times 3^{-1} \times 10^{-11}$ C m $\cong 3.33 \times 10^{-30}$ C m

[a] Unit of force constants; mdyn/Å = 10^2 N m^{-1}

[b] Unit of tensile strength or Young's modulus; dyn/cm^2 = 8.83×10^8 g/denier $\times$ ρ(g/cm^3), or kg/cm^2 = 9×10^2 g/denier $\times$ ρ(g/cm^3), where ρ is the density. The weight in grams of 9000 m of a filament equals 1 denier.

[c] The old definition of the liter (1.000 028 dm^3) was rescinded.

H.3 Values of the Fundamental Constants[3]

		Units	
Quantity and symbol	Value[a]	SI	cgs
Speed of light in vacuum c	2.997 924 58 (1.2)	10^8 m s^{-1}	10^{10} cm s^{-1}
Atomic mass unit u	1.660 565 5 (86)	10^{-27} kg	10^{-24} g
Proton rest mass m_p	1.672 648 5 (86)	10^{-27} kg	10^{-24} g
Electron rest mass m_e	9.109 534 (47)	10^{-31} kg	10^{-28} g
Elementary charge e	1.602 189 2 (46)	10^{-19} C	10^{-20} emu
	4.803 242 (13)		10^{-10} esu
Planck's constant h	6.626 176 (36)	10^{-34} J s	10^{-27} erg s
Avogadro's constant N_A	6.022 045 (31)	10^{23} mol^{-1}	10^{23} mol^{-1}
Molar gas constant R	8.314 41 (26)	J mol^{-1} K^{-1}	10^7 erg mol^{-1} K^{-1}
Boltzmann's constant $k = R/N_A$	1.380 662 (44)	10^{-23} J K^{-1}	10^{-16} erg K^{-1}

[a] Unified atomic mass unit defined as one-twelfth of the mass of the ^{12}C nuclide was used. The values in parentheses indicate the standard deviations and refer to the last decimal positions.

REFERENCES

1. Manual of Symbols and Terminology for Physicochemical Quantities and Units, *Pure Appl. Chem.*, **21,** 1 (1970).
2. M. L. McGlashan, *Physicochemical Quantities and Units*, 2nd ed., The Royal Institute of Chemistry, London, 1968.
3. Recommended Consistent Values of the Fundamental Physical Constants, 1973, *CODATA Bull.*, **11,** 1 (1973).

Author Index

Subject Index